Sinusoids

Theory and Technological Applications

MONOGRAPHS AND RESEARCH NOTES IN MATHEMATICS

Series Editors

John A. Burns
Thomas J. Tucker
Miklos Bona
Michael Ruzhansky
Chi-Kwong Li

Published Titles

Iterative Optimization in Inverse Problems, Charles L. Byrne

Modeling and Inverse Problems in the Presence of Uncertainty, H. T. Banks, Shuhua Hu, and W. Clayton Thompson

Sinusoids: Theory and Technological Applications, Prem K. Kythe

Forthcoming Titles

Stochastic Cauchy Problems in Infinite Dimensions: Generalized and Regularized Solutions, Irina V. Melnikova and Alexei Filinkov

Signal Processing: A Mathematical Approach, Charles L. Byrne

Monomial Algebra, Second Edition, Rafael Villarreal

Groups, Designs, and Linear Algebra, Donald L. Kreher

Geometric Modeling and Mesh Generation from Scanned Images, Yongjie Zhang

Difference Equations: Theory, Applications and Advanced Topics, Third Edition, Ronald E. Mickens

Set Theoretical Aspects of Real Analysis, Alexander Kharazishvili

Method of Moments in Electromagnetics, Second Edition, Walton C. Gibson

The Separable Galois Theory of Commutative Rings, Second Edition, Andy R. Magid

Dictionary of Inequalities, Second Edition, Peter Bullen

Actions and Invariants of Algebraic Groups, Second Edition, Walter Ferrer Santos and Alvaro Rittatore

Practical Guide to Geometric Regulation for Distributed Parameter Systems, Eugenio Aulisa and David S. Gilliam

Analytical Methods for Kolmogorov Equations, Second Edition, Luca Lorenzi

Handbook of the Tutte Polynomial, Joanna Anthony Ellis-Monaghan and Iain Moffat

Blow-up Patterns for Higher-Order: Nonlinear Parabolic, Hyperbolic Dispersion and Schrödinger Equations, Victor A. Galaktionov, Enzo L. Mitidieri and Stanislav Pohozaev

Application of Fuzzy Logic to Social Choice Theory, John N. Mordeson, Davendar Malik and Terry D. Clark

Microlocal Analysis on R^n and on NonCompact Manifolds, Sandro Coriasco

Cremona Groups and Icosahedron, Ivan Cheltsov and Constantin Shramov

MONOGRAPHS AND RESEARCH NOTES IN MATHEMATICS

Sinusoids
Theory and Technological Applications

Prem K. Kythe

University of New Orleans
Louisiana, USA

CRC Press
Taylor & Francis Group
Boca Raton London New York

CRC Press is an imprint of the
Taylor & Francis Group, an **informa** business
A CHAPMAN & HALL BOOK

CRC Press
Taylor & Francis Group
6000 Broken Sound Parkway NW, Suite 300
Boca Raton, FL 33487-2742

First issued in paperback 2019

© 2015 by Taylor & Francis Group, LLC
CRC Press is an imprint of Taylor & Francis Group, an Informa business

No claim to original U.S. Government works

ISBN-13: 978-1-4822-2106-0 (hbk)
ISBN-13: 978-0-367-37857-8 (pbk)

Visit the Taylor & Francis Web site at
http://www.taylorandfrancis.com

and the CRC Press Web site at
http://www.crcpress.com

TO THE MEMORY OF MY WIFE

Contents

Preface

The classical Fourier transform has been used in solving boundary value problems in a variety of applications including continuum mechanics, potential theory, geophysics, physics, biology, and mathematical economics. The discrete Fourier transform evolved into the fast Fourier transform technique that helped numerically investigate boundary value problems. In general, the Fourier transform carries over a function of time into a new function whose argument is the frequency in units of cycles per second or radians per second. This new function is the frequency spectrum of the given function, or vice versa. There are theorems that determine both of these functions, one from the other, i.e., one in the time domain and the other in the frequency domain. In fact, the Fourier transform is an extension of the Fourier series in the case of periodic functions where the period is allowed to tend to infinity.

Overview

The sinusoids, which are periodic sine or cosine functions, are explained in Chapter 1, with some well-known examples from wave theory, especially the traveling and standing waves, from continuous musical rhythms, and from the field of medicine in the human liver. In all cases the Fourier transform is used to calculate the discrete set of complex amplitudes that involve the Fourier series. After sampling a time-domain function to computer-processing, the discrete Fourier transform yields the original Fourier transform by applying the Poisson summation formula. The Fourier series and the Fourier transform are discussed in both continuous and discrete cases in Chapters 2 and 3, along with an analysis of the Dirichlet kernel and the Gibbs phenomenon. The amplitudes, phase, and frequency of periodic functions (signals) are studied in the case of deterministic continuous and discrete signals. Invertibility and periodicity of Fourier transforms are used in the development of signals and filters, which are discussed in Chapter 4. These two topics are not only useful in signal processing but also in subsequent different technological applications.

The general concept of communication systems is discussed in Chapter 5. It includes the process of quantization of analog/digital signals, interference, and data transmission. Space exploration, including the Mars project and SETI, are included. The general direction of current research in space travel seems to harness nuclear fusion and antimatter as an energy source to reach the nearest star in a few decades provided the machinery could last that long. The Alcuberre warp drive provides an interesting topic, which is based on the premise that it could travel faster than light if negative mass existed.

Although software-defined radio communication systems that use software for the modulation and demodulation of radio signals are known and have been used since 1995, the first Global Positioning System (GPS) was implemented in 1997. The Global Navigation Satellite Systems now include both GPS and Galileo systems. Although GPS is a complicated device, it can be understood as a single-frequency receiver, which describes its amplitude spectrum, auto-correlation function, and Fourier transform (power spectrum). This is done in Chapter 6. The linear time-invariant systems, already introduced in Chapter 4, together with L1 and L2 signals and binary offset carrier modulation, cyclic redundancy check, forward error correction, and block interleaving, are presented to explain the functioning of the GPS receiver, all based on the code division multiple access principle.

Fourier optics, which is the mathematical model for diffraction, uses Fourier transform in two dimensions. The theory is developed and examples in optical transform are presented in Chapter 7. This chapter includes the Helmholtz-Kirchhoff, Fresner-Kirchhoff and Rayleigh-Sommerfeld diffraction models, and scalar diffraction models of Fresnel and Fraunhofer approximations, which are solved in the case of quasi-optics. Electromagnetic radiation and its spectrum are discussed, and the adaptive additive algorithm is presented.

Chapter 8 on x-ray crystallography deals with the synchrotron light and x-ray diffraction. The electrons moving in a circular path are stored in the biggest machines in the world, where they are created by an electron gun (laser), such that the undulators (two rows of permanent magnets), while operating perpendicular to the electron beam and alternating in polarity, bend the electron beam up and down like a sinusoid with continuously varying radius. X-ray crystallography is based on Bragg's law for diffraction and interference, which explains why visible light and ultraviolet light cannot produce diffraction in crystals. The mathematical theory of electron distribution and the phase problem of sinusoids are applied to genetics and DNA, and to the mathematical solution for the hydrogen atom and its spectrum including the Bohr model. A discussion of laser and luminescence are included to present another useful property of light.

A short history of radioastronomy is presented in Chapter 9, and mathematical theory of radio waves and of image formation in a radiotelescope, including that of aperture synthesis, is developed. The Earth's rotation, quasars, and pulsars are described. This chapter also deals with the Big Bang theory and Olbers paradox. A mathematical theory of black holes is discussed and their solutions are presented.

In an effort to present a tangible definition of sinusoids, a brief description of certain aspects of sound and music is introduced in Chapter 10. An analysis of rhythm and music is presented in some detail, including certain mathematical structures in poetry and music. The problems of vibrating strings and drums are solved. A brief history of computerized music and music synthesizers is also presented.

Chapter 11 starts with a short history of computerized axial tomography and computerized tomographic scans, which takes the subject beyond classical radiology. Different types of tomographic procedures in medical science are presented, including sinogram, B-scan imaging, reflection tomography, and magnetic resonance imaging. Modern advances in tomography are explained, and nuclear medicine is discussed in some detail.

Wavelets are an extension of Fourier analysis. They are used to represent a function, including a sinusoid, by a series of orthogonal functions, but with the following differences: the wavelet series converge pointwise, are more localized, exhibit edge effects better, and use fewer coefficients to represent certain signals and images. However, the application of wavelet theory to different technological and tomographic cases is not covered, as this would require a separate volume to complete.

There are thirteen indices A through M, covering enclitic material that relates to certain words or concepts occurring in the main text. Since the book covers so many different and divergent topics, it is natural to supplement such issues in the form of indices, which would otherwise be left for the readers to search out.

Intended Readers

This book assumes a basic but thorough knowledge of advanced calculus, Fourier transforms, partial differential equations, and elementary complex analysis. Although it is intended at the graduate level for an effective study of different topics related to the sinusoids, and later to serve as a general literature for scientists, engineers, and technical experts in signal processing, and mathematicians in related industries, it will definitely provide an interesting reading material for scientifically inclined persons with sufficient knowledge of different technological applications. In most cases, the results are developed throughout the book from the basic level consistent with the mathematical background of the intended readers. In some instances the results are stated and the original references are cited. The majority of intended readers fall into one of the following three categories: firstly, they may be students ready for a graduate course in any of the topics in the book; secondly, they may be graduate students engaged in analytical and technical research in one or more topics discussed in the book; and thirdly, they may be scientists and researchers in varied areas of applied mathematics, engineering, physics, and technology. For them the book may become a source of information in some current trends in research where Fourier transforms and sinusoids are important.

Acknowledgments

I take this opportunity to thank Mr. Robert B. Stern, Executive Editor, Taylor & Francis/CRC Press, for his encouragement to complete this book, and to Project Editor Michele Dimont for doing a great job of editing the text. Thanks are due to the referees who took time to review the book proposal and made some valuable suggestions. I thank Mr. Charles F. Priore, Jr., Associate Librarian, Hustad Science Library, St. Olaf College, Northfield, MN, and Reference and Instruction Librarian, Lawrence McKinley Gould Library, Northfield. MN, for promptly providing logistic support whenever requested. I also thank Dr. Krishna K. Agarwal, Professor of Computer Science, LSU Shreveport, LA, for reading the manuscript and providing valuable suggestions. I am indebted to the reviewer who painstakingly edited the manuscript and made many significant suggestions to improve upon its quality and presentation. Finally, I thank my friend Michael R. Schäferkotter for help and advice freely given whenever needed.

Prem K. Kythe

Notations, Definitions, and Acronyms

A list of the notations, definitions, abbreviations, and acronyms used in this book is given below.

a, thermal conductivity

$a_0 = \dfrac{\hbar^2}{mc^2} = 0.529$ Å, first Bohr radius

a_0, a_n, b_n, Fourier coefficients, $n = 1, 2, \ldots$

a_e, b_e, major and minor axes of an ellipse; for Earth $a_e = 6378137 \pm 2$ m, and $b_e = 6356752.3142$ m, $e_c = 0.0033528106674$

$\arg\{z\}$, argument of a complex number z

arsis, unaccented part of a syllable

attosecond (as), 1 as $= 10^{-18}$ sec

$\overline{A}$, closure of a set A

A, amplitude; arithmetic mean

A_c, center amplitude

$A_t(\alpha/\lambda, \beta/\lambda)$, angular spectrum

AM, amplitude modulation

A^T, transpose of a matrix A

Å, angstrom unit is a unit of length equal to 0.1 nm; 1 Å $= 10^{-10}$ m

A_{peak}, peak amplitude

A_{RMS}, root mean square amplitude

AA Algorithm, Adaptive Additive Algorithm

ABCD matrix, ray-transfer matrix

ACF, auto-correction function

A, C, G, T, Adenine, Cytosine, Guanine, Thymine (four nucleic acids that are building blocks of DNA)

ADC, accurate digital correlation

ADM, acronym from the last names of the physicists Richard Arnowitt, Stanley Deser and Charles Misner

AFSCN, Air Force Satellite Control Network

ALARP, As Low As Reasonably Practicable

ALE, automatic link establishment

ALS, Advanced Photon Source

API, application programming interface

APT, atom probe tomography

ART, algebraic reconstruction technique

AWGN, additive white Gaussian noise

b, bias (or clock delay), time receiver's clock is off

$b(x)$, brilliance of an aperture field

ber, bei, ker, kei, Kelvin functions

B, bandwidth

$B(s)$, Fourier transform of $b(x)$

$\mathcal{B}\{f(x)\}$, Fourier-Bessel transform of f, also known as the Hankel transform of order zero $\mathcal{H}_0\{f(x)\} \equiv \hat{f}_0(\sigma)$

$\mathbf{B}$, magnetic field

BEC, binary erasure channel

BETA, billion channel extraterrestrial Assay

BP, backprojection (decoder)

BPM, beats per minute

BPSK, binary phase shift keying

BSC, binary symmetric channel

c, wave speed, or wave velocity; speed of light $(= 2.99792458 \times 10^8$ meter/sec$)$

c_n, Fourier coefficients

circle$(\sqrt{x^2 + y^2})$, circle function, equal to 1 for $r < 1$, equal to $\frac{1}{2}$ for $r = 1$, and equal to 0 otherwise, where $r = \sqrt{x^2 + y^2}$

comb(t), comb function of time t

const, constant

$C^k(D)$, class of real-valued functions continuous together with all the derivatives up to order k inclusive, $0 \leq k < \infty$, in a domain $D \in \mathbb{R}^n$

$C^\infty(D)$, class of functions infinitely differentiable in D, i.e., their continuous partial derivatives of all orders exist

$C_0^\infty(D)$, class of functions which are infinitely differentiable on D and vanish outside some bounded region (index 0 indicates compact support)

$C_0^\infty(\mathbb{R})$, class of functions with compact support on $\mathbb{R}$ and infinitely differentiable on $\mathbb{R}$

$\mathbf{c}(x)$, transmitted codeword; codeword polynomial

$\mathbb{C}$, complex plane

$\mathfrak{C}$, Shannon capacity of a transmission channel

CAT, computerized axial tomography

C/A, Course/Acquisition

CBR, cosmic (microwave) background radiation

CDMA, Code Division Multiple Access

CEP, circular error probable

CFT, complex Fourier transform

C, G, C, musical notes generated by strings of length 1, 3/2, and 2 units

CMB, cosmic microwave background radiation

CMYK, cyan-magenta-yellow-black color codes

CORBA, Common Object Request Broker Architecture

CPU, central processing unit

CRC, cyclic redundancy check

CT, computerized tomography

CTIS, computed tomography imaging spectrometer

Cyro-ET, cyro-electron

CW, continuous waveform; carrier wave

ds^2, metric

det $[A]$, determinant of a square matrix A

$D_{a,b}$, data

$D_N(k)$, Dirichlet kernel

$D_N(h,s)$, d-line

$\mathbf{d}(x)$, data polynomial

$\mathbf{D}$, electrical displacement field

dB, decibel; 0 dB is the pressure of 20 micropascal (μPa)

DC, direct current

DGPS, Differential GPS

DFT, discrete Fourier transform

DNA, DioxyriboNucleic Acid

DoD, Department of Defense (U.S.)

DR, dynamic range

DARPA, Defense Advanced Research Projects Agency

DRT, discrete Radon transform

DSSS, direct spectrum spread system

e, electron charge

e_c, eccentricity

e_p, ellipticity

E, energy; particle energy (eV); photon energy

$E(s)$, far field

E_b, energy per information bit

E_c, critical photon energy

E_s, average energy variance

$E(x,y,z) = u(x,y,z)\,e^{ikz}$, harmonic wave function

$E(r) = \Re\{\Psi(r)\}$, radial wave, where $\Psi(r)$ is a complex wave

$E(\mathbf{x})$, incident radiation

$E_N(r)$, radial wave for N slits

$\mathcal{E}_x(\omega)$, energy density spectrum

$e(x)$, error polynomial

$\mathbf{E}$, electric field

Eq(s), equation(s) (when followed by an equation number)

e.g., for example (Latin, *exempli gratia*)

et al., and others (Latin, *et alii*)

eV, electron volt

ECC, error correcting code

ECT, electric capacitance

EI, electron attenuation/scatter

EIT, electric impedance

EM, expectation minimization

EMR, electromagnetic radiation

ERT, electric resistivity

ESA, European Space Agency

f, cycle frequency

f_L, Larmor frequency

f_s, sampling rate

f_x, f_y, cycle frequencies for Fourier transform of two variables

$f \circ g$, composition of functions f and g, defined by $(f \circ g)(x) = f(g(x))$

$\|f\|$, norm of f, defined by $\langle f, f \rangle^{1/2}$

$f \star g$, convolution (*Faltung*) of f and g

$f(x) \rightleftharpoons F(\omega)$, Fourier transform pair, the relationship from a function $f(x)$ to its Fourier transform $F(\omega)$, and conversely

F, force of gravity ($F = 6.67384 \times 10^{-11}$ N)

F_{2^m}, field of binary numbers, m positive integer

$\mathcal{F}\{f(x)\} \equiv F(\omega)$, Fourier (complex) transform

$\mathcal{F}^{-1}\{F(\omega)\} \equiv f(x)$, inverse Fourier (complex) transform

$\mathcal{F}_s\{f(x)\} \equiv F_s(\omega)$, Fourier sine transform

$\mathcal{F}_c\{f(x)\} \equiv F_c(\omega)$, Fourier cosine transform

$\mathcal{F}\{f(x,y)\} \equiv F(f_x, f_y)$, Fourier transform of two variables (the cycle frequencies f_x, f_y can be replaced by the radian frequencies ω_x, ω_y, where $\omega_x = 2\pi f_x$ and $\omega_y = 2\pi f_y$)

$\mathcal{F}^{-1}\{F(f_x, f_y)\} \equiv f(x,y)$, inverse Fourier transform of two variables

$\mathcal{F}\{x(t)\} \equiv X(f)$, Fourier transform of a 1-D signal $x(t, \tau)$

$\mathcal{F}\{x(t, \tau)\} \equiv X(f, g)$, Fourier transform of a 2-D signal $x(t, \tau)$

fs, femtosecond, 1 fs = 10^{-15} sec

FBP, filtered backprojection

FDM, frequency division multiplexing

FEC, forward error correcting (code)

FFT, finite Fourier transform

FH/SS, frequency-hopping spread system

FM , frequency modulation

fMRI , functional MRI

FPGA , field programmable gate array

FPS , foot-pound-second units

FST , Fourier slice theorem

FT , Fourier transform

g , electric charge; total distribution charge density

$g_\varepsilon(x) = \dfrac{1}{\sqrt{2\pi\varepsilon}}\, e^{-x^2/(2\varepsilon)}$, Gaussian functions, or Gaussian distribution

$\mathbf{g}(x)$, generator polynomial

G , geometric mean; Newton's gravitation constant ($G = 6.67384 \times 10^{-11}$ N m^2 kg^{-2})

$G(P)$, Green's function at a point P

$G(t,t')$, or $G(x,x')$, also $G(t,s)$ or $G(x,s)$, Green's function

$G(u,v)$, Gold sequence

$G(\mathbf{x},\mathbf{x}')$, Green's function, also written as $G(\mathbf{x}-\mathbf{x}')$

$G(\mathbf{x},\mathbf{x}';t,t') \equiv G(\mathbf{x}-\mathbf{x}';t-t')$, Green's function for a space-time operator

gcd , greatest common divisor

GDOP , Geometric Dilution of Precision

GDPS , Global Data Processing System

GeV , gigaelectron volt (1 GeV = 10^9 eV)

GF , Galois field

GHz , gigahertz (1 GHz= 10^9 Hz)

GNSS , global navigational satellite system

GNU , not Unix (reverse definition)

GPS , global positioning system

h , Planck constant, $h = 2\pi\,\hbar$; altitude

$\hbar = 1.054 \times 10^{-34}$ joules-sec, or $= 6.625 \times 10^{-27}$ erg-sec , Planck constant

H , harmonic mean; entropy

H_0 , Hubble constant

H_n, harmonic sum $= \sum_{k=1}^{n} \frac{1}{k} = \ln k + \gamma + \varepsilon_n$, where $\gamma \approx 0.577215665$ is the Euler's constant and $\varepsilon_n \sim \frac{1}{2n} \to 0$ as $n \to \infty$

$H(t)$ or $H(x)$, Heaviside unit step function $H(t) = 1$ if $t > 0$, and $H(t) = 0$ if $t < 0$

$H_n(x) = (-1)^n e^{x^2} \dfrac{d^n}{dx^n} e^{-x^2}$, Hermite polynomials of degree n

$H_0^{(1)}(kr), H_0^{(2)}(kr)$, Hankel functions of the first and second kind, respectively, and of order n; $H_0^{(1,2)}(kr) = J_0(kr) \pm i\, Y_0(kr)$

$H(f_x, f_y)$, transfer function through the free space (in Talbot images)

$\mathcal{H}\{x(t)\} \equiv X_H(z)$, Hilbert transform of $x(t)$

$\mathcal{H}^{-1}\{X(z)\} = x(t)$, inverse Hilbert transform of $X(z)$

$\mathcal{H}_n\{f(x)\} \equiv \hat{f}_n(\sigma)$, Hankel transform of order n

$\mathcal{H}_0\{f(x)\} \equiv \hat{f}_0(\sigma)$, zero-order Hankel transform or Fourier-Bessel transform

$\mathcal{H}_n^{-1}\{\hat{f}_n(\sigma)\} \equiv f(x)$, inverse Hankel transform of order n

H, magnetic intensity

$$\mathbf{H} = -\frac{\hbar^2}{2m}\frac{d^2}{dt^2} + 2\pi^2\nu^2 m t^2 = -\frac{\hbar^2}{2m}\frac{d^2}{dt^2} + \tfrac{1}{2}m\omega^2 t^2 \,, \text{ Hamiltonian operator}$$

$\mathcal{H}(\mathfrak{p})$, binary entropy function

$\mathfrak{H}$, Hadamard matrix

HF, high frequency, 3 MHz to 300 MHz (radio)

HOW, handover word

HPLC, high-performance liquid chromatography

HPSDR, High-Performance Software Defined Radio

Hz, Hertz unit $= 1$ cycle per second

i, j, k, unit vectors along the rectangular coordinates axes x, y, and z, respectively

$i = \sqrt{-1}$, imaginary unity

I, beam current; photocurrent; intensity; modulation index

$I(x)$, $I(x,y)$, intensity of a wave

$I_0(z)$, modified Bessel function of the first kind and of order zero

$I_n(x)$, modified Bessel function of the first kind and order n, defined by
$I_n(x) = e^{-i\,n\pi/2}\, J_n\left(e^{i\,\pi/2}\,x\right)$

$\Im$, imaginary part of a complex quantity

$I(\mathbf{r})$, intensity of light

ictus, (Latin) beat or blow for rhythmic accent

i.e., that is (Latin *id est*)

iff, if and only if

ICO, intermediate circular orbit

ID, identification

IDFT, inverse discrete Fourier transform

IDRT, inverse discrete Radon transform

IFT, inverse Fourier transform

IFFT, inverse FFT

IR, iterative reconstruction

$J_n(x)$, Bessel function of first kind and order $n = 0, 1, 2, \ldots$

$J_0(x)$, cylindrical harmonic, Bessel function of first kind and zero order

J, current density

JND, just noticeable difference

JTRS, Joint Tactical Radio System

k, wavenumber

$K_0(z)$, modified Bessel function of the third kind and of order zero

$K_n(x)$, modified Bessel function of the third kind and order n, defined by

$$K_n(x) = \frac{\pi \left[I_{-n}(x) - I_n(x) \right]}{2 \sin n\pi}$$

$K_{1/3}$, $K_{2/3}$, modified Bessel functions of the second kind

kHz, kilohertz (1 kHz = 1,000 Hz)

keV, kiloelectron volt (1 keV= 10^3 eV)

L, loop-length of a harmonic; angular momentum; longitude

L_c, latitude

$L_k^\sigma(\rho)$, generalized Laguerre polynomials

$L_n(x)$, Laguerre polynomials of order n

$L_n^m(x)$, associated Laguerre polynomials of order n and degree m

$\mathbf{L}(x)$, error locator polynomial

Lambda-CDM, Lambda concordance model

LGM, Little Green Men (hypothesis)

$L_{u,\theta}$, line

L1C, L1 civilian (code)

LDPC, low-density parity-check (code)

LF, low frequency, 30 kHz to 300 kHz (radio)

LFSR, linear feedback shift register

LSB, least significant digit (rightmost digit)

LT, Luby transform

LTI , linear time-invariant

m, mass; mass of a quantum particle; magnetic quantum number

m_e, electron mass

micropascal , see μPa, dB, and pascal

microsecond , see μs

ms (millisecond), 1 ms = 10^{-3} sec

mm (millimeter), 1 mm = 10^{-3} meter

min (minute), 1 min = 10^{10} ns

M code, Military code

Mpc, Megaparsec (1 Mpc = 3.09×10^{19} km)

MCPS, megacycles per second

MCS, master control station

MD5, hash functions

MeV, megaelectron volt (1 MeV= 10^6 eV)

MEO, medium Earth orbit

MER, modulated error ratio

MF, medium frequency, 300 kHz to 3 MHz (radio)

MHz, megahertz (1 MHz = 1,000,000 Hz)

MIDI, Musical Instrument Digital Interface (logarithmic pitch scale)

MIP, maximum intensity projection

MIT, magnetic induction tomography

MLS, maximum length sequence

MRI, magnetic resonance imaging

MRO, Mars Reconnaissance Orbiter

MSB, most significant digit (leftmost digit)

MSPS, million samples per second (digital-to-analog converter)

$n_{1,2}$, refractive indices n_1 and n_2

$n(\mathbf{x})$, electron density

N, samples in discrete sinusoids; period (discrete case); noise power; principal quantum number

$N(E)$, number of photons of energy E

N_{ang}, discrete samplings of angles

$N(f)$, noise power

N_{pos}, discrete samplings of positions

$\mathbf{n}$, outward normal perpendicular to the boundary of a curve or surface; azimuthal (orbital angular) quantum number

$\mathbb{N} = \mathbb{Z}^+$, set of natural numbers

nm (nanometer), 1 nm $= 10^{-9}$ meter

ns (nanosecond), 1 ns $= 10^{-9}$ sec

NASA, National Astronautics and Space Administration

NAVSTAR, Navigational, Strategic, Tactical and Relay (GPS)

NDS, Nuclear Detonation Detection System Payload

NGA, National Geospatial-Intelligence Agency

NMR, nuclear magnetic resonance

NTSC, National Television Standards Committee

NUDET, Nuclear Detonation

OCT, optical coherence tomography

ODT, optical diffusion tomography

OPT, optical projection tomography

P, signal power; momentum; position; feedback gain

P_{ci}, carrier power

$P(s)$, angular power spectrum

$P_n(x)$, Legendre polynomials of degree $n = 0, 1, 2, \ldots$; Legendre's coefficients, or surface zonal harmonics

$P_n^m(\cos\theta)$ or $P_n^m(t)$, associated Legendre polynomials of order $m = 0, \pm1, \pm2, \ldots$ and degree $n = 0, 1, 2, \ldots$

$P_n^m(\theta)$, Legendre polynomials

$p_\theta(r)$, parallel projection in a direction θ

$P\{f(x)\}$, projection

$\mathbf{P}$, medium

$\mathbf{P}(x,t)$, polarization of the medium

$\mathfrak{p}$, probability of error; probability for the transmitted symbol to be received with error; $1 - \mathfrak{p} = \mathfrak{q}$ is the probability for it to be received with no error

$\mathfrak{P}(x|y)$, conditional probability of occurrence of x such that y occurs

p.v., Cauchy's principal value of an integral

ps (picosecond), 1 ps $= 10^{-12}$ sec

P code, Precision code

Pa (pascal), unit of atmospheric pressure, e.g., 1 atmospheric pressure is equal to 10^5 pascals

Pogo, dynamic coupling of the combustion process with structure. Pogo is not an acronym; it refers to the game of bouncing off a pogo stick

P(Y) code, encrypted precision code

PAT, photoacoustic tomography

PDOP, Position Dilution of Precision, or Precision Dilution of Position (GPS)

PET, position emission tomography

POSIX, Portable Operating System Interface

PRN, pseudorandom noise

PSK, phase-shift keying

q, deceleration factor

Q, charge of a black hole

$Q_n(x)$, Legendre function of the second kind of order n

quaver, 1/8

QAM, quadrature amplitude modulation

r, index of modulation

r_e, radius of an ideal spherical Earth or its average radius

$r_x(t)$, autocovariance

r_s, Schwarzschild radius

rect(t), rect function

rect$\left(\dfrac{t-x}{y}\right)$, general boxcar function

(r,θ), polar coordinates: $x = r\cos\theta$, $y = r\sin\theta$, $r = \sqrt{x^2+y^2}$, $\theta = \arctan\dfrac{y}{x}$

(r,θ,z), polar cylindrical coordinates: $x = r\cos\theta$, $y = r\sin\theta$, $r = \sqrt{x^2+y^2}$, $\theta = \arctan\dfrac{y}{x}$, $z = z$

(r,θ,ϕ), spherical coordinates: $x = r\sin\theta\cos\phi$, $y = r\sin\theta\sin\phi$, $z = r\cos\theta$, $r = \sqrt{x^2+y^2+z^2}$, $\theta = \arccos\dfrac{z}{r}$, $\phi = \arctan\dfrac{y}{x}$

$r^n P_n(\cos\theta)$ or $r^{-n-1}P_n(\cos\theta)$, solid zonal harmonics

$r^n Y_n^m(t,\phi)$ or $r^{-n-1}Y_n^m(t,\phi)$, solid spherical harmonics of degree n

$\mathbf{r}_{01}$, distance from a point P_0 to another point P_1

R, transmission rate

$R(t)$, scale factor of the universe

$R_\theta(t)$, Radon transform of a function $f(x,y)$

$\mathcal{R}\{R(\theta,\rho)\}$, inverse Radon transform of a function $R_\theta(t)$

$\mathcal{R}\{f(\rho,\theta)\}$, 2-D Radon transform of f

$\mathbb{R}$, real line

$\mathbb{R}^n$, Euclidean n-space; $\mathbb{R}^1 \equiv \mathbb{R}$

$\mathbb{R}^+$, set of positive real numbers

$\Re$, real part of a complex quantity

Raptor, rapid tornado (code)

RCA, Radio Corporation of America

RF, radio frequency

RGB, Red-Green-Blue (color model based on additive color primeries)

RMS, root mean square

RS, Reed-Solomon (code)

RTCM, Radio Technical Commission Marine

RTK, real-time kinematic

$s(t)$, signal; modulating signal

$s^*(x, y)$, modified reflectivity function

$s_{1/2}(t)$, square wave; two-level Rademacher function

$s_A(t)$, spin density

$\operatorname{sgn}(x) = \begin{cases} 1, & x > 0, \\ 0, & x = 0, \\ -1, & x < 0, \end{cases}$ signum function. Note that $\operatorname{sgn}(x) = 2H(x) - 1$,

    and $[\operatorname{sgn}(x)]' = 2H'(x) = 2\delta(x)$

$\operatorname{sinc}(x) = \dfrac{\sin x}{x}$, sinc function

$S(f)$, signal power

S_{ci}, envelope of line spectrum

SA, Selective availability

SCA, software communication architecture

SDR, software-defined radio

SETI, search for extraterrestrial intelligence

SEP, spherical error probable

SF, signal frequency

SNR, signal-to-noise ratio

SPECT, single photon emission CT

SPS, satellite positioning system

SRXTM, synchrotron x-ray tomographic microscopy

SS, spread-spectrum system

SV, space vehicle, or satellite

t, time

t', source point; singularity

$\operatorname{tri}(t) \equiv \wedge(t)$, triangle function

T, period; temperature

$T(x, y)$, transmittance function at a point (x, y)

$T_A(x, y)$, amplitude transmittance function of an aperture

T_s, sampling period

thesis, (Gk) accented part of a syllable

TAT, thermoacoustic tomography

TEM, transverse mode

TeV, teraelectron volt (1 TeV= 10^{12} eV)

TLM, telemetry (word)

THz, teraherz, 1 THz = 10^{12} Hz; wavelength 0.3 mm

TOW , truncated time of the week

u, dependent variable; displacement; temperature; volumetric energy density

$u(P, t)$, light disturbance function of position P and time t

$U_i(x, y; 0)$, incident field amplitude

$U_t(x, y; 0)$, transmitted field amplitude

$U(\mathbf{r})$, displacement of waves

$\mathbf{u}$, receiver data $(= [x\ y\ z\ b]^T)$

UCT, universal coordinated time

UHF, ultra-high frequency, 300 MHz to 3 GHz (radio)

UIUC, University of Illinois at Urbana-Champaign

UOT, ultrasound-modulated optical tomography

USB, universal serial bus

USRP, Universal Software Radio Peripheral

UTC , universal coordinated time

UV-Vis, visible ultraviolet

V, volume; potential energy

VHF, very high frequency, 30 MHz to 300 MHz (radio)

VLF, very low frequency, 3 kHz to 30 kHz (radio)

$w(x)$, weight function, $x \in (a, b)$

$\mathbf{w}(x)$, received codeword; received word polynomial

WGS, World Geodetic System

$\dot{x}$, time-derivative of x, i.e., dx/dt

$x(t)$, sinusoid; signal

$x(n)$, discrete sinusoid

$x_a(t)$, bandlimited signal

$\lfloor x \rfloor$, floor of x (largest integer not exceeding x)

$X_a(\omega)$, gate function

$\{x_n\}$, a sequence of real numbers

$\{x_{n_k}\}$, a subsequence of real numbers

$\{x(nT_s)\}$, sampling

$\mathbf{x}$, a point $(x_1, x_2, \ldots, x_n)$ in $\mathbb{R}^n$; a field point

$\mathbf{x}'$, source point; singularity

$\mathbf{x} \cdot \mathbf{y} = x_1 y_1 + x_2 y_2 + x_3 y_3$, scalar (dot) product of vectors $\mathbf{x} = x_1 \mathbf{i} + x_2 \mathbf{j} + x_3 \mathbf{k}$

and $\mathbf{y} = y_1\mathbf{i} + y_2\mathbf{j} + y_3\mathbf{k}$

$X(\omega)$, spectrum of a signal $x(t)$

X_i, error locations

$X(k)$, FT of a discrete sequence $x(n)$

$X_H(z)$, Hilbert transform of $x(t)$

$\mathbf{X}$, conducting medium

XOR, bitwise exclusive or, denoted by $\oplus$

XOR, logical exclusive or

$Y_0(x)$, Bessel function of second kind of order zero

$Y_n(x)$, Bessel function of the second kind of order n, defined by

$$Y_n(x) = \frac{\cos n\pi\, J_n(x) - J_{-n}(x)}{\sin n\pi}; \text{ sometimes denoted by } N_n(x)$$

Y_l^m, spherical harmonic functions of degree l and order m

$$y_n(x) = \sqrt{\frac{\pi}{2x}} Y_{n+1/2}(x) = (-1)^n \sqrt{\frac{\pi}{2x}} J_{n-1/2}(x), \text{ spherical Bessel function of}$$
order n

$Y_0(x)$, Bessel function of second kind of order zero

$Y_n(x)$, Bessel function of the second kind of order n, defined by

$$Y_n(x) = \frac{\cos n\pi\, J_n(x) - J_{-n}(x)}{\sin n\pi}; \text{ sometimes denoted by } N_n(x)$$

$Y_n^m(\theta, \phi)$, spherical harmonics

z, a complex number $z = x + iy$; redshift

$\bar{z} = x - iy$, complex conjugate of z

z^*, inverse (symmetrical) point of z with respect to the unit circle in $\mathbb{C}$

$|z| = \sqrt{x^2 + y^2}$, $z = x + iy$, modulus of a complex number z

z_n, correlation

Z, atomic number

$Z(k)$, DFT of correlation z_n

$\mathbb{Z}$, set of integers

$\mathbb{Z}^+$, set of positive integers

Zulu time, Greenwich meridian time

$\gamma = 0.577215665$, Euler gamma

$\delta(x)$, $\delta(x, x')$, $\delta(x, s)$, Dirac delta function; also denoted as $\delta(t)$, $\delta(t - t')$ and $\delta(t - \tau)$,

δ_{mn}, Kronecker delta, equal to 1 if $m = n$ and 0 if $m \neq n$

Δt, error in receiver's clock

ϵ, electric inductive capacity

λ, wavelength

λ_n, eigenvalues

μ, Earth's universal gravitational parameter ($\mu = 3.986005 \times 10^{14}$ meters^3/sec^2); magnetic inductive capacity

μm, micrometer, or micron (1 μm $= 10^{-6}$ meter)

μPa (micropascal) , 10^{-6} of 1 pascal; see dB and Pa

μs (microsecond), 1 $\mu s = 10^{-6}$ sec

ν, wave frequency; fundamental pitch

ϕ, phase

ϕ_n, eigenfunctions

$\langle \phi | \psi \rangle$, Dirac's bra-ket notation

$\Phi(\theta, \psi)$, photon flux

$\Phi(\omega)$, phase modifying function, or transfer function

$\Psi(r)$, complex wave

$\chi_D(x)$, characteristic function of a domain D

$\Psi(r)$, monochromatic complex wave

σ, covariance

σ_p, standard deviation of momentum

σ_x, standard deviation of position

$\theta_{1,2}$, angles of incidence; refractive index

$\hat{\theta}_{1,2}$, reduced angles of incidence

ω, radian frequency ($= 1/T$)

$\omega = 2\pi f_x$, radian frequency, where f_x is the circular (or circle) frequency
 (f_x is sometimes written as f)

ω_c, carrier frequency

ω_m, modulation frequency

Ω_M, density parameter for matter

Ω_λ, density parameter for dark energy

∂B, boundary of a set B

$\zeta = (\alpha, \beta, \gamma)$, variable of 3-D Fourier transform

1-D, 2-D, and 3-D, one-dimensional, two-dimensional, and three-dimensional,
 respectively

$\mathbf{0} = (0, 0, \dots, 0)$, zero vector, null vector, or origin in $\mathbb{R}^n$

$\mathbf{1} = (1, 1, \dots, 1)$, unit vector in $\mathbb{R}^n$

$\mathbf{i}, \mathbf{j}, \mathbf{k}$, unit vectors along the rectangular coordinate axes x, y, z

$n! = 1 \cdot 2 \cdot \dots (n-1)n$, $0! = 1! == (-1)! = 1$, factorial n

$\langle f, g \rangle$, inner product of functions f, g

$\|f_n\| = \sqrt{\langle f_n, f_n \rangle}$, norm of functions f_n

$\langle f, g \rangle = \int_a^b f(x)\, g(x)\, dx$, inner product of functions f and g

$\pm$, plus ($+$) or minus ($-$)

$\iint$, double (surface) integral

$\iiint$, triple (volume) integral

∇^2, Laplacian $\dfrac{\partial^2}{\partial x^2} + \dfrac{\partial^2}{\partial y^2} + \dfrac{\partial^2}{\partial z^2}$

$\nabla^2 + k^2$, Helmholtz operator

$\oplus$, bitwise XOR operation (addition mod 2)

$x(n) \otimes y(n)$, circular convolution operation between discrete signals $x(n)$ and $y(n)$; also, multiplication mod 2

$\cup$, operator joining left and right halves

$\blacksquare$, end of an example, or of a proof

§, section number, e.g., §7.3 means 'section 3 in chapter 7'

#, number of

REFERENCES are cited as 'author[year]', or 'author[year: page number(s)]'; 'ff' after a page number means the 'and the following page(s)'

Note that acronyms can be found at *http://www.acronymfinder.com.*

Multiples of Hertz, SI Units

10^{-1} Hz	dHz	decihertz	10^{1} Hz	daHz	decahertz
10^{-2} Hz	cHz	centihertz	10^{2} Hz	hHz	hectohertz
10^{-3} Hz	mHz	millihertz	10^{3} Hz	kHz	kilohertz
10^{-6} Hz	μHz	microhertz	10^{6} Hz	MHz	megahertz
10^{-9} Hz	nHz	nanohertz	10^{9} Hz	GHz	gigahertz
10^{-12} Hz	pHz	picohertz	10^{12} Hz	THz	terahertz
10^{-15} Hz	fHz	femtohertz	10^{15} Hz	PHz	pentahertz
10^{-18} Hz	aHz	attohertz	10^{18} Hz	EHz	exahertz
10^{-21} Hz	zHz	zeptohertz	10^{21} Hz	ZHz	zettahertz
10^{-24} Hz	yHz	yoctohertz	10^{24} Hz	YHz	yottahertz

Other SI units used are:

1 millisecond (ms) $= 10^{-3}$ s	1 centimeter (cm) $= 10^{-2}$ m
1 microsecond (μs) $= 10^{-6}$ s	1 millimeter (mm) $= 10^{-3}$ m
1 nanosecond (ns) $= 10^{-9}$ s	1 micrometer (μm) $= 10^{-6}$ m
1 picosecond (ps) $= 10^{-12}$ s	1 nanometer (nm) $= 10^{-9}$ m
1 femtosecond (fs) $= 10^{-15}$ s	1 picometer (pm) $= 10^{-12}$ m
1 attosecond (as) $= 10^{-18}$ s	1 femtometer (fm) $= 10^{-15}$ m

1 kiloelectron volt (keV) $= 10^{3}$ eV
1 megaelectron volt (MeV) $= 10^{6}$ eV
1 gigaelectron volt (GeV) $= 10^{9}$ eV
1 teraelectron volt (TeV) $= 10^{12}$ eV

1

Introduction

The concept of a continuous wave (CW), also called a continuous waveform, has at least two connotations: as an electromagnetic wave it has constant amplitude and frequency, and as a mathematical entity it has infinite duration together with a preassigned amplitude and frequency. A radio signal consists of a radio-frequency sinusoid, known as the carrier wave, that has undergone modulation in amplitude (AM) or in frequency (FM), such that the modulating signal creates variations in the carrier amplitude in the former case, while it does so in the carrier frequency in the latter. A carrier wave used to be called a CW in the early days of radio transmission when this wave was switched 'on' and 'off', to carry information in varying duration between the 'on' and 'off' periods of a signal. This was the case in the Morse code transmission. The CW waves were also called *undamped waves* in early wireless telegraphy, in order to distinguish them from the damped waves.

This chapter defines some elementary sinusoids, like $\sin x$ and $\cos x$, both of which are periodic functions of period 2π and generate continuous sine and cosine waves, known in general as the *sinusoids*. The discrete sinusoids consist of N samples and are defined as discrete periodic functions of period N. Since the sinusoids are progressive traveling waves, these waves and their interference are discussed in detail in this chapter. In contrast, the periodic vertical shifts of the complex exponential function e^z, where $z = x + iy \in \mathbb{C}$, are discussed. The real and imaginary parts of this function are not sinusoids. First, we will start with some useful definitions.

1.1 Definitions

A single-valued function $f(x)$ is said to be *smooth* in an interval $a < x < b$ if both f and its first derivative f' are continuous in this interval. A single-valued function $f(x)$ is said to be *piecewise continuous* in an interval $a < x < b$

if there exist finitely many points $a = x_1 < x_2 < \cdots < x_n = b$, such that f is continuous in the intervals $x_k < x < x_{k+1}$ and one-sided limit $f(x_{k+})$ or $f(x_{k+1-})$ exists for all $k = 1, 2, \dots, n$. Examples of piecewise continuous functions are $f(x) = 1/x$ and $f(x) = \sin(1/x)$ in the interval $[0, 1]$, for which the one-sided limit $f(0^+)$ does not exist. A graph of an arbitrary piecewise continuous function is shown in Figure 1.1.

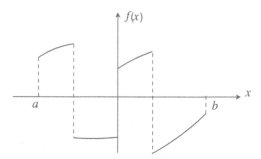

Figure 1.1 A Piecewise Continuous Function.

If f is piecewise continuous in an interval $[a, b]$ and if the first derivative f' is continuous in each of the subintervals $x_k < x < x_{k+1}$, and if the limits $f'(x_{k+})$ and $f'(x_{k-})$ exist, then f is said to be *piecewise smooth*; further, if the second derivative f'' is continuous in each of the subintervals $x_k < x < x_{k+1}$, and the limits $f''(x_{k+})$ and $f''(x_{k-})$ exist, then f is said to be *piecewise very smooth*.

A continuous or piecewise continuous function $f(x)$ on an interval $[a, b]$ is said to be *periodic* if there exists a real positive number p such that $f(x+p) = f(x)$ for all x, where p is the period of f. The smallest value of p is called the *fundamental period*. Let f be periodic with period p. Then $f(x + np) = f(x)$ for all $n \in \mathbb{Z}$. Also, if each of the functions $f_1, f_2, \dots, f_k$ has period p and c_k are constants, then $f = \sum_{j=1}^{k} c_j f_j$ has period p.

Example 1.1. The sinusoids ($\sin x$ and $\cos x$) are the well-known examples of periodic functions of period 2π. The complex exponential function $w = e^z$ is periodic with imaginary period $2\pi i$, since $e^{z+2n\pi i} = e^z$, $n \in \mathbb{Z}$. The function $w = e^z$ is an entire function that takes on every value in the complex plane except zero, i.e., $e^z \neq 0$ for any z, since $e^x > 0$ for every real number x (known as Picard's little theorem). Let $z = x + i y \in \mathbb{C}$. Then, by Euler's formula, the exponential function becomes $w = e^z = e^x (\cos y + i \sin y)$. Since the Taylor series for $e^x = \sum_{n=0}^{\infty} \frac{x^n}{n!}$, we have

$$e^{i y} = \left[1 + \frac{(i y)^2}{2!} + \frac{(i y)^4}{4!} + \cdots \right] + \left[\frac{(i y)^3}{3!} + \frac{(i y)^5}{5!} + \cdots \right] = \cos y + i \sin y. \quad (1.1)$$

The function $w = e^z$ is periodic under repeated shifts along the imaginary axis, i.e., $e^z = e^{z+2k\pi i}$, $k \in \mathbb{Z}$. Its behavior in the complex plane is determined by its behavior in any horizontal strip of width 2π, as shown in Figure 1.2. The functions $u = \Re\{e^z\}$ and $v = \Im\{e^z\}$ are not sinusoids. ∎

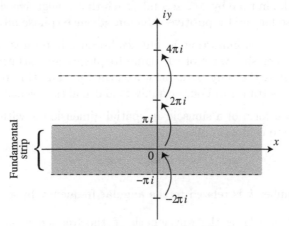

Figure 1.2 Behavior of e^z in $\mathbb{C}$.

Given a polynomial $P(z) = a_n(z - \zeta_1)^{m_1} \cdots (z - \zeta_k)^{m_k}$ for all $z \in \mathbb{C}$, the numbers ζ_k, generally complex-valued, are called the k zeros of $P(z)$ of multiplicity m_k (i.e., repeated m_k times). If $f(z)$ is analytic on a punctured disk $0 < |z - z_0| < r$, but is undefined at z_0, then z_0 is an *isolated singularity* of $f(z)$. For example, $f(z) = (\sin z)/z$ is analytic everywhere except at $z = 0$ where it is not well defined. Riemann's theorem states that if $f(z)$ is analytic and bounded on a punctured disk $0 < |z - z_0| < r$, then $z = z_0$ is a *removable singularity* for $f(z)$. An analytic function $f(z)$ with an isolated singularity at $z = z_0$ has a pole of order m iff $f(z)$ can be written in the form

$$f(z) = \frac{g(z)}{(z - z_0)^m} \quad \text{for all } z \neq z_0, \tag{1.2}$$

where $g(z)$ is analytic and nonzero at $z = z_0$.

1.2 Continuous Sinusoids

A *continuous sinusoid* is a sine wave that describes smooth repetitive oscillations. It occurs in pure and applied mathematics, in physics, engineering, signal processing, and many other fields. The most general form of a sinusoid is

$$x(t) = A\sin(2\pi f t + \phi) = A\sin(\omega t + \phi), \tag{1.3}$$

where A, the amplitude, is a peak deviation of the function from zero, f the frequency or the number of oscillations (cycles) per second, and $\omega = 2\pi f$ the angular frequency, which is the rate of change of the function argument per second, where units for ω and f are radian and cycle, respectively ($-\pi \leq \omega \leq$ π, and $-\frac{1}{2} \leq f \leq \frac{1}{2}$). The quantity ϕ represents the phase that specifies (in radians) when the oscillation is at $t = 0$ in its cycles. When $\phi \neq 0$ the entire waveform shifts in time by ϕ/ω seconds, such that a negative value represents a delay (phase lag) and a positive value an advance (phase advance).

The sine wave is important in physics because it retains its shape when added to another sine wave of the same frequency and arbitrary phase and magnitude. It is the only periodic waveform that has this property, which imparts its importance in Fourier analysis and makes it acoustically unique.

The general form of a sinusoid in spatial dimension x with wave number k and a nonzero center amplitude A_c is

$$y(x,t) = A\sin(\omega t - kx + \phi) + A_c. \tag{1.4}$$

The wave number k is related to the angular frequency by $k = \dfrac{\omega}{c} = \dfrac{2\pi f}{c} = \dfrac{2\pi}{\lambda}$, where $\lambda = c/f$ is the wavelength, f the frequency, and c the speed of propagation. This equation defines a one-dimensional sine wave. This gives the amplitude of the wave at a position x at time t along a single line. This could be considered the value of a wave on a wire. In two or three dimensions, the same equation describes a plane traveling wave if position x and wave number k are regarded as vectors, and their product as a dot (scalar) product.

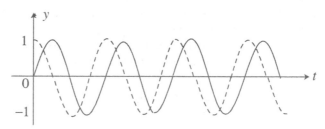

Figure 1.3 Sine Waves (solid) and Cosine Waves (dotted).

The sinusoidal wave pattern occurs often in nature, as in ocean waves, sound waves, and light waves. The cosine wave has a fundamental relation to the circle: it is the locus of a point moving in counter-clockwise direction along the circumference of the circle. A cosine curve is a sinusoid because $\cos(x) = \sin(x + \pi/2)$, which is also a sine wave with a phase shift of $\pi/2$. Because of this 'head start', it is often said that 'the cosine wave leads the sine wave or the sine legs the cosine.' This geometric feature is represented in Figure 1.3.

Since the sine waves propagate without changing form in distributed linear systems, they are often used to analyze wave propagation. The waveforms traveling in two directions can be represented as $y(t) = A\sin(\omega t - kx)$ and $y(t) = A\sin(\omega t + kx)$. When two forms having the same amplitude and frequency travel in opposite directions and superpose each other, then a standing wave pattern is created.

Besides the sine wave, there are other nonsinusoidal periodic waveforms with amplitude alternating at a given frequency between fixed minimum and maximum values. Such waves are, for example, the square wave, rectangle wave, triangle wave, and the sawtooth wave, shown in Figure 1.4. Each of these waves can be represented as an infinite sum of sinusoidal waves, and thus, as a Fourier series.

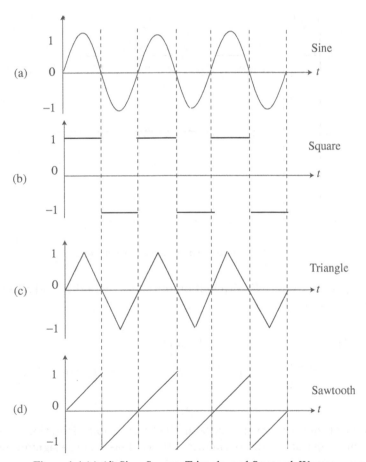

Figure 1.4 (a)-(d) Sine, Square, Triangle, and Sawtooth Waves.

In the case of ideal (mathematical) square waves the transition between minimum to maximum is instantaneous. Rectangle waves are similar to square waves, but they are not necessarily symmetrical and have different durations at minimum and maximum. Square waves are a particular case of rectangle waves. They are encountered in electronics and signal processing, especially in digital switching systems. They are generated by binary logic devices, and are used in timing references or 'clock signals', because they have fast transitions and are, therefore, very suitable for triggering synchronous logic circuits at preassigned precise intervals. An example of a square wave is a two-level Rademacher function, also known as a pulse wave, which is a periodic waveform consisting of instantaneous translations between two levels. The square wave $s_{1/2}(t)$, presented in Figure 1.5, has period 2 and levels $-1/2$ and $1/2$. Other common levels are $(-1,1)$ and $(0,1)$, which are digital signals.

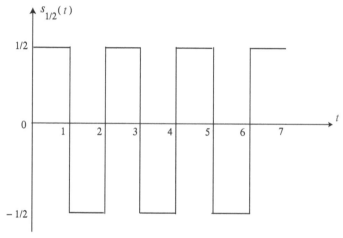

Figure 1.5 Two-Level Rademacher Function.

The square wave $s_{1/2}(t)$ with half-amplitude A, period T, and offset (phase shift) t_0 is defined by

$$s_{1/2}(t) = A(-1)^{\lfloor 2(t-t_0)/T \rfloor} = A \operatorname{sgn}\left[\sin\left(\frac{2(t-t_0)}{T}\right)\right]$$
$$= \frac{2i\,A}{\pi}\left[\tanh^{-1}\left(e^{-i\pi(t-t_0)/T}\right) - \tanh^{-1}\left(e^{i\pi(t-t_0)/T}\right)\right]. \tag{1.5}$$

The ideal square wave (in mathematics) has various definitions which are equivalent except at the discontinuities. It is defined via the signum functions of the sinusoids:

$$x(t) = \operatorname{sgn}\left(\sin(t)\right), \quad \text{or} \quad y(t) = \operatorname{sgn}\left(\cos(t)\right),$$

where sgn is ± 1 according as the sinusoid is positive or negative, and 0 at the discontinuities. It can also be defined in terms of the Heaviside step function

$H(t)$, or the rect function, as

$$x(t) = \sum_{n=-\infty}^{\infty} \left\{ H(t - nT + 1/2) - H(t - nT - 1/2) \right\}$$

$$= \sum_{n=-\infty}^{\infty} \text{rect}(t - nT), \tag{1.6}$$

where $T = 2$ for a half-cycle. It can be defined as a piecewise continuous function

$$x(t) = \begin{cases} 1 & \text{if } |t| < T_1, \\ 0 & \text{if } T_1 < |t| \le T/2, \end{cases} \tag{1.7}$$

where $x(t + T) = x(t)$, i.e., $x(t)$ is a periodic function of period T.

The square wave can be represented as a Fourier series with cycle frequency f over time t by

$$x_{\text{square}}(t) = \frac{4}{\pi} \sum_{k=1}^{\infty} \frac{\sin\left(2\pi(2k-1)ft\right)}{2k-1}$$

$$= \frac{4}{\pi} \left[\sin(2\pi ft) + \frac{1}{3}\sin(6\pi ft) + \frac{1}{5}\sin(10\pi ft) + \cdots \right]. \tag{1.8}$$

This result shows that the square wave contains only the components of odd-integer harmonic frequencies of the form $2\pi(2k-1)ft$. The convergence of the Fourier series expansion (1.8) leads to the Gibbs phenomenon, discussed in the next chapter.

The square wave changes between the maximum (high) and minimum (low) state instantaneously and without under- or over-shooting. This phenomenon does not exist in real physical systems as it would require infinite bandwidth. Square waves in all such physical systems have finite bandwidth and often generate ringing effects as in the Gibbs phenomenon, or the ripple effects as in the σ-approximation (see §2.6.1). The frequency graph (Figure 1.4(b)) shows that square waves contain a wide range of harmonics, which can generate electromagnetic radiation or pulses that interfere with other adjacent circuits and create noise or interference, resulting in errors. This problem is avoided in sensitive circuits, like the precision analog-to-digital converters, by using sine waves instead of square waves.

The rect function, also known as the Pi-function, unit impulse, gate function, or normalized boxcar function, is defined as

$$\text{rect}(t) = \Pi(t) = \begin{cases} 0 & \text{if } |t| > \frac{1}{2}, \\ \frac{1}{2} & \text{if } |t| = \frac{1}{2}, \\ 1 & \text{if } |t| < \frac{1}{2}. \end{cases} \tag{1.9}$$

An alternative definition is $\operatorname{rect}\left(\pm\frac{1}{2}\right) = 0, 1,$ or undefined. A graph of this function is shown in Figure 1.6(a). Let $H(t)$ be the Heaviside function, defined by $H(t) = \begin{cases} 0 \text{ if } t < 0 \\ 1 \text{ if } t > 0 \end{cases}$. Then $\operatorname{rect}(t)$ is a special case of the general boxcar function

$$\operatorname{rect}\left(\frac{t-x}{y}\right) = H\left(t - (x-y)/2\right) - H\left(t - (x+y)/2\right)$$
$$= H(t - x + y/2) - H(t - x - y/2),$$

where the function is centered at x and has duration $y \in (x - y/2, x + y/2)$. Another example is

$$\operatorname{rect}\left(\frac{t-T/2}{T}\right) = H(t) - H\left(\frac{t-T/2}{T}\right).$$

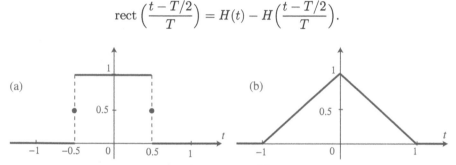

Figure 1.6 (a) Rect-Function $\operatorname{rect}(t)$; (b) Tri-Function.

The triangular function, also known as the triangle function, wedge function, or tent function, is defined either as:

$$\operatorname{tri}(t) \equiv \wedge(t) \overset{\text{def}}{=} \max\{1 - |t|, 0\}$$
$$= \begin{cases} 1 - |t| & \text{if } |t| < 1, \\ 0 & \text{otherwise;} \end{cases} \tag{1.10}$$

or, in terms of rect-functions, as

$$\operatorname{tri}(t) = \operatorname{rect}(t) \star \operatorname{rect}(t) \overset{\text{def}}{=} \int_{-\infty}^{\infty} \operatorname{rect}(u) \operatorname{rect}(t-u)\, du$$
$$= \int_{-\infty}^{\infty} \operatorname{rect}(u) \operatorname{rect}(u-t)\, du. \tag{1.11}$$

This function is also defined as $\operatorname{tri}(t) = \operatorname{rect}(t/2)\,(1 - |t|)$. Its graph is presented in Figure 1.6(b). This function, like the rect-function, is useful in signal processing, representing an idealized signal, and as a kernel from which more realistic signals are generated. It is also applied in pure code modulation as a pulse shape for transmitting digital signals, and as a matched filter for receiving signals.

In some cases, the tri-function is defined with a base of length t instead of $t/2$:

$$\text{tri}(t) \equiv \wedge(t) \stackrel{\text{def}}{=} \max\{1 - 2|t|, 0\}$$
$$= \begin{cases} 1 - 2|t| & \text{if } |2t| < 1, \\ 0 & \text{otherwise.} \end{cases} \tag{1.12}$$

This function is equivalent to the triangular window (because of its shape), sometimes called the *Bartlett window*.

1.3 Discrete Sinusoids

A *discrete sinusoid* which represents a digital signal is defined as

$$x(n) = a\cos(\omega n + \phi), \tag{1.13}$$

where the signals are denoted by a sequence of numbers (or *samples*)

$$\{\cdots, x(n-1), x(n), x(n+1), \cdots\},$$

the index $n \in \mathbb{Z}$ identifies the *sample number*, a is the amplitude, ω the angular frequency in radians, and ϕ the initial phase. In general, the phase depends on the sample number n, and is equal to $\omega n + \phi$, where $n = 0$ for the initial phase. A discrete sinusoid, as a signal 60 points long, is presented in Figure 1.7, with amplitude 1, angular frequency 0.25, and initial phase zero.

The relationship between digital signals and time is defined by the *sample rate R*, which is the number of samples that fit into one second, and where the relation between samples n and time t is given by $n = Rt$, or $t = n/R$. A sinusoidal signal with angular frequency ω has a real-time frequency f given by $f = \dfrac{\omega R}{2\pi}$ in Hertz (i.e., cycles per sec). The unit 'Hertz' is written in short as 'Hz'.

Since a cycle is 2π radians and one second covers n samples, the amplitude of a real-world signal is expressed as a time-varying voltage. But the samples of a digital signal are numbers that are independent of time. This feature allows the use of a range of hardware that is selected for an instrument. The interval, which the samples of a sinusoid $x(n)$ cover in a single period, is called the *window* of the signal.

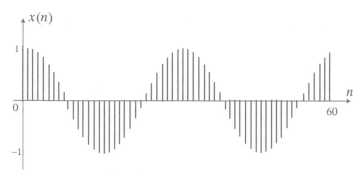

Figure 1.7 A Discrete Sinusoid.

1.4 Harmonic Series

In mathematics, the harmonic series is the divergent series

$$\sum_{n=1}^{\infty} \frac{1}{n} = 1 + \frac{1}{2} + \frac{1}{3} + \cdots . \tag{1.14}$$

Its name is derived from the concept of *overtones*, or harmonics, in music. The wavelengths of the overtone of a vibrating string (see §1.5.1) are $\frac{1}{2}, \frac{1}{3}, \frac{1}{4}, \cdots$ times the string's fundamental wavelength. Every term of this series after the first is the *harmonic mean* of the neighboring terms; the phrase 'harmonic mean' likewise derives from music (see §4.4.7). The pattern of the harmonic series as applied to music is discussed in §10.2.

The harmonic mean H of positive real numbers $a_1, a_2, \ldots , a_n$ is defined as the reciprocal of the arithmetic mean of the reciprocals of $a_1, a_2, \ldots , a_n$: $H = \left(\frac{1}{n} \sum_{k=1}^{n} a_k^{-1} \right)^{-1}$. For example, the harmonic mean of 1, 2 and 4 is $\frac{3}{\frac{1}{1} + \frac{1}{2} + \frac{1}{4}} = \frac{12}{7} = 1.\overline{714285}$. The harmonic mean of two numbers a and b is $H = \frac{2ab}{a+b}$, whereas their arithmetic mean $A = \frac{a+b}{2}$, and their geometric mean is $G = \sqrt{ab}$, thus $H = G^2/A$, or $G = \sqrt{AH}$. For a continuous distribution, the harmonic mean is $H = \frac{b-a}{\int_a^b (x(t))^{-1} \, dt}$.

1.5 Traveling and Standing Waves

A mechanical wave is a disturbance created by a vibrator. It travels through a medium from one location to another, and carries energy as it moves. The propagation of a wave through a medium involves particle interaction such that one particle pushes or pulls its adjacent neighbor, causing a displacement of the neighbor from its equilibrium position. During such a propagation, a crest appears to be moving along from particle to particle. This crest is followed by a trough that is in turn followed by the next crest. In fact, a distinct wave pattern in the form of a sine wave (or sinusoid) travels through the medium. This sinusoid continues to move uninterrupted until it encounters another wave in the medium or until it encounters a boundary with another medium. Waves that appear to move this way are sometimes called *traveling waves*. They are observed when a wave is not confined to a given space along the medium. The most common example of a traveling wave is an ocean wave. But if a wave is introduced into a bounded medium, it becomes confined to that region because it can travel only a finite distance. Once this wave reaches the boundary, it will reflect and travel back in the opposite direction. Any reflected portion of the wave will then interfere with the portion of the wave incident towards the boundary. This interference produces a new wave in the medium that hardly resembles a sinusoid. In fact, there are traveling waves in the region but they are not easily detectable because of their interference with each other. In such cases, instead of the sinusoid, an irregular and nonrepeating pattern is produced in the finite region that tends to change its shape over time. This shape is the result of interference of an incident sinusoid with a reflected sinusoid, such that both the incident and reflected waves continue to move through the medium, and encounter one another at different locations in different ways.

However, it is possible for a wave confined to a given region to produce a regular wave pattern. For example, if an elastic string is held end-to-end and vibrated at just the right frequency, a sinusoidal wave pattern is generated, which changes over time. But when a different frequency is used, the interference of the incident and the reflected wave occurs such that certain specific points along the medium will appear to be standing still. Such waves are called *standing waves*, in which there are other points along the medium whose displacement changes regularly over time. These points vibrate regularly back and forth from a positive displacement to a negative displacement. The first harmonic standing wave pattern is shown in Figure 1.8, which represents only one-half cycle of the motion of the standing wave.

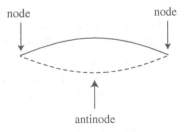

Figure 1.8 First Harmonic Standing Wave Pattern.

Standing wave patterns are produced in a medium when it is vibrated at certain frequencies and when two waves of the same frequency interfere. Then they generate points along the medium, which appear to be standing still. At these points each frequency is associated with a different and unique standing wave pattern.

A standing wave pattern is not actually a wave; rather it is the pattern resulting from the presence of two waves of the same frequency with different directions of travel within the same medium. It is an interference phenomenon formed as a result of perfectly timed interference of two waves passing through the same medium. The frequencies of standing waves and their associated wave patterns are referred to as *harmonics*.

Consider the first harmonic standing wave pattern for a vibrating string as shown in Figure 1.9. The pattern for the first harmonic shows a single antinode in the middle of the string. This antinode on the string vibrates up and down to a maximum upward displacement from rest to a maximum downward displacement. A continuous vibration of the string in this manner creates the appearance of a loop within the string. A complete wave pattern can be described as starting at the rest position, rising upward to a peak displacement, then returning back down to a rest position, and then descending to a peak downward displacement, and finally returning back to the original rest position. As shown in Figure 1.10, one complete wave in a standing wave pattern consists of two loops. Thus, each loop is equivalent to one-half of a wavelength. This wave pattern can be described in terms of the following list, where L is the loop-length and λ is the wavelength:

1st Harmonic: $L = \frac{1}{2}\lambda$;   2nd Harmonic: $L = \frac{2}{2}\lambda$;
3rd Harmonic: $L = \frac{3}{2}\lambda$;   4th Harmonic: $L = \frac{4}{2}\lambda$;
5th Harmonic: $L = \frac{5}{2}\lambda$, and so on; nth Harmonic: $L = \frac{n}{2}\lambda$.

Some of their graphs are presented in Figure 1.9.

The standing wave patterns and the length-wavelength relationships for the first three harmonics, as shown in Figure 1.10, are as follows: the number of antinodes in the pattern is equal to the harmonic number of that pattern. The first harmonic has one antinode; the second harmonic has two antinodes; and the third harmonic has three antinodes. Thus, in general, the nth harmonic has n antinodes, where n is the harmonic number. Also, there are n halves of wavelengths present within the length of the string.

Example 1.2. Let the length of a string be L units. Then, according to the above relationships, for the first harmonic, $L = \lambda/2$, or $\lambda = 2L$. For the second harmonic, $\lambda = L$, and for the third harmonic, $L = \frac{3}{2}\lambda$. Further, if this string is vibrating up and down as the first harmonic and completes 36 cycles in 10 secs, then the frequency $f = 36/10 = 3.6$ Hz, and the period is

$T = 1/(3.6)$ Hz $= 0.2\overline{7}$ secs. The speed v of the wave is given by $v = f\lambda$. ∎

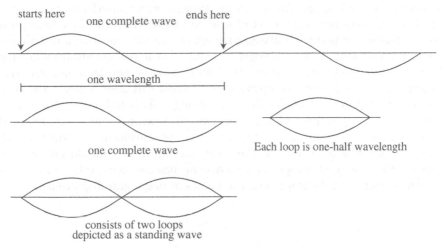

Figure 1.9 Wave Patterns.

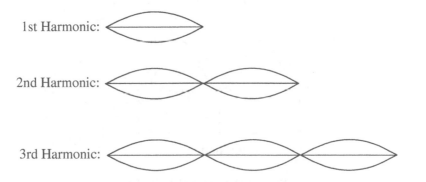

1st Harmonic:

2nd Harmonic:

3rd Harmonic:

Figure 1.10 First Three Harmonics: Loop-lengths and Wavelengths.

Every standing wave pattern is such that there are points along the medium that appear to be standing still. These points are called *nodes*. Then there are other points along the medium that undergo maximum displacement during each vibrational cycle of the standing wave. That is, they are the opposite of nodes, and hence, they are called *antinodes*. Every standing waveform always consists of an alternating pattern of nodes and antinodes, as shown in Figure 1.11, where the nodes and antinodes are labeled.

The location of nodes and antinodes in a standing waveform can be explained using the interference of the two waves. The nodes are located where interference occurs. Thus, nodes at locations where a crest of one wave meets a trough of a second wave, or a half-crest of one wave meets a half-trough of a second wave, or a quarter-crest of one wave meets a quarter-trough of a second wave, and so on. Antinodes are located where interference occurs: a crest of one wave meeting a crest of a second wave will result in a large positive displacement point. Similarly, a trough of one wave meeting a trough of a second wave results in a large negative point. Antinodes are always vibrating back and forth between these points of large positive and large negative displacement, since during a complete cycle, a crest will meet a crest, and then one-half cycle later, a trough will meet a trough. Although antinodes vibrate between a large positive and large negative displacement and are represented as the graph of a standing wave by drawing the waveform at a time and at one-half vibrational cycle later, the nodes and antinodes should not be confused with crests and troughs of a waveform, because the crests and troughs are the extrema on its graph, i.e., the points of maximum displacement.

nodes

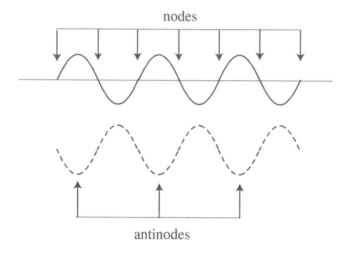

antinodes

Figure 1.11 Nodes and Antinodes.

When the motion of a traveling wave is discussed, it is customary to refer to a point of maximum positive displacement as a crest and a point of maximum negative displacement as a trough in a traveling wave. These points travel along the wave from one location to another through the medium. On the other hand, an antinode is a point on the medium that is staying in the same location and does not travel, but rather vibrates back and forth between upward and downward displacements. Thus, nodes and antinodes are not actually part of a wave, since a standing wave is not actually a wave but rather a pattern that results from the interference of two or more waves. A standing wave is formed when two identical waves moving in different directions along the same medium interfere.

A transverse wave moves to the right while the medium vibrates at right angles, up and down, as shown in Figure 1.12(a). A compressional wave moves to the right while the medium vibrates in the same direction, right to left (Figure 1.12(b)).

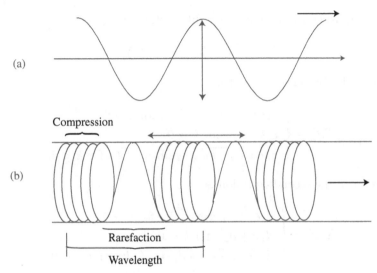

Figure 1.12 (a) Transverse and (b) Compressional Waveform.

1.5.1 Vibrating String.

The problem of a vibrating string is governed by the one-dimensional wave equation. Consider the boundary value problem

$$\frac{\partial^2 u}{\partial t^2} = c^2 \frac{\partial^2 u}{\partial x^2}, \qquad 0 < x < l,$$
$$u(0,t) = 0 = u(l,t), \qquad t > 0, \tag{1.15}$$
$$u(x,0) = f(x), \quad u_t(x,0) = h(x), \quad 0 < x < l,$$

where $f \in C^1(0,l)$ is a given function. Using Bernoulli's separation method, we seek a solution of the form $u(x,t) = X(x)T(t)$, where X is a function of x only and T a function of t only. Since $\frac{\partial^2 u}{\partial t^2} = X\ddot{T}$, and $\frac{\partial^2 u}{\partial x^2} = X''T$, where the primes denote the derivative with respect to its corresponding independent variable and the dot denotes the time derivative, Eq (1.15) reduces to $X\ddot{T} = c^2 X''T$, or, after separating the variables,

$$\frac{\ddot{T}}{T} = c^2 \frac{X''}{X}. \tag{1.16}$$

Since the left side of Eq (1.16) is a function of t only and the right side a function of x only, the only situation where $X(x)$ and $T(t)$ have solutions for all x and all t is when $\dfrac{c^2\,X''(x)}{X} = \dfrac{\ddot{T}(t)}{T} = \text{const}$. Hence,

$$\frac{1}{c^2}\frac{\ddot{T}}{T} = \frac{X''}{X} = k, \quad k = \text{const},$$

which is equivalent to two ordinary differential equations:

$$\ddot{T} - k\,c^2\,T = 0, \quad X'' - k\,X = 0.$$

The general solution of the first equation is

$$T(t) = \begin{cases} c_1 e^{c\sqrt{k}t} + c_2 e^{-c\sqrt{k}t} & \text{for } k > 0 \\ c_1 t + c_2 & \text{for } k = 0 \\ c_1 \cos c\sqrt{-k}t + c_2 \sin c\sqrt{-k}t & \text{for } k < 0, \end{cases}$$

while that of the second equation is

$$X(x) = \begin{cases} d_1 e^{\sqrt{k}x} + d_2 e^{-\sqrt{k}x} & \text{for } k > 0 \\ d_1 x + d_2 & \text{for } k = 0 \\ d_1 \cos \sqrt{-k}x + d_2 \sin \sqrt{-k}x & \text{for } k < 0. \end{cases}$$

Using the boundary conditions in (1.15), the solution for $X(x)$ is

$$X_n(x) = d_{2,n} \sin \frac{n\pi x}{l},$$

where each eigenfunction $\sin \dfrac{n\pi x}{l}$ corresponds to the eigenvalue $k = -\dfrac{n^2\pi^2}{l^2}$.
The solutions for $T(t)$ for the choice of $k < 0$ are then obtained as

$$T_n(t) = c_{1,n} \cos \frac{n\pi ct}{l} + c_{2,n} \sin \frac{n\pi ct}{l}.$$

Thus, the infinite set of solutions is

$$u_n(x,t) = X_n(x)T_n(t) = \left[A_n \cos \frac{n\pi ct}{l} + B_n \sin \frac{n\pi ct}{l} \right] \sin \frac{n\pi x}{l},$$

where the constants A_n and B_n are determined from the initial conditions.
Next, after satisfying the initial conditions in (1.15) and using the superposition principle, which holds since the boundary conditions are homogeneous, the solution of the problem is

$$u(x,t) = \sum_{n=1}^{\infty} X_n(x)T_n(t) = \sum_{n=1}^{\infty} \left[A_n \cos \frac{n\pi ct}{l} + B_n \sin \frac{n\pi ct}{l} \right] \sin \frac{n\pi x}{l}, \quad (1.17)$$

where

$$A_n = \frac{1}{l} \int_{-l}^{l} f(x) \sin \frac{n\pi x}{l} \, dx = \frac{2}{l} \int_{0}^{l} f(x) \sin \frac{n\pi x}{l} \, dx,$$

and

$$B_n = \frac{2}{n\pi c} \int_{0}^{l} h(x) \sin \frac{n\pi x}{l} \, dx, \quad n = 1, 2, \ldots .$$

Hence, the solution is completely determined. The details of the solution can be found in any textbook on partial differential equations, or in Kythe et al. [2003: 121].

The *d'Alembert solution* for this problem is

$$u(x, t) = \phi(x + ct) + \psi(x - ct), \tag{1.18}$$

where c is the wave velocity, and

$$\phi(x + ct) = \frac{1}{2} \left[f(x + ct) + g(x - ct) \right]$$

$$\psi(x - ct) = \frac{1}{2c} \left[f(x + ct) - g(x - ct) \right].$$

An interpretation of the solution of this problem is as follows: at each point x of the string we have

$$u(x, t) = \sum_{n=1}^{\infty} \alpha_n \cos \frac{n\pi c}{l} (t + \delta_n) \sin \frac{n\pi x}{l}.$$

This equation describes a harmonic motion with amplitudes $\alpha_n \sin \frac{n\pi x}{l}$. Each such motion of the string is a *standing wave*, which has its nodes at the points where $\sin(n\pi x/l) = 0$; these points remain fixed during the entire process of vibration. But the string vibrates with maximum amplitudes α_n at the points where $\sin(n\pi x/l) = \pm 1$. For any t the structure of the standing wave is described by

$$u(x, t) = \sum_{n=1}^{\infty} C_n(t) \sin \frac{n\pi x}{l},$$

where

$$C_n(t) = \alpha_n \cos \omega_n (t + \delta_n), \quad \omega_n = \frac{n\pi c}{l}.$$

At times t when $\cos \omega_n (t + \delta_n) = \pm 1$, the velocity becomes zero and the displacement reaches its maximum value.

Example 1.3. We will consider the case of a vibrating string with nonuniform density. In particular, let the density of the string be proportional to $p(x) = (1+x)^{-2}$. Then the governing equation is

$$\frac{1}{(1+x)^2}\frac{\partial^2 u}{\partial t^2} - \frac{\partial^2 u}{\partial x^2} = 0, \quad 0 < x < 1, \ t > 0, \tag{1.19}$$

subject to the initial conditions $u(x,0) = f(x)$, $\dfrac{\partial u}{\partial t}(x,0) = 0$, and the boundary conditions $u(0,t) = 0 = u(1,t)$. By Bernoulli's separation method $(u = X(x)\,T(t))$ we get

$$X'' + \frac{\lambda}{(1+x)^2}X = 0, \qquad \ddot{T} + \lambda T = 0,$$
$$X(0) = 0 = X(1), \qquad T(0) = f(x), \ \dot{T}(0) = 0, \tag{1.20}$$

where λ is the separation constant. This is an eigenvalue problem. The equation in X has the solutions of the form $(1+x)^a$, where $a(a-1)+\lambda = 0$, or $a = \frac{1}{2}\left(1 \pm \sqrt{1-4\lambda}\right)$. Then for $X(0) = 0$ we get

$$X(x) = (1+x)^{1+\sqrt{1-4\lambda})/2} - (1+x)^{1-\sqrt{1-4\lambda})/2}.$$

The condition $X(1) = 0$ gives $2^{1+\sqrt{1-4\lambda})/2} - 2^{1-\sqrt{1-4\lambda})/2} = 0$, or $2^{\sqrt{1-4\lambda}} = 1$. If $\lambda < 1/4$, then $\sqrt{1-4\lambda}$ is real, and this equation has no solution in this case. If $\lambda = 1/4$, the two solutions are $(1+x)^{1/2}$ and $(1+x)^{1/2}\ln(1+x)$, which are no longer linearly independent; although the latter solution satisfies the condition $X(0) = 0$, it fails to satisfy $X(1) = 0$, thus showing that $\lambda = 1/4$ is not an eigenvalue. Finally, if $\lambda > 1/4$, the expression $\sqrt{1-4\lambda}$ is imaginary, but it has the solutions

$$(1+x)^{\left(1+\sqrt{1-4\lambda}\right)/2} = (1+x)^{1/2}\exp\left\{\frac{i}{2}\sqrt{4\lambda-1}\,\ln(1+x)\right\}$$
$$= (1+x)^{1/2}\left[\cos\left(\sqrt{\lambda-1/4}\,\ln(1+x)\right) + i\,\sin\left(\sqrt{\lambda-1/4}\,\ln(1+x)\right)\right],$$

and the two linearly independent solutions are given by the real and imaginary parts of this expression. To satisfy the condition $X(0) = 0$ we take

$$X(x) = (1+x)^{1/2}\sin\left(\sqrt{\lambda-1/4}\,\ln(1+x)\right).$$

Then the condition $X(1) = 0$ gives $2^{1/2}\sin\left(\sqrt{\lambda-1/4}\,\ln 2\right) = 0$, so $\sqrt{\lambda-1/4}\,\ln 2$ must be an integral multiple of π, i.e., $\sqrt{\lambda-1/4}\,\ln 2 = n\pi$, which gives the eigenvalues as

$$\lambda_n = \left(\frac{n\pi}{\ln 2}\right)^2 + \frac{1}{4}, \quad n = 1, 2, \ldots,$$

and thus, the corresponding eigenfunctions are

$$X_n(x) = (1+x)^{1/2} \sin \frac{n\pi \ln(1+x)}{\ln 2}.$$

We take the Fourier series $f(x) = \sum_{n=1}^{\infty} c_n X_n(x)$, where, since $p(x) = (1+x)^{-2}$,

$$c_n = \frac{\int_0^1 f(x)(1+x)^{-3/2} \sin \dfrac{n\pi \ln(1+x)}{\ln 2} \, dx}{\int_0^1 (1+x)^{-1} \sin^2 \dfrac{n\pi \ln(1+x)}{\ln 2} \, dx}$$

$$= \frac{2}{\ln 2} \int_0^1 f(x)(1+x)^{-3/2} \sin \frac{n\pi \ln(1+x)}{\ln 2} \, dx,$$

assuming these integrals exist. The Fourier series converges absolutely and uniformly if $f(0) = f(1) = 0$ and $\int_0^1 f'^2 \, dx < +\infty$. Since Eq (1.20) with $T(0) = 1$, $T'(0) = 0$ has the solution $T(t) = \cos \sqrt{\lambda}\, t$, the problem (1.19) has the solution

$$u(x,t) = \sum_{n=1}^{\infty} c_n \cos\left(\sqrt{\frac{n^2\pi^2}{(\ln 2)^2} + \frac{1}{4}}\, t\right)(1+x)^{1/2} \sin \frac{n\pi \ln(1+x)}{\ln 2}. \quad (1.21)$$

This series converges uniformly, but to ensure continuous derivatives of u we must assume that $f \in C^2$, and $f(0) = f(1) = 0$, $f''(0) = f''(1) = 0$, and $\int_0^1 f'''^2 \, dx < +\infty$. This solution is available in Kythe [2011: 177]. ∎

1.6 Wave Propagation and Dispersion

The d'Alembert solution of the problem of a vibrating string, defined in §1.5.1 by the wave equation $\dfrac{\partial^2 u}{\partial t^2} = c^2 \dfrac{\partial^2 u}{\partial x^2}$, and subject to the boundary conditions $u(0,t) = 0 = u(l,t)$ and the initial conditions $u(x,0) = f(x)$, $u_t(x,0) = g'(x)$, is

$$u(x,t) = \frac{f(x+ct) + f(x-ct)}{2} + \frac{g(x+ct) - g(x-ct)}{2c}, \quad (1.22)$$

which can be represented as $u(x,t) = u_1(x,t) + u_2(x,t)$, where

$$u_1(x,t) = \frac{f(x+ct) + f(x-ct)}{2},$$

$$u_2(x,t) = \frac{g(x+ct) - g(x-ct)}{2c}. \quad (1.23)$$

The function $u_1(x,t)$ represents the path of the propagation of the initial displacement without the initial velocity, i.e., for $g'(x) = 0$. The other function $u_2(x,t)$ contains the initial velocity (initial impulse) with zero initial displacement. Geometrically, the function $u(x, t)$ represents a surface in the (u, x, t)-space (Figure 1.13(a)). The intersection of this surface by a plane $t = t_0$ is described analytically by $u = u(x, t_0)$, which exhibits the profile of the string at time t_0. However, the intersection of the surface $u(x, t)$ by a plane $x = x_0$ is given by $u = u(x_0, t)$, which represents the path of the motion of the point x_0.

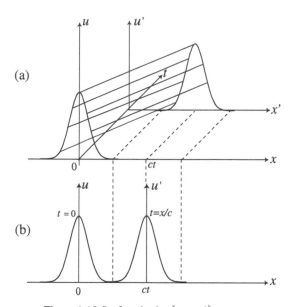

Figure 1.13 Surface in the (u, x, t)-space.

The function $u = f(x - ct)$ represents a *propagating* wave. The structure of a propagating wave at different times t is described as follows: Assume that an observer moves parallel to the x-axis with velocity c (Figure 1.13(b)). If the observer is at the initial time $t = 0$ at the position $x = 0$, then he (she) has moved along the path toward the right until some time t. Let a new coordinate system, defined by $x' = x - ct$, $t' = t$, move along with the observer. Then $u(x, t) = f(x - ct)$ is defined in this new coordinate system by

$$u(x', t') = f(x'),$$

which means that the observer sees one and the same profile $f(x')$ during the entire time t. Thus, $f(x - ct)$ represents a fixed profile $f(x')$, which moves to the right with velocity c (hence, a propagating wave). In other words, in the (x, t)-plane the function $u = f(x - ct)$ remains constant on the line $x - ct =$ const.

The other function $f(x+ct)$ similarly represents a wave propagating toward the left with velocity c. For this wave we have a similar explanation. Thus, the initial form of both waves is characterized by the function $f(x')/2$, which is equal to one-half of the original displacement.

Next, consider the one-dimensional wave equation $u_{tt} = c^2 u_{xx}$, subject to the initial conditions $u(x,0) = f(x)$, $u_t(x,0) = g'(x)$. This problem has the solution

$$u(x,t) = \frac{1}{2}\left[f(x - ct) + f(x + ct)\right] + \frac{1}{2c}\left[g(x + ct) - g(x - ct)\right].$$

To arrive at this solution, we note that the d'Alembert solution for the one-dimensional wave equation is

$$u(x,t) = F(x - ct) + G(x + ct), \quad x \in \mathbb{R}^1, \tag{1.24}$$

where F represents a disturbance (wave) traveling in the positive x direction with velocity c, while G is a disturbance traveling in the negative x direction with the same velocity. The initial conditions give

$$f(x) = F(x) + G(x), \quad g'(x) = -c\,F'(x) + c\,G'(x). \tag{1.25}$$

If we integrate the second equation in (1.25), we obtain

$$g(x) = -c\,F(x) + c\,G(x) + A,$$

where A is an arbitrary constant. This equation, along with the first equation in (1.25), gives

$$2F(x) = f(x) + \frac{1}{c}\,g(x) + A, \quad 2G(x) = f(x) - \frac{1}{c}\,g(x) - A. \tag{1.26}$$

Substituting (1.26) into (1.24), we find the solution to be

$$u(x,t) = \frac{1}{2}\left[f(x - ct) + f(x + ct) + \frac{1}{c}\,g(x + ct) - \frac{1}{c}\,g(x - ct)\right]. \tag{1.27}$$

A geometrical interpretation of Eq (1.27) is obvious from Figure 1.14, where we draw through the point P in the (x,t)-plane two lines PA and PB, which satisfy the equation $x \pm ct = \text{const}$. The disturbance at the point P at the time T is caused by those initial data that lie on the segment $AB = (x_1, x_2)$. In optics, the triangle APB is called the *retrograde light cone*. If the initial velocities are zero, i.e., if $g = 0$, then

$$u(x,t) = \frac{1}{2}\left[f(x - ct) + f(x + ct)\right]. \tag{1.28}$$

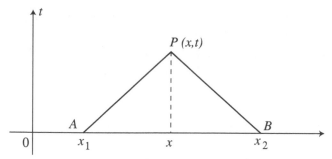

Figure 1.14 Huygens Principle in $\mathbb{R}^1$.

This represents the *Huygens principle*, which states that each point of an advancing wavefront becomes a new source of secondary waves such that the envelope tangent to all these secondary waves forms a new wavefront. The secondary waves have the same frequency and speed as the primary advancing waves. Thus, the disturbance at x at time t, originating from the sources at A and B at time zero, needs the time $\tau = (x_2 - x)/c = (x - x_1)/c$ to reach the point x.

Green's functions for the wave operator in $\mathbb{R}^1$, $\mathbb{R}^2$, and $\mathbb{R}^3$ are as follows (Kythe [2011: 179–182]):

$$G_1(x,t;x',t') = \frac{1}{2c} H\big[c(t - t') - (|x - x'|)\big], \tag{1.29}$$

$$G_2(r,t;t') = \frac{H(c(t - t') - r)}{2\pi c\sqrt{c^2(t - t')^2 - r^2}}, \tag{1.30}$$

$$G_3(x,y,z,t;x',y',z',t') = \frac{1}{4\pi\rho c}\,\delta\left(c(t - t') - \rho\right). \tag{1.31}$$

In $\mathbb{R}^1$, the solution (1.29) shows that the wave originating instantaneously at a point source $\delta(x,t)$ at time $t > 0$ covers the interval $-ct \le x \le ct$, where there exist two edges defined by $x = \pm ct$ that move forward with velocity c. This wave is observed behind the front edge and has amplitude $1/2c$. Hence, wave dispersion occurs in this case. A three-dimensional representation of Green's function in $\mathbb{R}^1$ is shown in Figure 1.15. It can be viewed as that of a wave starting at the point source and propagating as a plane wave $|x| \le ct$ whose front edge $|x| = ct$ moves with the velocity c perpendicular to the plane $x = 0$. There does not exist a rear edge of the wave in this case.

In $\mathbb{R}^2$, Green's function defined by (1.30), with $(\mathbf{x}',t') = (\mathbf{0},0)$, shows that the disturbance originates instantaneously at the point source $\delta(\mathbf{x},t)$, and at time $t > 0$ it occupies the entire circle $|\mathbf{x}| \le ct$ (see Figure 1.16(a)). The wavefront at $|\mathbf{x}| = ct$ propagates throughout the plane with velocity c, but wave propagation exists behind the front edge at all subsequent times, and the wave has no rear edge. The wave diffusion occurs in this case, and the Huygens principle does not apply.

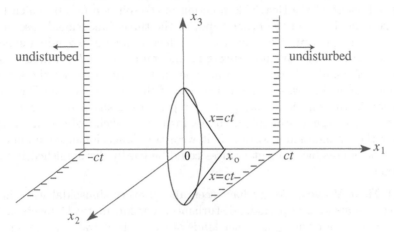

Figure 1.15 Wave propagation in $\mathbb{R}^1$.

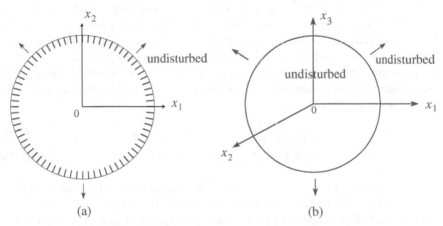

Figure 1.16 Wave propagation in (a) $\mathbb{R}^2$, and (b) in $\mathbb{R}^3$.

In $\mathbb{R}^3$, the Green's function (1.31), with $(\mathbf{x}', t') = (\mathbf{0}, 0)$, implies that the disturbance originating at a point source $\delta(\mathbf{x}, t)$ at time $t > 0$ occupies a spherical surface of radius ct and center at the origin. The wave propagates as a spherical wave with wavefront at $|\mathbf{x}| = ct$ and velocity c, and after the wave has passed, there is no disturbance (see Figure 1.16(b)). The Huygens principle applies in this case. The amplitude of the wave decays like r^{-1} as the radius increases.

There is a significant difference between the two- and three-dimensional cases. If a stone is dropped in a calm shallow pond, the leading water wave spreads out in a circular form with its radius increasing uniformly with time, but the water contained by this wave continues to move after its passage. This is because of the Heaviside function in the solution (1.30), which leaves a wake behind it. On the other hand, in the three-dimensional case, a shot fired suddenly at time $t = t'$ in still air is heard only on expanding spherical surfaces with center at the firing gun and radius $c|t - t'|$, where c is the velocity of sound. However, the air does not continue to reverberate after the passage of this wave. This is because of the presence of the Dirac delta function in the solution (1.31), which represents a sharp bang and no tail effect. The Huygens principle accounts for the simplicity of communications in our three-dimensional world. If it were two-dimensional, communication would be impossible since utterances could be hardly distinguished from one another.

1.6.1 Heat Waves. As we have seen above, every sinusoidal wave has its source in some kind of periodic disturbance. The kinematical aspects of wave motion are important. Like other kinds of waves, heat waves possess similar kinematics. We will discuss the problem of propagation of heat waves in the earth.

Changes in the earth's surface occur daily (day-night cycle) as well as annually (summer-winter cycle). Disregarding the earth's inhomogeneity, we will assume that the periodic temperature distribution in the earth is homogeneous. Since the effect of the initial temperature becomes small after several temperature variations, we have to solve a steady-state boundary value problem without initial conditions. Thus, we seek a bounded solution $u(y, t)$ for the following problem:

$$u_t = k\, u_{yy}, \qquad\qquad 0 \le y < \infty, \qquad\qquad (1.32)$$

$$u(0, t) = A\, \cos\omega t, \quad -\infty < t. \qquad\qquad (1.33)$$

Note that the range $0 \le y < \infty$ represents the medium from the earth's surface to its interior. We assume that the solution is of the form $u(y, t) = A\, e^{\alpha y + \beta t}$, where α and β are constants to be determined, and A is a preassigned constant. Then on substituting this solution into Eq (1.32) and using the complex form of the boundary condition $u(0, t) = A\, e^{i\omega t}$, we find that $\beta = i\omega$, and $\alpha^2 = \omega/k = i\omega/k$, which gives $\alpha = \pm(1 + i)\sqrt{\omega/2k}$. Hence,

$$u(y, t) = A\, e^{\pm y\,\sqrt{\omega/2k} + i\left(\pm y\,\sqrt{\omega/2k} + \omega t\right)}. \qquad\qquad (1.34)$$

The bounded solution is obtained by choosing the minus sign in α. Thus, taking the real part of (1.34), we obtain the required solution as

$$u(y, t) = A\, e^{-y\,\sqrt{\omega/2k}} \cos\left(y\,\sqrt{\omega/2k} - \omega t\right). \qquad\qquad (1.35)$$

It is known that when the temperature of the earth's surface changes periodically over a long period of time, the temperature fluctuations in its interior develop with the same period. The structure of the heat waves is as follows:

(i) The amplitude of these waves, given by $A\,e^{-y\,\sqrt{\omega/2k}}$, decreases exponentially with the depth y. Thus, an increase in depth results in a decay of the amplitude.

(ii) The temperature fluctuations in the earth occur with a phase lag of $y/\sqrt{2k\omega}$, which denotes the time that lapses between the temperature maximum and minimum inside the earth, and the corresponding time point on the surface is proportional to the depth y.

(iii) The change in temperature amplitude is given by $e^{-y\,\sqrt{\omega/2k}}$, which means that the depth of penetration of the temperature is smaller when the period $2\pi/\omega$ is smaller. Thus, for two distinct temperature distributions at time t_1 and t_2, the corresponding depths y_1 and y_2 at which the relative temperature change is the same are related by

$$y_2 = y_1\,\sqrt{t_2/t_1}.$$

For example, a comparison between the daily $(t_1 = 1)$ and the annual variation with $t_2 = 365\,t_1$ yields $y_2 = \sqrt{365}\,y_1 \approx 19.104\,y_1$, i.e., the depth of penetration of the annual temperature distribution with the same amplitude as on the surface is about 19 times larger than the depth of penetration of the daily temperature distribution.

1.7 Applications of Sinusoids

Sinusoids have found applications in various fields, which include, *inter alia*, signal processing, software-defined radio (SDR) communication systems, global positioning system (GPS) and global navigation satellite systems (GNSS), Fourier optics, spectroscopy and synchrotron light, x-rays, lasers, x-ray crystallography, computerized axial tomography (CAT), magnetic resonance imaging (MRI), radio waves, radiotelescope, Earth's rotation, quasars and pulsars, nanoseconds, and wavelet analysis. The areas of research are also increasing rapidly, so much so that it is quite a task to keep pace with them.

An interesting application comes from the existence of a human liver sinusoid, which is a type of small sinusoidal blood vessel, fenestrated or discontinuous (i.e., open-pore capillary as opposed to fenestrated) endothelium that serves as a location for the oxygen-rich blood from the hepatic artery and the nutrient-rich blood from the postal vein, reported in UIUC Histology Subject 589 [2007] and Sellaro et al. [2007]. A scanned copy of the micrograph of a liver sinusoid with fenestrated endothelial cells, whose fenestratae are about 100 nm in diameter is shown in Figure 1.17(a) ; its Fourier transform is presented in Figure 1.17(b).

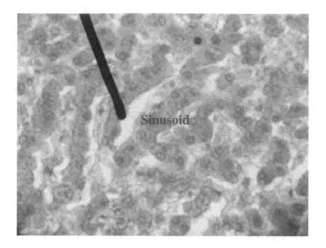

Figure 1.17(a) Liver Sinusoid.

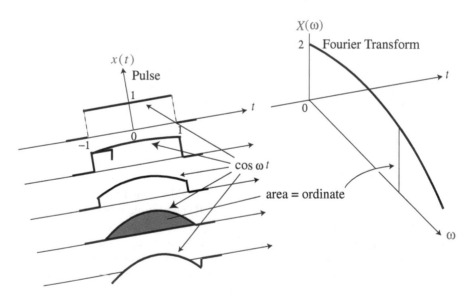

Figure 1.17(b) Fourier Transform of the Liver Sinusoid.

1.8 Historical Notes

Terras [1985] has noted: "Since its beginnings with Fourier (and as far back as the Babylonian astronomers), harmonic analysis has been developed with the goal of unraveling the mysteries of the physical world of quasars, brain tumors, and so forth, as well as the mysteries of the nonphysical, but no less concrete, world of prime numbers, diophantine equations, and zeta function." The reference to the Babylonian astronomers relates to the discovery by Neugebauer [1952] that the Babylonians used a primitive kind of Fourier series to predict celestial events.

As remarked by Zygmund [1968] and Edwards [1967], during Fourier's lifetime and thereafter, the Fourier theory had already started gaining importance in one-dimensional and two-dimensional cases (as noted in Stein and Weiss [1971]), and later in groups and discrete structures. It was used in solving partial differential equations, and this was the very incentive that started the Fourier revolution.

It was Fourier's work on the analytic theory of heat, published in 1822 (reprinted, Fourier [1978]), that developed into a method of representing a function as a sum of infinitely many harmonics. The so-called Fourier's method arose from his solution of the boundary value problem $u_{xx} + u_{yy} = 0$ subject to the conditions $u(x, \pm 1) = 0$ and $u(0, y) = 1$ for $-1 < y < 1$, which defines the heat diffusion in a lamina of semi-infinite length and a finite width centered on the x-axis, such that the temperature is held at zero along the edges $y = \pm 1$ and unity along the width $-1 \leq y \leq 1$. While imposing the boundary condition $u(0, y) = 1$, Fourier found the solution as

$$1 = a_1 \cos \frac{\pi y}{2} + a_2 \cos \frac{3\pi y}{2} + a_3 \cos \frac{5\pi y}{2} + \cdots . \tag{1.36}$$

To determine the coefficients a_n, $n \geq 1$, he differentiated the above result term-by-term infinitely many times and, after substituting $y = 0$, he obtained a linear system of infinitely many algebraic equations in infinitely many unknowns a_n. Then, using a long complicated method he determined that $a_1 = 4/\pi$, and so on. After analyzing these values he found, in terms of modern terminology, that his solution was the 'Fourier series' of the *square wave* (see Example 2.9). He also calculated the trigonometric Fourier series for other particular periodic functions, such as the triangular wave, sawtooth function, and others discussed above and in Chapter 2.

Fourier's method requires successive sinusoids that are in a harmonic relationship. They synthesize the mean value of variation of a quantity for a given location that requires the first sinusoid to have a period of, say, p per unit, the second sinusoid a period of $p/2$, the third sinusoid a period of $p/3$, and so on. This technique of summation, if stopped after a few terms, and sinusoids replaced by triangles (which are easier to calculate), would describe the Babylonian method for calculating approximate corrections to planetary positions. Similar approximating schemes of summation by triangles were adopted by Ptolemy in his system of deferents and epicycles used for motions of planets around the Earth.

Some years before Fourier's research, an astronomical computation of the orbit of a comet as a sum of sinusoids was performed by Gauss (1777–1855) who noted the harmonic relationship of the chosen periods and fixed the amplitudes and phases of the waves using a numerical algorithm that is now known as the *fast Fourier transform*. This algorithm was perfected by Cooley and Tukey [1965] and is known as the FFT.

The synthesis of time-varying signals from sinusoids is the process most commonly used to understand how the imaging technique works in technological applications, some of which have been mentioned above. To highlight the wide range of technological applications, a statement from Bernhard Riemann (1826–1866) is quoted:

"Nearly fifty years had passed without any progress on the question of analytic representation of an arbitrary function, when an assertion of Fourier threw new light on the subject. Thus a new era began for the development of this part of Mathematics and this was heralded in a stunning way by major developments in mathematical Physics."

In his work '*Partial Differential Equations in Physics*', Sommerfeld expressed the importance of Fourier analysis as follows: "Fourier's *Théorie Analytique de la Chaleur* is the bible of the mathematical physicist. It contains not only an exposition of the trigonometric series and integrals named after Fourier, but the general boundary value problem is treated in an exemplary fashion for the typical case of heat conduction."

It is generally believed that the concept of integral transform originated from Fourier's integral theorem as published in his above-mentioned research paper in 1816. But in fact Augustin Louis Cauchy (1789–1857) had already obtained in 1817 what is now called Fourier's integral theorem contained in the exponential form in his treatise entitled *Memoire sur l'emploi des equations symboliques*:

$$f(x) = \int_{-\infty}^{\infty} \int_{-\infty}^{\infty} f(t)\, e^{i\,(x-t)u}\, dt\, du, \qquad (1.37)$$

as well as the formula for functions of the operator D:

$$\phi(D)\{f(x)\} = \frac{1}{2\pi} \int_{-\infty}^{\infty} \int_{-\infty}^{\infty} \phi(i\,u) f(t)\, e^{i\,(x-t)u}\, dt\, du, \qquad (1.38)$$

which is the modern form of the operational calculus.

However, it was Oliver Heaviside (1850–1925) who, realizing the power and usefulness of operational calculus, provided methods to solve the telegraph equation and the second-order partial differential equation with constant coefficients, published in his two papers [Heaviside, 1892/1893]. He developed the operational methods, and published a book in 1899 that contained the application of the operational methods to electrical circuits and networks. His method was simple and consisted in replacing the differential operator $D = d/dt$ by p which was then treated like an ordinary algebraic quantity. However, this approach did not muster much support at that time and created a lot of controversy. Almost equally controversial was a brilliant idea of P. M. M. Dirac (1902–1984), Nobel laureate (1933), on his introduction of the δ-function that arose out of his logical formulation of quantum mechanics [Dirac, 1964], in which he states: 'All electrical engineers are familiar with the idea of a pulse, and the δ-function is just a way of expressing a pulse mathematically.' Dirac also applied Heaviside's operational calculus in electromagnetic theory. The work of these two scientists, both trained as electrical engineers, has left a great impact not only on the development of quantum mechanics but also on the study and applications of the sinusoidal theory.

Heaviside's work, even though it might lack rigor, is hailed as a remarkable achievement by all except for a few pure mathematicians. Richard P. Feynman (1918–1988), the famous theoretical physicist and Nobel laureate (1965), has remarked on Heaviside's work: 'However, the emphasis should be somewhat more on how to do mathematics quickly and easily, and what formulas are true, rather than the mathematicians' interest in methods of rigorous proof.' From the applied mathematical point of view, Heaviside's operational calculus was an important achievement. In one of Heaviside's obituaries, the famous mathematician E. T. Whittaker said: 'Looking back ..., we should place the operational calculus with Poincaré's discovery of automorphic functions and Ricci's discovery of the tensor calculus as the three most important mathematical advances of the last quarter of the nineteenth century.'

Besides sinusoids, the electromagnetic force is equally important. Writing about the four basic forces, namely, the electromagnetic force, strong nuclear force, weak nuclear force, and gravitational force, Kaku [1994: 13–14] recounts the benefits of the electromagnetic force as follows: "The electromagnetic force takes a variety of forms, including electricity, magnetism, and light itself. The electromagnetic force lights our cities, fills the air with music from radios and stereos, entertains us with television, reduces housework with electrical appliances, heats our food with microwaves, tracks our planes and space probes with radar, and electrifies our power plants. More recently, the power of the electromagnetic force has been used in electronic computers ..., and in lasers .... More than half the gross national product of the earth, representing the accumulated wealth of our planet, depends in some way on the electromagnetic force."

2

Fourier Series

Fourier series and Fourier transforms are of fundamental importance for the mathematical theory of signals. We will discuss both continuous and discrete cases of Fourier series in this chapter. The continuous case is very well known and can be found in many textbooks on integral transforms and boundary value problems of partial differential equations. The convergence of Fourier series is discussed, and Heaviside unit step function, Dirac δ-function, Dirichlet kernel, and Gibbs phenomenon are presented. The discrete case related to signals and filters is discussed in detail in Chapter 4.

2.1 Orthogonality

It is assumed that the reader is familiar with the concepts of convergence and uniform convergence of infinite series. Other definitions are as follows: the *inner product*, or dot product, of two real-valued Lebesgue-measurable functions $f_1(x)$ and $f_2(x)$, $a \leq x \leq b$, denoted by $\langle f_1, f_2 \rangle$, is defined as $\langle f_1, f_2 \rangle = \int_a^b f_1(x) f_2(x) \, dx$, provided the integral exists. The functions f_1 and f_2 are said to be *orthogonal* on the interval $I = [a, b]$ if $\langle f_1, f_2 \rangle = 0$, i.e., the above integral vanishes. Similarly, a set of real-valued functions $\{f_1(x), f_2(x), \dots\}$ defined on the interval I is said to be an *orthogonal set of functions* on this interval if $\langle f_m, f_n \rangle = \int_a^b f_m(x) f_n(x) \, dx = 0$, $m, n = 1, 2, \dots$, for each m and n, $m \neq n$, provided each integral exists.

The *norm* of the functions $f_n(x)$ is denoted by $\|f_n\|$ and defined as $\|f_n\| = \sqrt{\langle f_n, f_n \rangle} = \left(\int_a^b f_n^2(x) \, dx \right)^{1/2} \geq 0$. An orthogonal set of functions $\{f_n(x)\}$ is called an *orthonormal set of functions* on the interval I if $\|f_n\| = 1$ for all $n = 1, 2, \dots$. Thus, if $\{f_n(x)\}$ is an orthonormal set of functions, then

$\langle f_m, f_n \rangle = \int_a^b f_m(x) f_n(x)\, dx = \delta_{mn}$, where $\delta_{mn} = \begin{cases} 1 & \text{if } m = n \\ 0 & \text{if } m \neq n \end{cases}$ is the

Kronecker delta. If an orthogonal set of functions $\{f_n(x)\}$ is defined on the interval I with $\|f_n\| \neq 0$, we can always construct an orthonormal set of functions $\{g_n(x)\}$ by the formula

$$g_n(x) = \frac{f_n(x)}{\|f_n\|}, \quad x \in I. \tag{2.1}$$

In fact, $\langle g_m, g_n \rangle = \displaystyle\int_a^b \frac{f_m(x)}{\|f_m\|} \frac{f_n(x)}{\|f_n\|}\, dx = \frac{1}{\|f_m\| \|f_n\|} \langle f_m, f_n \rangle = \delta_{mn}$, and hence $\|g_n\| = 1$ for all n.

The sequence of functions $\{1, \cos x, \sin x, \cos 2x, \sin 2x, \ldots, \cos nx, \sin nx\}$ forms an orthogonal system on $[-\pi, \pi]$, since for positive integers m and n

$$\int_{-\pi}^{\pi} \sin mx \sin nx\, dx = \begin{cases} 0 & \text{if } m \neq n, \\ \pi & \text{if } m = n, \end{cases}$$

$$\int_{-\pi}^{\pi} \sin mx \cos nx\, dx = 0 \quad \text{for all } m, n, \tag{2.2}$$

$$\int_{-\pi}^{\pi} \cos mx \cos nx\, dx = \begin{cases} 0 & \text{if } m \neq n, \\ \pi & \text{if } m = n, \end{cases}$$

Example 2.1. The sequence of functions $\{f_n(x) = \sin nx\}$, $n = 1, 2, \ldots$, form an orthogonal set of functions on the interval $-\pi \leq x \leq \pi$, since

$$\langle f_m, f_n \rangle = \int_{-\pi}^{\pi} \sin mx \sin nx\, dx = 0, \quad m \neq n,$$

and $\langle f_n, f_n \rangle = \pi$. Thus, $\|f_n\| = \sqrt{\pi}$, and the orthonormal set of functions is given by

$$\{g_n(x)\} = \left\{ \frac{\sin nx}{\sqrt{\pi}} \right\}, \quad n = 1, 2, \ldots. \blacksquare$$

Example 2.2. The set of functions

$$\left\{ \frac{1}{\sqrt{2\pi}}, \frac{\cos x}{\sqrt{\pi}}, \frac{\sin x}{\sqrt{\pi}}, \frac{\cos 2x}{\sqrt{\pi}}, \frac{\sin 2x}{\sqrt{\pi}}, \ldots \right\}$$

forms an orthonormal set on the interval $-\pi \leq x \leq \pi$. We have seen that $\dfrac{\sin nx}{\sqrt{\pi}}$ is orthonormal on the interval $-\pi \leq x \leq \pi$. Now, to verify the orthonormality of other functions, we have

$$\int_{-\pi}^{\pi} \frac{\cos mx}{\sqrt{\pi}} \frac{\cos nx}{\sqrt{\pi}}\, dx$$

$$= \begin{cases} \dfrac{1}{2\pi} \int_{-\pi}^{\pi} [\cos(m-n)x + \cos(m+n)x]\, dx = 0, & \text{for } m \neq n, \\[2mm] \dfrac{1}{2\pi} \int_{-\pi}^{\pi} [1 + \cos 2mx]\, dx = 1, & \text{for } m = n. \end{cases}$$

Also,

$$\int_{-\pi}^{\pi} \frac{\sin mx}{\sqrt{\pi}} \frac{\cos nx}{\sqrt{\pi}} \, dx = \frac{1}{2\pi} \int_{-\pi}^{\pi} [\sin(m-n)x + \sin(m+n)x] \, dx = 0.$$

Hence,

$$\int_{-\pi}^{\pi} \frac{1}{\sqrt{2\pi}} \frac{\sin nx}{\sqrt{\pi}} \, dx = -\frac{1}{\pi\sqrt{2}} \frac{\cos nx}{n} \bigg|_{-\pi}^{\pi} = 0,$$

$$\int_{-\pi}^{\pi} \frac{1}{\sqrt{2\pi}} \frac{\cos nx}{\sqrt{\pi}} \, dx = \frac{1}{\pi\sqrt{2}} \frac{\sin nx}{n} \bigg|_{-\pi}^{\pi} = 0,$$

$$\int_{-\pi}^{\pi} \left(\frac{1}{\sqrt{2\pi}}\right)^2 \, dx = \frac{x}{2\pi} \bigg|_{-\pi}^{\pi} = 1. \ \blacksquare$$

Consider the problem of approximating a given function $f(x)$ in an interval $[a, b]$ by a sum $s_N(x) = \sum_{n=1}^{N} c_n g_n(x)$, where $g_n(x)$, $n = 1, 2, \dots, N$, are fixed functions of x. In this process, known as *pointwise approximation*, the coefficients c_n are chosen such that the value of $s_N(x)$ comes near to $f(x)$ for each value of $x \in [a, b]$. Thus, if there is an infinite sequence $\{g_1, g_2, \dots\}$, if $s_N(x) = \sum_{n=1}^{N} c_n g_n(x)$, and if $\lim_{N \to \infty} s_N(x) = f(x)$, then the series $\sum_{n=1}^{\infty} c_n g_n(x)$ converges to $f(x)$ pointwise (i.e., at each point x). Moreover, if for each $\varepsilon > 0$ there is an N_ε, independent of x, such that $|f(x) - s_N(x)| < \varepsilon$ for all x in $[a, b]$, whenever $N \geq N_\varepsilon$, then the series $\sum_{n=1}^{N} c_n g_n(x)$ converges uniformly to $f(x)$.

We can also find the coefficients c_n so that $s_N(x)$ only approximates $f(x)$ in the sense of least squares, i.e., in the mean. Thus, for some given positive weight function $w(x)$ the quantity $\int_a^b [f(x) - s_N(x)]^2 \, w(x) \, dx$ approaches zero as $N \to \infty$. Here $s_N(x)$ is assumed to be close to $f(x)$ except for a set of intervals of small total length. If $\int_a^b (f - s_N)^2 \, w(x) \, dx = 0$, then the sequence $s_N(x)$ converges in the mean to $f(x)$, and the integral itself is called the *mean-square deviation* of s_N from f.

We will consider the problem of approximating $f(x)$ in the sense of least squares for a restricted class of functions $g_n(x)$. Two functions $g(x)$ and $h(x)$ are said to be orthogonal on the interval (a, b) with respect to a positive weight function $w(x)$ if

$$\langle g(x), h(x) \rangle_w = \int_a^b g(x) h(x) w(x) \, dx = 0.$$

Let the functions $g_n(x)$, $n = 1, 2, \dots$, be mutually orthogonal with respect to a fixed positive weight function $w(x)$: $\langle g_m, g_n \rangle = 0$ for $m \neq n$. Because of this orthogonality relation, we find by squaring and integrating that

$$\int_a^b \left[f(x) - \sum_{n=1}^N c_n g_n(x) \right]^2 w(x)\, dx$$

$$= \int_a^b f^2\, dx - 2 \sum_{n=1}^N c_n \int_a^b f g_n w\, dx + \sum_{n=1}^N c_n^2 \int_a^b g_n^2 w\, dx$$

$$= \sum_{n=1}^N \int_a^b g_n^2 w\, dx \left[c_n - \frac{\int_a^b f g_n w\, dx}{\int_a^b g_n^2 w\, dx} \right]^2 + \int_a^b f^2 w\, dx - \sum_{n=1}^N \frac{\left[\int_a^b f g_n w\, dx \right]^2}{\int_a^b g_n^2 w\, dx}, \quad (2.3)$$

on completing the square. The coefficients c_n occur only in the first sum, and since this is a sum of squares, it is obviously minimized by taking all its terms to be zero, i.e., by choosing

$$c_n = \frac{\int_a^b f g_n w\, dx}{\int_a^b g_n^2 w\, dx}. \quad (2.4)$$

This choice of the coefficients c_n minimizes the left side of (2.3). Thus, the coefficients (2.4) provide the best approximation in the sense of least squares, and these best coefficients are independent of N because of the orthogonality of $g_n(x)$. The coefficients c_n given by (2.4) are called the *Fourier coefficients* of $f(x)$ with respect to the orthogonal functions $g_n(x)$. The series $\sum_{n=1}^\infty c_n g_n(x)$ is called the *Fourier series* of $f(x)$. Since this series may or may not converge to $f(x)$, we cannot write $f(x) = \sum_{n=1}^\infty c_n g_n(x)$. However, we use the notation

$$f(x) \sim \sum_{n=1}^\infty c_n g_n(x) \quad (2.5)$$

to simply denote the fact that c_n are the Fourier coefficients defined by (2.4).

Example 2.3. Consider the function $f(x) = x^2$, and find the best approximation in the mean to $f(x)$ by a linear combination of two orthogonal functions $g_1 = 1$ and $g_2 = x^2$ on the interval $-1 \le x \le 1$. We have

$$c_1 = \frac{\int_{-1}^1 x^2\, dx}{\int_{-1}^1 dx} = \frac{1}{3}, \quad c_2 = \frac{\int_{-1}^1 x^2 \cdot x\, dx}{\int_{-1}^1 x^2\, dx} = 0.$$

Thus the function $s_2(x) = \frac{1}{3}$ is the best approximation in the sense of least squares on the given interval to x^2 among all first-degree polynomials. ∎

Note that formula (2.4) for the Fourier coefficients c_n can be derived directly if the series $\sum_{n=1}^\infty c_n g_n(x)$ converges to $f(x)$ uniformly. In that case we may multiply the series by $g_m(x)w(x)$ and integrate term-by-term from a to b. Because of orthogonality, all integrals will be zero except when $n = m$, which gives

$$\int_a^b f g_m w \, dx = c_m \int_a^b g_m^2 \, w \, dx,$$

which yields (2.4). Also, note that if no weight function $w(x)$ is specified, it is understood that $w(x) = 1$. This case is used in §2.3.

2.2 Completeness and Uniform Convergence

If we substitute the Fourier coefficients (2.5) into (2.3), we see that

$$\int_a^b \left[f(x) - \sum_{n=1}^N c_n g_n(x) \right]^2 w(x) \, dx = \int_a^b f^2 w \, dx - \sum_{n=1}^N c_n^2 \int_a^b g_n^2 w \, dx. \quad (2.6)$$

Since the left side is nonnegative, we have

$$\sum_{n=1}^N c_n^2 \int_a^b g_n^2 w \, dx \le \int_a^b f^2 w \, dx \quad (2.7)$$

for any N. This is called *Bessel's inequality*. The sum on the left side in (2.7) is nondecreasing in N and is bounded by the right side. Hence, it converges provided that the right side is finite. The functions f or w may be discontinuous or even infinite at some points as long as the integral on the right side is finite. The limit of the sum as $N \to \infty$ is denoted by the term on the left side of (2.7), and the convergence of the series therein means, in particular, that its terms approach zero. Therefore, if $f(x)$ has the property the $\int_a^b f^2 w \, dx$ is finite, then

$$c_n^2 \int_a^b g_n^2 w \, dx = \frac{\left(\int_a^b f g_n^2 w \, dx \right)^2}{\int_a^b g_n^2 w \, dx} \to 0 \quad \text{as } n \to \infty. \quad (2.8)$$

If the sequence $\left\{ \sum_{n=1}^N c_n^2 g_n^2 \right\}$ converges in the mean to f, the left side of (2.6) approaches zero as $N \to \infty$. Hence,

$$\sum_{n=1}^\infty c_n^2 \int_a^b g_n^2 w \, dx = \int_a^b f^2 w \, dx. \quad (2.9)$$

This is called *Parseval's equation*; it holds iff

$$\lim_{N \to \infty} \int_a^b \left(f - \sum_{n=1}^N c_n g_n \right)^2 w \, dx = 0.$$

If this limit holds for every function f for which $\int_a^b f^2 w\, dx$ is finite, then the set of functions $\{g_1, g_2, \dots\}$ is said to be *complete*. In that case the function f is continuous and $\int_a^b f^2 w\, dx$ is finite, and if $\int_a^b f g_n w\, dx = 0$ for all n, then $f \equiv 0$, because in this case $c_n = 0$ for all n, and (2.9) gives $\int_a^b f^2 w\, dx = 0$, which implies that $f \equiv 0$. This property means that two continuous functions having the same Fourier coefficients with respect to a complete set of functions must be equal, since their difference has Fourier coefficients zero. Conversely, if the set of functions $\{g_1, g_2, \dots\}$ is such that $f \equiv 0$ is the only function with the property that $\int_a^b f^2 w\, dx$ is finite and that all its Fourier coefficients are zero, then the set $\{g_1, g_2, \dots\}$ is complete.

2.3 Fourier Series

Let the weight function be $w(x) = 1$. Then some important types of series expansions obtained from orthogonal sets of functions are as follows. Let $\{g_1(x), g_2(x), \dots\}$ be an orthogonal set of functions on an interval $a \leq x \leq b$, and let a function $f(x)$ be represented in terms of the functions $g_n(x)$, $n = 1, 2, \dots$, by a uniformly convergent series

$$f(x) = \sum_{n=1}^{\infty} c_n\, g_n(x). \tag{2.10}$$

This series is called a *generalized Fourier series* of $f(x)$, and the coefficients c_n, $n = 1, 2, \dots$, are called the *Fourier coefficients* of $f(x)$ with respect to the orthogonal set of functions $\{g_n(x)\}$, $n = 1, 2, \dots$. As in §2.2, to determine the coefficients c_n, we multiply both sides of (2.10) by $g_m(x)$ for a fixed m, integrate with respect to x over the interval $[a, b]$, and assume term-by-term integration, which is justified in the case of uniform convergence. Then we obtain

$$\langle f, g_m \rangle = \int_a^b f g_m\, dx = \int_a^b \left(\sum_{n=1}^{\infty} c_n\, g_n(x) \right) g_m(x)\, dx$$
$$= \sum_{n=1}^{\infty} c_n \langle g_n(x), g_m(x) \rangle. \tag{2.11}$$

Since $\langle g_n, g_m \rangle = \delta_{nm}$, we find that $\langle g_n, g_m \rangle = \|g_n\|^2$ for $n = m$, and then (2.11) yields $\langle f, g_n \rangle = c_n \|g_n\|^2$, or

$$c_n = \frac{\langle f, g_n \rangle}{\|g_n\|^2} = \frac{1}{\|g_n\|^2} \int_a^b f(x) g_n(x)\, dx.$$

Example 2.4. The functions $1, \cos x, \sin x, \cos 2x, \sin 2x, \dots$ are orthogonal on the interval $[-\pi, \pi]$ with respect to $w(x) = 1$. In the absence of any other specified functions $g_n(x)$, we obtain a Fourier series expansion using these functions. Since $\int_{-\pi}^{\pi} 1^2\, dx = 2\pi$, and $\int_{-\pi}^{\pi} \cos^2 nx\, dx = \int_{-\pi}^{\pi} \sin^2 nx\, dx = \pi$ for $n = 1, 2, \dots$, so if we define the Fourier coefficients as

$$a_n = \frac{1}{\pi} \int_{-\pi}^{\pi} f(t) \cos nt\, dt, \quad n = 0, 1, 2, \dots,$$

$$b_n = \frac{1}{\pi} \int_{-\pi}^{\pi} f(t) \sin nt\, dt, \quad n = 1, 2, \dots, \tag{2.12}$$

then the Fourier series of $f(x)$ is

$$f(x) = \frac{a_0}{2} + \sum_{n=1}^{\infty} (a_n \cos nx + b_n \sin nx), \tag{2.13}$$

where the Fourier coefficients a_n and b_n are defined by (2.12). ∎

In this example, we have introduced the *trigonometric Fourier series* of a periodic function $f(x)$ of period 2π, under the assumption that the series converges and represents the function $f(x)$. The coefficients a_0, a_n and b_n for $n = 1, 2, \dots$ are called the *Fourier coefficients*.

A function $f(x)$ is said to be *periodic* of period p if $f(x + p) \equiv f(x)$, $c \le x \le c + p$, where c is a constant, and p a positive real number for which the identity holds. For example, the functions $\sin x$ and $\cos x$ have period 2π. More generally, each of the functions $\sin \dfrac{2n\pi x}{p}$ and $\cos \dfrac{2n\pi x}{p}$ is periodic of period p, where n is a positive integer. Hence, a function $f(x)$ of period p is represented by the infinite series

$$f(x) = \frac{a_0}{2} + \sum_{n=1}^{\infty} \left(a_n \cos \frac{2n\pi x}{p} + b_n \sin \frac{2n\pi x}{p} \right) \tag{2.14}$$

provided this series is convergent.

Theorem 2.1. (Fourier Theorem I for Periodic Functions) *Let $f(x)$ be a single-valued, piecewise continuous, periodic function of period p on a finite interval $I = [c, c+p]$, where c is a constant. Then the series (2.14) converges to $f(x)$ at all points of continuity and to*

$$\frac{1}{2} \left[f(x+) + f(x-) \right] \tag{2.15}$$

at the points of discontinuity (and also at all points of continuity). The coefficients a_0, a_n and b_n are given by

$$a_n = \frac{2}{p} \int_{c}^{c+p} f(x) \cos \frac{2n\pi x}{p}\, dx, \quad n = 0, 1, 2, \dots,$$

$$b_n = \frac{2}{p} \int_{c}^{c+p} f(x) \sin \frac{2n\pi x}{p}\, dx, \quad n = 1, 2, \dots. \tag{2.16}$$

2.3.1 Some Special Cases of the Fourier series expansion (2.14) are:

(a) If we set $p = 2L$ and $c = -L$ in (2.16), then these formulas become

$$a_n = \frac{1}{L} \int_{-L}^{L} f(x) \cos \frac{n\pi x}{L} \, dx, \quad n = 0, 1, 2, \ldots,$$

$$b_n = \frac{1}{L} \int_{-L}^{L} f(x) \sin \frac{n\pi x}{L} \, dx, \quad n = 1, 2, \ldots.$$

(2.17)

(b) If the period $p = 2\pi$, then

$$f(t) = \frac{a_0}{2} + \sum_{n=1}^{\infty} (a_n \cos n\omega t + b_n \sin n\omega t), \quad \text{where } a_0 = \int_{\pi} f(t) \, dt,$$

$$a_n = \frac{1}{\pi} \int_{-\pi}^{\pi} f(t) \cos nt \, dt, \quad b_n = \frac{1}{\pi} \int_{-\pi}^{\pi} f(t) \sin nt \, dt \quad \text{for } n = 1, 2, \ldots.$$

(2.18)

(c) For even and odd f, we have

$$a_n = \frac{4}{p} \int_0^{p/2} f(t) \cos n\omega t \, dt, \;\; b_n = 0 \quad \text{if } f(t) \text{ is even;}$$

$$a_n = 0, \;\; b_n = \frac{4}{p} \int_0^{p/2} f(t) \sin n\omega t \, dt \quad \text{if } f(t) \text{ is odd.}$$

(2.19)

2.3.2 Amplitude and Phase Form. Let $f(t)$ be be real, and $\omega = 2\pi/t$. Then

$$f(t) = \frac{A_n}{2} + \sum_{n=1}^{\infty} A_n \cos(n\omega t + \alpha_n), \quad A_n \geq 0 \text{ for } n \geq 1.$$

(2.20)

The amplitudes A_n are defined in sin, cos form as

$$A_0 = \frac{a_0}{2}, \; A_n = \sqrt{a_n^2 + b_n^2}, \; \alpha_n = \arg\{a_n + i\, b_n\} = \begin{cases} -\arctan(b_n/a_n) \text{ if } a_n > 0, \\ \pi - \arctan(b_n/a_n) \text{ if } a_n < 0. \end{cases}$$

Example 2.5. Consider the wave equation $u_{tt} = c^2 u_{xx}$, where c is the wave velocity, subject to the boundary conditions $u(0, t) = 0 = u(l, t)$ for $t \geq 0$, and the initial conditions $u(x, 0) = x$, and $u_t(x, 0) = 0$. We have already seen that the d'Alembert solution of the wave equation is of the form

$$u(x, t) = f(x + ct) + g(x - ct).$$

From the boundary condition $u(0, t) = 0$ we find that

$$f(ct) + g(-ct) = 0,$$

or, setting $ct = z$, we get $f(z) = -g(-z)$. Thus, the solution reduces to

$$u(x, t) = f(ct + x) - f(ct - x).$$

Using the other boundary condition $u(l, t) = 0$, we have

$$f(l + ct) + g(l - ct) = 0,$$

which implies that $f(ct + l) - f(ct - l) = 0$, i.e., $f(z) = f(z + 2l)$. This last equation means that f is a periodic function of period $2l$. If we apply the initial conditions, we get

$$f(x) - f(-x) = x, \quad \text{and} \quad f'(x) - f'(-x) = 0,$$

which means that $f'(x)$ is an even function, and therefore $f(x)$ an odd function, i.e., $f(x) = -f(-x)$. Hence, $2 f(x) = x$. Since $f(x)$ is an odd periodic function of period $2l$, it can be expressed as a Fourier sine series

$$f(x) = \frac{x}{2} = \frac{l}{\pi} \sum_{n=1}^{\infty} \frac{(-1)^{n+1}}{n} \sin \frac{n \pi x}{l},$$

which yields

$$u(x, t) = \frac{l}{\pi} \sum_{n=1}^{\infty} \frac{(-1)^{n+1}}{n} \left[\sin \frac{n\pi(ct + x)}{l} - \sin \frac{n\pi(ct - x)}{l} \right]$$

$$= \frac{2l}{\pi} \sum_{n=1}^{\infty} \frac{(-1)^{n+1}}{n} \sin \frac{n\pi x}{l} \cos \frac{n\pi ct}{l}. \ \blacksquare$$

The partial differential equation $u_{tt} = c^2 u_{xx}$ was discovered by d'Alembert [1747/1749] for the general case of n beads on a string of length l, and letting $n \to \infty$, where $c = \sqrt{T/m}$ has the dimension of velocity, T being the tension and m the total mass of the beads. It was further assumed that the ratio m/T is negligible enough to ignore gravity, thus giving tension the prominent role in the formulation of this problem. The solution was found to be

$$u(x, t) = \sum_{n=1}^{\infty} A_n \sin \frac{n\pi x}{l} \cos \frac{n\pi ct}{l},$$

where the coefficients A_n are determined by $f(x) = \sum_{n=1}^{\infty} A_n \sin \frac{n\pi x}{l}$. The d'Alembert solutions for $t = 1/6$ and $t = 2/3$ are presented in Figure 2.1, where the dotted lines show the position at different times as marked.

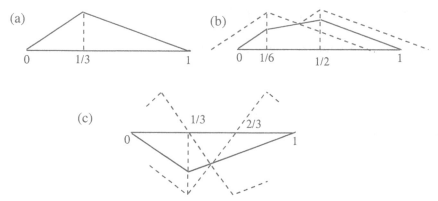

Figure 2.1 (a) Plucked String of Length 1 for $c = 1$;
d'Alembert Solutions for (b) $t = 1/6$, and (c) $t = 2/3$. ∎

Theorem 2.2. (Fourier Theorem II) *Let $f(x)$ denote a periodic function of period 2π, which is integrable on $-\pi < x < \pi$. If this integral is improper, let it be absolutely convergent. Then at each point x which is interior to an interval on which f is piecewise continuous, the Fourier series for the function f converges to the average value (2.10).*

The asymptotic behavior of the Fourier coefficients of a periodic function $f(x)$ is given by the following theorem:

Theorem 2.3. *As $n \to \infty$, the Fourier coefficients a_n and b_n always approach zero at least as rapidly as α/n, where α is a constant independent of n. If the function $f(x)$ is piecewise continuous, then either a_n or b_n, and in general both, decrease no faster than α/n. In general, if $f(x)$ and its first $(k-1)$ derivatives satisfy the conditions of the Fourier theorems I and II, then the Fourier coefficients a_n and b_n approach zero as $n \to \infty$ at least as rapidly as α/n^{k+1}. Moreover, if $f^{(k)}(x)$ is not continuous everywhere, then either a_n or b_n, and in general both, approach zero no faster than α/n^{k+1}.*

This theorem implies that the smoother the function f is, the faster its Fourier series converges. Note that the analysis of the Fourier series for functions in $\mathbb{R}^2$ and $\mathbb{R}^3$ is similar. Proofs of these theorems are available in Davis [1963], Churchill and Brown [1978], and Walker [1988].

2.3.3 Fourier Sine Series. Let f be a function defined on the interval $0 \le x \le L$ such that the integrals $\displaystyle\int_0^L f(x) \sin \frac{n\pi x}{L}\, dx$, $n = 1, 2, \ldots$, exist.

Then the series

$$\sum_{n=1}^{\infty} b_n \sin \frac{n\pi x}{L}, \qquad (2.21)$$

where

$$b_n = \frac{2}{L} \int_0^L f(x) \sin \frac{n\pi x}{L} \, dx, \quad n = 1, 2, \ldots, \qquad (2.22)$$

is called the *Fourier sine series* of f on the interval $0 \le x \le L$. Note that the series expansion (2.22) is identical to the trigonometric Fourier series (2.14) of an *odd* function defined on the interval $-L \le x \le L$, which coincides with $f(x)$ on the interval $0 \le x \le L$.

2.3.4 Fourier Cosine Series. Let f be a function defined on the interval $0 \le x \le L$ such that the integrals

$$\int_0^L f(x) \cos \frac{n\pi x}{L} \, dx, \quad n = 0, 1, 2, \ldots,$$

exist. Then the series

$$\frac{a_0}{2} + \sum_{n=1}^{\infty} a_n \cos \frac{n\pi x}{L}, \qquad (2.23)$$

where

$$a_n = \frac{2}{L} \int_0^L f(x) \cos \frac{n\pi x}{L} \, dx, \quad n = 0, 1, 2, \ldots, \qquad (2.24)$$

is called the *Fourier cosine series* of f on the interval $0 \le x \le L$.

Note that the series expansion (2.23) is identical to the trigonometric Fourier series (2.14) of an *even* function defined on the interval $-L \le x \le L$, which coincides with $f(x)$ on the interval $0 \le x \le L$.

2.3.5 Uniform Convergence. It is shown in (2.8) that if $f(x)$ is an arbitrary function for which $\int_a^b f^2 w \, dx$ is finite and of the set $\{g_n(x)\}$ is any orthogonal set of functions with respect to the positive weight function $w(x)$, then

$$\lim_{n \to \infty} \frac{\int_a^b f g_n w \, dx}{\sqrt{\int_a^b g_n^2 w \, dx}} = 0. \qquad (2.25)$$

The *Riemann-Lebesgue lemma* is as follows: If $g_n(x)/\sqrt{\int_a^b g_n^2 w \, dx}$ is uniformly bounded, i.e, if

$$\frac{|g_n(x)|}{\sqrt{\int_a^b g_n(x)^2 w(x) \, dx}} \le K, \quad K > 0,$$

and if $f(x)$ is such that the (possibly improper) integral $\int_a^b |f(x)|w(x)\,dx$ is finite, then the limit (2.25) holds. A proof of this lemma can be found, e.g., in Weinberger [1965: 77]. Now, since the functions $\cos nx$, $\sin nx$, and

$$s_N(x) = \frac{1}{2\pi} \int_{-\pi}^{\pi} f(t) \frac{\sin\left(N+\frac{1}{2}\right)(x-t)}{\sin\frac{1}{2}(x-t)}\,dt$$

are periodic of period 2π, we can expand the function $f(x)$ outside the interval $[-\pi, \pi]$ as a periodic function by defining $f(x + 2n\pi) = f(x)$ for $n = 0, \pm1, \pm2, \ldots$. Then the integrand in the integral for $s_N(x)$ is periodic, and the integral is over one period. Thus,

$$s_N(x) = \frac{1}{2\pi} \int_{-\pi}^{\pi} f(t+\tau) \frac{\sin\left(N+\frac{1}{2}\right)\tau}{\sin\frac{1}{2}\tau}\,d\tau. \qquad (2.26)$$

Integrating $\dfrac{1}{2} + \displaystyle\sum_{n=1}^{N} \cos n\tau = \dfrac{\sin\left(N+\frac{1}{2}\right)\tau}{\frac{1}{2}\sin\frac{1}{2}\tau}$ from $-\pi$ to π, we get

$$\pi = \frac{1}{2} \int_{-\pi}^{\pi} \frac{\sin\left(N+\frac{1}{2}\right)\tau}{\frac{1}{2}\sin\frac{1}{2}\tau},$$

which, when multiplied by $\dfrac{1}{\pi}f(x)$ and when subtracted from (2.26), yields

$$s_N(x) - f(x) = \frac{1}{2\pi} \int_{-\pi}^{\pi} \frac{f(t+\tau) - f(x)}{\sin\frac{1}{2}\tau} \sin(N+\tfrac{1}{2})\tau\,d\tau. \qquad (2.27)$$

The set of functions $\{\sin(N+\frac{1}{2})\tau\}$, where $N = 0, 1, 2, \ldots$, is orthogonal and satisfies the Riemann-Lebesgue lemma. Hence, if

$$\int_{-\pi}^{\pi} \left| \frac{f(t+\tau) - f(x)}{\sin\frac{1}{2}\tau} \sin(N+\tfrac{1}{2})\tau \right| d\tau < \infty,$$

the right side of (2.27) approaches zero, i.e., $s_N(x) \to f(x)$ as $N \to \infty$. This is known as *Dini's test*, and the Fourier series (2.13) or (2.14) converges uniformly to $f(x)$ where this test is satisfied.

2.4 Dirac Delta Function

The Heaviside unit step function $H(t)$, sometimes also denoted by $u(t)$, is defined as

$$H(t) = \begin{cases} 1 & \text{if } t > 0, \\ 0 & \text{if } t < 0, \end{cases}$$

and its plot is presented in Figure 2.2.

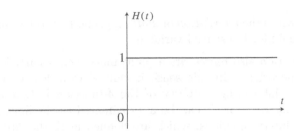

Figure 2.2 Heaviside Function $H(t)$.

It is known that the Laplace transform $\mathcal{L}\{\delta(t)\} = 1$. We will explain the existence of the δ-function heuristically. Consider the Heaviside unit step function $H(t)$. Since the Laplace transform of $H(t)$ is $\overline{H}(s) = \dfrac{1}{s}$, then $\mathcal{L}H'(t) = s\overline{H}(s) = 1$. Obviously, $H'(t)$ vanishes for $|t| > 0$ and does not exist for $t = 0$. From the graph of $H(t)$ it is clear that there is a vertical jump at $t = 0$. Therefore, it seems reasonable to assume that $\lim\limits_{t\to 0} H'(t) \to \infty$. But since $\int_{-\varepsilon}^{\varepsilon} H'(t)\, dt = 1$, it is obvious that a function like $H'(t)$ exists only in the generalized sense. The function $H'(t)$ is known as the Dirac delta function[1], and is denoted by $\delta(t)$ such that

$$\delta(t) = \begin{cases} 0 & \text{for } |t| > 0, \\ \infty & \text{for } t = 0, \end{cases} \quad \text{and} \quad \int_{-\varepsilon}^{\varepsilon} \delta(t)\, dt = 1 \quad \text{for any } \varepsilon > 0. \qquad (2.28)$$

This function is not a function in the ordinary sense; it is called a *generalized function*.

Physically, the δ-function signifies a force of strength unity, a sudden impulse, at time $x = 0$ which is called the *source point* or *singularity* of the δ-function. If the source point is translated to t', then this function is $\delta(t - t')$, which will signify a sudden impulse of unit strength at the source point $t = t'$. The δ-function in either case is also written as $\delta(x - x')$ or $\delta(t - t')$.

There are other descriptions of this function, but they all lead to the same properties, which are

$$\delta(t-t') = \begin{cases} 0 & \text{if } t \neq t', \\ +\infty & \text{if } t = t', \end{cases} \quad \text{and} \quad \int_{t'-\varepsilon}^{t'+\varepsilon} \delta(t - t')\, dt = 1, \qquad (2.29)$$

where $\varepsilon > 0$ is an arbitrarily small real number. This property implies that

$$\int_{t'-\varepsilon}^{t'+\varepsilon} f(t)\, \delta(t - t')\, dt = f(t'), \qquad (2.30)$$

---

[1] This function was introduced by P. A. M. Dirac in 1926–1927. The development of the analysis in this and the next two sections can be traced back to examples and ideas found in Dirac [1939; 1947]. In fact, δ is the weak derivative of H via integration by parts, and thus δ is a measure.

which is known as the *translation* or *shifting property* of the δ-function. These results also hold for the spatial variable x.

The δ-function also represents a point mass, i.e., a particle of unit mass located at the origin. In this sense it can be regarded as a mass-density function. An interesting property of the δ-function is that although it is not an ordinary function, it can be approximated as the limit of a sequence of certain ordinary functions, which are defined in Kythe [2011: 44ff]. The important properties of the δ-function are:

$$\text{(i) } \delta(x) = 0 \text{ for } x \neq 0, \quad \text{(ii) } \int_{-\infty}^{\infty} \delta(x)\,dx = 1,$$

$$\text{(iii) } \int_{-\infty}^{\infty} f(x)\delta(x)\,dx = f(0). \tag{2.31}$$

By shifting the singularity (source point) from $x = 0$ to $x = x' \neq 0$, the third result in (2.31) becomes

$$\int_{-\infty}^{\infty} f(x)\delta(x - x')\,dx = f(x'). \tag{2.32}$$

which, as mentioned above, is the *translation or shifting property* of the δ-function. In the 3-D case and using spherical coordinates we have for the singularity (or source) at the origin

$$\int_0^\infty \int_0^\pi \int_0^{2\pi} \delta(r)r^2 \sin\theta\, dr\, d\theta\, d\phi = \int_{-\infty}^\infty \int_{-\infty}^\infty \int_{-\infty}^\infty \delta(x)\delta(y)\delta(z)\,dx\,dy\,dz = 1,$$
$$\tag{2.33}$$

whereas for the singularity at $r = r_1$, Eq (2.33) becomes

$$\int_0^\infty \int_0^\pi \int_0^{2\pi} \delta(r_2 - r_1)r_2^2 \sin\theta_2\, dr_2\, d\theta_2\, d\phi_2 = 1.$$

Note that $\delta(r_1 - r_2) = \delta(r_2 - r_1)$. The Dirac δ-function is used in electronics to represent the unit impulse and sampled signals.

To summarize, the important properties of the δ-function with source point at t' are as follows:

1. $\displaystyle\int_{-\infty}^{\infty} f(t)\,\delta(t - t')\,dt = f(t')$ (Fundamental property);

2. $\displaystyle\int_{-\infty}^{\infty} f(t')\,\delta(t - t')\,dt = f(t') \int_{-\infty}^{\infty} \delta(t - t')\,dt = f(t')$;

3. $f(t)\,\delta(t - t') = f(t')\,\delta(t - t')$;

4. $t\,\delta(t) = 0$;

5. $\delta(t - t') = \delta(t' - t)$;

6. $\displaystyle\int_{-\infty}^{\infty} \delta(s)\,ds = \begin{cases} 1 & \text{for } t > 0, \\ 0 & \text{for } t < 0 \end{cases} = H(t);$

7. $\dfrac{d}{dt}H(t) = \delta(t),$

where $H(t)$ is the Heaviside unit step function. Note that the last property is not true in the classical sense since $H(t)$ is not differentiable at $t = 0$.

The δ-function can be similarly defined for the space variable $\mathbf{x}$, and all the results given above hold for $\delta(\mathbf{x})$, where $\mathbf{x} \in \mathbb{R}^n$, or for $\delta(\mathbf{x} - \mathbf{x}')$ where $\mathbf{x}'$ is the source point for the δ-function. As in the 1-D case, the δ-function in $\mathbb{R}^2$ is formally written as $\delta(\mathbf{x}) = \delta(x)\,\delta(y)$, and $\delta(\mathbf{x}; t) = \delta(x)\,\delta(y)\,\delta(t)$; similarly, in $\mathbb{R}^3$ it is formally written as $\delta(\mathbf{x}) = \delta(x)\,\delta(y)\,\delta(z)$, and $\delta(\mathbf{x}; t) = \delta(x)\,\delta(y)\,\delta(z)\,\delta(t)$.

The dimension of the δ-function, which is important to physicists and engineers but mostly ignored by mathematicians, can be determined from the fact that this function is one of those few self-similar functions whose argument can be a space variable or a time variable, and therefore, depending on the dimension of its argument, the δ-function has a nonzero dimension. For example, the dimension of the δ-function of time $\delta(t)$ is equal to the inverse time, i.e., the dimension of frequency because, by definition, $\int_{-\infty}^{\infty} \delta(t)\,dt = 1$, which is dimensionless. In other words, the dimension of $\delta(t)$ is equal to the dimension of the inverse function $1/t$. A detailed analysis of the δ-function is also available in Kythe [2011].

2.5 Delta Function and Dirichlet Kernel

Consider the 1-D transient problem: $\dfrac{\partial^2 u}{\partial x^2} = \dfrac{1}{k}\dfrac{\partial u}{\partial t}$, $-a < x < a$, with the initial and boundary conditions as $u(x,0) = k(x)$ for $-a < x < a$, and $u(-a, t) = 0 = u(a, t)$ for $t > 0$. Using Bernoulli's separation method, i.e., by assuming $u(x, t) = X(x)T(t)$, we get

$$\frac{1}{X}\frac{d^2 X}{dx^2} = \frac{1}{kT}\frac{dT}{dt} = -\lambda^2,$$

where λ is a real parameter. The boundary conditions and the initial condition become: $X(-a) = 0 = X(a)$, and $T(0) = k(x)$. It will be found that only the choice of $\lambda > 0$ gives meaningful results, leading to the solution of the following two equations

$$\frac{d^2 X}{dx^2} + \lambda^2 X = 0, \qquad \frac{dT}{dt} + k\lambda^2 T = 0,$$

which eventually leads to the solution

$$u(x, t) = \sum_{n=-\infty}^{\infty} A_n \sin\frac{n\pi x}{a}\, e^{-n^2\pi^2 kt/a^2}, \tag{2.34}$$

where

$$2aA_n = \int_{-\infty}^{\infty} k(x) \sin \frac{n\pi x}{a}\, dx. \qquad (2.35)$$

The orthonormal set of eigenpairs for this problem are $\left(\dfrac{n\pi}{a}, \dfrac{1}{2a} \sin \dfrac{n\pi x}{a} \right)$, and the orthonormal eigenfunctions are $\phi_n(x) = \frac{1}{2a} e^{in\pi x/a}$. Hence, the Dirac δ-function in the region $-a < x < a$ for the steady-state (as $t \to \infty$) 1-D transient problem involving Laplace's equation over the interval $-a < x < a$ is

$$\delta(x, x') = \frac{1}{4a^2} \sum_{n=-\infty}^{\infty} e^{in\pi x/a} e^{-in\pi x'/a} = \frac{1}{4a^2} \sum_{-\infty}^{\infty} e^{in\pi(x-x')/a}. \qquad (2.36)$$

Note that in (2.35) we have assumed that the eigenfunction expansion is periodic with period $2a$. Hence, the Dirac δ-function (2.36) also has period $2a$.

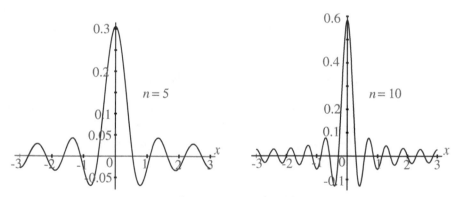

Figure 2.3 Dirac δ-Function with 5, 10, and 25 Terms.

The solution (2.34) can also be obtained by using the Laplace transform method. We have graphed the real part of this function for the basic interval $-a < x < a$ in Figure 2.3, using $n = 5$ and 10 terms in (2.36), with $a = 3$, and $x' = 0$. The graphs, however, repeat outside this interval with a period $2a$. From the graphs in Figure 2.3, it is obvious that the peak becomes infinitely higher and narrower at $x = 0$ as n increases.

The *Dirichlet kernel* $D_n(x) = \sum_{n=-N}^{N} e^{inx}$ is represented for different values of N in Figure 2.4. It shows peaks of heights $2N+1$ and width of the order of $2\pi/(2N+1)$ due to periodicity at $0, \pm 2\pi, \pm 4\pi, \ldots$. The height of the peaks is equal to the number of the elements in the lattice.

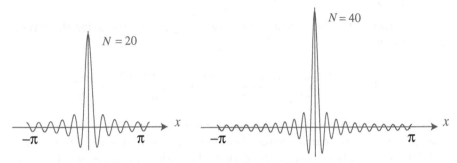

Figure 2.4 Dirichlet Kernel D_N for $N = 20, 40$.

2.6 Gibbs Phenomenon

In the case of a jump discontinuity the Fourier series leads to what is known as the *Gibbs phenomenon*[2]. For example, consider the Fourier sine series for

$$f(x) = \begin{cases} 1, & 0 < x < \pi, \\ 0, & x = 0, \\ -1, & -\pi < x < 0. \end{cases} \tag{2.37}$$

Then the coefficients b_n are determined by (2.10), and we get

$$f(x) = \frac{4}{\pi} \sum_{n=0}^{\infty} \frac{1}{2n+1} \sin(2n+1)x. \tag{2.38}$$

The partial sums

$$f_N(x) = \frac{4}{\pi} \sum_{n=0}^{N} \frac{1}{2n+1} \sin(2n+1)x \tag{2.39}$$

define the N harmonics, which approximate the jump as shown in Figure 2.5. Notice the sharp peaks in the harmonics $f_1(x)$, $f_2(x)$, $f_{10}(x)$, $f_{20}(x)$, and $f_{40}(x)$ near 0, which is the discontinuity of $f(x)$. Gibbs [1899] showed that the height or *overshoot* of these peaks is greater than $f(0+)$ by about 9%. The width of the overshoot goes to zero as $N \to \infty$, but the height remains at 9% both at the top and the bottom such that

$$\lim_{N \to \infty} \max |f(x) - f_N(x)| \neq 0.$$

---

[2] Maxime Bôchner [1905-1906] gave it the name Gibbs phenomenon. See also Carslaw [1925].

This phenomenon does not go away even when the number of harmonics is increased. ∎

Example 2.6. The Fourier sine series for $f(x) = 10 + x$ for the interval $0 < x < k$ is

$$10 + x = \frac{2}{\pi} \sum_{n=1}^{\infty} \left[\frac{10\left(1 - (-1)^n\right)}{n} + \frac{(-1)^{n+1} k}{n} \right] \sin \frac{n\pi x}{k}.$$

This series is valid everywhere except the end points where it becomes zero. The graphs of the partial sums $f_N(x)$ are given in Figure 2.6 for $N = 1, 5, 10, 20, 40$. These graphs show that the series converges to $10 + x$ for $0 < x < k$. For these finitely many partial sums $f_N(x)$ the solution starts from zero at $x = 0$, shoots up beyond $10 + x$ and then reduces to zero at $x = k$ (we have taken $k = 10$ in this figure). This overshoot, as seen in Figure 2.6, exhibits the Gibbs phenomenon.

2.6.1 Sigma-Approximation. The σ-approximation adjusts a Fourier sum to eliminate the Gibbs phenomenon, since otherwise this phenomenon will occur at discontinuities. Let $f(x)$ be a piecewise smooth function on $[-\pi, \pi]$ and let $f_N(x)$ be its Fourier partial sum. Set

$$S_N(x) = \frac{a_0}{2} + \sum_{n=1}^{N} \left(a_n \sigma_n \cos(nx) + b_n \sigma_n \sin(nx) \right), \qquad (2.40)$$

where

$$\sigma_n = \mathrm{sinc}\left(\frac{b\pi}{N}\right), \quad n = 1, 2, \ldots, N, \qquad (2.41)$$

are known as the σ-factors. Since the sequence of sums $\{S_N(x)\}$ approximates the function $f(x)$ better than the Fourier partial sums $\{f_N(x)\}$, the following result is used:

Theorem 2.4. *We have*

$$S_N(x) = \int_{-\pi/N}^{\pi/N} f_N(x + t)\, dt. \qquad (2.42)$$

PROOF. Since

$$\frac{N}{2\pi} \int_{-\pi/N}^{\pi/N} \cos\left(n(x + t)\right)\, dt = \frac{N}{2\pi} \cos(nx) \frac{2 \sin(n\pi/N)}{n} = \cos(nx)\,\sigma_n,$$

and similarly

$$\frac{N}{2\pi} \int_{-\pi/N}^{\pi/N} \sin\left(n(x + t)\right)\, dt = \sin(nx)\,\sigma_n,$$

we get

$$\frac{N}{2\pi} \int_{-\pi/N}^{\pi/N} S_N(x+t)\, dt = \frac{a_0}{2} + \sum_{n=1}^{N} \big(a_n \sigma_n \cos(nx) + b_n \sigma_n \sin(nx)\big).$$

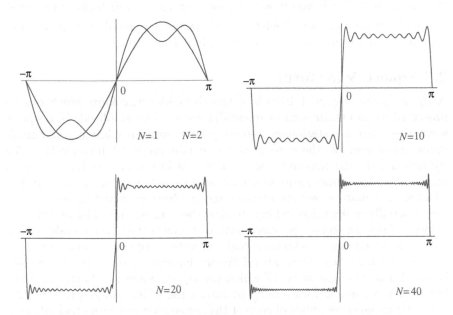

Figure 2.5 Gibbs phenomenon.

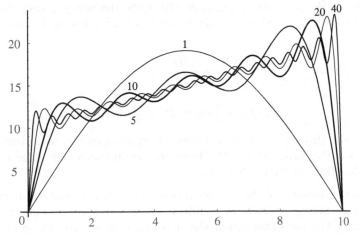

Figure 2.6 Graphs of partial sums S_n for $n = 1, 5, 10, 20, 40$. ∎

Obviously, the sums $\{S_N(x)\}$ approximate the function $f(x)$ in a very smooth manner. In the case of the function $f(x)$ defined by (2.37), the partial sums (2.39) have the σ-approximation

$$S_N(x) = \frac{4}{\pi}\left[\sin(x)\sin\left(\frac{\pi}{N}\right) + \cdots + \frac{1}{(N-1)^2}\sin((n-1)x)\sin\left(\frac{(N-1)\pi}{N}\right)\right].$$

(2.43)

The term sinc $\left(\frac{n\pi}{N}\right)$ is known as the *Lanczos σ-factor*, which eliminates most of the Gibbs phenomenon. Sometimes this factor is squared or even cubed to serially attenuate the Gibbs phenomenon in extreme cases.

2.7 Square Waveform

A square wave (Figure 1.4(b)) is a non-sinusoidal waveform which can be presented as an infinite sum of sinusoidal waves. The amplitude in a square wave alternates at a stationary frequency between fixed minimum and maximum values, and has the same period at maximum or minimum value. The transition from the maximum to minimum is instantaneous for ideal (i.e., mathematically defined) square waves; it does not happen in real physical situations. Square waves are encountered in electronics and signal processing, especially in digital switching circuits; they are generated by binary logic devices. They are known as 'clock signals' because their fast transitions are suited for triggering synchronous logic circuits at precisely preassigned intervals. A similar waveform with different duration at maxima and minima is called a *rectangular wave*, of which the square wave is a particular case. Square waves contain a wide range of harmonics, which can generate electromagnetic radiation of pulses of current that often interfere with other adjacent circuits, such as analog-to-digital converters. Therefore, sine waves are used instead of square waves as timing references.

Using a Fourier series expansion with cycle frequency f over time t, the ideal square wave (1.8) with a peak amplitude of 2 has an infinite series of the form

$$x(t) = \frac{4}{\pi}\sum_{k=1}^{\infty}\frac{\sin\left(2\pi(2k-1)ft\right)}{2k-1}$$

$$= \frac{4}{\pi}\left[\sin(2\pi ft) + \tfrac{1}{3}\sin(6\pi ft) + \tfrac{1}{5}\sin(10\pi ft) + \cdots\right],$$

(2.44)

Notice that this representation contains only components of odd-integer frequencies of the form $2\pi(2k-1)$. However, the sawtooth waves and real-world signals contain all integer harmonics.

The convergence of the Fourier series (2.44) leads to Gibbs phenomenon, where the ringing artefacts in nonideal square waves are related to this phenomenon. On the other hand, the mathematical square waves change between high and low states instantaneously without under- or over-shooting.

This cannot be achieved in physical systems since it would require an infinite bandwidth. The times taken by a square-wave signal to rise from the low to the high level and back again are called the *rise time* and *fall time*, respectively. If the system is overdamped, the square waveform may never reach the rise and fall times, and if the system is underdamped, it will oscillate about the high and low levels before it settles down. In these cases the bandwidth of the system is related to the transition times of the waveform.

A square waveform is also defined in different but equivalent forms. For example, it can be defined as a sine function of a periodic sinusoid form as

$$x(t) = \text{sgn}\,(\sin(t)), \quad y(t) = \text{sgn}\,(\cos(t)), \tag{2.45}$$

which is 1 when the sinusoid is positive, -1 when the sinusoid is negative, and 0 at the discontinuities. The sinusoid in the definition (2.45) can be replaced by any other periodic function, like the modulo operation[3]

$$x(t) = \text{sgn}\,\left(\text{mod}\left(t - \frac{\lambda}{2}, \lambda\right) - \frac{\lambda}{2}\right), \tag{2.46}$$

It can also be defined in terms of the Heaviside function, or the rect function as

$$x(t) = \sum_{n=-\infty}^{\infty} \left[H\left(t - nT + \tfrac{1}{2}\right) - H\left(t - nT - \tfrac{1}{2}\right)\right] = \sum_{n=-\infty}^{\infty} \text{rect}(t - nT), \tag{2.47}$$

where $t = 2$ for a 50% duty cycle[4]. For $t \geq 0$, the square wave function is defined in terms of the Heaviside function as

$$x(t) = H(t) - 2H(t - a) + 2H(t - 2a) - 2H(t - 3a) + \cdots. \tag{2.48}$$

The square wave can also be defined as a piecewise continuous function

$$x(t) = \begin{cases} 1 & \text{if } |t| < T_1, \\ 0 & \text{if } T_1 < |t| < T/2, \end{cases} \tag{2.49}$$

when $x(t + T) = x(t)$, i.e., $x(t)$ is a periodic function of period T.

---

[3] The modulo operation, or simply modulus, is the process of finding the remainder of division of one quantity by another.

[4] A duty cycle is the percent of time that a system spends in an active state as a fraction of the total time. This term is often used in electrical devices, such as switching power supplies. Mathematically, the duty cycle d in a periodic event is the ratio of the duration (τ) of the event to the total period T of a signal, i.e., $d = \tau/T$.

2.8 Examples of Fourier Series

Example 2.7. Consider the Rademacher function $s_{1/2}(x)$ (square wave), defined by Eq (1.5), and suppose that it has period $2L$. This square wave function is odd, so the Fourier series has $a_0 = a_n = 0$, and

$$b_n = \frac{2}{L} \int_0^L \sin\left(\frac{n\pi x}{L}\right) dx = \frac{4}{n\pi} \sin^2 \frac{n\pi}{2} = \frac{2}{n\pi}[1-(-1)^n] = \begin{cases} 0 & \text{if } n \text{ is even,} \\ \dfrac{4}{n\pi} & \text{if } n \text{ is odd.} \end{cases}$$

The Fourier series for the square wave with period $2L$, phase shift 0, and half-amplitude 1 is

$$s_{1/2}(x) = \frac{4}{\pi} \sum_{n=1,3,5,\ldots} \frac{1}{n} \sin \frac{n\pi x}{L}. \blacksquare \qquad (2.50)$$

Example 2.8. The function $f(t)$ defined in Figure 2.7 has the Fourier series expansion $f(t) = ah + \dfrac{2h}{\pi} \sum_{n=1}^{\infty} \dfrac{\sin n\pi a}{n} \cos \dfrac{n\pi t}{L}. \blacksquare$

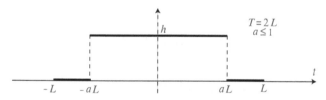

Figure 2.7 Function $f(t)$.

Example 2.9. The function $f(t)$ defined in Figure 2.8 has the Fourier series expansion $f(t) = \dfrac{2h}{\pi} \sum_{n=1}^{\infty} \dfrac{1 - \cos n\pi a}{n} \sin \dfrac{n\pi t}{L}. \blacksquare$

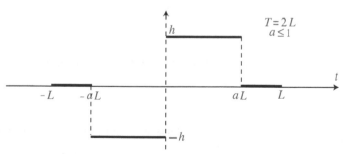

Figure 2.8 Function $f(t)$.

Example 2.10. The distribution $\delta(t - a)$ defined in Figure 2.9 has the

Fourier series expansion $\delta(t-a) = \dfrac{1}{T} \displaystyle\sum_{n=-\infty}^{\infty} e^{2in\pi(t-a)/T} = \dfrac{1}{T} + \dfrac{2}{T} \displaystyle\sum_{n=1}^{\infty} \cos \dfrac{2n\pi(t-a)}{T}$. ■

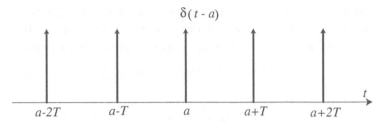

$$\delta(t-a)$$

Figure 2.9 The Distribution $\delta(t-a)$.

Example 2.11. (Sawtooth wave). A sawtooth wave of period 1 is defined as $x(t) = t - \lfloor t \rfloor$, where $\lfloor t \rfloor$ is the floor of t. A general definition is $x(t) = a\,\mathrm{frac}\!\left(\dfrac{t}{p} + \phi\right)$, where $\mathrm{frac}(t) = t - \lfloor t \rfloor$ is the fractional part of t, a is the amplitude, p the period, and ϕ the phase. Thus, in the interval $[-1,1]$ with period p this function is defined as $x(t) = 2\!\left(\dfrac{t}{p} - \left\lfloor \dfrac{1}{2} + \dfrac{t}{p} \right\rfloor\right)$. This wave has the same phase as the sine function. A graph of this function is given in Figure 1.4(d). If $a = 1, p = 2L$ and $\pi = 0$, a couple of expressions for the Fourier series of a sawtooth wave are

$$x(t) = \frac{1}{2} - \frac{1}{\pi} \sum_{n=0}^{\infty} \frac{1}{n} \sin\left(\frac{n\pi t}{L}\right) = \frac{1}{2} - \arctan\left(\cot\left(\frac{\pi t}{2L}\right)\right).\ \blacksquare$$

2.9 Complex Fourier Coefficients

A function $f(x)$ is said to satisfy *Dirichlet conditions* in the interval $I : -a < x < a$, if (i) $f(x)$ has only finitely many finite discontinuities in the interval I and has no infinite discontinuities, and (ii) $f(x)$ has only finitely many maxima and minima in the interval I. Then the theory of Fourier series implies that if $f(x)$ satisfies these two Dirichlet conditions (i) and (ii) in the interval I, it can be represented as the complex Fourier series

$$f(x) = \sum_{n=-\infty}^{\infty} a_n\, e^{n\pi i x/a}, \tag{2.51}$$

where the coefficients a_n are defined as

$$a_n = \frac{1}{2a} \int_{-a}^{a} f(t)\, e^{-n\pi i t/a}\, dt. \tag{2.52}$$

This representation is periodic of period $2a$ in the interval I. However, the right side of (2.51) cannot represent $f(x)$ outside the interval I unless $f(x)$

itself is periodic of period $2a$. Thus, we have Fourier series on finite intervals, but on infinite intervals $-\infty < x < \infty$ we obtain Fourier integrals.

An integral representation of a nonperiodic function $f(x)$ on $(-\infty, \infty)$ can be found by letting $a \to \infty$. Notice that as the interval I grows, i.e., as $a \to \infty$, the values $\dfrac{n\pi}{a} \equiv u_n$ become closer together into a dense set. Let $\delta u = (u_{n+1} - u_n) = \dfrac{\pi}{a}$. Then, substituting a_n from (2.52) into (2.51) we get

$$f(x) = \frac{1}{2\pi} \sum_{n=-\infty}^{\infty} (\delta u) \left[\int_{-a}^{a} f(t)\, e^{-i\, t u_n}\, dt \right] e^{i\, x u_n}, \qquad (2.53)$$

where in the limit as $a \to \infty$, u_n becomes a continuous variable u and δu becomes du. Thus, formally the sum in (2.53) can be replaced by the integral in the limit and (2.53) reduces to the *Fourier integral formula*:

$$f(x) = \frac{1}{2a} \int_{-\infty}^{\infty} \left[\int_{-\infty}^{\infty} f(t)\, e^{-i\, ut}\, dt \right] e^{i\, ux}\, du. \qquad (2.54)$$

Although the above limiting process may not be very rigorous, the formula (2.54) is nevertheless correct and can be used for functions $f(x)$ that are piecewise continuously differentiable in every finite interval and absolutely integrable on the whole real line $(-\infty, \infty)$, i.e., the integral $\int_{-\infty}^{\infty} |f(x)|\, dx < \infty$ exists on $(-\infty, \infty)$. In fact, formula (2.54) is valid under more general conditions, as the following theorem, known as the *Fourier integral theorem* reveals.

Theorem 2.5. *If $f(x)$ satisfies Dirichlet conditions on $(-\infty, \infty)$, and is absolutely integrable on $(-\infty, \infty)$, then the Fourier integral (2.54) converges to the function $\frac{1}{2}[f(x+0) + f(x-0)]$ at a finite discontinuity at x. In other words,*

$$\frac{1}{2}[f(x+0) + f(x-0)] = \frac{1}{2a} \int_{-\infty}^{\infty} e^{i\, ux} \left[\int_{-\infty}^{\infty} f(t)\, e^{-i\, ut}\, dt \right] du. \qquad (2.55)$$

If the function $f(x)$ is continuous at x, then $f(x+0) = f(x-0) = f(x)$, and (2.55) reduces to (2.54).

The exponential factor $e^{i\, u(x-t)}$ in (2.54) can be expressed in terms of trigonometric functions. Thus, using the even and odd nature of the cosine and sine functions, respectively, as functions of u, formula (2.54) can be written for a real valued f as

$$f(x) = \frac{1}{\pi} \int_{0}^{\infty} du \int_{-\infty}^{\infty} f(t) \cos u(x - t)\, dt. \qquad (2.56)$$

This is another version of the Fourier integral formula.

If $f(x)$ is an even function, then expanding the cosine function in (2.56) we get *Fourier cosine integral formula*:

$$f(x) = f(-x) = \frac{2}{\pi} \int_0^\infty \cos ux \int_0^\infty f(t) \cos ut \, dt \, du.$$

Similarly, if $f(x)$ is an odd function, we obtain *Fourier cosine integral formula*:

$$f(x) = -f(-x) = \frac{2}{\pi} \int_0^\infty \sin ux \int_0^\infty f(t) \sin ut \, dt \, du.$$

The Fourier series expansion can be formulated in two ways, as follows.

FIRST FORMULATION. The orthogonality relation is

$$\int_0^L e^{ik\omega x} e^{in\omega x} \, dx = \begin{cases} L & k = n, \\ 0 & k \neq n. \end{cases}$$

Then the Fourier expansion for $f(x)$ is

$$f(x) = \sum_{-\infty}^\infty c_n e^{in\omega x}, \quad c_n = \frac{1}{L} \int_a^{a+L} f(x) e^{-in\omega x} \, dx.$$

The totality of the Fourier coefficients c_n combined is called the *spectrum* while the indices n that indicate the precise coefficients are called the *frequencies*. The amplitudes A_n defined in the sin, cos form are

$$c_0 = \frac{a_0}{2}, \; c_n = \tfrac{1}{2}(a_n - ib_n), \; c_{-n} = \tfrac{1}{2}(a_n + ib_n), \; n \geq 1,$$
$$a_0 = 2c_0, \; a_n = c_n + c_{-n}, \; b_n = i(c_n - c_{-n}), \; n \geq 1,$$

where $f(x)$ is real, $a_0 = 2c_0$, $a_n = 2\Re\{c_n\}$, $b_n = -2\Im\{c_n\}$, and $c_{-n} = \bar{c}_n$ for $n \geq 1$.

The amplitudes A_n defined in the complex form are

$$A_0 = c + 0, \; A_n e^{i\alpha_n} = 2c_n, \; A_n = 2|c_n|, \; \alpha_n = \arg\{c_n\}, n \geq 1,$$
$$c_0 = A_0, \; c_n = \frac{A_n}{2} e^{i\alpha_n} = \frac{A_n}{2}(\cos \alpha_n + i \sin \alpha_n), \; c_{-n} = \bar{c}_n = \frac{A_n}{2} e^{-i\alpha_n}, n \geq 1.$$

SECOND FORMULATION. This formulation uses the real numbers and is defined by

$$f(x) = \frac{A_0}{2} + \sum_{n=1}^\infty A_n \cos(nx + \phi_n). \tag{2.57}$$

Note that the formulations (2.29) and (2.36) are equivalent since, using Euler's formula $e^{inx} = \cos nx + i \sin nx$ we have $c_{\pm n} = \frac{1}{2}(a_n \mp i b_n)$, where $n \geq 0$ is an integer. This equivalence also allows us to switch to polar coordinates A_n and $-\phi_n$ by taking $\left\{ \begin{matrix} a_n \\ b_n \end{matrix} \right\} = \left\{ \begin{matrix} A_n \cos \phi_n \\ -A_n \sin \phi_n \end{matrix} \right\}$. The terms in the series (2.36) are called *harmonics*. Thus, for $n = 1$ there is the *fundamental harmonic* with frequency $1/2\pi$ which is equal to the inverse of the period 2π. For higher values of n there are *higher harmonics* with frequency $n/2\pi$. Thus, the fundamental harmonic makes one complete oscillation over one period of 2π and is called *one cycle*, whereas the higher harmonics that make n complete oscillations are called the n *cycles*. The numbers A_n and ϕ_n are called *amplitudes* and *phases*, respectively. Geometrically, a phase generates a translation; for example, $\sin x = \cos(x - \pi/2)$ (phase $\pi/2$) means that the graph of $\sin x$ is obtained by translating the graph of $\cos x$ by $\pi/2$ to the right (see Figure 2.10).

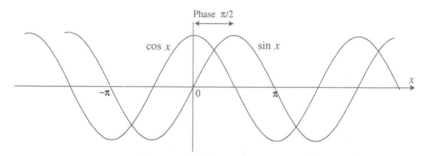

Figure 2.10 Translation of the Graph of $\cos x$ by the Phase $\pi/2$.

Example 2.12. A function $f(x)$ defined on the interval $(-T/2, T/2)$ which is periodic of period T has the Fourier series

$$F(x) = \frac{a_0}{2} + \sum_{n=1}^{\infty} a_n \cos \frac{2\pi n x}{T} + b_n \sin \frac{2\pi n x}{T}, \qquad (2.58)$$

where the coefficients a_n and b_n are given by

$$\left\{ \begin{matrix} a_n \\ b_n \end{matrix} \right\} = \frac{2}{\pi} \int_{-T/2}^{T/2} f(x) \left\{ \begin{matrix} \cos \dfrac{2\pi n x}{T} \\ \sin \dfrac{2\pi n x}{T} \end{matrix} \right\} dx.$$

Here, the coefficients a_n and b_n are, in fact, $a_{2\pi n/T}$ and $b_{2n\pi/T}$, respectively. Thus, if f is defined periodic on a larger interval T the spectrum becomes denser, although the spectrum remains discrete. ∎

3

Fourier Transforms

Fourier transforms (FTs) constitute the central theme of this book. They are used in every technological application which starts from Chapter 4 onward. The purpose of the Fourier transform is to represent an image in terms of sine and cosine functions (sinusoids). Using these two basic functions, it is easy to perform processing operations, especially frequency domain filtering, which after inversion using inverse Fourier transform (IFT) is converted back to the spatial domain.

Using Fourier transform, any digital image can be represented as a weighted sum of sine and cosine functions. These weights are used to reconstruct the image using the inverse Fourier transform. Linear image processing operations can be implemented using convolution in the spatial domain or by filtering in the frequency domain. Some image processing operations perform better if frequency domain solutions are used.

3.1 Definitions

Let $f(x)$, $-\infty < t < \infty$, be a piecewise differentiable and absolutely integrable function. Then the Fourier transform of $f(x)$, denoted by $\mathcal{F}\{f(x)\} = F(\omega)$, is defined by

$$\mathcal{F}(f(x)) \equiv F(\omega) = \int_{-\infty}^{\infty} f(x)\, e^{-i\,\omega x}\, dx, \qquad (3.1)$$

and its inverse by

$$\mathcal{F}^{-1}\{F(\omega)\} \equiv f(x) = \frac{1}{2\pi} \int_{-\infty}^{\infty} F(\omega) e^{i\omega x}\, d\omega, \qquad (3.2)$$

where ω is the radian frequency, also called the *variable of the transform*. The integrals converge if $\int_{-\infty}^{\infty} |f(x)|\,dx$ does. If $f(x)$ is differentiable at x, then the relationship between $f(x)$ and $F(\omega)$ can be expressed as

$$f(x) \rightleftharpoons F(\omega). \tag{3.3}$$

Since $\omega = 2\pi f_x$, where f_x called the *circular* or *cycle frequency*, the above definitions in the frequency domain become

$$F(f) = \int_{-\infty}^{\infty} f(x)\, e^{-2\pi i\, f_x x}\, dx, \quad f(x) = \int_{-\infty}^{\infty} F(f)\, e^{2\pi i\, f_x x}\, df_x, \tag{3.4}$$

respectively, so that

$$f(x) \rightleftharpoons F(2\pi f_x). \tag{3.5}$$

The definition (3.1) first appeared in Fourier [1816]. The definitions (3.1)–(3.2) also appear in the literature in the following two different forms:

$$F(\omega) = \int_{-\infty}^{\infty} f(x)\, e^{-i\omega t}\, dt \qquad \rightleftharpoons f(x) = \int_{-\infty}^{\infty} F(\omega)\, e^{i\omega x}\, dt, \tag{3.6}$$

$$F(\omega) = \frac{1}{\sqrt{2\pi}} \int_{-\infty}^{\infty} f(x)\, e^{-i\omega x}\, d\omega \rightleftharpoons f(x) = \frac{1}{\sqrt{2\pi}} \int_{-\infty}^{\infty} F(\omega)\, e^{i\omega x}\, d\omega. \tag{3.7}$$

However, we will use the definitions (3.1)–(3.2) instead, because the other set of definitions (3.4) can be easily obtained by setting $\omega = 2\pi f_x$ in (3.1)–(3.2).

Example 3.1. (i) $\mathcal{F}\{\delta(x)\} = \int_{-\infty}^{\infty} \delta(x)\, e^{-i\omega x}\, dx = 1$, and $\delta(x) = \mathcal{F}^{-1}\{1\}$ $= \frac{1}{2\pi} \int_{-\infty}^{\infty} e^{i\,\omega x}\, d\omega$.

(ii) Since $\operatorname{sgn}(x) = \begin{cases} 1 & \text{if } x > 0, \\ -1 & \text{if } x < 0, \end{cases}$, then $H(x) = \frac{1}{2}\,[1 + \operatorname{sgn}(x)]$, or $\operatorname{sgn}(x)$ $= 2H(x) - 1$, or $\frac{d}{dx}\,[\operatorname{sgn}(x)] = 2H'(x) = 2\delta(x)$, which gives $\mathcal{F}\{e^{-ax} H(x)\} =$ $\int_{0}^{\infty} e^{-(i\omega + a)x}\, dt = \frac{1}{i\omega + a}.$ ∎

The Fourier inversion theorem holds for all continuous functions that are absolutely integrable with absolutely integrable Fourier transform. However, there is a slight variation where the condition that the function $f(x)$ be continuous is dropped, but still required is the condition that the function $f(x)$ and its Fourier transform be absolutely integrable. Then $f(x) = g(x)$ almost everywhere where g is a continuous function, and $\mathcal{F}^{-1}\mathcal{F}\{f(x)\} = g(x)$ for every real x.

Theorem 3.1. (Fourier integral theorem.) *Let $f(x)$ be an absolutely integrable function on $(-\infty, \infty)$ and satisfy the Dirichlet conditions, defined* *in §2.9, on the entire real line. Then*

$$\frac{1}{2}\left[f(x+0) + f(x-0)\right] = \frac{1}{2\pi}\int_{-\infty}^{\infty} e^{-i\omega x}\int_{-\infty}^{\infty} x(u)e^{i\omega u}\, du\, d\omega. \qquad (3.8)$$

3.1.1 Fourier Integral The complex Fourier coefficients for a function f over a finite interval $(-T/2, T/2)$ are given by

$$c_n = \int_{-T/2}^{T/2} f(x)\, e^{-2\pi i\, nx/T}\, dx.$$

If we let $T \to \infty$, then the result so obtained will include the functions defined on the entire real axis. This extension leads to a continuous spectrum and is known as the Fourier integral, which is defined by

$$F(\omega) = \int_{-\infty}^{\infty} f(x)\, e^{i\omega x}\, dx.$$

This is the same as (3.1), where $F(\omega)$ is the Fourier transform of f.

3.1.2 Properties of Fourier Transforms. Let $\mathcal{F}\{f(x)\} = F(\omega)$. Then

(1) $\mathcal{F}\{f(x-a)\} = e^{i\omega a}\, F(\omega)$ (Shifting).

(2) $\mathcal{F}\{f(ax)\} = \dfrac{1}{|a|}F(\omega/a)$ (Scaling).

(3) $\mathcal{F}\{e^{iax}\, f(x)\} = F(\omega + a)$ (Translation).

(4) $\mathcal{F}\{\overline{-f(x)}\} = \overline{\mathcal{F}\{f(x)\}}$ (Conjugate).

(5) $\mathcal{F}\{F(x)\} = f(-\omega)$ (Duality).

(6) $\mathcal{F}\{x^n f(x)\} = (-i)^n \dfrac{d^n}{d\omega^n}\, F(\omega)$.

(7) $\mathcal{F}\{f(ax)e^{i\, bx}\} = \dfrac{1}{|a|}F\left(\dfrac{\omega + b}{a}\right)$.

Property (1) shows that the product of $f(x)$ by an exponential $e^{i\omega a}$, a real, corresponds to a shifting of $F(\omega)$ by $e^{i\omega a}$. This property is useful in interference. For example, a function $f(x)$ as shown in Figure 3.1 can be regarded as the sum of two rectangular pulses, one translated to A and the other to $-A$, defined by $f(x - A) + f(x + A)$. This yields the corresponding Fourier transform as $F(\omega)\left(e^{iA\omega} + e^{-iA\omega}\right) = 2F(\omega)\cos A\omega$, where $\cos A\omega$ is the oscillating factor, which is responsible for the so-called *interference fringes*. If A is very large, then $\cos A\omega$ oscillates very rapidly creating very dense fringes practically indistinguishable from one another; however, if A is small (of the order of 1), the fringes are easily discernible. This is known as the *principle of interference* which is useful in certain fields, such as interferometry, radiotelescopes and radioastronomy.

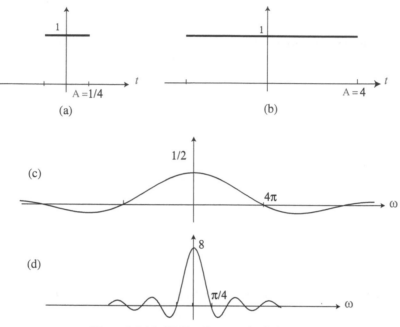

Figure 3.1 (a)–(b) Two Rectangular Pulses;
(c)–(d) Fourier Transforms of the above Functions.

Property (2) implies that a function and its Fourier transform cannot both be of "short duration," since if the details of $f(x)$ as small as a are to be reconstructed, then $F(\omega)$ must be known on an interval as large as $1/a$. The meaning of "duration" is significant and depends on the context. For example, in quantum mechanics the context involves both position and momentum. Property (2) is known as the *uncertainty principle* and its various interpretations depend on the meaning of the term 'duration'. For Heisenberg's uncertainty principle, see Appendix L.

3.1.3 Fourier Transforms of the Derivatives of a Function. We assume that $f(x)$ is differentiable n-times, and that this function and its derivatives approach zero as $|x| \to \infty$. Then

$$F^{(k)}(\omega) = (i\omega)^k F^{(k-1)}(\omega), \tag{3.9}$$

where $F^{(k)}$ is the Fourier transform of $f^{(k)}(x)$, which is the kth derivative of

$f(x)$ for $0 \le k \le n$. Thus, in particular,

$$\mathcal{F}\{f'(t)\} = \int_{-\infty}^{\infty} e^{-i\omega x} f'(x) \, dx$$

$$= \left[f(x) \, e^{-i\omega x} \right]_{-\infty}^{\infty} + i\omega \int_{-\infty}^{\infty} e^{-i\omega x} f(x) \, dx = i\omega F(\omega).$$
(3.10)

An extension of this property for functions of two variables is given in §3.3.

3.1.4 Convolution Theorems for Fourier Transform. The convolution (or *Faltung*) of $f(x)$ and $g(x)$ over $(-\infty, \infty)$ is defined by

$$f \star g = \int_{-\infty}^{\infty} f(u) \, g(x-u) \, du = \int_{-\infty}^{\infty} f(x-u) \, g(u) \, du,$$
(3.11)

and we have the transform pair

$$f \star g \rightleftharpoons F(\omega)G(\omega).$$
(3.12)

Theorem 3.2. If $\mathcal{F}\{f(x)\} = F(\omega)$ and $\mathcal{F}\{g(x)\} = G(\omega)$, then

$$\mathcal{F}\{f(x) \star g(t)\} = F(\omega)G(\omega), \quad f(x) \star g(x) = \mathcal{F}^{-1}\{F(\omega)G(\omega)\}.$$
(3.13)

Some algebraic properties of convolution are:

$$
\begin{array}{ll}
f \star g = g \star f & \text{Commutative,} \\
f \star (g \star h) = (f \star g) \star h & \text{Associative,} \\
f \star (g + h) = f \star g + f \star h & \text{Distributive,} \\
f \star \delta = f = \delta \star f & \text{Identity.}
\end{array}
$$

The above commutative property, using (3.12), can be written as

$$\int_{-\infty}^{\infty} f(u) \, g(x-u) \, du = \int_{-\infty}^{\infty} e^{i\omega x} \, F(\omega)G(\omega) \, d\omega.$$

Since this is valid for all real x, set $x = 0$, which gives

$$\int_{-\infty}^{\infty} f(u) \, g(-u) \, du = \int_{-\infty}^{\infty} f(x) \, g(-x) \, dx = \int_{-\infty}^{\infty} F(\omega)G(\omega) \, d\omega.$$

Substitute $g(x) = \overline{f(-x)}$ to obtain

$$G(\omega) = \mathcal{F}\{g(x)\} = \mathcal{F}\{\overline{f(-x)}\} = \int_{-\infty}^{\infty} F(\omega) \, \overline{F(\omega)} \, d\omega,$$

which gives *Parseval's formula*:

$$\int_{-\infty}^{\infty} |f(x)|^2 \, dx = \int_{-\infty}^{\infty} |F(\omega)|^2 \, d\omega. \qquad (3.14)$$

Using the inner product $\langle f, g \rangle$, and the norm, as defined in §2.1, Parseval's formula can be written as

$$\|f\| = \|F\| = \|\mathcal{F}\{f\}\|.$$

This shows that the action of Fourier transforms is unitary. Physically, the quantity $\|f\|$ is a measure of energy, and $\|F\|$ the power spectrum of $f(x)$.

If $f(x)$ and $g(x)$ are two square-integrable functions, then $f(x) \pm g(x)$ are also square-integrable. Then Parseval's formula (3.14) gives

$$\int_{-\infty}^{\infty} |F \pm G|^2 \, d\omega = \int_{-\infty}^{\infty} |f \pm g|^2 \, dx.$$

Example 3.2. From Example 3.1 (ii) we know that $\mathcal{F}\left\{ \dfrac{d}{dx} \operatorname{sgn}(x) \right\} = \mathcal{F}\{2H'(x)\} = 2\mathcal{F}\{\delta(x)\} = 2$. Using formula (3.9), we find that $\mathcal{F}\{\operatorname{sgn}(x)\} = \dfrac{2}{i\omega}$. Also, using other properties of Fourier transform,

$$\mathcal{F}\{H(x)\} = \frac{1}{2}\mathcal{F}\{1 + \operatorname{sgn}(x)\} = \frac{1}{2}\left[\mathcal{F}\{1\} + \mathcal{F}\{\operatorname{sgn}(x)\}\right] = \left[\delta(\omega) + \frac{2}{i\omega}\right]. \quad \blacksquare$$

Convolution is important in the mathematical theory of filters (Chapter 4) and resolving power of radiotelescopes. Mathematically, the action of a filter is a convolution if the signal is defined in terms of frequency, i.e., by its Fourier transform. Convolution can be regarded as a kind of multiplication in the class of absolutely integrable functions; under the pointwise addition it is associative, commutative, and distributive. The δ-function behaves like the unity, since $f(x) \star \delta(x) = f(x)$, provided x is continuous and absolutely integrable. Care should be taken to remember that while $1 \in \mathbb{R}$ is a number, the delta function is not a function but a distribution.

Example 3.3. Find the Fourier transform of $f(x) = e^{-a|x|}, a > 0$.

$$\mathcal{F}\{f(x)\} = F(\omega) = \int_{-\infty}^{\infty} e^{-a|x|} e^{i\omega x} \, dx$$

$$= \left[\int_{-\infty}^{0} e^{ax} e^{i\omega x} \, dx + \int_{0}^{\infty} e^{-ax} e^{i\omega x} \, dx \right]$$

$$= \left(\frac{1}{a + i\omega} - \frac{1}{-a + i\omega} \right) = \frac{2a}{a^2 + \omega^2}. \quad \blacksquare$$

Example 3.4. Find the Fourier transform of $f(x) = e^{-ax^2}$, $a > 0$.

$$F(\omega) = \int_{-\infty}^{\infty} e^{-ax^2} e^{i\omega x}\, dx$$

$$= \int_{-\infty}^{\infty} e^{-a(x^2 - i\omega x/a - \omega^2/(4a^2) + \omega^2/(4a^2))}\, dx$$

$$= \int_{-\infty}^{\infty} e^{-a((x - i\omega/a)^2 - \omega^2/(4a^2))}\, dx$$

$$= e^{-\omega^2/(4a)} \int_{-\infty}^{\infty} e^{-u^2}\, du = \sqrt{\frac{\pi}{a}}\, e^{-\omega^2/(4a)}. \quad \blacksquare$$

Example 3.5. Find the Fourier transform of $f(x) = \begin{cases} 0 & \text{if } x < b, \\ e^{-a^2 x^2} & \text{if } 0 < b < x. \end{cases}$

The solution is

$$F(\omega) = \int_{b}^{\infty} e^{-a^2 x^2} e^{i\omega x}\, dx = \int_{b}^{\infty} e^{-a^2 (t - i\omega/(2a^2))^2 - \omega^2/(4a^2)}\, dx$$

$$= \frac{e^{-\omega^2/(4a^2)}}{a} \int_{(ab - i\omega/2a)}^{\infty} e^{-u^2}\, du = \frac{\sqrt{\pi}}{2a}\, e^{-\omega^2/(4a^2)}\, \mathrm{erfc}\left(ab - \frac{i\omega}{2a}\right). \quad \blacksquare$$

A list of Fourier transforms is provided in Appendix A. Some typical signals with respect to the circular frequency are presented in the next section.

3.1.5 Fourier Transforms of Some Typical Signals. Using the definitions (3.4), the Fourier transform of a signal $x(t)$, $-\infty < t < \infty$, is defined in the frequency domain as[1]

$$\mathcal{F}\{x(t)\} \equiv X(f) = \int_{-\infty}^{\infty} x(t)\, e^{-2\pi i f t}\, dt, \quad x(t) = \int_{-\infty}^{\infty} X(f)\, e^{2\pi i f t}\, df, \quad (3.15)$$

respectively, so that $x(t) \rightleftharpoons X(f)$. Some Fourier transforms are as follows.

(i) **Impulse:** $\mathcal{F}\{\delta(t)\} = 1$.

(ii) **Unit Step:** $\mathcal{F}\{u(t)\} \equiv \mathcal{F}\{H(t)\} = \dfrac{1}{i\,f} + \pi\delta(f)$.

(iii) **Constant** $x(t) = c$: $\mathcal{F}\{c\} = \displaystyle\int_{-\infty}^{\infty} c\, e^{-i f t}\, dt = 2\pi c\delta(f)$.

(iv) **Complex Exponential** (due to frequency shift): $\mathcal{F}\{e^{iat}\} = 2\pi\delta(f - a)$.

---

[1] Note that in this definition we have used the frequency component f instead of x_t for the sake of simplicity since it does not give rise to any confusion. Similarly, in the case of a two-dimensional signal $x(t, \tau)$, as in §3.3.1, we will use f and g instead of x_t and x_τ, respectively, to denote the circular frequencies.

(v) **Sinusoids** $x(t) = \cos(at) = \frac{1}{2}\left(e^{iat} + e^{-iat}\right)$:

$\mathcal{F}\{\cos(at)\} = \frac{1}{2}\left(\mathcal{F}\{e^{-iat}\} + \mathcal{F}\{e^{iat}\}\right) = \pi\left[\delta(f-a) + \delta(f+a)\right]$.

Similarly, $\mathcal{F}\{\sin(at)\} = \frac{\pi}{i}\left[\delta(f-a) - \delta(f+a)\right]$.

(vi) **Exponential Decay, Right-Sided:** $x(t) = e^{-at}H(t)$, $a > 0$, where $H(t)$ is the Heaviside function. Then

$$\mathcal{F}\{e^{-at}H(t)\} = \int_0^\infty e^{-at}e^{-ift}\,dt = -\frac{1}{a+if}\left[e^{-(a+if)t}\right]_0^\infty = \frac{1}{a+if}.$$

Exponential Decay, Left-Sided: $a > 0$. Then $\mathcal{F}\{e^{at}H(-t)\} = \dfrac{1}{a-if}$, or $\mathcal{F}\{-e^{at}H(-t)\} = \dfrac{1}{if-a}$.

Exponential Decay, Two-Sided: $x(t) = e^{-a|t|}$, $a > 0$. Since the spectrum $x(t)$ is the sum of right- and left-sided exponential decays due to linearity, we have

$$\mathcal{F}\{e^{-a|t|}\} = \frac{1}{a+if} + \frac{1}{a-if} = \frac{2a}{a^2+f^2}.$$

(vii) **Square Wave** or rectangular function of width a is defined as the difference between two Heaviside functions:

$$\text{rect}_a(t) = H(t+a) - H(t-a) = \begin{cases} 1 & \text{if } -a \le t < a, \\ 0 & \text{elsewhere.} \end{cases}$$

Since this function is linear, its FT is the difference between the two corresponding FTs:

$$\mathcal{F}\{\text{rect}_a(t)\} = \mathcal{F}\{H(t+a)\} - \mathcal{F}\{H(t-a)\} = \frac{1}{if}e^{iaf} - \frac{1}{if}e^{-iaf} = \frac{2\sin(af)}{f}.$$

(viii) **Sinc Function:** The spectrum of an *ideal lowpass filter* is $x(t) = \text{sinc}(x)$. Then

$$X(f) = \begin{cases} 1 & \text{if } -a \le t < a, \\ 0 & \text{elsewhere,} \end{cases}$$

and its impulse response can be found by IFT:

$$x(t) = \frac{1}{2\pi}\int_{-\infty}^\infty X(f)e^{ift}\,df = \frac{1}{2\pi i t}\left(e^{iat} - e^{-iat}\right) = \frac{\sin(at)}{\pi t}.$$

(ix) **Triangle Function:** This function is defined as

$$x(t) = \text{tri}(t) = \begin{cases} 1 - |t| & \text{if } |t| < 1, \\ 0 & \text{if } |t| \ge 1. \end{cases} \tag{3.16}$$

Since $x(t)$ is an even function, its FT is

$$\mathcal{F}\{x(t)\} = \int_{-\infty}^{\infty} x(t)\, e^{-i\,ft}\, dt = 2 \int_{0}^{1} (1-t) \cos(ft)\, dt$$

$$= 2 \left\{ \int_{0}^{1} \cos(ft)\, dt - \int_{0}^{1} t \cos(ft)\, dt \right\}$$

$$= \frac{2}{f} \left\{ \sin(f) - \int_{0}^{1} t\, d\sin(ft) \right\}$$

$$= \frac{2}{f} \left\{ \sin(f) - t\sin(ft) \Big|_{0}^{1} - \int_{0}^{1} \sin(ft)\, dt \right\}$$

$$= \frac{2}{f^2} \cos(ft) \Big|_{0}^{1} = \frac{2}{f^2}(1 - \cos(f))$$

$$= \left(\frac{2}{f}\right)^2 \sin^2(f/2) = \operatorname{sinc}^2\left(\frac{f}{2}\right). \tag{3.17}$$

Alternatively, since the triangle function is convolution of two square functions for $a = 1/2$, its FT can be obtained using the convolution theorem as

$$\mathcal{F}\{\operatorname{tri}(t)\} = \mathcal{F}\{\operatorname{rect}(t) \star \operatorname{rect}(t)\} = \operatorname{sinc}\left(\frac{f}{2}\right) \operatorname{sinc}\left(\frac{f}{2}\right) = \operatorname{sinc}^2\left(\frac{f}{2}\right).$$

(x) **Gaussian Function:** The FT of the Gaussian function $x(t) = e^{-\pi t^2}$ is

$$X(f) = \mathcal{F}\{x(t)\} = \int_{-\infty}^{\infty} e^{-\pi t^2}\, e^{-i\,ft}\, dt$$

$$= \int_{-\infty}^{\infty} e^{-\pi\left(t^2 + 2i\,ft\right)}\, dt$$

$$= e^{\pi(i\,f)^2} \int_{-\infty}^{\infty} e^{-\pi\left[t^2 + 2i\,ft + (i\,f)^2\right]}\, dt$$

$$= e^{-\pi f^2} \int_{-\infty}^{\infty} e^{-\pi(t+i\,f)^2}\, d(t + i\,f) = e^{-\pi f^2},$$

since $\int_{-\infty}^{\infty} e^{-\pi t^2}\, dt = 1$. Thus, the FT of a bell-shaped function is also a bell-shaped function. Note that the area under either $x(t)$ or $X(f)$ is 1. Also, due to the scaling property of time and frequency, we have

$$\mathcal{F}\{a e^{-\pi(at)^2}\} = e^{-\pi(f/a)^2}.$$

Also, in the case when $a = 1/\sqrt{2\pi\sigma^2}$, then $a x(t)$ is a normal distribution with variance σ^2 and mean $\mu = 0$. As $a \to \infty$, $x(t)$ becomes taller and approaches $\delta(t)$, and its spectrum $e^{-\pi(f/a)^2}$ becomes wider and approaches 1. On the other hand, if we write $\mathcal{F}\{e^{-\pi(at)^2}\} = \dfrac{1}{a} e^{-\pi(f/a)^2}$, and let $a \to 0$, then $x(t)$ approaches 1 and $X(f)$ approaches $\delta(f)$.

(xi) **Comb Function:** This function is defined in terms of the δ-function as

$$\text{comb}(t) = \sum_{m=-\infty}^{\infty} \delta(t - mT),$$

where T is some given period, $m \in \mathbb{Z}$, and $1/T$ is the sampling rate. Since the delta function is periodic, this definition can be represented as a Fourier series

$$\text{comb}(t) = \frac{1}{T} \sum_{m=-\infty}^{\infty} e^{2\pi i \, mt/T}.$$

The expansion is derived as follows: the complex Fourier series expansion is $\text{comb}(t) = \sum\limits_{m=-\infty}^{\infty} c_m \, e^{2\pi i \, mt/T}$, where

$$\begin{aligned}
c_m &= \frac{1}{T} \int_{t_0}^{t_0+T} \text{comb}(t) \, e^{-2\pi i \, mt/T} \, dt, \quad -\infty < t_0 < +\infty \\
&= \frac{1}{T} \int_{-T/2}^{T/2} \text{comb}(t) \, e^{-2\pi i \, mt/T} \, dt \\
&= \frac{1}{T} \int_{-T/2}^{T/2} \delta(t) \, e^{-2\pi i \, mt/T} \, dt = \frac{1}{T} e^{-2\pi i \, 0/T} = \frac{1}{T}.
\end{aligned}$$

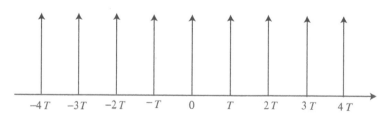

Figure 3.2 Comb Function comb(t).

The FT of the comb function in the frequency domain (in Hz units) is

$$\mathcal{F}\{\text{comb}(t)\} = \text{comb}\left(f - \frac{m}{T}\right) = \sum_{m=-\infty}^{\infty} e^{-2\pi i \, mfT},$$

and in the angular frequency domain (in radian/sec) is

$$\mathcal{F}\{\text{comb}(t)\} = \frac{\sqrt{2\pi}}{T} \text{comb}\left(f - \frac{m}{T}\right) = \frac{1}{\sqrt{2\pi}} \sum_{m=-\infty}^{\infty} e^{-\pi i \, fmT}, \tag{3.18}$$

Thus, the FT of a comb function is a comb function. The graph of the comb function[2] is given in Figure 3.2.

3.1.6 Poisson's Summation Formula. Consider a continuous function $f(x)$ on $-\infty < x < \infty$ which vanishes for large $|x| > N$. Then for each x, the series $\sum\limits_{n=-\infty}^{\infty} f(x + 2na)$ converges absolutely and uniformly for $-a \leq x \leq a$ to a continuous function $g(x)$, which is a periodic function of period $2a$. The complex Fourier series for $g(x)$ is

$$g(x) = \sum_{n=-\infty}^{\infty} c_k \, e^{k\pi i x/a},$$

where

$$c_k = \frac{1}{2a} \int_{-a}^{a} g(x) \, e^{k\pi i x/a}, \quad k = 0, \pm 1, \pm 2, \dots .$$

Since $g(x) = \lim\limits_{N \to \infty} \sum\limits_{-N}^{N} f(x + 2na)$, then

$$c_k = \frac{1}{2a} \lim_{N \to \infty} \int_{-a}^{a} f(x + 2na) \, e^{-k\pi i x/a} \, dx$$

$$= \frac{1}{2a} \lim_{N \to \infty} \int_{-(2n-1)a}^{(2n+1)a} f(u) \, e^{-k\pi i u/a} \, du$$

$$= \frac{1}{2a} \lim_{N \to \infty} \int_{-(2N-1)a}^{(2N+1)a} f(x) \, e^{-k\pi i x/a} \, dx = \frac{\pi}{a} F\left(\frac{k\pi}{a}\right).$$

Hence,

$$g(x) = \sum_{n=-\infty}^{\infty} f(x + 2na) = \sum_{k=-\infty}^{\infty} \frac{\pi}{a} F\left(\frac{k\pi}{a}\right) e^{k\pi i x/a}.$$

Setting $x = 0$ in this result, we obtain the *Poisson's summation formula*:

$$\sum_{n=-\infty}^{\infty} f(2na) = \sum_{k=-\infty}^{\infty} \frac{\pi}{a} F\left(\frac{k\pi}{a}\right), \tag{3.19}$$

---

[2] Bracewell [1986] calls it the 'Shah' function, based on the form of the cyrillic letter 'sh, Sh' (ш, Ш). This letter is not pronounced as 'Shah', which may also be confused for the last name of a person as there are over 8 million persons with this last name, many of whom are scientist and engineers. A better choice would have been to name this function as 'Sha' function.

which for $2a = 1$ becomes

$$\sum_{n=-\infty}^{\infty} f(n) = \sum_{k=-\infty}^{\infty} 2\pi F(2n\pi).$$

These formulas can be used to obtain the following results:

$$\sum_{n=-\infty}^{\infty} \frac{1}{n^2 + a^2} = \frac{\pi}{a} \coth(\pi a), \quad \sum_{n=-\infty}^{\infty} e^{-\pi n^2 x} = \frac{1}{\sqrt{x}} \sum_{n=-\infty}^{\infty} e^{-\pi n^2 x}. \quad (3.20)$$

The details of proofs are easy and can be found in any standard textbook on integral transforms.

3.1.7 Sine and Cosine Transforms. Fourier sine and cosine transforms are forms of the Fourier transforms that do not use complex quantities. They are the forms originally used by Fourier and are still in use in signal processing and statistics. The Fourier sine transform of $f(x)$ is defined as

$$\mathcal{F}_s\{f(x)\} = F_s(\omega) = 2 \int_{-\infty}^{\infty} f(x) \sin(\omega x) \, dx. \quad (3.21)$$

Note that $F_s(\omega)$ is an odd function of the radian frequency ω, i.e., $F_s(\omega) = -F_s(-\omega)$ for all ω. The numerical factors are uniquely defined only by their product, and they vary in mathematical and physical situations. In definition (3.21), the factor 2 appears because the sine function has L^2-norm of $1/\sqrt{2}$. The Fourier cosine transform of $f(x)$ is similarly defined as

$$\mathcal{F}_c\{f(x)\} = F_c(\omega) = 2 \int_{-\infty}^{\infty} f(x) \cos(\omega x) \, dx. \quad (3.22)$$

Note that $F_c(\omega)$ is an even function of the radian frequency ω, i.e., $F_c(\omega) = F_c(-\omega)$ for all ω.

   Theorem 3.3. (Fourier sine theorem) *If $f(x)$ satisfies the Dirichlet conditions on the nonnegative real line and is absolutely integrable on $(0, \infty)$, then*

$$\frac{1}{2}[f(x+0) + f(x-0)] = 2 \int_0^{\infty} d\omega \int_0^{\infty} f(u) \sin(\omega u) \sin(\omega x) \, du.$$

If $f(x)$ is continuous, then $\frac{1}{2}[f(x+0) + f(x-0)] = f(x)$.

   Theorem 3.4. (Fourier cosine theorem) *If $f(x)$ satisfies the Dirichlet conditions on the nonnegative real line and is absolutely integrable on $(0, \infty)$, then*

$$\frac{1}{2}[f(x+0) + f(x-0)] = 2 \int_0^{\infty} d\omega \int_0^{\infty} f(u) \cos(\omega u) \cos(\omega x) \, du.$$

Using the definition (3.1), the Fourier transform $F(\omega)$ can be expressed as

$$F(\omega) = \int_{-\infty}^{\infty} f(x)\, e^{-i\omega x}\, dx = \int_{-\infty}^{\infty} f(x) \big[\cos(\omega x) - i\,\sin(\omega t)\big]\, dx$$

$$= \int_{-\infty}^{\infty} f(x)\cos(\omega x)\, dx - i \int_{-\infty}^{\infty} f(x)\sin(\omega x)\, dx$$

$$= \frac{1}{2} F_c(\omega) - \frac{i}{2} F_s(\omega).$$

Properties of Fourier sine and cosine transforms are as follows:

(a) $\mathcal{F}_c\{f(x)\} = F_c(\omega), \quad \mathcal{F}_s\{f(x)\} = F_s(\omega),$

(b) $\mathcal{F}_c\{F_c(x)\} = f(\omega), \quad \mathcal{F}_s\{F_s(x)\} = f(\omega),$

(c) $\mathcal{F}_c\{f(kx)\} = \dfrac{1}{k} F_c\left(\dfrac{\omega}{k}\right), \quad k > 0,$

(d) $\mathcal{F}_s\{f(kt)\} = \dfrac{1}{k} F_s\left(\dfrac{\omega}{k}\right), \quad k > 0,$

(e) $\mathcal{F}_c\{f(kx)\cos bx\} = \dfrac{1}{2k}\left[F_c\left(\dfrac{\omega+b}{k}\right) + F_c\left(\dfrac{\omega-b}{k}\right) \right], \quad k > 0,$

(f) $\mathcal{F}_c\{f(kx)\sin bx\} = \dfrac{1}{2k}\left[F_s\left(\dfrac{\omega+b}{k}\right) - F_s\left(\dfrac{\omega-b}{k}\right) \right], \quad k > 0,$

(g) $\mathcal{F}_s\{f(kx)\cos bt\} = \dfrac{1}{2k}\left[F_s\left(\dfrac{\omega+b}{k}\right) + F_s\left(\dfrac{\omega-b}{k}\right) \right], \quad k > 0,$

(h) $\mathcal{F}_s\{f(kx)\sin bx\} = \dfrac{1}{2k}\left[F_c\left(\dfrac{\omega-b}{k}\right) - F_c\left(\dfrac{\omega+b}{k}\right) \right], \quad k > 0,$

(i) $\mathcal{F}_c\{t^{2n} f(x)\} = (-1)^n \dfrac{d^{2n} F_c(\omega)}{d\omega^{2n}},$

(j) $\mathcal{F}_c\{t^{2n+1} f(x)\} = (-1)^n \dfrac{d^{2n+1} F_s(\omega)}{d\omega^{2n+1}},$

(k) $\mathcal{F}_s\{t^{2n} f(x)\} = (-1)^n \dfrac{d^{2n} F_s(\omega)}{d\omega^{2n}},$

(l) $\mathcal{F}_s\{t^{2n+1} f(x)\} = (-1)^{n+1} \dfrac{d^{2n+1} F_c(\omega)}{d\omega^{2n+1}}.$

(m) $\mathcal{F}_s\{f'(x)\} = -\omega\, F_c(\omega),$

(n) $\mathcal{F}_s\{f''(x)\} = -\omega^2\, F_s(\omega) + 2\omega f(0),$

(o) $\mathcal{F}_c\{f'(x)\} = \omega\, F_s(\omega) - 2f(0),$ and

(p) $\mathcal{F}_c\{f''(x)\} = -\omega^2\, F_c(\omega) - 2f'(0).$

Convolution theorems for Fourier sine and cosine transforms are as follows:

Theorem 3.5. *Let $F_c(\omega)$ and $G_c(\omega)$ be the Fourier cosine transforms of $f(x)$ and $g(x)$, respectively, and let $F_s(\omega)$ and $G_s(\omega)$ be the Fourier sine transforms of $f(x)$ and $g(x)$, respectively. Then*

$$\mathcal{F}_c^{-1}\{F_c(\omega)\,G_c(\omega)\} = \int_0^\infty g(s)\,[f(|t-s|)+f(x+s)]\,ds$$

$$= \int_0^\infty f(s)\,[g(|t-s|)+g(x+s)]\,ds.$$

Theorem 3.6. *If $F_s(\omega)$ and $F_c(\omega)$, and $G_s(\omega)$ and $G_s(\omega)$ are the Fourier sine and cosine transforms of $f(x)$ and $g(x)$, respectively, then the following results hold:*

(i) $\displaystyle\int_0^\infty F_c(\omega)\,G_s(\omega)\,\sin\omega x\,d\omega = \frac{1}{2}\int_0^\infty g(s)[f(|x-s|)-f(x+s)]\,ds,$

(ii) $\displaystyle\int_0^\infty F_s(\omega)\,G_c(\omega)\,\sin\omega x\,d\omega = \frac{1}{2}\int_0^\infty f(s)[g(|x-s|)-g(x+s)]\,ds,$

(iii) $\displaystyle\int_0^\infty F_s(\omega)\,G_s(\omega)\,\cos\omega x\,d\omega$

$$= \frac{1}{2}\int_0^\infty g(x)\left[H(x+s)\,f(x+s)+H(x-s)\,f(x-s)\right]dx$$

$$= \frac{1}{2}\int_0^\infty f(x)\left[H(x+s)\,g(x+s)+H(x-s)\,g(x-s)\right]dx,$$

(iv) $\displaystyle\int_0^\infty F_c(\omega)\,G_c(\omega)\,d\omega = \int_0^\infty x(s)y(s)\,ds = \int_0^\infty F_s(\omega)G_s(\omega)\,d\omega.$

Proofs of all above theorems can be found in Sneddon [1978] or Kythe et al. [2003].

3.2 Finite Fourier Transforms

When the domain of a physical problem is finite, it is generally not convenient to use the transforms with an infinite range of integration. In many cases finite Fourier transfoms can be used with advantage The finite sine and cosine transforms are defined, respectively, by

$$\mathcal{F}_s\{f(x)\} = F_s(n) = \int_0^a f(x)\sin\left(\frac{n\pi x}{a}\right)dx,$$

$$\mathcal{F}_c\{f(x)\} = F_c(n) = \int_0^a f(x)\cos\left(\frac{n\pi x}{a}\right)dx, \tag{3.23}$$

and their inverse transforms, respectively, by

$$f(x) = \frac{2}{a}\sum F_s(n)\sin\left(\frac{n\pi x}{a}\right), \quad f(x) = \frac{F_c(0)}{a}+\frac{2}{a}\sum F_c(n)\cos\left(\frac{n\pi x}{a}\right). \tag{3.24}$$

Example 3.6. (a) For $f(x) = 1$ we have

$$\mathcal{F}_s\{1\} = F_s(n) = \int_0^a \sin \frac{n\pi x}{a} \, dx = \frac{a}{n\pi} \left[1 - (-1)^n\right];$$

$$\mathcal{F}_c\{1\} = F_c(n) = \int_0^a \cos \frac{n\pi x}{a} \, dx = \begin{cases} a, & n = 0, \\ 0, & n \neq 0. \end{cases}$$

(b) For $f(x) = x$, we have

$$\mathcal{F}_s\{x\} = F_s(n) = \int_0^a x \sin \frac{n\pi x}{a} \, dx = \frac{(-1)^{n+1} a^2}{n\pi};$$

$$\mathcal{F}_c\{x\} = F_c(n) = \int_0^a x \cos \frac{n\pi x}{a} \, dx = \begin{cases} \dfrac{a^2}{2}, & n = 0, \\ \left(\dfrac{a}{n\pi}\right)^2 \left[1 - (-1)^n\right], & n \neq 0. \end{cases} \blacksquare$$

The *discrete* finite Fourier transform is defined as follows: Let $f(j)$, $j = 0, 1, \ldots, N-1$, denote a sequence of N finite valued complex numbers. Then the finite Fourier transform $F(n)$ of $f(j)$ is defined by

$$F(n) = \frac{1}{N} \sum_{j=0}^{N-1} f(j) \, e^{2\pi i \, nj/N} = \frac{1}{N} \sum_{j=0}^{N-1} f(j) \, W_N^{nj}, \qquad (3.25)$$

where $W_N = e^{2\pi i/N}$ [Cooley et al., 1969]. Note that W_N^{nj} is periodic of period N, i.e., $W_N^{nj} = W_N^{(n+N)j}$. The inverse finite Fourier transform is defined by

$$f(j) = \sum_{j=0}^{N-1} F(n) W_N^{nj}, \qquad j = 0, 1, \ldots, N-1, \qquad (3.26)$$

so that $f(j)$ and $F(n)$ form a transform pair, i.e., $f(j) \rightleftharpoons F(n)$. Since $f(-j) = F(N-j)$ and $F(-n) = F(N-n)$, we have $f(-j) \rightleftharpoons F(-n)$. Note that

(i) $f(j)$ is real iff $F(n) = \tilde{F}(-n) = \tilde{F}(N-n)$;

(ii) $F(n)$ is real iff $f(j) = \tilde{f}(-j) = \tilde{f}(N-j)$;

(iii) $f(j)$ is purely imaginary iff $F(n) = -\tilde{F}(-n) = -\tilde{F}(N-m)$; and

(iv) $F(n)$ is purely imaginary iff $f(j) = -\tilde{f}(-j) = -\tilde{f}(m-j)$, where $\tilde{f}(j) = -\tilde{f}(-j) = -\tilde{f}(n-j)$, and $m = -n$. Also,

$f(j)$ is real and even iff $F(n)$ is real and even;

$f(j)$ is real and odd iff $F(n)$ is purely imaginary and odd;

$f(j)$ is purly imaginary and even iff $F(n)$ is purely imaginary and even;

$f(j)$ is purly imaginary and odd iff $F(n)$ is real and odd.

Since $f(j) \rightleftharpoons F(n)$, we have the following results:

$$f(j - k) \rightleftharpoons W_N^{-nk} F(n), \qquad W_N^{mj} f(j) \rightleftharpoons F(n - m).$$

Example 3.7. Define the sequence $\delta(j)$ by

$$\delta(j) = \begin{cases} 1 & \text{if } j = 0 \;(\text{mod } N) \\ 0 & \text{otherwise.} \end{cases}$$

Then $\delta(j) \rightleftharpoons \dfrac{1}{N}$, and $1 \rightleftharpoons \delta(n)$. This result is very useful. For example, suppose we have the transform of $f(j)$ and want to find the transform of $f(j) - a$, where a is a constant, i.e., we want to move $f(j)$ up or down. Then, if $f(j) \rightleftharpoons F(n)$, we have $f(j) - a \rightleftharpoons F(n) - a\delta(n)$. Now, $F(n) - a\delta(n)$ is just $F(n)$ with $F(0)$ replaced by $F(0) - a$. Hence, subtracting a constant from all the values of $f(j)$ is equivalent, in the frequency domain, to subtracting the same constant from $F(0)$. ∎

Finally, note that $f(j) \rightleftharpoons F(n)$, then $f(0) = \sum_{n=0}^{N-1} F(n)$ and $F(0) = \dfrac{1}{N} \sum_{j=0}^{N-1} f(j)$. This result is useful in statistical applications, since if $f(j)$ is a set of samples of a random variable, then $F(0)$ is the sample mean.

3.2.1 Properties. The following properties are useful:

(i) $\mathcal{F}_s\{f'(x)\} = -\left(\dfrac{n\pi}{a}\right) F_c(n),$

(ii) $\mathcal{F}_s\{f''(x)\} = -\left(\dfrac{n\pi}{a}\right)^2 F_s(n) + \left(\dfrac{n\pi}{a}\right) [f(0) + (-1)^{n+1} f(a)],$

(iii) $\mathcal{F}_c\{f'(x)\} = \left(\dfrac{n\pi}{a}\right) F_s(n) + (-1)^n f(a) - f(0),$ and

(iv) $\mathcal{F}_c\{f''(x)\} = -\left(\dfrac{n\pi}{a}\right)^2 F_c(n) + (-1)^n f'(a) - f'(0).$

3.2.2 Finite Fourier Transforms of Derivatives

$$\begin{aligned}
\mathcal{F}_s\{f'(x)\} &= -\left(\frac{n\pi}{a}\right) F_c(n), \\
\mathcal{F}_s\{f''(x)\} &= -\left(\frac{n\pi}{a}\right)^2 F_s(n) + \left(\frac{n\pi}{a}\right) [x(0) + (-1)^{n+1} x(a)], \\
\mathcal{F}_c\{f'(x)\} &= -\left(\frac{n\pi}{a}\right) F_s(n) + (-1)^n x(a) - x(0), \\
\mathcal{F}_c\{f''(x)\} &= -\left(\frac{n\pi}{a}\right)^2 F_c(n) + (-1)^n x'(a) - x'(0).
\end{aligned}$$

(3.27)

3.2.3 Periodic Extensions. A function $g(x)$ is said to be an *odd periodic extension* of a periodic function $f(x)$ of period 2π if

$$g(x) = \begin{cases} f(x) & \text{for } 0 < x < \pi, \\ -f(-x) & \text{for } -\pi < x < 0. \end{cases}$$

That is, $g(x) = f(x)$ for $0 < x < \pi$, and $g(-x) = -g(x)$ for $x \in (-\pi, \pi)$. Use periodicity to extend g, that is, $g(x + 2\pi) = g(x)$ for $-\infty < x < \infty$.

A function $h(x)$ is said to be an *even periodic extension* of a periodic function $f(x)$ of period 2π if

$$h(x) = \begin{cases} f(x) & \text{for } 0 < x < \pi, \\ f(-x) & \text{for } -\pi < x < 0. \end{cases}$$

That is, $h(x) = f(x)$ for $0 < x < \pi$, and $h(-x) = h(x)$. Extend g by periodicity, i.e., $g(x + 2\pi) = h(x)$ for $-\infty < x < \infty$.

A theorem on odd and even extensions of a periodic function f is given below without proof (Churchill [1972]).

Theorem 3.7. *(i) If $g(x)$ is the odd periodic extension of $f(x)$ of period 2π, then for any constant ω*

$$\mathcal{F}_s\{g(x - \omega) + g(x + \omega)\} = 2\cos n\omega\,\mathcal{F}_s\{f(x)\},$$
$$\mathcal{F}_c\{g(x + \omega) + g(x - \omega)\} = 2\sin n\omega\,\mathcal{F}_s\{f(x)\}.$$

(ii) If $h(t)$ is an even periodic extension of $f(x)$ of period 2π, then for any constant ω

$$\mathcal{F}_c\{h(x - \omega) + h(x + \omega)\} = 2\cos n\omega\,\mathcal{F}_c\{f(x)\},$$
$$\mathcal{F}_c\{h(x - \omega) + h(x + \omega)\} = 2\sin n\omega\,\mathcal{F}_c\{f(x)\}.$$

3.2.4 Convolution. Let $f(x)$ and $g(x)$ be two piecewise continuous periodic functions defined on $-\pi < x < \pi$. Then their *convolution* is defined by

$$f(x) \star g(x) = \int_{-\pi}^{\pi} f(x - u)\,y(u)\,du, \tag{3.28}$$

which is a continuous periodic function of period 2π. The following results hold (Churchill [1972]):

Theorem 3.8. *Let $f_1(x)$ and $g_1(x)$ be the odd periodic extensions of $f(x)$ and $g(x)$, respectively on $0 < x < \pi$, and $f_2(x)$ and $g_2(x)$ their even periodic extensions on $0 < x < \pi$. Then*

$$\mathcal{F}_s\{f_1(x) \star g_2(t)\} = 2F_s(n)\,G_c(n), \quad \mathcal{F}_s\{f_2(t) \star g_1(x)\} = 2F_c(n)\,G_s(n),$$
$$\mathcal{F}_c\{f_1(x) \star g_1(t)\} = -2F_s(n)\,G_s(n), \quad \mathcal{F}_c\{f_2(t) \star g_2(x)\} = 2F_c(n)\,G_c(n),$$

or inversely,

$$\mathcal{F}_s^{-1}\{F_s(n)\,G_c(n)\} = \frac{1}{2}\{f_1(x) \star g_2(x)\},$$
$$\mathcal{F}_s^{-1}\{F_c(n)\,G_s(n)\} = 2\{f_2(x) \star g_1(x)\},$$
$$\mathcal{F}_s^{-1}\{F_s(n)\,G_s(n)\} = -\frac{1}{2}\{f_1(x) \star g_1(x)\},$$

$$\mathcal{F}_c^{-1}\{F_c(n)\,G_c(n)\} = \frac{1}{2}\{f_2(x) \star g_2(x)\}.$$

Example 3.8. To determine $\mathcal{F}^{-1}\{(n^2 - a^2)^{-1}\}$, $n \neq 0$, let $\dfrac{1}{n2 - a^2} =$

$F_s(n)\,G_s(n)$, where $F_s(n) = \dfrac{n(-1)^{n+1}}{n^2 - a^2}$ and $G_s(n) = \dfrac{(-1)^{n+1}}{n}$, which gives

$f(x) = \dfrac{\sin ax}{\sin a\pi}$, and $g(x) = \dfrac{x}{\pi}$, so that

$$\frac{1}{n^2 - a^2} = F_s(n)G_s(n) = \mathcal{F}_s\left\{\frac{\sin ax}{\sin a\pi}\right\}\mathcal{F}_s\left\{\frac{x}{\pi}\right\}; \quad \text{thence}$$

$$= \mathcal{F}^{-1}\left\{\frac{1}{n^2 - a^2}\right\} = \mathcal{F}^{-1}\{F_s(n)G_s(n)\} = -\frac{1}{2}\,f_1(x) \star g_1(x),$$

where $f_1(x)$ is the periodic extension of the odd function $f(x)$ and $g_1(x) = \dfrac{x}{\pi}$.
Thus, using Example 3.8, we get

$$\mathcal{F}^{-1}\left\{\frac{1}{n^2 - a^2}\right\} = -\frac{1}{2\pi}\int_{-\pi}^{\pi} f_1(x - u)\,g_1(u)\,du$$

$$= -\frac{1}{2\pi}\int_{-\pi}^{\pi} f_1(x - u)\,u\,du = -\frac{\cos[a(\pi - x)]}{a\sin a\pi}. \quad \blacksquare$$

3.2.5 Differentiation with Respect to a Parameter

$$f(x, a) \rightleftharpoons F(\omega, a) \implies \frac{\partial}{\partial a}f(x, a) \rightleftharpoons \frac{\partial}{\partial a}F(\omega, a).$$

Example 3.9. Consider the rectangular pulse $f(x)$, equal to h on the interval $[-aL, aL]$ and zero elsewhere (Figure 2.9). The Fourier transform of this pulse is given by

$$F(\omega) = \int_{-aL}^{aL} h\,e^{-i\omega x}\,dx = \frac{h}{i\omega}\left[e^{i\omega aL} - e^{-i\omega aL}\right] = \frac{h}{2\omega}\sin\omega aL,$$

which in the simple case of a pulse with $a = 1 = L = h$ reduces to

$$F(\omega) = \int_{-1}^{1} \cos\omega x\,dx = \frac{2\sin\omega}{\omega} = 2\,\text{sinc}\,\omega = \int_{-1}^{1} \cos\omega x\,dx.$$

Geometrically, it means that the area under $F(\omega)$, for a fixed ω, varies continuously with a. Thus, $F(\omega)$ is continuous although $f(x)$ is not. Assuming that $f(x)$ is absolutely continuous, the Fourier transform $F(\omega)$ defined by (3.1) is always continuous and so is f. Figure 1.17(b) shows the graph of the rectangular pulse $f(x)$, and the surface $f(x)\cos\omega t$ in slices along with the graph of $F(\omega)$ of the above rectangular pulse. Note that $\text{sinc}(0) = \lim_{\omega \to 0} \dfrac{\sin\omega}{\omega} = 1$, and $\text{sinc}(\infty) = \lim_{\omega \to \infty} \dfrac{\sin \omega}{\omega} = 0$. For a given value of ω the area of the region under the curve $\cos \omega x$ from $x = -1$ to $x = 1$ is equal to the ordinate at that value of ω. Part (b) of Figure 1.17 is quite similar to that in Bracewell [2000].

3.3 Fourier Transforms of Two Variables

The Fourier transform of a function $f(x, y)$ of two independent spatial variables x and y is denoted by $\mathcal{F}\{f(x, y)\}$ and defined by

$$\mathcal{F}\{f(x,y)\} \equiv F\left(f_x, f_y\right) = \int_{-\infty}^{\infty} \int_{-\infty}^{\infty} f(x,y) \, e^{-2\pi i \, (f_x x + f_y y)} \, dx \, dy. \qquad (3.29)$$

The Fourier transform so defined is a complex-valued function of two independent variables f_x and f_y, which are generally called the *cycle frequencies*. The inverse transform is defined by

$$f(x,y) \equiv \mathcal{F}^{-1}\{F\left(f_x, f_y\right)\} = \int_{-\infty}^{\infty} \int_{-\infty}^{\infty} F\left(f_x, f_y\right) \, e^{2\pi i \, (f_x x + f_y y)} \, df_x \, df_y. \qquad (3.30)$$

The inverse Fourier transform is also called the *Fourier integral representation* of the function $f(x, y)$. There are three sufficient conditions that the function $f(x, y)$ must satisfy in order for the definitions (3.29) and (3.30) to be meaningful. They are: (i) $f(x, y)$ must be absolutely integrable over the (x, y)-plane; (ii) $f(x, y)$ must have only finitely many discontinuities and finitely many maxima and minima in any finite rectangle; and (iii) $f(x, y)$ must not have any infinite discontinuities.

3.3.1 Local Spatial Frequency. Since each component of a transform is a complex exponential of a unique frequency, the frequency components f_x and f_y extend over the entire (x, y)-plane. It is, therefore, impossible to associate a location to a particular frequency. However, in practice a certain portion of an image may contain parallel grid lines at some fixed spacing, and the frequency (or frequencies) represented by these grid lines may be localized to some region of the image. Thus, consider a complex-valued function

$$g(x,y) = A(x,y) \, e^{i \, \phi(x,y)}, \qquad (3.31)$$

where $A(x, y)$ is a real-valued and nonnegative amplitude, and $\phi(x, y)$ a real-valued phase distribution. Since our interest is in the phase function $\phi(x, y)$, let us further assume that $A(x, y)$ is a slowly varying function. We will consider two examples:

Example 3.10. Define the local frequency of the function $g(x, y)$ as a pair $\left(f_x^l, f_y^l\right)$ by

$$f_x^l = \frac{1}{2\pi} \frac{\partial}{\partial x} \phi(x,y), \quad f_y^l = \frac{1}{2\pi} \frac{\partial}{\partial y} \phi(x,y), \qquad (3.32)$$

where the superscript l signifies the 'local' frequency, and both frequencies f_x^l and f_y^l are zero in regions where the function $g(x,y)$ is zero. If the definitions (3.32) are applied to a given complex-valued function $g(x,y) = e^{2\pi i\,(f_x x + g_y y)}$ that represents linear-phase exponential frequencies (f_x, f_y), we get

$$f_x^l = \frac{1}{2\pi}\frac{\partial}{\partial x}\{2\pi(f_x x + g_y y)\} = f_x, \quad f_y^l = \frac{1}{2\pi}\frac{\partial}{\partial y}\{2\pi(f_x x + g_y y)\} = f_y. \quad (3.33)$$

In this case of a single Fourier component, the local frequencies reduce to the frequencies of that component, and they are constant over the entire (x,y)-plane. ∎

Example 3.11. Consider a quadratic-phase exponential function, known as a *finite chirp function*,

$$g(x,y) = e^{\pi i\,\beta(x^2 + y^2)}\,\mathrm{rect}\left(\frac{x}{2L_x}\right)\mathrm{rect}\left(\frac{y}{2L_y}\right). \quad (3.34)$$

Carrying out differentiations as required by definitions (3.33), we obtain

$$f_x^l = \beta x\,\mathrm{rect}\left(\frac{x}{2L_x}\right), \quad f_y^l = \beta y\,\mathrm{rect}\left(\frac{y}{2L_y}\right). \quad (3.35)$$

It is obvious that in this case the local frequencies depend on the location in the (x,y)-plane, within a rectangle $R = \{2L_x \times 2L_y\}$, where f_x^l varies linearly with x and f_y^l varies linearly with y. The function (3.34) is one of the examples where the local frequency depends on position in the (x,y)-plane. Also note that the dimensions of f_x^l and f_y^l are cycles/meter, and of β are meters^{-2}.

The Fourier transform of the function $g(x,y)$ in the rectangle R is given by

$$G(f_x, f_y) = \int_{-L_x}^{L_x}\int_{-L_y}^{L_y} e^{\pi i\,\beta(x^2 + y^2)}\,e^{-2\pi i\,(f_x x + g_y y)}\,dx\,dy.$$

Since this integral is separable in rectangular coordinates, it would suffice to evaluate the one-dimensional spectrum:

$$\begin{aligned}
G(f_x) &= \int_{-L_x}^{L_x} e^{\pi i\,\beta(x^2 + y^2)}\,e^{-2\pi i\,f_x x}\,dx \\
&= \frac{1}{\sqrt{2\beta}}\,e^{-i\pi f_x/\beta}\int_{-\sqrt{2\beta}(L_x + f_x/\beta)}^{\sqrt{2\beta}(L_x - f_x/\beta)} e^{i\,\pi w^2/2}\,dw \\
&= \frac{e^{-i\pi f_x/\beta}}{\sqrt{2\beta}}\left[C\left\{\sqrt{2\beta}\,(L_x - f_x/\beta)\right\} - C\left\{\sqrt{2\beta}\,(-L_x - f_x/\beta)\right\} \right. \\
&\quad \left. + i\,S\left\{\sqrt{2\beta}\,(L_x - f_x/\beta)\right\} - i\,S\left\{\sqrt{2\beta}\,(-L_x - f_x/\beta)\right\}\right],
\end{aligned}$$

$$(3.36)$$

where we have made a change of variable from t to $w = \sqrt{2\beta}\,(x - f_x/\beta)$ in the second step; and the function C and S are the Fresnel integrals defined by

$$C(z) = \int_0^z \cos\left(\pi t^2/2\right)\,dt, \quad S(z) = \int_0^z \sin\left(\pi t^2/2\right)\,dt.$$

The graph of $|G(f_x)|$ for $\beta = 1, L_x = 5$ is shown in Figure 3.3. ∎

Local frequencies are important in Fourier optics. They appear as a comprehensive optical wavefront and signify the ray directions of geometric optics related to that wavefront (see Chapter 7).

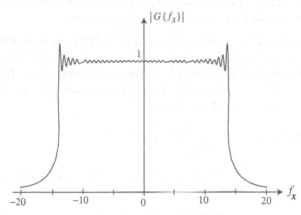

Figure 3.3 Spectrum of $|G(f_x)|$.

3.3.2 Fourier Transform of Signals. In the two-dimensional case of a signal $x(t, \tau)$, we define the Fourier transform (FT) as

$$\mathcal{F}\{x(t,\tau)\} \equiv X(f,g) = \int_{-\infty}^{\infty} \int_{-\infty}^{\infty} x(t,\tau)\,e^{-2\pi i\,(ft+g\tau)}\,dt\,d\tau, \qquad (3.37)$$

where we have used the cycle frequencies f and g instead of x_t and x_τ as remarked in footnote 1 in §3.1.4. In definition (3.37) the function $x(t,\tau)$ must satisfy the same type of three conditions mentioned in §3.3. However, any one of these conditions can be weakened at the cost of strengthening the others. For example, the two-dimensional Dirac delta function, defined by

$$\delta(t,\tau) = \lim_{N\to\infty} N^2\,e^{-N^2\pi(t^2+\tau^2)},$$

is infinite at the origin and zero elsewhere, has an infinite discontinuity, and therefore fails to satisfy the condition (iii). Similarly, the functions $x(t,\tau) = 1$ and $x(t,\tau) = \cos(2\pi ft)$ both fail to satisfy the existence condition (i). However, the Dirac delta function, for example, does satisfy the existence condition (i) and has Fourier transform given by

$$\mathcal{F}\{\delta(t,\tau)\} = \lim_{N\to\infty}\left\{\mathcal{F}\{N^2 e^{-N^2\pi(t^2+\tau^2)}\}\right\} = \lim_{N\to\infty}\left\{e^{-\pi(f^2+g^2)/N^2}\right\} = 1.$$

The Fourier cosine and sine transforms for a signal $x(t,\tau)$ are given by

$$\Re\{X(f,g)\} = \int_{-\infty}^{\infty}\int_{-\infty}^{\infty} x(t,\tau)\cos\left(-2\pi\cos(ft+g\tau)\right)\,dt\,d\tau,$$

$$\Im\{X(f,g)\} = \int_{-\infty}^{\infty}\int_{-\infty}^{\infty} x(t,\tau)\sin\left(-2\pi\cos(ft+g\tau)\right)\,dt\,d\tau.$$

The 2-D FT basic functions are of the form of waves with different frequencies and orientations which are given by the (f,g) values.

The purpose of inverse Fourier transform (IFT) is to reconstruct an image $x(t,\tau)$ using a weighted sum of sine and cosine functions. The only major difference between the FT and IFT is the sign on the basic function. This symmetry simplifies the implementation of the discrete FT and IFT. The IFT of a continuous function $x(t)$ is defined by

$$\mathcal{F}^{-1}\{X(f)\} \equiv x(t) = \int_{-\infty}^{\infty} X(f)\,e^{2\pi i\,ft}\,df$$

$$= \int_{-\infty}^{\infty} X(f)\cos(2\pi ft)\,df + i\int_{-\infty}^{\infty} X(f)\sin(2\pi ft)\,df. \tag{3.38}$$

The IFT of a continuous function (signal) $x(t,\tau)$ of two variables is defined as

$$\mathcal{F}^{-1}\{X(f,g)\} \equiv x(t,\tau) = \int_{-\infty}^{\infty}\int_{-\infty}^{\infty} X(f,g)\,e^{2\pi i\,(ft+g\tau)}\,df\,dg$$

$$= \int_{-\infty}^{\infty} X(f,g)\int_{-\infty}^{\infty} X(f,g)\cos\left(2\pi(ft+g\tau)\right)\,df\,dg$$

$$+ i\int_{-\infty}^{\infty}\int_{-\infty}^{\infty} X(f,g)\sin\left(2\pi(ft+g\tau)\right)\,df\,dg. \tag{3.39}$$

3.3.3 Circle Function. $\operatorname{circ}\left(\sqrt{x^2+y^2}\right) = \begin{cases} 1 & \text{for } \sqrt{x^2+y^2}=1 \\ 0 & \text{otherwise} \end{cases}$. Note

that the function $e^{-\pi(x^2+y^2)}$ is separable in rectangular coordinates and is circularly symmetric. However, the two-dimensional Fourier transform of most circularly symmetric functions cannot be found from products of known one-dimensional transforms. For example, consider a slightly different circularly symmetric function

$$\operatorname{circ}(r) = \begin{cases} 1 & \text{for } r < 1, \\ \frac{1}{2} & \text{for } r = 1, \\ 0 & \text{otherwise} \end{cases}$$

The Fourier-Bessel transform of the circle function is

$$\mathcal{B}\{\text{circ}(r)\} = 2\pi \int_0^1 r J_0(2\pi r \rho)\, dr.$$

If we change the variable by setting $r' = 2\pi r \rho$ and use the identity $\int_0^x t J_0(t)\, dx = x J_1(x)$, where J_0 and J_1 are Bessel functions of the first kind and order zero and 1, respectively, then the above transform can be written as

$$\mathcal{B}\{\text{circ}(r)\} = \frac{1}{2\pi \rho^2} \int_0^{2\pi \rho} r' J_0(r')\, dr' = \frac{J_1(2\pi \rho)}{\rho}. \tag{3.40}$$

A normalized form of this transform with value is 1 at the origin is $\dfrac{2J_1(2\pi \rho)}{2\pi \rho}$.

This function is known as the *besinc* function or the *jinc* function.

3.3.4 Amplitude Transmittance Function. The amplitude transmittance function $T_A(x, y)$ of the aperture is defined as the ratio of the transmitted field amplitude $U_t(x, y; 0)$ and the incident field amplitude $U_i(x, y; 0)$ at each point (x, y) in the $z = 0$ plane (i.e., the (x, y)-plane):

$$T_A(x, y) = \frac{U_t(x, y; 0)}{U_i(x, y; 0)}, \tag{3.41}$$

where the subscripts 't' and 'i' stand for the transmitted and incident field amplitude, respectively. This implies that $U_t(x, y; 0) = U_i(x, y; 0)\, T_A(x, y)$. We apply the convolution theorem to this equation, and relate the angular spectrum $A_i\,(\alpha/\lambda, \beta/\lambda)$ of the incident field and the angular spectrum $A_t\,(\alpha/\lambda, \beta/\lambda)$ of the transmitted field by

$$A_t\left(\frac{\alpha}{\lambda}, \frac{\beta}{\lambda}\right) = \left[A_i\left(\frac{\alpha}{\lambda}, \frac{\beta}{\lambda}\right) \star T\left(\frac{\alpha}{\lambda}, \frac{\beta}{\lambda}\right)\right], \tag{3.42}$$

where

$$T\left(\frac{\alpha}{\lambda}, \frac{\beta}{\lambda}\right) = \int_{-\infty}^{\infty} \int_{-\infty}^{\infty} T_A(x, y) \exp\left[-2\pi i \left(\frac{\alpha}{\lambda}x + \frac{\beta}{\lambda}y\right)\right] dx\, dy. \tag{3.43}$$

Thus, the angular spectrum of the transmitted disturbance is the convolution of the angular spectrum of the incident disturbance with a second angular spectrum. This defines the diffraction structure.

In particular, if a unit amplitude plane wave illuminates the diffraction structure, we have

$$A_i\left(\frac{\alpha}{\lambda}, \frac{\beta}{\lambda}\right) = \delta\left(\frac{\alpha}{\lambda}, \frac{\beta}{\lambda}\right),$$

and the amplitude is

$$A\left(\frac{\alpha}{\lambda}, \frac{\beta}{\lambda}\right) = \delta\left(\frac{\alpha}{\lambda}, \frac{\beta}{\lambda}\right) \star T\left(\frac{\alpha}{\lambda}, \frac{\beta}{\lambda}\right) = T\left(\frac{\alpha}{\lambda}, \frac{\beta}{\lambda}\right).$$

Thus, the diffraction structure is determined directly using the Fourier transform of the amplitude transmittance function of the aperture.

3.3.5 Convolution. According to the convolution theorem (§3.1.4), convolution in the frequency domain is equivalent to multiplication in the spatial domain:

$$F(f) \star G(f) = \text{IFT}\left\{f(x) \cdot g(x)\right\}, \quad \text{or} \quad \text{FT}\left\{F(f) \star G(f)\right\} = f(x) \cdot g(x). \quad (3.44)$$

This result manifests itself in the fact that both of these transforms become periodic when discrete samples are taken, since sampling is equivalent to multiplication by the *comb function* (see Example 3.12).

In the case of a signal, by calculating FT, the image processing in the frequency domain can be performed by filtering the FT coefficients. Thus,

$$x(t, \tau) \xrightarrow{\text{FT}} X(f, g),$$

$$X(f, g) \xrightarrow{\text{filter}} X'(f, g),$$

$$X'(f, g) \xrightarrow{\text{IFT}} x'(t, \tau).$$

Note that there are linear and nonlinear filters in both frequency and spatial domain image processing, which perform a wide variety of operations.

The separability of FT is justified by considering the 2-D FT, defined by (3.37), which can be written as a sequence of 1-D FTs:

$$\begin{aligned}
X(f, g) &= \int_{-\infty}^{\infty} \int_{-\infty}^{\infty} x(t, \tau) e^{-2\pi i (ft + g\tau)} \, dt \, d\tau \\
&= \int_{-\infty}^{\infty} \left[\int_{-\infty}^{\infty} x(t, \tau) e^{-2\pi i ft} \, dt\right] e^{-2\pi i y\tau} \, d\tau \\
&= \int_{-\infty}^{\infty} X(t, g) e^{-2\pi i g\tau} \, d\tau.
\end{aligned} \quad (3.45)$$

Thus, the algorithm for separability of FT can be given as follows:

STEP 1: For each row of the matrix $x(t, \tau)$, perform FT to get $X(f, g)$.

STEP 2: For each column of the matrix $X(t, g)$, perform FT to get $X(f, g)$.

This algorithm is schematically presented in Figure 3.4.

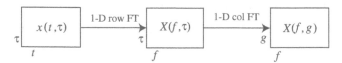

Figure 3.4 Separability of 2-D FT.

This process is much faster than the traditional $O(N^2M^2)$ 2-D computation, because of the following facts: 1-D FT of N rows is of the order $O(NM^2)$; 1-D FT of M columns is $O(MN^2)$; and 2-D FT is $O(NM^2 + MN^2)$.

3.3.6 Symmetry of Fourier Transform. The FT $X(f)$ of a signal $x(t)$ has even-odd symmetry: when $x(t)$ is symmetric, so is $X(f)$, and conversely. Thus, $x(t)$ even implies $X(f)$ even, and $x(t)$ odd implies $X(f)$ odd, and conversely. Similarly, an imaginary $x(t)$ implies anti-hermitian $X(f)$, and conversely, where anti-hermitian means real odd, or imaginary even. Schematically,

$$x(t) \text{ even} \rightleftarrows X(f) \text{ even};$$
$$x(t) \text{ odd} \rightleftarrows X(f) \text{ odd};$$
$$x(t) \text{ imaginary} \rightleftarrows X(f) \text{ anti-hermitian};$$
$$x(t) \text{ anti-hermitian} \rightleftarrows X(f) \text{ imaginary}.$$

These relations are useful when computing the FT of a real image.

3.3.7 Image Rotation. The effects of image rotation are better explained using the polar coordinates system. Thus, using $t = r\cos(\alpha)$, $\tau = r\sin(\alpha)$, $f = \gamma\cos(\beta)$, and $g = \gamma\sin(\beta)$, we have $\mathcal{F}\{x(t,\tau)\} = X(f,g)$, and $\mathcal{F}\{x(r,\alpha)\} = X(\gamma,\beta)$. Rotating $x(r,\alpha)$ by an angle θ yields $x(r,\alpha+\theta)$. Then substituting it we get $\mathcal{F}\{x(r,\alpha+\theta)\} = X(\gamma,\beta+\theta)$. Hence, a rotation of an image also rotates the FT.

The amplitude of FT is given by $|X(f,g)| = \sqrt{\Re\{X(f,g)\}^2 + \Im\{X(f,g)\}^2}$. The phase of the FT is given by $\phi = \arctan\dfrac{\Im\{X\}}{\Re\{X\}}$. An image can be reconstructed using either the FT amplitude or FT phase.

3.3.8 Other Properties of 2-D Fourier Transform

(i) LINEARITY: $\mathcal{F}\{ax(t,\tau) + by(t,\tau)\} = aX(f,g) + bX(f,g)$.

(ii) SCALING: $\mathcal{F}\{x(at,b\tau)\} = \dfrac{1}{|ab|}X\left(\dfrac{f}{a}, \dfrac{g}{b}\right)$.

(iii) TRANSLATION: $\mathcal{F}\{x(t+a, \tau+b)\} = X(f,g)\, e^{-2\pi i\,(fa/M + gb/M)}$.

(iv) DERIVATIVE: $\mathcal{F}\left\{\left(\partial_t^m\right)\left(\partial_\tau\right)^n x(t,\tau)\right\} = (2\pi i\, f)^m (2\pi i\, g)^n\, X(f,g)$.

(v) UNIT FUNCTION: For the unit step function $u(t,\tau)$[†]

$$\mathcal{F}\left\{\frac{\partial u}{\partial t}\right\} = i\,fU(f,\tau), \qquad \mathcal{F}\left\{\frac{\partial u}{\partial \tau}\right\} = \frac{dU}{d\tau},$$
$$\mathcal{F}\left\{\frac{\partial^2 u}{\partial t^2}\right\} = -f^2U(f,\tau), \qquad \mathcal{F}\left\{\frac{\partial^2 u}{\partial \tau^2}\right\} = \frac{d^2U}{d\tau^2}.$$

[†] The unit step function u is the same as the Heaviside step function H; but u is used here on purpose so that its FT can be written as U.

The convolution theorem enables us to perform any linear operation by filtering (multiplication) in the frequency domain instead of performing it in the spatial domain.

Example 3.12. (2-D comb function) This function is defined as $\text{comb}(t,\tau)$ $= \text{comb}(t)\,\text{comb}(\tau) = \sum_{m=-\infty}^{\infty} \delta(t-m) \sum_{n=-\infty}^{\infty} \delta(\tau-n)$. We have

$$\mathcal{F}\{\text{comb}(t)\} = \sum_{m=-\infty}^{\infty} e^{\pm 2\pi i m f} = \text{comb}(f),$$

which is obtained using the shift property, which translates into multiplication by a linear phase factor (i.e., $\mathcal{F}\{x(t \pm t_0)\} = e^{\pm 2\pi i t_0 f} X(f)$. Similarly, the scaling property of FT that transforms magnification into reduction: $\mathcal{F}\{x(t/b)\} = |b| X(bf)$. Using both shifting and scaling on the comb function, we get

$$\text{comb}\left(\frac{t-t_0}{b}\right) = |b| \sum_{n=-\infty}^{\infty} \delta(t-t_0-nb).$$

Hence,

$$\mathcal{F}\left\{\text{comb}\left(\frac{t-t_0}{b}\right)\right\} = |b|\,\text{comb}(bf)\,e^{-2\pi i t_0 f}.$$

Also, multiplying $x(t)$ and the δ function we get

$$x(t)\,\text{comb}(t) = x(t) \sum_{m=-\infty}^{\infty} \delta(t-m) = \sum_{m=-\infty}^{\infty} x(t)\delta(t-m) = \sum_{m=-\infty}^{\infty} x_m\delta(t-m)$$

and their convolution is

$$x(t) \star \text{comb}(t) = x(t) \star \sum_{m=-\infty}^{\infty} \delta(t-m) = \sum_{m=-\infty}^{\infty} x(t)\star\delta(t-m) = \sum_{m=-\infty}^{\infty} x(t-m).$$

A function $u_0(t)$ of bounded support can be used to create a periodic function using convolution as follows. Let $u(t) = \Delta f\,\text{comb}(\Delta ft) \star u_0(t)$. Then its FT is

$$U(f) = \Delta U_0(f)\,\text{comb}\left(\frac{f}{\Delta f}\right), \tag{3.46}$$

where U_0 is the FT of u_0. Now, by breaking the FT in (3.46) into periods and then summing the periods, we get

$$U(f) = \sum_{m=-\infty}^{\infty} \int_{-1/(2\Delta f)}^{1/(2\Delta f)} u_0(t)\, e^{-2\pi i\, f(t-m/\Delta f)}\, dt$$

$$= \left(\sum_{m=-\infty}^{\infty} e^{2\pi i\, fm/\Delta f} \right) \int_{-1/(2\Delta f)}^{1/(2\Delta f)} u_0(t)\, e^{-2\pi i\, ft}\, dt$$

$$= \mathrm{comb}(f/\Delta f) \int_{-1/(2\Delta f)}^{1/(2\Delta f)} u_0(t)\, e^{-2\pi i\, ft}\, dt$$

$$= \Delta f\, \mathrm{comb}(f/\Delta f) \int_{-1/(2\Delta f)}^{1/(2\Delta f)} u_0(t)\, e^{-2\pi i\, \Delta f\, t}\, dt,$$

where we have used the relation $\mathrm{comb}(f/\Delta f) = \sum_{m=-\infty}^{\infty} e^{2\pi i\, mf/\Delta f}$, and the factor Δf in the last expression comes from the scaling property when a delta function is applied. Geometrically, the effect is that of the tooth-thickening in the comb.

Now we consider the effect of the weights applied to the δ functions in the comb which will be the FT evaluated at specific positions:

$$U(m\Delta f) = \delta f \int_{-1/(2\Delta f)}^{1/(2\Delta f)} u(t)\, e^{-2\pi i\, m\Delta f t}\, dt.$$

But the Fourier series coefficients are also

$$C_m = \Delta f \int_{-1/(2\Delta f)}^{1/(2\Delta f)} u(t)\, e^{-2\pi i\, m\Delta f t}\, dt.$$

Comparing these two relations we find that

$$C_m = U(m\Delta f).$$

DISCRETE CASE. Again, consider a function with bounded support, and use it to create another function that is both periodic and discrete with a period N times the point spacing. Let

$$u(t) = \Delta f\, \mathrm{comb}(\Delta f\, t) \left\{ \frac{\Delta f}{N}\, \mathrm{comb}\left(\frac{\Delta f}{N} t \right) \right\} \star u_0(t),$$

while taking FT of $u(t)$ we get

$$U(f) = \int_{-\infty}^{\infty} u(t)\, e^{-2\pi i\, ft}\, dt.$$

Then we obtain the Fourier series relation

$$C_k = U(k\Delta f) = \frac{\Delta f}{N} \int_{-N/(2\Delta f)-\varepsilon}^{N/(2\Delta f)+\varepsilon} \Delta f \, \mathrm{comb}(\Delta ft)u_0(t) \, e^{-2\pi i k\Delta f t/N}$$

$$= \frac{\Delta f}{N} \sum_{n=-N/2}^{N/2-1} u_0(n/\Delta f) \, e^{-2\pi i kn/N}, \tag{3.47}$$

where, because of the FT evaluated at N specific positions, the integration range has been modified accordingly.

Now, the discrete FT coefficients are

$$U_k = \frac{1}{\sqrt{N}} \sum_{n=-N/2}^{N/2-1} u_0(n/\Delta f) \, e^{-2\pi i kn/N}. \tag{3.48}$$

Comparing these coefficients in (3.48) to the weights of the δ function in the FT (3.47), we find that

$$U_k = \frac{1}{\sqrt{N}}U(k\Delta f)\frac{N}{\Delta f} = \frac{\sqrt{N}}{\Delta f}U(k\Delta f).$$

3.3.9 Grating. In the three-dimensional coordinate system, the complex-valued amplitudes of two waves can be represented by

$$U^{\mathrm{r}}(\mathbf{r}) = A \, e^{i\mathbf{k}_{\mathrm{r}}\cdot\mathbf{r}}, \quad U^{\mathrm{o}}(\mathbf{r}) = a \, e^{i\mathbf{k}_{\mathrm{o}}\cdot\mathbf{r}}, \tag{3.49}$$

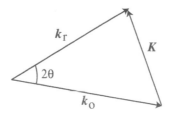

Figure 3.5 Grating Vector $\mathbf{K}$.

where $\mathbf{k}_{\mathrm{r}}$ and $\mathbf{k}_{\mathrm{o}}$ are the wave vectors of the reference and object waves, respectively, and $\mathbf{r} = (x, y, z)$ is a position vector. Then the intensity distribution, generated by superimposing these waves, is given by

$$I(\mathbf{r}) = |A|^2 + |a|^2 + 2|A|\,|a|\cos\left((\mathbf{k}_{\mathrm{r}} - \mathbf{b})\cdot\mathbf{r} + \phi\right), \tag{3.50}$$

where A and a are the fringe periods, $\mathbf{b}$ is a vector of magnitude $2\pi/A$ which points in the direction of the difference between $\mathbf{k}_{\mathrm{r}}$ and $\mathbf{k}_{\mathrm{o}}$, and ϕ is the phase difference between the phases A and a. Now, we define a grating vector $\mathbf{K}$ as the difference of the two waves

$$\mathbf{K} = \mathbf{k_r} - \mathbf{k_o}. \tag{3.51}$$

This geometry is presented in Figure 3.5, which shows that

$$A = \frac{2\pi}{|\mathbf{K}|} = \frac{\lambda}{2\sin\theta}. \tag{3.52}$$

3.4 Fourier Transforms: Discrete Case

The discrete Fourier transform (DFT) that transforms one function into another is called the *frequency domain representation*; it requires a discrete input function whose nonzero real or complex values are finite. Unlike the discrete-time Fourier transform, the DFT only evaluates enough frequency components to reconstruct the finite segment that was analyzed. The DFT is used in processing information (data) stored in computers, and in signal processing and related fields where the frequencies in a signal are analyzed. An efficient computation of the DFT is provided by a fast Fourier transform (FFT) algorithm. For more details, see Kythe and Schäferkotter [2005: 299 ff]. The DFT is defined as follows: let a sequence of n nonzero complex numbers $\{x(0), \ldots, x(N-1)\}$ be transformed into a sequence of N complex numbers $\{X(0), \ldots, X(N-1)\}$ by the formula

$$X(k) = \sum_{n=0}^{N-1} x(n)\, e^{-2\pi nk\, i/N} \quad \text{for } 0 \le k \le N-1, \tag{3.53}$$

where

$$x(n) = \frac{1}{N} \sum_{k=0}^{N-1} X(k)\, e^{2\pi kn\, i/N} \quad \text{for } 0 \le n \le N-1 \tag{3.54}$$

Formula (3.53) is known as the *DFT analysis equation* and its inverse (3.54) the *DFT synthesis equation*, or the *inverse discrete Fourier transform* (IDFT). The DFT pair can be represented as

$$x(n) \;\rightleftharpoons\; X(k). \tag{3.55}$$

The DFT in matrix form is represented as follows: expand (3.53) in terms of the time and frequency indices (k, n), and we get

$$X(0) = x(0) + x(1) + \cdots + x(N-1),$$

$$X(1) = x(0) + x(1)\, e^{-2\pi\, i/N} + \cdots + x(N-1)\, e^{-2(N-1)\pi\, i/N},$$

$$X(2) = x(0) + x(1)\, e^{-4\pi\, i/N} + \cdots + x(N-1)\, e^{-4(n-1)\pi\, i/N},$$

$$\vdots$$

$$X(N-1) = x(0) + x(1)\, e^{-2(N-1)\pi\, i/N} + \cdots + x(N-1)\, e^{-2(N-1)^2\pi\, i/N},$$

which can be written in matrix form as

$$
\left\{
\begin{array}{c}
X(0) \\
X(1) \\
X(2) \\
\vdots \\
X(N-1)
\end{array}
\right\}
=
\begin{bmatrix}
1 & 1 & \cdots & 1 \\
1 & e^{-2\pi i/N} & \cdots & e^{-2(N-1)\pi i/N} \\
1 & e^{-4\pi i/N} & \cdots & e^{-4(N-1)\pi i/N} \\
\vdots & \vdots & \vdots & \vdots \\
1 & e^{-2(N-1)\pi i/N} & \cdots & e^{-2(N-1)^2\pi i/N}
\end{bmatrix}
\left\{
\begin{array}{c}
x(0) \\
x(1) \\
x(2) \\
\vdots \\
x(N-1)
\end{array}
\right\}.
$$

$$(3.56)$$

Similarly, for IDFT in matrix form we have

$$
\left\{
\begin{array}{c}
x(0) \\
x(1) \\
x(2) \\
\vdots \\
x(N-1)
\end{array}
\right\}
=
\frac{1}{N}
\begin{bmatrix}
1 & 1 & \cdots & 1 \\
1 & e^{2\pi i/N} & \cdots & e^{2(N-1)\pi i/N} \\
1 & e^{4\pi i/N} & \cdots & e^{4(N-1)\pi i/N} \\
\vdots & \vdots & \vdots & \vdots \\
1 & e^{2(N-1)\pi i/N} & \cdots & e^{2(N-1)^2\pi i/N}
\end{bmatrix}
\left\{
\begin{array}{c}
X(0) \\
X(1) \\
X(2) \\
\vdots \\
X(N-1)
\end{array}
\right\}.
$$

$$(3.57)$$

The properties of the N-point DFT are as follows:

(i) PERIODICITY. $X(k) = X(k + aN)$ for $0 \le k \le N - 1$, where $a \in \mathbb{R}^+$. In other words, the N-point DFT of an aperiodic sequence of length N, $N \le N$, is periodic with period N.

(ii) LINEARITY. If $x(n)$ and $y(n)$ are two DT sequences with the N-point DFT pairs: $x(n) \xrightarrow{DFT} X(k)$ and $y(n) \xrightarrow{DFT} Y(k)$, then for any arbitrary constants a and b (which may be complex)

$$
a\,x(n) + b\,y(n) \xrightarrow{DFT} a\,X(k) + b\,Y(k).
$$

(iii) HERMITIAN SYMMETRY. This symmetry implies that for the N-point DFT $X(k)$ of a real-valued aperiodic sequence $x(n)$

$$
X(k) = \overline{X}(N - k),
$$

where $\overline{X}$ is the complex conjugate of X. In other words, $X(k)$ is conjugate-symmetric about $k = N/2$. The magnitude $|X(N - k)| = |X(k)|$, and the phase $\phi(X(N - k)) = -\phi(X(k))$, i.e., the phase of the spectrum is odd.

(iv) TIME SHIFTING. If $x(n) \xrightarrow{DFT} X(k)$, then for an N-point DFT and an arbitrary integer n_0

$$
x(n - n_0) \xrightarrow{DFT} e^{2n_0 k\pi i/N} X(k).
$$

(v) CIRCULAR CONVOLUTION. For two DT sequences $x(n)$ and $y(n)$ with the N-point DFT pairs where $x(n) \xrightarrow{DFT} X(k)$ and $y(n) \xrightarrow{DFT} Y(k)$, the circular convolution is defined by

$$
x(n) \otimes y(n) \xrightarrow{DFT} X(k)Y(k),
$$

$$x(n)\, y(n) \xrightarrow{DFT} \frac{1}{N}\, [X(k) \otimes Y(k)]\,,$$

where $\otimes$ denotes the circular convolution operation[4]. In this operation the two sequences must be of equal length.

3.4.1 Discrete Signal. The discrete FT for a discrete signal $x(t)$ with N values (samples) is given by

$$X(f) = \frac{1}{N} \sum_{k=0}^{N-1} x(k)\, e^{-2\pi i\, kt/N}$$

$$= \frac{1}{N} \sum_{k=0}^{N-1} x(k)\, \cos(-2\pi i\, kt/N) + i\, \frac{1}{N} \sum_{k=0}^{N-1} x(k)\, \sin(-2\pi i\, kt/N).$$
$$(3.58)$$

Example 3.13. Let $x(t)$ be a discrete signal with $N = 8$ samples. Then for $f = 1$

$\cos(2\pi t/N)$	1	0.7	0	-0.7	-1	-0.7	0	0.7
$\sin(2\pi t/N)$	0	0.7	1	0.7	0	-0.7	-1	-0.7
$x(t)$	1	1	1	0	0	0	0	1
t	0	1	2	3	4	5	6	7

Computation of discrete FT for $f = 1$ yields

$$\Re\{x(1)\} = (1 + 0.7 + 0 + 0 + 0 + 0 + 0 + 0.7)/8 = 0.3,$$
$$\Im\{x(1)\} = (0 + 0.7 + 1 + 0 + 0 + 0 + 0 - 0.7)/8 = 0.12.$$

---

[4] To improve the resolution of the frequency axis ω in the DFT domain, a common practice is to append additional zero-valued samples to the DT sequences. This process is known as *zero padding*. The linear convolution $x(n) \star y(k)$ between two time-limited DT sequences $x(n)$ and $y(k)$ of lengths n_1 and n_2, respectively, can be expressed in terms of the circular convolution $x(n) \otimes y_k$ by zero-padding both $x(n)$ and y_k such that each sequence has length $N \geq (n_1 + n_2 - 1)$. It is known that the circular convolution of the zero-padded sequences is the same as that of the linear convolution. The algorithm for implementing the linear convolution of two sequences $x(n)$ and y_k is available in Kythe and Kythe [2012: 452].

Similarly, for $f = 2$

$\cos(4\pi t/N)$	1	0	-1	0	1	0	-1	0
$\sin(4\pi t/N)$	0	1	0	-1	0	1	0	-1
$x(t)$	1	1	1	0	0	0	0	1
t	0	1	2	3	4	5	6	7

Computation of discrete FT for $f = 2$ yields

$$\Re\{x(2)\} = (1 + 0 - 1 + 0 + 0 + 0 + 0 + 0)/8 = 0,$$
$$\Im\{x(2)\} = (0 + 1 + 0 + 0 + 0 + 0 + 0 - 1)/8 = 0.$$

Again, for $f = 3$

$\cos(6\pi t/N)$	1	-0.7	0	0.7	-1	0.7	0	-0.7
$\sin(6\pi t/N)$	0	0.7	-1	0.7	0	-0.7	1	-0.7
$x(t)$	1	1	1	0	0	0	0	1
t	0	1	2	3	4	5	6	7

Computation of discrete FT for $f = 3$ yields

$$\Re\{x(3)\} = (1 - 0.7 + 0 + 0 + 0 + 0 + 0 - 0.7)/8 = -0.05,$$
$$\Im\{x(3)\} = (0 + 0.7 - 1 + 0 + 0 + 0 + 0 - 0.7)/8 = -0.12.$$

The values of 1-D discrete FT and IFT for $N = 8$ are presented in Table 3.1.

In the 2-D case, the discrete FT for a discrete $N \times M$ image $x(t, \tau)$ is given by

$$X(f, g) = \frac{1}{N} \frac{1}{M} \sum_{t=0}^{N-1} \sum_{\tau=0}^{M-1} x(t, \tau) \, e^{-2\pi i \, (ft/N + g\tau/M)},$$

where

$$\Re\{X(f, g)\} = \frac{1}{N} \frac{1}{M} \sum_{t=0}^{N-1} \sum_{\tau=0}^{M-1} x(t, \tau) \, \cos\left(-2\pi i \left(ft/N + g\tau/M\right)\right),$$

$$\Im\{X(f, g)\} = \frac{1}{N} \frac{1}{M} \sum_{t=0}^{N-1} \sum_{\tau=0}^{M-1} x(t, \tau) \, \sin\left(-2\pi i \left(ft/N + g\tau/M\right)\right).$$

The 2-D discrete IFT for a discrete $N \times M$ image $X(f, g)$ is given by

$$x(t, \tau) = \sum_{f=0}^{N-1} \sum_{g=0}^{M-1} X(f, g) \, e^{2\pi i \, (ft/N + g\tau/M)},$$

where
$$\Re\{x(t,\tau)\} = \sum_{f=0}^{N-1}\sum_{g=0}^{M-1} X(f,g)\,\cos\left(2\pi(ft/N + g\tau/M)\right),$$

$$\Im\{x(t,\tau)\} = \sum_{f=0}^{N-1}\sum_{g=0}^{M-1} X(f,g)\,\sin\left(2\pi(ft/N + g\tau/M)\right).$$

Table 3.1 Values of Discrete FT [†]

	t	0	1	2	3	4	5	6	7	
	$x(t)$	1	1	1	0	0	0	0	1	
f	$2\pi t/8$	0.0	0.8	1.6	2.4	3.1	3.9	4.7	5.5	$X(f)$
-1	$\cos(2\pi ft/8)$	1.0	0.7	0.0	-0.7	-1.0	-0.7	0.0	0.7	0.30
-1	$\sin(2\pi ft/8)$	0.0	-0.7	-1.0	-0.7	0.0	0.7	1.0	0.7	-0.12
0	$\cos(2\pi ft/8)$	1.0	1.0	1.0	1.0	1.0	1.0	1.0	1.0	0.50
0	$\sin(2\pi ft/8)$	0.0	0.0	0.0	0.0	0.0	0.0	0.0	0.0	0.00
1	$\cos(2\pi ft/8)$	1.0	0.7	0.0	-0.7	-1.0	-0.7	0.0	0.7	0.30
1	$\sin(2\pi ft/8)$	0.0	0.7	1.0	0.7	0.0	-0.7	-1.0	-0.7	0.12
2	$\cos(2\pi ft/8)$	1.0	0.0	-1.0	0.0	1.0	0.0	-1.0	0.0	0.00
2	$\sin(2\pi ft/8)$	0.0	1.0	0.0	-1.0	0.0	1.0	0.0	-1.0	0.00
3	$\cos(2\pi ft/8)$	1.0	-0.7	0.0	0.7	-1.0	0.7	0.0	-0.7	-0.05
3	$\sin(2\pi ft/8)$	0.0	1.7	-1.0	0.7	0.0	-0.7	1.0	-0.7	-0.12
4	$\cos(2\pi ft/8)$	1.0	-1.0	1.0	-1.0	1.0	-1.0	1.0	-1.0	0.00
4	$\sin(2\pi ft/8)$	0.0	0.0	0.0	0.0	0.0	0.0	0.0	0.0	0.00
5	$\cos(2\pi ft/8)$	1.0	-0.7	0.0	0.7	-1.0	0.7	0.0	-0.7	-0.05
5	$\sin(2\pi ft/8)$	0.0	-0.7	1.0	-0.7	0.0	0.7	-1.0	0.7	0.12
6	$\cos(2\pi ft/8)$	1.0	0.0	-1.0	0.0	1.0	0.0	-1.0	0.0	0.00
6	$\sin(2\pi ft/8)$	0.0	-1.0	0.0	1.0	0.0	-1.0	0.0	1.0	0.00
7	$\cos(2\pi ft/8)$	1.0	0.7	0.0	-0.7	-1.0	-0.7	0.0	0.7	0.30
7	$\sin(2\pi ft/8)$	0.0	-0.7	-1.0	-0.7	0.0	0.7	1.0	0.7	-012
8	$\cos(2\pi ft/8)$	1.0	1.0	1.0	1.0	1.0	1.0	1.0	1.0	0.50
8	$\sin(2\pi ft/8)$	0.0	0.0	0.0	0.0	0.0	0.0	0.0	0.0	0.00

[†] The discrete FT values repeat with a period of 8, i.e., $X(-1)$ is the same as $X(7)$, and $X(8)$ is the same as $X(0)$. Also, $\Re\{X(0)\}$ is the average of all pixels, while $\Im\{X(f)\}$ is always zero.

Note that

(i) 1-D FT requires N computations for each value, so it is $O(N^2)$.

(ii) 2-D requires $N \times M$ computations for each value, so it is $O(N^2 M^2)$. However, since 2-D FT is separable, it is then possible to reduce the computations down to $O(NM^2 + N^2 M)$.

(iii) The differences in the normalizing weights in the FT and IFT formulas are important.

(iv) The convolution in the spatial domain is equivalent to multiplication in the frequency domain. Thus, $x(t) \star y(t) = \text{IFT}\{X(f) \cdot Y(f)\}$, or $\text{FT}\{x(t) \star y(t)\} = X(f) \cdot Y(f)$. This means that the computation time of convolution can be reduced by implementing linear operations using FT and IFT.

3.5 Fast Fourier Transform

The DFT, defined by (3.53), represents a system of N equations. For the sake of convenience, set $W = e^{-2\pi i/N}$. Then Eq (3.53) can be written as

$$X(0) = x(0)W^0 + x(1)W^0 + x(2)W^0 + \cdots + x(N-1)W^0,$$
$$X(1) = x(0)W^0 + x(1)W^1 + x(2)W^2 + \cdots + x(N-1)W^{N-1},$$
$$X(2) = x(0)W^0 + x(1)W^2 + x(2)W^4 + \cdots + x(N-1)W^{2(N-1)},$$
$$\vdots$$
$$X(N-1) = x(0)W^0 + x(1)W^{N-1} + x(2)W^{2(N-1)} + \cdots + x(N-1)W^{(N-1)(N-1)},$$

or, in matrix form, as

$$
\begin{Bmatrix} X(0) \\ X(1) \\ X(2) \\ \vdots \\ X(N-1) \end{Bmatrix} =
\begin{bmatrix}
W^0 & W^0 & W^0 & \cdots & W^0 \\
W^0 & W^1 & W^2 & \cdots & W^{N-1} \\
W^0 & W^2 & W^4 & \cdots & W^{2(N-1)} \\
\vdots & & & & \\
W^0 & W^{N-1} & W^{2(N-1)} & \cdots & W^{(N-1)(N-1)}
\end{bmatrix}
\begin{Bmatrix} x(0) \\ x(1) \\ x(2) \\ \vdots \\ x(N-1) \end{Bmatrix}.
$$
$$(3.59)$$

This algorithm was introduced by Cooley and Tukey [1965][5].

It is further simplified by considering the complex roots of W. Since $W = e^{-2\pi i/N}$ is the Nth root of unity, we have $W^N = 1$, which helps in simplifying other powers of W. Consider the case for $N = 4$. Then $W^6 = W^4 W^2 = W^2$, $W^9 = W^4 W^4 W = W$, $W^3 = -W$, $W^2 = -W^0$, and although $W^0 = 1$, we

---

[5] A detailed historical description of the FFT algorithm, tracing it back to Carl Fredrick Gauss (1777–1855), is given in Prestini [2004].

will retain W^0 for reasons of uniformity. Then Eq (3.59) reduces to

$$\left\{\begin{array}{c} X(0) \\ X(1) \\ X(2) \\ X(3) \end{array}\right\} = \begin{bmatrix} 1 & 1 & 1 & 1 \\ 1 & W & W^2 & W^3 \\ 1 & W^2 & W^0 & W^2 \\ 1 & W^3 & W^2 & W \end{bmatrix} \left\{\begin{array}{c} x(0) \\ x(1) \\ x(2) \\ x(3) \end{array}\right\},$$

which can be written as product of two matrices as

$$\left\{\begin{array}{c} X(0) \\ X(1) \\ X(2) \\ X(3) \end{array}\right\} = \begin{bmatrix} 1 & W^0 & 0 & 0 \\ 1 & W^2 & 0 & 0 \\ 0 & 0 & 1 & W^1 \\ 0 & 0 & 1 & W^3 \end{bmatrix} \begin{bmatrix} 1 & 0 & W^0 & 0 \\ 0 & 1 & 0 & W^0 \\ 1 & 0 & W^2 & 0 \\ 0 & 1 & 0 & W^2 \end{bmatrix} \left\{\begin{array}{c} x(0) \\ x(1) \\ x(2) \\ x(3) \end{array}\right\}, \qquad (3.60)$$

This algorithm involves $(N/2)\log_2 N$ multiplications, and saves about 99% of computing time. It is known (see Press [2006]) that the FFT approximates the one-dimensional continuous transform in $O(N \log N)$ operations and the two-dimensional in $O(N^2 \log N)$ operations.

3.5.1 Radix-2 Algorithm for FFT. The following result is useful for developing a radix-2 algorithm for FFT.

Theorem 3.9. *For even values of N, the N-point DFT of a real-valued sequence x_k of length $m \leq N$ can be computed from the DFT coefficients of two subsequences: (i) x_{2k}, which contains the even-valued samples of x_k, and (ii) x_{2k+1}, which contains the odd-valued samples of x_k.*

This theorem leads to the following algorithm to determine the N-point DFT:

STEP 1. Determine the $(N/2)$-point DFT G_j for $0 \leq j \leq N/2 - 1$ of the even-numbered samples of x_k.

STEP 2. Determine the $(N/2)$-point DFT H_j for $0 \leq j \leq N/2 - 1$ of the odd-numbered samples of x_k.

STEP 3. The N-point DFT coefficients X_j for $0 \leq j \leq k-1$ of x_k are obtained by combining the $(N/2)$ DFT coefficients of G_j and H_j using the formula $X_j = G_j + W_k^j H_j$, where $W_k^j = e^{-2\pi i/N}$ is known as the *twiddle factor*. Note that although the index $j = 0, \ldots, N - 1$, we only compute G_j and H_j over $0 \leq j \leq (N/2-1)$, and any outside value can be determined using the periodicity properties of G_j and H_j, which are defined by $G_j = G_{j+N/2}$ and $H_j = H_{j+N/2}$.

The flow graph for the above method for $N = 8$-point DFT is shown in Figure 3.6.

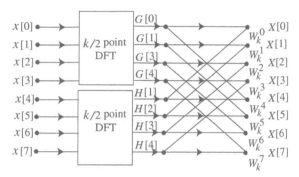

Figure 3.6 Flow Graph for 8-Point DFT.

In general, this figure computes two $(N/2)$-point DFTs along with N complex additions and N complex multiplications. Thus, $(N/2)^2 + N$ complex additions and $(N/2)^2 + N$ complex multiplications are needed. Since $(N/2)^2 + N < N^2$ for $N > 2$, the above algorithm is considerably cost effective. MATLAB library contains a variety of programs on CFT and DFT (see also Harris [1978]).

3.6 Multiple Fourier Transform

Let $\mathbf{t} = (t_1, \dots, t_n) \in \mathbb{R}^n$. Then under the same assumptions on $x(\mathbf{t})$ as on $x(t) \in \mathbb{R}$, the multi-dimensional Fourier transform of $x(\mathbf{t})$ is defined by

$$\mathcal{F}\{x(\mathbf{t})\} = X(\varsigma) = \frac{1}{(2\pi)^{n/2}} \int_{-\infty}^{\infty} \cdots \int_{-\infty}^{\infty} \exp\{-i\,(\varsigma \cdot \mathbf{t})\}\, x(\mathbf{t})\, dt, \qquad (3.61)$$

where $\varsigma = (\varsigma_1, \dots, \varsigma_n)$ denotes the n-dimensional transform variable and $\varsigma \cdot \mathbf{t} = (\varsigma_1 t_1 + \cdots + \varsigma_n t_n)$.

The inverse Fourier transform is defined by

$$\mathcal{F}^{-1}\{X(\varsigma)\} = x(\mathbf{t}) = \frac{1}{(2\pi)^{n/2}} \int_{-\infty}^{\infty} \cdots \int_{-\infty}^{\infty} \exp\{-i\,(\varsigma \cdot \mathbf{t})\}\, x(\mathbf{t})\, dt, \qquad (3.62)$$

In 2-D and 3-D cases we will use $\varsigma = (\omega, \beta)$ and $\varsigma = (\omega, \beta, \gamma)$, respectively. Thus, the *double Fourier transform* is defined by

$$\mathcal{F}\{x(t, u)\} = \frac{1}{2\pi} \int_{-\infty}^{\infty} \int_{-\infty}^{\infty} e^{-i\,(\omega t + \beta u)}\, x(t, u)\, dt\, du,$$

and its inverse by

$$\mathcal{F}^{-1}\{F(\omega, \beta)\} = \frac{1}{2\pi} \int_{-\infty}^{\infty} \cdots \int_{-\infty}^{\infty} e^{-i\,(\omega t + \beta u)}\, x(t, u)\, d\omega\, d\beta,$$

and the 3-D Fourier transform by

$$\mathcal{F}\{x(t,u,v)\} = \frac{1}{2\pi} \int_{-\infty}^{\infty} \int_{-\infty}^{\infty} e^{-i\,(\omega t + \beta u + \gamma v)}\, x(t,u,v)\, dt\, du\, dv,$$

and its inverse by

$$\mathcal{F}^{-1}\{F(\omega,\beta,\gamma)\} = \frac{1}{2\pi} \int_{-\infty}^{\infty} \cdots \int_{-\infty}^{\infty} e^{-i\,(\omega t + \beta u + \gamma v)}\, x(t,u,v)\, d\omega\, d\beta\, d\gamma.$$

Example 3.14. For $x(t,u,v) = \delta(t)\delta(u)\delta(v)$, the 3-D Fourier transform is $X(\zeta) = (2\pi)^{-3/2}$. ∎

Other examples are available in Kythe [2011: §7.10.1].

3.7 Fourier Slice Theorem

The Fourier slice theorem involves line integrals and projections. Let $R_\theta(t)$ denote the Radon transform of the function $f(x,y)$, and let L denote the (θ,t)-line. Then the Radon transform in a two-dimensional domain (see Appendix C) is defined as

$$R_\theta(t) = \int_L f(x,y)\, ds, \tag{3.63}$$

or

$$R_\theta(t) = \int_{-\infty}^{\infty} \int_{-\infty}^{\infty} f(x,y)\, \delta(x\cos\theta + y\sin\theta, t)\, dx\, dy. \tag{3.64}$$

A projection is formed by combining a set of line integrals. The simplest projection, shown in Figure 3.7, is a collection of parallel ray integrals (for constant c). The one-dimensional case of a Fourier slice is shown in Figure 3.8 (for details, see Appendix C, Figure C.3).

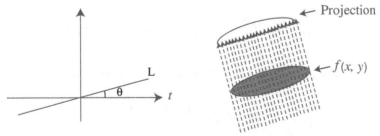

Figure 3.7 Simplest Projection of a 2-D Domain $f(x,y)$.

The two-dimensional Fourier slice theorem, also known as the *Fourier projection-slice theorem*, states that the following two methods of calculations yield the same result: (i) take a two-dimensional function $f(\mathbf{r})$, project it onto a one-dimensional line, and perform a FT of that projection; or (ii) take the same function $f(\mathbf{r})$, but do a two-dimensional FT first, and then slice it through the origin, which is parallel to the projection line. In terms of the operators, this theorem can be stated as follows.

Theorem 3.10. *Let F_1 and F_2 be the one- and two-dimensional FT operators. Let P_1 be the projection operator that projects a two-dimensional function onto a one-dimensional line, and let S_1 be the slice operator that extracts a one-dimensional central slice from a function. Then*

$$F_1 P_1 = S_1 F_2. \tag{3.65}$$

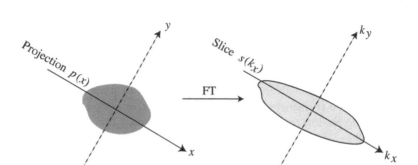

Figure 3.8 One-Dimensional Simple Fan Beam Projection.

Figure 3.8 shows the one-dimensional Fourier transform pair $f(\mathbf{r}) \rightleftharpoons F(\mathbf{k})$, where the function $p(\mathbf{x})$ is the projection of $f(\mathbf{r})$ onto the x-axis, defined by the integral of $f(\mathbf{r})$ along the lines parallel to the y-axis. The slice $s(k_x)$ through $F(\mathbf{k})$ is situated on the k_x-axis, which is parallel to the x-axis.

PROOF. (Two-dimensional case) Define a two-dimensional Fourier transform of the object function as

$$F(u, v) = \int_{-\infty}^{\infty} \int_{-\infty}^{\infty} f(x, y)\, e^{-2\pi i\,(ux+vy)}\, dx\, dy. \tag{3.66}$$

Now, define the projection R_θ at an angle θ and its transform by

$$S_\theta(\omega) = \int_{-\infty}^{\infty} R_\theta(t)\, e^{-2\pi i t}\, dt. \tag{3.67}$$

For the sake of simplicity, take $v = 0$. Then

$$F(u, 0) = \int_{-\infty}^{\infty} \int_{-\infty}^{\infty} f(x, y)\, e^{-2\pi i u x}\, dx\, dy = \int_{-\infty}^{\infty} \left[\int_{-\infty}^{\infty} f(x, y)\, dy \right] e^{-2\pi i u x}\, dx,$$
$$\tag{3.68}$$

since the phase factor does not depend on y, so the integral in (3.68) is split. The integral within the square brackets in (3.68) is the Radon transform $R_\theta(x)$, which defines a projection along a line $x = \text{const}$. Then (3.68) reduces to

$$F(u,0) = \int_{-\infty}^{\infty} R_\theta(x) \, e^{-2\pi i \, ux} \, dx,$$

and the vertical projection and the two-dimensional transform of the object function are replaced by $F(u,0) = S_{\theta=0}(u)$. This proves the Fourier slice theorem along a radial line. ■

The inverse Radon transform $\mathcal{R}^{-1}\{R(\rho,\theta)\}$ (see Appendix C) recovers the original function $f(x,y)$ from its projections. The existence of this inverse for suitably well-behaved functions f is a direct consequence of the Fourier slice theorem, which states that the one-dimensional Fourier transform of a projection at an angle θ has values identical to a radial slice through the origin of the two-dimensional Fourier transform of the original image. For proof see Kak and Slaney [1988].

4

Signals and Filters

Signals and Fourier transforms are related to one another. They play a significant role in the development of all technological innovations. Filters are important in that they eliminate or reduce noise, and hence, error.

4.1 Description of Signals

A signal in the context of communications is a quantitative function that varies in time and conveys information about some phenomenon from source to destination. Some common examples are traffic signals, railway signals, whistle of a train, or the blinking of an automobile's left or right turn signals. The beacon is used in navigation for defensive communication. The native Americans used smoke signals during colonial times, although this technique is known to have been prevalent throughout the world since ancient times. Distress signals are internationally recognized as a means of obtaining help when a group of people, ship, aircraft or other vehicle is threatened by grave and imminent danger and requests immediate assistance. There are maritime distress signals, by transmitting voice signal 'May Day', or pressing the distress key, or transmitting SOS in Morse code, or burning red flame, or emitting orange smoke from a canister, or flying international maritime signal flags. In modern times the semaphore flag signaling system has been used for sending messages, although always repeated to ensure error-free delivery. Details of the semaphore system can be found at the International Scouts Association's website: http://inter.scoutnet.org/semaphore/.

The term 'signal' includes audio, video, speech and image communication. Every signal must be preceded by a code. By a *code* we mean one thing that stands for another. Naturally, a code must be smaller than the thing it stands for, and this is the case except for Morse code. A code consists of *characters* which are symbols that are either hand-guided or machine-generated. Finally, there is a *function code* that causes the machine to do something, such as ring a bell, carriage return, and so on, and is given a mnemonic name to distinguish it from the characters used in the code. Telegraphy and telephony are other examples where signals are realized and manipulated. Details of these topics are available in Kythe and Kythe [2012: Chapter 1].

In the area of telephone and radio, Faraday's law of induction states that any change in a magnetic field generates an electric field. Two years after Faraday's publication, James Clerk Maxwell (1831–1879) formulated his famous Maxwell equations which established a quantitative relationship between magnetic and electric fields: a dielectric behaves like an elastic membrane that transmits the electrical disturbance as a transverse wave through the dielectric medium with finite velocity. Based on this theory the physicist Heinrich Hertz (1857–1894) generated, transmitted, and received electromagnetic waves in his laboratory in 1887. He found that the electromagnetic waves are periodic in both space and time, just like the transverse waves on a pond. In space, the period is λ, which is the wavelength (or distance between two crests). If the period is T, then the radian frequency is $\omega = 1/T$. The following list provides the frequency terms that are in use: 1 hertz (Hz) = 1 cycle per second; 1 kilohertz (kHz) = 1,000 Hz; 1 megahertz (MHz) = 1,000,000 Hz; and 1 gigahertz (GHz) = 1,000,000,000 Hz.

Hertz conducted an experiment to determine the oscillations of a cork on the surface of a disturbed pond, in which the cork remains stationary but moves up and down (transverse motion). It was found that the motion of all electromagnetic waves is governed by the formula $\lambda = vT$, where v is the velocity, which is about 300,000 km (186,410 miles) per second.

The signal in telephony is transmitted by means of two cables (wires) as well as by satellite, using radio waves (signals). The telephone is not an example of high fidelity transmission because the bandwidth of its signal (the width of the range of frequencies) is 4,000 Hz. The human ear can perceive sound waves from about 30 Hz up to 18,000 Hz. This band is known as the *telephone channel*. Higher frequencies, if present in the original conversation, are cut off by suitable filters for reasons of economy. It has been determined that for adequate telephone conversation the frequency range of 300 Hz to 3,400 Hz is sufficient.

A *radio signal* consists of a radio-frequency sinusoidal (sine or cosine) wave, known as the *carrier wave*, that has undergone modulation, which is a variation in amplitude (volume), frequency (pitch), or phase (timing) of a carrier signal. There are two types: (i) Amplitude modulation (AM) in which the carrier amplitude is varied by the modulating signal, and (ii) frequency modulation (FM) in which the carrier frequency is varied by the modulating sign. For an acceptable audio quality the AM radio requires a bandwidth as small as 5 kHz, but for musical sound quality the frequency bandwidth extends to about 10 kHz in commercial broadcasting. For high-fidelity music transmission the FM frequency range is extended up to 15 kHz (e.g., for cymbals sounds) and to 30 Hz (for organ pipes).

In television both human ears and eyes are involved. The eye distinguishes the color, brightness, size, shape, details, and position of objects. A picture is split into elementary details by superimposing a fine grid. This enables a single frame (picture) to be realized as a set of lines each created by a string of details which are associated to each shades of gray for black-and-white and RGB and CMYK color transmissions supported by RCA and NTSC. The phenomenon that works in this process is known as the *persistence of vision*, based on the fact that the brain retains the impression of an image for about 0.1 second after the light source is shut off. Thus, if the process of image synthesis occurs before that 0.1 second interval, the eye is deceived about the piecewise reassembly of images. Although each picture (frame) is broken into at least 300,000 elementary details, about 25 to 30 picture frames per second are required for a successful rapid smooth motion. The high frequency required for picture details is at a rate of 4,000,000 per second, and such requirements are limited by the frequency-allocation authorities to 6 MHz for video (color) and audio signals.

The band of frequencies used in radar systems extends from about 400 MHz to about 40 GHz. A widely used radar, known as the *pulsed radar* which detects and locates distant objects, transmits a short signal made by a carrier wave of frequency 1 GHz, modulated by a sequence of rectangular pulses with a low repetition interval; it hits the target and a part of it is reflected back (see Figure 4.1(a)). The receiver in the radar apparatus computes the distance of the target from the time interval between the transmitted signal and the detection of the reflected signal (called *echo*). The 'lidar' also uses the radar echo-ranging technique at optical frequencies ('li' short for 'light' and 'dar' from 'radar').

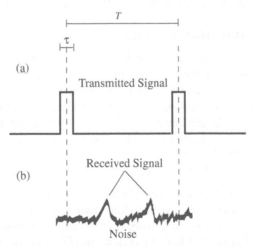

Figure 4.1 (a) Modulation Pulse Transmitted by a Pulsed Radar;
(b) Received Signal and Noise.

Sometimes the echoes in radio waves pose serious problems for the airport traffic control system. Such situations arise under certain atmospheric conditions or with certain orientation of a distant large reflecting object. A solution is the choice of repetition interval. Another problem that disturbs a signal received by a radar is noise (Figure 4.1(b)). Noise generally gives a false alarm, and therefore, is a serious problem in radar technology. However, filters allow elimination of most noise, but not all.

Very short pulses of the order of milliseconds or less are used for accuracy of transmitted signal. The pulse duration is defined to be the time interval of one microsecond (μs), for example, and the repetition interval of one millisecond (ms). In practice, a transmitter pulse duration of one millisecond requires a receiver bandwidth of one MHz, which is precisely the reciprocal, because the frequency content of a rectangular pulse of one millisecond duration is mostly confined to a one MHz interval, centered on plus or minus the modular frequency (for rectangular pulse, see Examples 4.4 and 4.5). For example, if the frequency of a carrier wave is $f = 1$ GHz, then the modulating pulse will have duration $\tau = 1\,\mu$s and the repetition period $T = 1$ ms.

The bands of the radio spectrum are allocated by the International Telecommunication Union in Geneva depending on the kind of usage, including mobile radio (aeronautical, marine, land), radio navigation (aeronautical, marine), broadcasting (AM, FM, TV), amateur radio, space communications (satellite, communication between stations), and radioastronomy. For example, in North America, AM broadcasting is allocated 535 – 1,605 kHz band, FM 88 – 108 MHz band, and TV with three major bands: 54 – 88 MHz (VHF lowband), 170 – 220 MHz (VHF highband), and 470 – 890 MHz (UHF Band).

4.2 Convolution and Signals

Convolution describes how an output image can be formed as a weighted sum of pixels in an input image. These convolution weights are known as the *impulse responses* of the linear system. The convolution weights are 'flipped and shifted' to different locations and multiplied by the input image to yield the output image. Although the range of the convolution summation is usually infinite, in practice it is limited to finite images. If the values outside discrete images are assumed to be zero, a convolution of an $N \times N$ image by an $M \times M$ image will produce an $(N + M - 1) \times (N + M - 1)$ output image.

The purpose of the Fourier transform is to represent an image in terms of sine and cosine functions. Using these two basic functions, it is easy to carry out processing operations, especially frequency domain filtering, which after inversion using inverse Fourier transform (IFT) is converted back to the spatial domain.

Example 4.1. (Synthesis of a signal) A *bandlimited signal* $x(t)$ is a continuous function of time t, $-\infty < t < \infty$, such that its Fourier transform $X(\omega) = 0$ outside a finite frequency band, i.e., for $|\omega| > a$. In particular, let

$$X(\omega) \equiv X_a(\omega) = \begin{cases} 1 & \text{if } |\omega| < a, \\ a & \text{if } \omega > 0 \end{cases}, \text{ where } X_a(\omega) \text{ in the frequency domain is}$$

called a *gate function*, and the corresponding signal $x_a(t)$ is given by

$$x_a(t) = \frac{1}{2\pi} \int_{-\infty}^{\infty} X_a(\omega) e^{i\omega t} \, d\omega = \frac{1}{2\pi} \int_{-a}^{a} e^{i\omega t} \, d\omega = \frac{\sin at}{\pi t}. \qquad (4.1)$$

The plots of both $X_a(\omega)$ and $x_a(t)$ are presented in Figure 4.2. Taking the limit as $a \to \infty$ in (4.1) we find that

$$1 = \lim_{a \to \infty} \int_{-\infty}^{\infty} e^{-i\omega t} x_a(t) \, dt = \lim_{a \to \infty} \int_{-\infty}^{\infty} e^{-i\omega t} \frac{\sin at}{\pi t} \, dt$$

$$= \int_{-\infty}^{\infty} e^{-i\omega t} \left[\lim_{a \to \infty} \frac{\sin at}{\pi t} \right] \, dt = \int_{-\infty}^{\infty} e^{-i\omega t} \delta(t) \, dt.$$

Thus, $\delta(t)$ can be regarded as the limit of sequences of functions $x_a(t)$, i.e.,
$\delta(t) = \lim_{a \to \infty} \dfrac{\sin at}{\pi t}.$

Now, for the bandlimited signal

$$x_a(t) = \frac{1}{2\pi} \int_{-a}^{a} X(\omega) e^{i\omega t} \, d\omega = \frac{1}{2\pi} \int_{-\infty}^{\infty} X(\omega) x_a(t) e^{i\omega t} \, d\omega$$

$$= \int_{-\infty}^{\infty} x(u) x_a(t - u) \, du = \int_{-\infty}^{\infty} \frac{\sin a(t - u)}{\pi(t - u)} \, du, \text{ by Property (5), §3.1.2.} \qquad (4.2)$$

This is the sampling integral representation of the bandlimited signal $x_a(t)$.

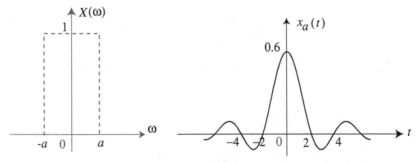

Figure 4.2 Gate Function and its Inverse Fourier Transform.

As an example, let $x(t) = e^{i\omega t}$ be the signal with a single frequency ω in a simple electrical device such as an amplifier, with an output signal $y(t) = e^{i\omega t}$, also with a single frequency ω. The amplifier will change the amplitude and possibly the phase as well, so that the output can be expressed in terms of the input. Let the amplitude and the phase modifying function be $\Phi(\omega)$, known as the *transfer function* of the real variable ω, defined as the Fourier transform of some function $\phi(t)$, i.e., $\Phi(\omega) = \int_{-\infty}^{\infty} \phi(t) e^{-i\omega t} \, dt$. Then the total output is given by

$$
\begin{aligned}
y(t) &= \frac{1}{2\pi} \int_{-\infty}^{\infty} X(\omega) \Phi(\omega) e^{i\omega t} \, d\omega \\
&= \mathcal{F}^{-1}\{X(\omega)\Phi(\omega)\} = (x \star \phi)(t), \quad \text{using the convolution (3.12)} \\
&= \int_{-\infty}^{\infty} x(u)\phi(t-u) \, du.
\end{aligned}
$$

Physically, this result defines the output signal $y(t)$ as the integral superposition of the input signal $x(t)$ modified by $\phi(t-u)$. This is an example where an output (effect) function in terms of an input (cause) function is modified by the amplifier. ∎

Example 4.2. (Series sampling expansion of a bandwidth signal) Consider a bandlimited signal $x_a(t)$ with Fourier Transform $X_a(\omega) \equiv X(\omega) = 0$ for $|\omega| > a$. The Fourier series expansion of $X(\omega)$ on the interval $|\omega| < a$ in terms of the orthogonal set of functions $\{e^{-n\pi i \omega/a}\}$ is

$$
X(\omega) = \sum_{n=-\infty}^{\infty} c_n e^{-n\pi i \omega/a}, \tag{4.3}
$$

where

$$
c_n = \frac{1}{2a} \int_{-a}^{a} X(\omega) e^{n\pi i \omega/a} \, d\omega = \frac{1}{2a} x_a(n\pi/a).
$$

Thus, the expansion (4.3) becomes

$$
X(\omega) = \frac{1}{2a} \sum_{n=-\infty}^{\infty} x_a\left(\frac{n\pi}{a}\right) e^{-n\pi i \omega/a}, \tag{4.4}
$$

The input signal $x_a(t)$ is then obtained by multiplying (4.4) by $e^{i\omega t}$ and integrating over $(-a, a)$:

$$
\begin{aligned}
x_a(t) &= \int_{-a}^{a} X(\omega) e^{i\omega t} \, d\omega \\
&= \frac{1}{2a} \int_{-a}^{a} e^{i\omega t} \left[\sum_{n=-\infty}^{\infty} x_a\left(\frac{n\pi}{a}\right) e^{-n\pi i \omega/a} \right] d\omega \\
&= \frac{1}{2a} \sum_{n=-\infty}^{\infty} x_a\left(\frac{n\pi}{a}\right) \int_{-a}^{a} e^{i\omega(t-n\pi/a)} \, d\omega \\
&= \sum_{n=-\infty}^{\infty} x_a\left(\frac{n\pi}{a}\right) \frac{\sin a(t-n\pi/a)}{a(t-n\pi/a)} = \sum_{n=-\infty}^{\infty} x_a\left(\frac{n\pi}{a}\right) \operatorname{sinc} a(t-n\pi/a).
\end{aligned}
\tag{4.5}
$$

Instead of this mathematical formulation in terms of the continuous bandlimited signal $x_a(t)$, in practice a discrete set of samples $\{x_a(t_n)\}$ is used as input. The result (4.5) can be obtained from the convolution theorem by using discrete input samples

$$\sum_{n=-\infty}^{\infty} \frac{\pi}{a} x_a\left(\frac{n\pi}{a}\right)\delta\left(t - \frac{n\pi}{a}\right) = x(t). \tag{4.6}$$

Then the sampling equation (4.2) gives the bandlimited signal

$$x_a(t) = \int_{-\infty}^{\infty} \frac{\sin a(t-u)}{\pi(t-u)}\left[\sum_{n=-\infty}^{\infty} \frac{\pi}{a} x_a\left(\frac{n\pi}{a}\right)\delta\left(t - \frac{n\pi}{a}\right)\right] du$$

$$= \sum_{n=-\infty}^{\infty} x_a\left(\frac{n\pi}{a}\right)\int_{-\infty}^{\infty} \frac{\sin a(t-u)}{a(t-u)}\delta\left(t - \frac{n\pi}{a}\right) du$$

$$= \sum_{n=-\infty}^{\infty} x_a\left(\frac{n\pi}{a}\right)\operatorname{sinc} a\left(t - \frac{n\pi}{a}\right), \quad \text{which is the same as (4.6).} \blacksquare$$

4.3 Theory of Signals

If the signal is given in terms of frequencies, that is by its Fourier transform, then the filter acts by convolution. An ideal filter stops certain frequencies while leaving others unaltered. Ideal filters are classified as highpass, lowpass, and bandpass, as presented in Figure 4.3.

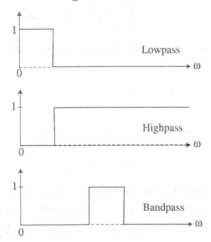

Figure 4.3 Ideal Highpass, Lowpass, and Bandpass Filters.

4.3.1 Types of Signals. There are two types of signals: deterministic and random. Deterministic systems are predefined by explicit expressions, like

$x(t) = 5\sin(5t)$ or $x(t) = 8e^{15t}$. Random signals have a certain degree of uncertainty; for example, a received communication signal is a random signal that contains not only the information but noise as well from the atmosphere, or internal circuitry of the receiver.

4.3.2 Continuous Deterministic Signals. Let $x(t)$, real or complex, denote a continuous deterministic signal with finite energy $\mathcal{E}$, defined by $\mathcal{E} = \int_{-\infty}^{\infty} |x(t)|^2 \, dt$. In the frequency domain the Fourier transform pair for this signal is

$$x(t) \rightleftharpoons X(\omega), \qquad (4.7)$$

where ω is the angular (radian) frequency (see §3.1). The Fourier transform $X(\omega)$ is a complex number, so we can write

$$X(\omega) = \Re\{X(\omega)\} + i\Im\{X(\omega)\} = |X(\omega)|\, e^{i\,\arg\{X(\omega)\}}. \qquad (4.8)$$

Since the Fourier transform measures the frequency content, the quantity $X(\omega)$ is also called the *spectrum* of the signal $x(t)$, $|X(\omega)|$ the *magnitude of the spectrum of* $x(t)$, and $\arg\{X(\omega)\} = \dfrac{\Im\{X(\omega)\}}{\Re\{X(\omega)\}}$ the *phase spectrum of* $x(t)$. The quantity $|X(\omega)|^2$, denoted by $\mathcal{E}_x(\omega)$, is known as the *energy density spectrum* of $x(t)$, since it represents the distribution of signal energy as a function of the frequency. The spectrum $\mathcal{E}_x(\omega)$ of a deterministic continuous signal $x(t)$ can be determined using the average auto-correlation function (ACF), also known as *autocovariance*, $r_x(t)$ of the finite energy signal $x(t)$, which is defined in terms of statistical expected value $E = \overline{x(t)}\, x(t+\tau)$, i.e.,

$$r_x(\tau) = \int_{-\infty}^{\infty} \overline{x(t)}\, x(t+\tau)\, dt, \qquad (4.9)$$

where the line over an expression denotes its complex conjugate. Then the energy density spectrum is

$$\mathcal{E}_x(\omega) = \int_{-\infty}^{\infty} r_x(\tau)\, e^{-i\omega\tau}\, d\tau. \qquad (4.10)$$

The definitions (4.9) and (4.10) together define another Fourier transform pair:

$$r_x(\tau) \rightleftharpoons \mathcal{E}_x(\omega). \qquad (4.11)$$

4.3.3 Discrete Deterministic Signals. Let $\{x(n)\}_{n\in\mathbb{Z}}$ denote a real or complex deterministic sequence which is obtained by uniformly sampling the continuous signals $x(t)$. If $x(n)$ has finite energy $\mathcal{E} = \sum_{-\infty}^{\infty} |x(n)|^2$, then it has the frequency domain representation (discrete Fourier transform (DFT))

$$X(\omega) = \sum_{-\infty}^{\infty} x(n)\, e^{-i\omega n}, \quad \text{or} \quad X(f) = \sum_{-\infty}^{\infty} x(n)\, e^{-2\pi i f n}, \qquad (4.12)$$

where both definitions are periodic, the former with period 2π and the latter with period 1. The inverse DFT $x(n)$ obtained from $X(\omega)$ or $X(f)$ is given by

$$x(n) = \frac{1}{2\pi} \int_{-\pi}^{\pi} X(f) e^{i\omega n} \, d\omega = \int_{-1/2}^{1/2} X(f) e^{i 2\pi fn} \, df. \tag{4.13}$$

The energy density spectrum of $x(n)$ is defined by $\mathcal{E}_x(f) = |X(f)|^2$, where $f = \omega/(2\pi)$. The auto-correlation sequence (ACS) of $x(n)$ is defined by DFT as

$$r_x(k) = \sum_{n=-\infty}^{\infty} \overline{x(n)} \, x(n+k),$$

so that

$$\mathcal{E}_x(f) = \sum_{k=-\infty}^{\infty} r_x(k) e^{-i 2\pi fk}. \tag{4.14}$$

Then we obtain the discrete Fourier transform pair

$$r_x(k) \rightleftharpoons \mathcal{E}_x(f). \tag{4.15}$$

Example 4.3. (Unit Impulse) In view of the translation (shifting) property of the Dirac delta function (§2.4), we have $\int_{-\infty}^{\infty} \delta(t) x(t) \, dt = x(0)$, where $x(t)$ is any signal continuous at $t = 0$. In discrete time the unit impulse is defined by $\delta(n) = \begin{cases} 1, & n = 0, \\ 0, & n \neq 0 \end{cases}$. If a continuous signal is defined as

$$x(t) = \int_{-\infty}^{\infty} x(u) \, \delta(t-u) \, du \quad \text{for all } t,$$

then the discrete signal is represented by

$$x(n) = \sum_{k=-\infty}^{\infty} x(k) \, \delta(n-k) \quad \text{for all } n.$$

The Fourier transform of the unit impulse is

$$\int_{-\infty}^{\infty} \delta(t) e^{-i 2\pi ft} \, dt = 1. \tag{4.16}$$

Thus, we get the Fourier transform pair

$$\delta(t) \rightleftharpoons 1. \tag{4.17}$$

In the discrete case, the spectrum of the unit impulse is given by

$$\sum_{n=-\infty}^{\infty} \delta(n)\, e^{-i\, 2\pi f n} = 1, \tag{4.18}$$

thus yielding the Fourier transform pair

$$\delta(n) \; \rightleftharpoons \; 1. \; \blacksquare \tag{4.19}$$

Example 4.4. (Rectangular Pulse) Consider a single rectangular pulse $x(t)$ of amplitude and pulse-width T (Figure 4.4) defined by

$$x(t) = \begin{cases} 1, & |t| \leq T/2, \\ 0, & |t| > T/2 \end{cases}.$$

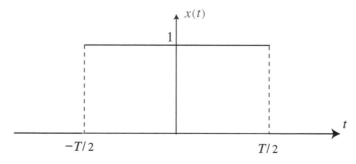

Figure 4.4 Rectangular Pulse.

Suppose that the frequency is f Hz (cycles/sec) and, using the radian frequency $\omega = 2\pi f$, we obtain the Fourier transform of the above rectangular pulse as

$$X(\omega) = T\, \frac{\sin(\omega T/2)}{\omega T/2} = T \operatorname{sinc}\left(\frac{\omega T}{2}\right).$$

The magnitude spectrum $|X(\omega)|$ and the phase spectrum $\arg\{X(\omega)\}$ are presented in Figure 4.5.

The auto-correlation function (ACF) $r_x(t)$ of the finite energy signal and the energy density spectrum $\mathcal{E}_x(\omega)$ of this pulse are defined, respectively, by

$$r_x(\tau) = \begin{cases} T\left(1 - \dfrac{|\tau|}{T}\right) & \text{for } |\tau| \leq T, \\ 0 & \text{otherwise}, \end{cases}$$

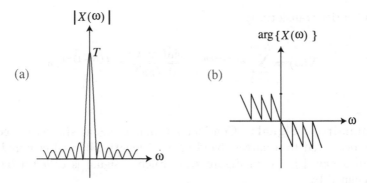

Figure 4.5 (a) Magnitude, and (b) Phase Spectrum of a Rectangular Pulse.

and

$$\mathcal{E}_x(\omega) = \int_{-\infty}^{\infty} r_x(\tau)\, e^{-i\omega\tau}\, d\tau = \int_{-\infty}^{\infty} r_x(\tau)\, \cos(\omega\tau)\, d\tau,$$

since $\int_{-\infty}^{\infty} r_x(\tau) \sin(\omega\tau) = 0$, because $r_x(\tau)$ is even and $\sin(\omega\tau)$ is odd. Hence, the energy density spectrum for this rectangular pulse simplifies to

$$\mathcal{E}_x(\omega) = \int_{-T}^{T} (T-\tau) \cos(\omega t)\, d\tau = T^2 \left(\frac{\sin(\omega T/2)}{\omega T/2} \right)^2 = T^2 \operatorname{sinc}^2 \left(\frac{\omega T}{2} \right). \quad (4.20)$$

This energy density spectrum $\mathcal{E}_x(\omega)$ is presented in Figure 4.6. ∎

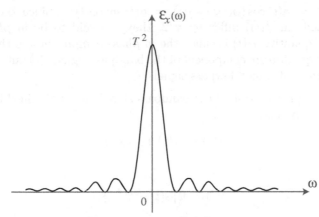

Figure 4.6 Energy Density Spectrum $\mathcal{E}_x(\omega)$ of a Rectangular Pulse.

Example 4.5. (Discrete Rectangular Pulse) The rectangular pulse in discrete time is defined by

$$x(n) = \begin{cases} 1 & \text{for } 0 \le n \le N-1,\ N \in \mathbb{Z}, \\ 0 & \text{otherwise,} \end{cases}$$

and its Fourier transform is

$$X(\omega) = \sum_{n=0}^{N-1} e^{-i\,2\pi x n} = \frac{\sin(\pi x N)}{\sin(\pi x)} e^{-i\,(N-1)\pi x}. \ \blacksquare$$

4.3.4 Bandpass Signals. Consider an analog signal $s(t)$ with frequency content restricted to a narrow band of total length $2W$ and centered about some frequency $\pm f_c$ (as in Figure 4.7). Such a signal is called a *bandpass signal* defined by

$$s(t) = a(t)\cos\left(2\pi f_c t + \phi(t)\right), \tag{4.21}$$

where $a(t)$ is the *amplitude* or *envelope*, $\phi(t)$ the *phase* of the signal, and f_c is called the *carrier frequency*. Eq (4.21) represents a hybrid form of amplitude-phase modulation that includes amplitude modulation, frequency modulation, and phase (angle) modulation as special cases. If we expand the cosine function in (4.21), the following alternative representation of the bandpass signal is obtained:

$$\begin{aligned} s(t) &= a(t)\left[\cos\left(\phi(t)\right)\cos\left(2\pi f_c t\right) - \sin\left(\phi(t)\right)\sin\left(2\pi f_c t\right)\right] \\ &= s_I(t)\cos\left(2\pi f_c t\right) - s_Q(t)\sin\left(2\pi f_c t\right), \end{aligned} \tag{4.22}$$

where $s_I(t) = a(t)\cos\left(\phi(t)\right)$, $s_Q(t) = a(t)\sin\left(\phi(t)\right)$. Since the sinusoids $\cos\left(\phi(t)\right)$ and $\sin\left(\phi(t)\right)$ differ by $\pi/2$, they are said to be in *phase quadrature*. The quantity $s_I(t)$ is called the *in-phase component* and the quantity $s_Q(t)$ as the *quadrature component* of the bandpass signal $s(t)$, and both $s_I(t)$ and $s_Q(t)$ are real-valued lowpass signals.

Another representation of the bandpass signal $s(t)$ is obtained by defining the *complex envelope*

$$\tilde{s}(t) = s_I(t) + i\,s_Q(t). \tag{4.23}$$

Then

$$s(t) = \Re\left\{\tilde{s}(t)\,e^{i\,2\pi f_c t}\right\}. \tag{4.24}$$

Eq (4.23) can be regarded as the Cartesian form of the complex envelope $\tilde{s}(t)$, which in polar form can be written as $\tilde{s}(t) = a(t)\,e^{i\,\phi(t)}$, where $a(t) = \sqrt{s_I^2(t) + s_Q^2(t)}$ and $\phi(t) = \arctan\dfrac{s_Q(t)}{s_I(t)}$ are both real-valued lowpass signals. Thus, the information carried in the bandpass signal $s(t)$ is preserved in $\tilde{s}(t)$, and therefore, we can represent $s(t)$ in terms of its in-phase and quadrature components as in (4.23).

In the frequency domain the bandpass signal $s(t)$ is represented by its Fourier transform

$$S(f) = \int_{-\infty}^{\infty} s(t)\, e^{-2\pi i\, ft}\, dt = \int_{-\infty}^{\infty} \Re\{\tilde{s}(t)\, e^{2\pi i\, f_c t}\}\, e^{-2\pi i\, ft}\, dt$$

$$= \tfrac{1}{2} \int_{-\infty}^{\infty} \left[\tilde{s}(t)\, e^{2\pi i\, f_c t} + \overline{\tilde{s}(t)}\, e^{-2\pi i\, f_c t} \right] e^{-2\pi i\, ft}\, dt$$

$$= \tfrac{1}{2} \left[\int_{-\infty}^{\infty} \tilde{s}(t)\, e^{2\pi i\, (f - f_c) t}\, dt + \int_{-\infty}^{\infty} \overline{\tilde{s}(t)}\, e^{-2\pi i\, (f + f_c) t}\, dt \right].$$

$$(4.25)$$

Let $S(f)$ and $\tilde{S}(f)$ denote the Fourier transform of $s(t)$ and $\tilde{s}(t)$, respectively. Then (4.25) can be written as

$$S(f) = \tfrac{1}{2} \left[\tilde{S}(f - f_c) + \overline{\tilde{S}(-f - f_c)} \right]. \qquad (4.26)$$

This result is presented in Figure 4.7.

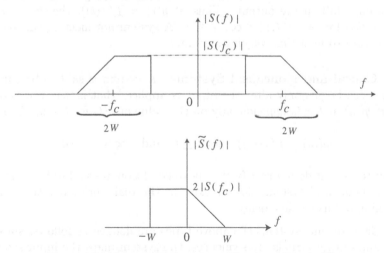

Figure 4.7 (a) Magnitude Spectrum of Bandpass Signal $s(t)$;
(b) Complex Envelope $\tilde{s}(t)$.

4.3.5 Continuous-Time Domain. Let input and output signals be denoted by $x(t)$ and $y(t)$. Then the input-output relation of a system at a finite time t_0 is represented by

$$y(t_0) = f(x(t_0)), \quad -\infty < t, t_0 < \infty.$$

Schematically, this relation is represented as block diagram in Figure 4.8.

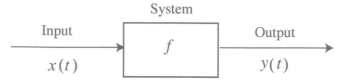

Figure 4.8 Block Diagram for a System.

Based on this relationship the systems are classified as follows.

4.3.6 Linear and Nonlinear Systems. A system is said to be *linear* if the principle of superposition holds. Thus, if $y_1(t) = f(x_1(t))$ and $y_2(t) = f(x_2(t))$, then

$$a_1 y_1(t) + a_2 y_2(t) = f(a_1 x_1(t) + a_2 x_2(t)).$$

If the superposition does not hold, then the system is said to be *nonlinear*.

4.3.7 Time-Invariant and Time-Varying Systems. A system is said to be *time-invariant* if a time shift (translation) in the input results in a corresponding shift in the output. Thus, if $y(t) = f(x(t))$, then $y(t - t_0) = f(x(t - t_0))$ for $-\infty < t, t_0 < \infty$, $t_0 \in \mathbb{R}$. A system not meeting this requirement is said to be a *time-varying* system.

4.3.8 Causal and Noncausal Systems. A system is said to be *causal* if its response begins only after the input is applied, that is, the values of the output $y(t_0)$ at $t = t_0$ depends only on the value of $x(t)$ for $t \leq t_0$. Thus,

$$y(t_0) = f(x(t_0)) \quad \text{for } t \leq t_0 \text{ and } -\infty < t < \infty.$$

The systems that do not satisfy the above condition are said to be *noncausal*. In fact, noncausal systems do not exist in the real world, but they can be approximated using time delay.

In discrete-time systems the above types are defined as follows: since the input and output signals are sequences, the system maps the input sequence $\{x(n)\}$ into the output sequence $\{y(n)\}$. Thus, a *linear discrete system* can, in general, be defined by

$$y(n) = x(n) + a_1 x(n - 1) + a_2 x(n - 2). \tag{4.27}$$

This system is presented in Figure 4.9.

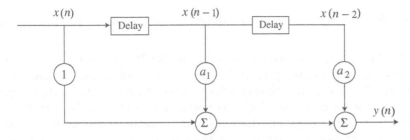

Figure 4.9 Linear Discrete System defined by Eq (4.27).

4.3.9 Linear Time-Invariant Systems. In the case of continuous-time, linear, time-invariant (LTI) systems that are represented by an impulse function $h(t)$ such that $h(t)$ is equivalent to the output response $y(t)$ obtained from the LTI system to a unit impulse $\delta(t)$, we have

$$h(t) \equiv y(t) \quad \text{when } x(t) = \delta(t).$$

The response output is $y(t) = (x \star \delta)(t)$, where $\star$ denotes the convolution in the time domain, i.e.,

$$y(t) = x(t) \star h(t) = \int_{-\infty}^{\infty} h(u)\, x(t-u)\, du. \qquad (4.28)$$

For causal systems $h(t) = 0$ for $t < 0$, and thus Eq (4.28) reduces in this case to

$$y(t) = \int_{0}^{\infty} h(u)\, x(t-u)\, du.$$

For example, in the case of a continuous exponential signal $x(t) = e^{i\, 2\pi f t}$, we obtain from (4.28)

$$y(t) = h(t) \star x(t) = h(t) \star e^{i\, 2\pi f t} = \int_{0}^{\infty} h(u)\, e^{i\, 2\pi f(t-u)}\, du$$

$$= e^{i\, 2\pi f t} \left[\int_{0}^{\infty} h(u)\, e^{-i\, 2\pi f u}\, du \right] = [H(f)]\, e^{i\, 2\pi f t}, \qquad (4.29)$$

where $H(f)$ is the Fourier transform of $h(t)$.

In the discrete-time, linear, time-invariant case, define an exponential sequence by $x(n) = e^{i\, 2\pi f n} = e^{i \omega n}$. Then the discrete-time LTI is represented by its frequency response to $x(n)$. Using the convolution sum formula, the discrete form of (4.29) is the response

$$y(n) = \sum_{k=-\infty}^{\infty} e^{i \omega k}\, h(n-k) = \left(\sum_{k=-\infty}^{\infty} h(k)\, e^{-i \omega k} \right) e^{i \omega n} = H(\omega)\, e^{i \omega n}, \qquad (4.30)$$

where $H(\omega)$ is the discrete-time Fourier transform of the impulse response $h(n)$.

A LTI system has direct application to NMR spectroscopy, seismology, circuits, signal processing, control theory, and other areas. It analyses the response of a linear and time-invariant system to an arbitrary input signal. Linearity means that the relation between the input and the output of the system is a linear map, and time invariance means that whether we apply an input to the system now or T seconds from now, the output will be the same except for a time delay of T secs. For more information, see Oppenheim [2009], and Oppenheim and Schäfer [2009].

4.4 Random Signals

A *random process* can be regarded as a mapping of the outputs of random signals $x(t)$. A random signal is said to be *stationary* if the density functions $\rho\{x(t)\}$ that describe this signal are invariant under translation in time t. Since a random stationary process is an infinite energy signal, i.e., its inputs and outputs are not square-integrable, its Fourier transform does not exist. Instead, we compute the Fourier transform of the auto-correlation function (ACF) of this signal, using the Wiener-Khinchin theorem[1], and obtain the distribution of the signal power as a function of frequency:

$$S_X(\omega) = \int_{-\infty}^{\infty} r_X(\tau) e^{-i\omega\tau} d\tau, \tag{4.31}$$

where $S_X(\tau)$ is called the *power density spectrum* of $X(t)$. The ACF of this stationary process $X(t)$ is defined as

$$r_X(\tau) = E[\overline{X(t)}\, X(t+\tau)],$$

where E denotes the expectation operator and τ the time lag. Then the inverse Fourier transform is given by

$$r_X(\tau) = \int_{-\infty}^{\infty} S_X(\omega) e^{i\omega\tau} d\tau. \tag{4.32}$$

---

[1] This theorem was published by Norbert Wiener in 1930 and by Aleksandr Khinchin in 1934 (see Couch [2001], Krzysztof [2007], and Engelberg [2007]). It states that the power spectrum density of a wide-sense-stationary random process is the Fourier transform of the corresponding auto-correlation function (see Eq (4.11)). This theorem is used in analyzing linear time-invariant (LTI) systems, when the inputs and outputs are not square-integrable, and so their Fourier transforms do not exist. A corollary is that the Fourier transform of the ACF of the output of an LTI system is equal to the product the Fourier transform of the ACF of the input of the system and the square of the magnitude of the Fourier transform of the system impulse response (see Engelberg [2007]). Since the Fourier transform of the ACF of a signal is the power spectrum of the signal, this corollary implies that the power spectrum of the output is equal to the power spectrum of the input times the power transform function. This corollary is useful in the parametric method for power spectrum estimation.

4.4.1 Discrete Case. Although the discrete-time random process is a sequence with infinite energy, it has a finite average power given by $E[X^2(n)] = r_X(0)$. In view of the above Wiener-Khinchin theorem, the spectral property of the discrete-time random process $X(t)$ is obtained by the Fourier transform of the auto-correlation sequence $r_X(k)$, that is,

$$S_X(f) = \sum_{k=-\infty}^{\infty} r_X(k)\, e^{-i\,2\pi f k}, \tag{4.33}$$

and the inverse transform by

$$r_X(k) = \int_{-1/2}^{1/2} S_X(f)\, e^{i\,2\pi f k}\, df. \tag{4.34}$$

Compare these definitions with those given in §4.6.2.

4.4.2 Random Sequence of Pulses. We will derive some properties of random sequences of pulses of duration T, each with amplitude ± 1. As in the case of the rectangular pulse (Examples 4.4 and 4.5), the ACF of a random function $x(t)$ of a process $X(t)$ that consists of a random sequence of pulses with amplitudes ± 1 with equal probability for the output $+1$ or -1 is

$$r_X(\tau) = \begin{cases} \left(1 - \dfrac{|\tau|}{T}\right) & \text{for } |\tau| \leq T, \\ 0 & \text{otherwise,} \end{cases}$$

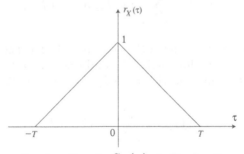

Figure 4.10 Power Spectrum Density $S_X(\omega)$ of a Random Sequence of Pulses.

This ACF is presented in Figure 4.10. Then the power spectral density is

$$S_X(\omega) = \int_{-T}^{T} \left(1 - \frac{|\tau|}{T}\right) e^{-i\,\omega\tau}\, d\tau = T\, \text{sinc}^2\, (\omega/2)\,.$$

This is plotted in Figure 4.11, where it will be noticed that the power spectral density $S_X(\omega)$ shows a main loop bounded by well-defined spectral nulls (or 0s), which can be inserted at any frequency using a filter in the transmitter.

Thus, the null-to-null bandwidth is a simple measure for the bandwidth of $X(t)$. Figure 4.11 shows that a sequence of pulses with amplitude ± 1 differs from the energy spectral density $\mathcal{E}_f(\omega)$, defined by (4.20), for a single rectangular pulse only by a factor of T.

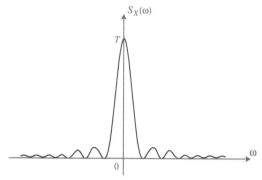

Figure 4.11 Power Spectrum Density $S_X(\omega)$ of a Random Sequence of Pulses.

4.4.3 Sampling. If we sample a continuous signal $x(t)$ at a uniform rate, say every T_s seconds, then we obtain an (infinite) sequence of samples. This sequence is denoted by $\{x(nT_s)\}$, where $n \in \mathbb{Z}$, and is called *sampling*, while the quantity T_s is called the *sampling period*, such that its reciprocal $1/T_s = f_s$ is the *sampling rate*. Analytically, for a *sampled* signal $x(t)$ this sampling operation is defined by

$$x_\delta(t) = \sum_{n=-\infty}^{\infty} x(nT_s)\, \delta(t - nT_s), \qquad (4.35)$$

where $x_\delta(t)$ is the sampled signal that consists of sequence of impulses separated in time by the sampling period T_s, and the term $\delta(t - nT_s)$ is the Dirac delta function at the time location $t = nT_s$. The Fourier transform of $x_\delta(t)$ is

$$X_\delta(f) = f_s \sum_{n=-\infty}^{\infty} X(f - nT_s), \qquad (4.36)$$

where f_s is the sampling rate, and X is the FT of x. The signals $X(f)$ and $X_\delta(f)$ for $f_s < 2B$ are presented in Figure 4.13.

The signal $x(t)$ and its sampled signal $x_\delta(t)$ are presented in Figure 4.12 (a) and (b), respectively. The signal $X(f)$ with bandwidth B and the signal $x_\delta(f)$ for $f_s > 2B$ are shown, respectively, in Figure 4.12 (a) and (b).

Note that in the case when the sampling rate $f_s < 2B$, the frequency-shifted components of $X(f)$ overlap, and the spectrum of the sampled signal is different from that of the original signal $x(t)$. This overlapping phenomenon is known as *aliasing*, and the sampling rate $f_s = 2B$ is called the *Nyquist rate*.

The aliasing is avoided by using a lowpass anti-aliasing filter to attenuate frequency components in excess of B (overlapping), and sample the signal such that its rate is greater than the Nyquist rate (keeping $f_s > 2B$).

4.4.4 Amplitude. The amplitude is the most important property of a digital signal. As defined above, the samples in a digital signal are the amplitudes, each one relocated to a sinusoid. There are two mostly used measures of amplitudes of a digital signal: the *peak amplitude*, and the *root mean square amplitude* (RMS). Let the sequence at a window starting at a sample M of length N of a signal $x(n)$ be denoted by $\{x(M), x(M+1), \dots, x(M+N-1)\}$. For this sequence the peak amplitude, which is the greatest sample in absolute value over the window, is

$$A_{\text{peak}}\{x(N)\} = \max_n |x(n)| \quad \text{for } n = M, \dots, M+N-1,$$

while the root mean square (RMS) amplitude is $A_{\text{RMS}}\{x(n)\} = \sqrt{P\{x(n)\}}$, where $P\{x(n)\} = \dfrac{1}{N}\left(|x(M)|^2 + \cdots + |c(M+N-1)|^2\right)$ is the *mean power*. Recall that an amplitude is always non-negative. Also, $\dfrac{1}{\sqrt{N}}A_{\text{peak}} \leq A_{\text{RMS}} \leq A_{\text{peak}}$. These two measures are presented in Figure 4.14 for the Sinusoid of Figure 4.14(a), where the peak and RMS amplitudes are the same in Figure 4.14(b) but different in Figure 4.14(c).

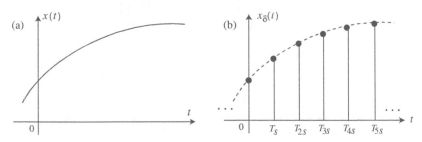

Figure 4.12 (a) Signal $x(t)$, and (b) Sampled Signal $x_\delta(t)$.

A better method of comparing these two measures is to use their ratio. Thus, for example, to say that a signal's amplitude is greater than another's by a factor of 2 gives more information than saying that it is greater by 25 millivolts. Thus, using ratio, the amplitudes are expressed in their logarithmic units called *decibels*. Let a denote the amplitude of a signal (peak or RMS). Then this amplitude's decibel (dB) level d is defined as $d = 20 \cdot \log_{10}(a/a_0)$, where a_0 is a reference amplitude. This definition simply implies that if the signal power is increased by 10, the amplitude is increased by a factor of $\sqrt{10}$, i.e., the logarithm increases by $1/2$, and the value in decibels increases (additively) by 3.01 dB. A graph showing the relationship of the amplitudes to their decibels is shown in Figure 4.15, where the linear amplitude 1 is assigned to 0 dB.

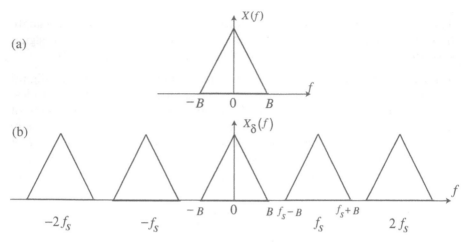

Figure 4.13 (a) Signal $X(f)$, and (b) Signal $X_\delta(f)$ for $f_s < 2B$.

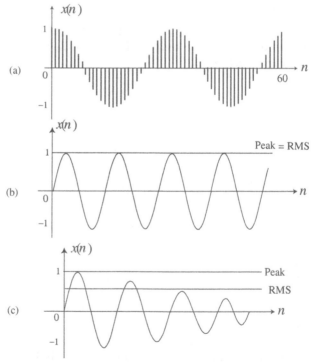

Figure 4.14 Peak and RMS Measures of the Sinusoid (a).

If a_0 is the reference amplitude, a signal with linear amplitude less than a_0 will have negative amplitude in decibels. For example, $a_0/10$ yields -20 dB;

$a_0/100$ yields -40 dB; and so on. For most of the hardware, $a_0 = 10^{-5} = 0.00001$, so that the maximum possible amplitude is 100 dB. Of course, 0 dB is always inaudible no matter what the maximum amplitude is assumed, and 100 dB is chosen as a range for the human ear, 0 dB being inaudible and 100 dB dangerously loud. However, the perception of loudness is not always related to the amplitude. For example, two signals each with the same peak or RMS amplitude may have different loudness, but amplifying a signal by 3 dB will make the sound a 'bit' louder. The mystery of adopting the logarithmic scale of decibels may speak about the human nature of hearing.

Control or change of amplitudes plays an important role in electronic music, where a simple strategy for synthesizing sounds is by combining sinusoids, sample by sample. But since sinusoids have a constant amplitude a, a technique to vary it in time is as follows: to multiply the amplitude a of a signal $x(n)$ by a factor $y \geq 0$, just multiply each sample by y. This will yield a new signal $y \cdot x(n)$. Then the measure of the peak or RMS amplitude of $x(n)$ will be greater or less by the factor y. In general, the amplitude can be changed by the (variable) amount $y(n)$ which varies sample by sample. If $y(n) \geq 0$ and if it varies slowly enough so that the values of $n = M, M+1, M+N-1$ in the window do not change, then the amplitude of the product $y(n) \cdot x(n)$ will be equal to the amplitude of $x(n)$ multiplied by the value of $y(n)$ for each value of N. However, sometimes the values of $x(n)$ and $y(n)$ are allowed to take negative or positive values and may be allowed to change quickly; in such cases the effect of multiplying $y(n) \cdot x(n)$ cannot be precisely described.

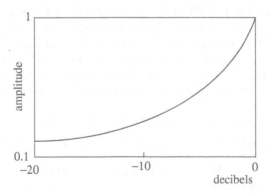

Figure 4.15 Amplitude vs. Decibels.

4.4.5 Frequency. To compare two signals, the frequency f is also measured on the logarithmic scale. The frequency ratio between them determines the interval between them. In music, the *octave* defines the musical interval associated with a ratio of 2 : 1. The musical scale divides the octave into 12 equal subintervals, each of which then corresponds to a ratio of $2^{1/12}$. A subinterval is called a *half-step*. A logarithmic scale for pitch is simply to count the number of half-steps (subintervals) from a reference pitch, allowing fractions so that the pitches that do not fall on a note can be specified. The most popular logarithmic pitch scale is 'MIDI pitch' m, in which the pitch 69 is assigned to the frequency $f = 440$ cycles/sec, i.e., A above middle C. The *Pitch-Frequency Conversion* formulas are:

$$m = 69 + 12 \cdot \log_{10}(f/440), \quad \text{and} \quad f = 440 \cdot 2^{(m-69)/12}.$$

For example, middle C, corresponding to MIDI pitch $m = 60$, has the frequency $f = 261.626$ cycles/sec. In hardware, MIDI allows only integer pitches between 0 and 127. A half-step has a ratio of about 1.059 to 1, or about 6% increase in frequency. Half-steps are further divided into *cents*, each cent being one hundredth of a half-step. The general rule is that it takes about 3 cents to make a real change in the pitch of a musical note. At middle C this amounts to a difference of about 0.5 cycle/sec. The relationship between MIDI pitch and frequency in cycles/sec (Hz) is given in Table 4.1.

Table 4.1 MIDI Pitch and Frequency

Pitch	Frequency
45	0
57	220
69	440

4.4.6 Superposition of Signals. If a signal $x(n)$ has a peak or RMS amplitude A in some fixed window, then the scaled signal $k \cdot x(n)$, $k \geq 0$, has the amplitude kA. The mean power of the scaled signal changes by a factor of k^2. When two different signals are added, the amplitude of the sum is not equal to the sum of the amplitudes. Let $x(n)$ and $y(n)$ be two different signals. Then the sum of their amplitudes obeys the triangle inequality, i.e.,

$$A_{\text{peak}}\{x(n)\} + A_{\text{peak}}\{y(n)\} \geq A_{\text{peak}}\{x(n) + y(n)\},$$
$$A_{\text{RMS}}\{x(n)\} + A_{\text{RMS}}\{y(n)\} \geq A_{\text{RMS}}\{x(n) + y(n)\}.$$

Let a window $[M, \ldots, M + N - 1]$ be fixed. Then the mean power of the sum of two signals is given by

$$P\{x(n) + y(n)\} = P\{x(n)\} + P\{y(n)\} + 2\sigma\{x(n), y(n)\},$$

where σ is the *covariance* of the two signals, defined by

$$\sigma\{x(n), y(n)\} = \frac{x(M)y(M) + \cdots + x(M + N - 1)y(M + N - 1)}{N}.$$

Note that the covariance σ is a measure of how much two signals change together. If the greater values of one signal mainly correspond with the greater values of the other, and the same holds for the smaller values, then $\sigma > 0$. On the other hand, when the greater values of one signal mainly correspond to the smaller values of the other, i.e., the signals tend to show opposite behavior, then $\sigma < 0$. The sign of σ therefore shows the tendency in the linear relationship between the signals. Thus, if the window is sufficiently large, the covariance of the two sinusoids with different frequencies is negligible compared to the mean power. Two signals that have no covariance are called *uncorrelated*, where the correlation is defined as covariance normalized over the interval $(-1, 1)$. In general, for two uncorrelated signals, the power of the sum is equal to the sum of the powers, i.e., the following Pythagorean relationship holds:

$$P\{x(n) + y(n)\} = P\{x(n)\} + P\{y(n)\} \quad \text{whenever} \quad \sigma\{x(n), y(n)\} = 0.$$

Geometrically, two uncorrelated signals can be regarded as orthogonal vectors, such that the positive correlated ones have an acute angle between them while the negative correlated have an obtuse angle between them. For example, the sum of two uncorrelated signals, both having RMS amplitude a, will have RMS amplitude $a\sqrt{2}$. However, for two equal signals, which is the most correlated case, their sum will have amplitude $2a$ which is the maximum value under the triangle inequality.

4.4.7 Periodic Signals. A signal $x(n)$ is said to be *periodic* with period τ if $x(n + \tau) = x(n)$ for all n. This signal repeats at $\tau = n$, $n \in \mathbb{N}$, but the smallest period $(n = 1)$ for which it repeats is the signal's period. The situation of a non-integer period arises in some cases, like the audio digital signals, where it is defined as follows: a sinusoid has a period (in samples) of $2\pi/\omega$, where ω is the angular frequency. In general, any sum of sinusoids with frequencies $2k\pi/\omega$, $k \in \mathbb{N}$, will repeat after $2\pi/\omega$ samples. Such a sum is the Fourier series in discrete form:

$$x(n) = a_0 + a_1 \cos(\omega n + \phi) + a_2 \cos(2\omega n + \phi) + \cdots + a_\nu \cos(\nu \omega n + \phi),$$

where it is assumed that the signals contain frequencies of finite bound, and $\nu \in \mathbb{N}$. The angular frequencies of the sinusoids are all integer multiples of ω and are called the *harmonics* of ω, called the *fundamental harmonic*. In terms of pitch, the harmonics $\omega, 2\omega, \ldots$ are at intervals of $0, 1200, 1902, 2400$, $2786, 3102, 3369, 3600, \ldots$ cents above the fundamental. This sequence of pitches is also known as the *harmonic series*. The first six harmonics of a pitch are almost multiples of 100, i.e., they land close to, but not exactly on, other pitches of the same scale, where the third and sixth miss only by 2 cents and the fifth by 14 cents. In other words, the frequency ratio $3 : 2$ (called a *perfect fifth*) is almost 7 half-steps; the ratio $4 : 3$ (called a *perfect fourth*) is almost 5 half-steps, and the ratios $5 : 4$ and $6 : 5$ (called *perfect major* and *minor thirds*) are almost 4 and 3 half-steps, respectively.

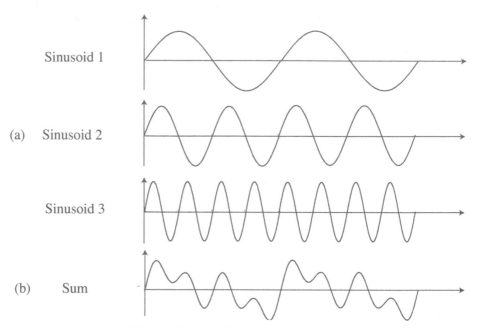

Figure 4.16 Three Sinusoids and Their Sum.

A Fourier series with three sinusoids and their sum are shown in Figure 4.16. The frequencies of the three sinusoids are in a 1 : 2 : 3 ratio. The common period is marked on the horizontal axis. Each sinusoid has different amplitude and initial phase. However, the sum of these three sinusoids in Figure 4.16(b) is not a sinusoid, but still preserves the periodicity shared by the three sinusoids.

4.5 Discrete Fourier Analysis

Two observations about the symmetry in waveforms of Fourier series of classical waveforms are as follows:

(i) A Fourier series consists of only even- or odd-numbered harmonics; this is due to symmetry comparing a waveform to its displacement by half a cycle.

(ii) A Fourier series may contain only real-valued or purely imaginary-valued coefficients corresponding to cosine or sine functions; this is due to symmetry comparing waveform to its reversal in time.

Let a waveform have an integer period f (cyclic frequency) which, for the sake of simplicity, is assumed to be even; if not, just up-sample, e.g., by a factor of 2. A real- or complex-valued waveform $x(n)$ can be written as a

Fourier series, with coefficients $a(k)$, $k = 0, 1, \ldots, f - 1$:

$$x(n) = a(0) + a(1)u^n + \cdots + a(f-1)u^{(f-1)n},$$

or equivalently

$$\begin{aligned}x(n) = a(0) + a(1)\,(\cos(\omega n) + i\,\sin(\omega n)) + \cdots \\ + a(f-1)\,(\cos(\omega(f-1)n) + i\,\sin(\omega(f-1)n)),\end{aligned}$$

where $\omega = 2\pi/f$ is the fundamental frequency of the waveform, and $u = \cos(\omega) + i\,\sin(\omega) = e^{i\omega}$ is a complex number of magnitude unity and argument ω.

To analyze the first symmetry, we delay the signal $x(n)$ by a half-cycle. Since $u^{n/2} = -1$, we get

$$x(n+f/2) = a(0) - a(1)u^n + a(2)u^{2n} - \cdots + a(f-2)u^{(f-2)n} - a(f-1)u^{(f-1)n}.$$

Notice that a half-cycle delay changes the sign of every other term in the Fourier series. We combine this series with the original series in two ways:

METHOD 1: Let x^* denote half the sum of the two, i.e.,

$$x^*(n) = \frac{x(n) + x(n+f/2)}{2} = a(0) + a(2)u^{2n} + \cdots + a(f-2)u^{(f-2)n},$$

and x^{**} half the difference of the two, i.e.,

$$x^{**}(n) = \frac{x(n) - x(n+f/2)}{2} = a(1)u^n + a(3)u^{3n} + \cdots + a(f-1)u^{(f-1)n}.$$

Note that x^* contains only the even-numbered harmonics while x^{**} only the odd-numbered ones. Also, if $x(n)$ happens to be equal to itself shifted a half-cycle, i.e., if $x(n) = x(n+f/2)$, then obviously $x^*(n) = x(n)$ and $x^{**}(n) = 0$, which implies that in this case $x(n)$ has only even-numbered harmonics. Similarly, if $x(n) = -x(n+f/2)$, then $x(n)$ has only odd-numbered harmonics. This method can be used to split any given waveform into its even- and odd-numbered harmonics, or equivalently, to use a comb filter to extract even or odd harmonics.

METHOD 2: Compare $x(n)$ with its time reversal $x(-n)$, or equivalently, compare $x(n)$ with $x(f-n)$ since $x(n)$ repeats every f samples. Then the Fourier series becomes

$$\begin{aligned}x(-n) = a(0) + a(1)\,(\cos(\omega n) - i\,\sin(\omega n)) + \cdots \\ + a(f-1)\,(\cos(\omega(f-1)n) - i\,\sin(\omega(f-1)n)),\end{aligned}$$

and, as in Method 1, we obtain the cosines by forming $x^*(n)$ as half the sum of the two, i.e.,

$$x^*(n) = \frac{x(n) + x(-n)}{2} = a(0) + a(1)\cos(\omega n) + \cdots + a(f-1)\cos(\omega(f-1)n),$$

and the series by forming x^{**} half the difference of the two divided by i, i.e.,

$$x^{**}(n) = \frac{x(n) - x(n + f/2)}{2\,i} = a(1)\sin(\omega n) + \cdots + a(f-1)\sin(\omega(f-1)n).$$

Thus, if $x(-n) = x(n)$, the Fourier series consists of cosine terms only, and if $x(-n) = -x(n)$, it consists of sine terms only. Hence, we can decompose any $x(n)$ as sum of the two provided $x(n)$ repeats every f samples, that is, it is periodic of period f.

Example 4.6. (Sawtooth wave) Figure 4.17, where part (a) shows the original sawtooth wave; part (b) shows the result of shifting by a half-cycle; part (c) sum of parts (a) and (b), which drops discontinuously whenever either one of the part (a) or part (b) does so, and traces a line segment whenever the two component sawtooth waves do, so that it becomes a single sawtooth wave of half the original period (twice the fundamental frequency); and part (d) shows the waveform obtained after subtracting the two sawtooth waves, and in particular shows zero slope except at the jump discontinuities, which arising out of the original sawtooth wave jump in the same direction (negative to positive) while those coming from the shifted one jump from positive to negative. The result, shown in part (d), is a square wave which is a particular rectangle wave in which the two component segments have the same duration. ∎

This symmetry is used in the design of the Buchla analog synthesizers. Buchla designed an oscillator that outputs the even and odd harmonic components separately, so that cross-fading between the two allows a continuous control over the relative strengths of even and odd harmonics in the analog waveform.

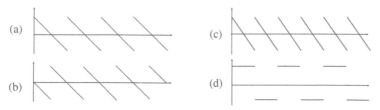

Figure 4.17 Sawtooth Waves.

Example 4.7. (Parabolic waves) Consider the parabolic waveform with corners over a single period from 0 to N given by (Figure 4.18)

$$p(n) = \frac{1}{2}\left(\frac{n}{N} - \frac{1}{2}\right)^2 - \frac{1}{24}.$$

It is a quadratic polynomial in n with a maximum halfway through the cycle at $n = N/2$, and its average value over one cycle of the waveform is zero, so that the slope changes discontinuously by $-1/N$ at the beginning of the cycle.

Figure 4.18 Parabolic Waveform.

To construct a waveform with any desired number of corners, say at the points $M_i, \dots, M_m$, with changes equal to $c_1, \dots, c_m$, add all the necessary parabolic waves, i.e.,

$$x(n) = -Nc_1 p\,(n - M_1) - \cdots - Nc_m p\,(n - N_m). \quad\blacksquare \qquad (4.37)$$

4.5.1 Fourier Series of Elementary Waveforms.
Given a periodic (repeating) waveform $x(n)$ we can determine its Fourier series coefficients $c(k)$, $k = 0, 1, \dots, N - 1$, directly from the Fourier transform using the formula:

$$c(k) = \frac{1}{N}\, \mathcal{F}\{x(n)\}(k) = \frac{1}{N}\left[x(0) + u^{-1}x(1) + \cdots + u^{-(N-1)}x(N-1)\right], \quad (4.38)$$

where $u = e^{i\omega}$ is a unit-magnitude complex number. But this sometimes leads to a very tedious mathematical process, especially in the case of sawtooth and parabolic waves. Instead, we use the properties of the Fourier transform of a signal $x(n)$ and its first difference defined by $x(n) - x(n-1)$. As noted above, the first difference of the parabolic wave will result in a sawtooth wave, and the Fourier transform of the sawtooth wave will be simple to determine, thus obtaining the desired Fourier series. In general, to determine the strength of the kth harmonic, we will assume that $N \gg k$, so that k/N is negligible. Using the time shift formula for Fourier transform applied to one sample we get

$$\mathcal{F}\{x(n-1)\} = [\cos(k\omega) - i\,\sin(k\omega)]\,\mathcal{F}\{x(n)\}$$
$$\approx (1 - i\,k\omega)\,\mathcal{F}\{x(n)\},$$

where $k\omega = 2\pi k/N \ll 1$ since $N \gg k$. Thus, $\cos(k\omega) \approx 1$, $\sin(k\omega) \approx k\omega$, with an error of the order of $(k/N)^2$. Then substituting these values we get

$$\mathcal{F}\{x(n) - x(n-1)\} \approx i\,k\omega \mathcal{F}\{x(n)\}.$$

Example 4.8. (Sawtooth Wave, see Example 2.11) First, we apply this method to the sawtooth wave $s(n)$. For $0 \leq n < N$ we get

$$s(n) - s(n-1) = -\frac{1}{N} + \begin{cases} 1 & \text{if } n = 0, \\ 0 & \text{otherwise.} \end{cases}$$

Ignoring the constant offset of $-\frac{1}{N}$, this gives an impulse, zero everywhere except one sample per cycle. The summation in the Fourier transform has only one term, which yields

$$\mathcal{F}\{s(n)\}(k) \approx \frac{1}{i\,k\omega} = -\frac{i\,N}{2k\pi},$$

where $\omega = 2\pi/N$. This result is valid for integer values of nonzero k which are smaller than N. However, this method does not give the component $\mathcal{F}\{s(n)\}(0)$, since we cannot divide by zero. But this term can be evaluated directly as the sum of all the points of the waveform: it is approximately zero by symmetry. To obtain a Fourier series in terms of real-valued sine and cosine functions, we add corresponding terms for negative and positive values of k. The first harmonic corresponds to $k = \pm 1$ and is given by

$$\frac{1}{N} \left[\mathcal{F}\{s(n)\}(1)u^n + \mathcal{F}\{s(n)\}(-1)u^{-n} \right] \approx \frac{-i}{2\pi} \left[u^n - u^{-n} \right]$$
$$= \frac{\sin(n\omega)}{\pi};$$

Similarly, the kth harmonic is $\dfrac{\sin(n\omega)}{k\pi}$. Hence, the entire Fourier series is

$$s(n) \approx \frac{1}{\pi} \left[\sin(n\omega) + \frac{\sin(2n\omega)}{2} + \frac{\sin(3n\omega)}{3} + \cdots \right]. \blacksquare \qquad (4.39)$$

Example 4.9. (Parabolic Wave $p(n)$) First, using the above method we determine the difference

$$p(n) - p(n-1) = \frac{\left(\dfrac{n}{N} - \dfrac{1}{2} \right)^2 - \left(\dfrac{n-1}{N} - \dfrac{1}{2} \right)^2}{2}$$

$$= \frac{\left(\dfrac{n}{N} - \dfrac{N}{2N} \right)^2 - \left(\dfrac{n}{N} - \dfrac{N-2}{2N} \right)^2}{2}$$

$$= \frac{\dfrac{2n}{N^2} - \dfrac{1}{N} + \dfrac{1}{N^2}}{2} \approx -\frac{s(n)}{N}.$$

For nonzero $k \ll N$, we get

$$\mathcal{F}\{p(n)\}(k) \approx \frac{-1}{N} \cdot \frac{-i\,N}{2k\pi} \mathcal{F}\{s(n)\}(k)$$

$$\approx \frac{-1}{N} \cdot \frac{-i\,N}{2k\pi} \cdot \frac{-i\,N}{2k\pi} = \frac{N}{4k^2\pi^2}.$$

Hence, we obtain the Fourier series

$$p(n) \approx \frac{1}{2\pi^2} \left[\cos(n\omega) + \frac{\cos(2n\omega)}{4} + \frac{\cos(3n\omega)}{9} + \cdots \right]. \ \blacksquare \qquad (4.40)$$

Example 4.10. (Symmetric and Nonsymmetric Triangle Waves) A symmetric triangle wave, obtained by superposing parabolic waves with (M, c) pairs equal to $(0, 8)$, is shown in Figure 4.19.

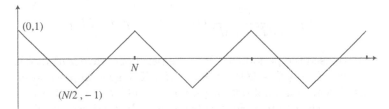

Figure 4.19 Symmetric Triangle Wave.

To obtain the Fourier series we will first consider the square wave

$$x(n) = s(n) - s(n - N/2) = \begin{cases} \dfrac{1}{2} & \text{for the first half cycle } 0 \le n < N/2, \\ 0 & \text{otherwise.} \end{cases}$$

We obtain the Fourier series by substituting the Fourier series for $s(n)$ twice:

$$x(n) \approx \frac{1}{\pi} \left[\sin(n\omega) + \frac{\sin(2n\omega)}{2} + \frac{\sin(3n\omega)}{3} + \cdots - \sin(n\omega) + \frac{\sin(2n\omega)}{2} - \frac{\sin(3n\omega)}{3} \pm \cdots \right]$$

$$= \frac{2}{\pi} \left[\sin(n\omega) + \frac{\sin(3n\omega)}{3} + \frac{\sin(5n\omega)}{5} + \cdots \right].$$

Hence, the symmetric triangle wave (Figure 4.19), defined by $x(n) = 8p(n) - 8p(n - N/2)$, where $p(n)$ is the parabolic wave (Example 4.7), is given by

$$x(n) \approx \frac{8}{\pi^2} \left[\cos(n\omega) + \frac{\cos(3n\omega)}{9} + \frac{\cos(5n\omega)}{25} + \cdots \right]. \ \blacksquare \qquad (4.41)$$

Example 4.11. (Nonsymmetric Triangle Wave) This wave with vertices at $(M, 1)$ and $(N - M, -1)$ is shown in Figure 4.20, where the cycles are set such that the two vertices have equal and opposite heights and the mid-point of the shorter segment passes through the point $(0,0)$. The two line segments have the slopes $1/M$ and $-2/(N-2M)$.

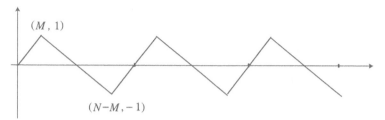

Figure 4.20 Nonsymmetric Triangle Wave.

Thus, the decomposition into parabolic waves is given by

$$x(n) = \frac{N^2}{MN - 2M^2} \{p(n-M) - p(n+M)\},$$

where we have used the periodicity N of the parabolic wave $p(n)$ (see Example 4.9) to replace $p(n-(N-M))$ by $p(n+M)$. The general method has so far been to use linear combinations of parabolic and/or sawtooth waves which takes us back to the complex Fourier transform. Instead, we will first use the real-valued Fourier series

$$x(n) = \frac{N^2}{2\pi^2(MN - 2M^2)} \left[\cos((n-M)\omega) - \cos((n+M)\omega) \right.$$
$$\left. - \frac{\cos(2(n-M)\omega) - \cos(2(n+M)\omega)}{4} + \cdots \right].$$
$$(4.42)$$

Then, using the identity $\cos A - \cos B = 2\sin\left(\frac{B-A}{2}\right)\sin\left(\frac{B+A}{2}\right)$, and letting $\omega = 2\pi/N$, we get

$$x(n) = a(1)\sin(n\omega) + a(2)\sin(2n\omega) + \cdots,$$

where the amplitudes $a(k)$, $k = 0, 1, \ldots, N-1$, are given by

$$a(k) = \frac{1}{(M/N - 2(M/N)^2)\,\pi^2} \cdot \frac{\sin(2\pi k M/N)}{k^2}.$$

This result depends on both M and N together since they appear in the form M/N. This fact conforms to the shape of the waveform, which depends on this ratio. By taking small values of $k < \dfrac{1}{4M/N}$ and using the approximation

$\sin\theta \approx \theta$, we find that the argument of the sine function is $\pi/2$. Thus, using $\sin\theta \approx \theta$ in the first quadrant, $a(k)$ decays at rate $1/k$, as in the case of sawtooth wave. However, for larger values of k, the sine term oscillates between 1 and -1 so that the amplitudes decay irregularly at the rate $1/k^2$.

4.6 Fourier Resynthesis: Discrete Case

We will discuss methods to reconstruct $x(n)$ from its Fourier Transform $X(k)$. It is known that Fourier analysis takes as input any periodic signal (of period N) and yields as output the complex-valued amplitudes of its N possible sinusoidal components. These N complex amplitudes can be theoretically used to exactly reconstruct the original signal. This reconstruction is called *Fourier Resynthesis*.

Let $x(n)$ be a complex-valued signal that repeats every N samples. Since the period is N, we determine the values of $x(n)$ for $n = 0, 1, \ldots, N-1$. Let $x(n)$ be written as a sum of complex sinusoids of frequency $0, 2\pi/N, 4\pi/N, \ldots$, $2(N-1)\pi/N$, where frequency of the kth partial is $2\pi k/N$, $k = 0, 1, \ldots, N-1$. We can find its complex amplitude by modulating x downward $2\pi k/N$ radians per sample in frequency, so that the kth partial is modulated to frequency zero. Then we pass the signal through a lowpass filter with such a low cutoff frequency that nothing but the zero-frequency partial remains. This can be done by averaging over a very large number of samples; but since the signal repeats every N samples, this very large average is the same as the average over the first N samples. Hence, to measure a sinusoidal component of a periodic signal, just modulate it down to DC and then average over one period.

Let $\omega = 2\pi/N$ be the fundamental frequency for the period N, and let u be the unit-magnitude complex number with argument ω, i.e., $u = e^{i\omega} = e^{2\pi i/N}$, with $\arg\{u^k\} = k \arg\{u\} = k\omega = 2k\pi/N$. Assuming that the signal $x(n)$ can be written as a sum of the n partials, i.e.,

$$x(n) = A_0 \left[u^0\right]^n + A_1 \left[u^1\right]^n + \cdots + A_{N-1} \left[u^{N-1}\right]^n,$$

where A_k is the complex amplitude of the partial, each A_k can be determined by multiplying the sinusoid of frequency $-k\omega$ and averaging over one period, i.e.,

$$A_k = \frac{1}{N}\left\{ \left[W^k\right]^0 x(0) + \left[W^k\right]^1 x(1) + \cdots + \left[W^k\right]^{N-1} x(N-1)\right\},$$

where $W = u^{-1} = e^{-2\pi i/N}$ is defined in §3.5. Thus, the *Fourier transform* of a signal $x(n)$ over N samples (i.e., of period N) is defined as

$$X(k) = \left[W^k\right]^0 x(0) + \left[W^k\right]^1 x(1) + \cdots + \left[W^k\right]^{N-1} x(N-1). \qquad (4.43)$$

The Fourier transform is a function of the variable k, equal to N times the amplitude of the input's kth partial, where k has so far taken integer values.

But the above definition can be extended for any value of k if we define W as above. Also, if $x(n)$ is a signal of period N samples (i.e., it repeats every N samples), the Fourier Transform of $x(n)$ also repeats itself every N units of frequency:

$$X(k+N) = X(k) \quad \text{for all real values of } k. \tag{4.44}$$

This follows from the definition (4.43) since the factor W^k remains unchanged when we add N, or any multiple of N, to k.

4.6.1 Additive Synthesis.

There is no direct way to obtain the Fourier series for a signal $x(n)$. We have assumed that the signal $x(n)$ can be obtained as a sum of sinusoids, but we have not yet found out whether every periodic $x(n)$ can be so obtained. Now, consider an arbitrary signal $x(n)$ that repeats every N samples. Let $X(k)$ be the Fourier Transform of $x(n)$ for $k = 0, 1, \ldots, N - 1$. Then by (4.43)

$$X(k) = \left[W^0\right]^k x(0) + \left[W^1\right]^k x(1) + \cdots + \left[W^{(N-1)}\right]^k x(N-1), \tag{4.45}$$

where the exponents are rearranged such that $X(k)$ is a sum of complex sinusoids with complex amplitudes $x(m)$ and frequencies $-m\omega$ for $m = 0, 1, \ldots, N-1$. Thus, $X(k)$ can be also regarded as a Fourier series, in which the mth component (term) has strength $x(-m)$. Since $x(n)$ is a periodic signal, the amplitude of the partials of $X(k)$ can be expressed in terms of its own Fourier transform. By equating the two expressions we get $\frac{1}{N}X(m) = x(-m)$. Thus, $x(-m)$ can be obtained by summing sinusoids with amplitudes $X(k)/N$. Set $n = -m$, and we get

$$x(n) = \frac{1}{N}X(-n) = \left[W^0\right]^n X(0) + \left[W^1\right]^n X(1) + \cdots + \left[W^{-(N-1)}\right]^n X(N-1).$$

Thus, any periodic $x(n)$ can be obtained as a sum of sinusoids, which shows how to reconstruct $x(n)$ from its Fourier transform $X(k)$ if we know its values for $k = 0, 1, \ldots, N - 1$.

Example 4.12. (Fourier Transform of DC) The simplest case of a sinusoid is direct current (DC) which is a sinusoid of frequency zero. Let $x(n) = 1$ for all n. This signal repeats with any desired integer period $N > 1$, and consists of all 1s. Its Fourier transform is given by

$$X(k) = \begin{cases} N & \text{for } k = 0, \\ 0 & \text{for } k = 1, \ldots, N - 1. \end{cases}$$

For noninteger k, we use (4.43) whose right side is a geometric series if $W \neq 1$. Thus,

$$X(k) = \frac{(W^k)^N - 1}{W^k - 1} = \frac{V^{2N} - 1}{V^2 - 1},$$

where $V = e^{-k\pi i/N}$, so that $V^2 = W$. This is simplified to yield

$$X(k) = V^{N-1} \frac{V^N - V^{-N}}{V - V^{-1}} = e^{-k(N-1)i\,\pi/N} \frac{\sin k\pi}{\sin k\pi/N}$$
$$= e^{-k(N-1)i\,\pi/N} D_N(k), \qquad (4.46)$$

where $D_N(k)$ is the Dirichlet kernel, defined by

$$D_N(k) = \begin{cases} N & \text{for } k = 0, \\ \dfrac{\sin(k\pi)}{\sin(k\pi/N)} & \text{for } k \neq 0,\ -N < k < N. \end{cases} \qquad (4.47)$$

Thus, the Fourier transform of $x(n) = 1$ repeats every N samples, with a peak at $k = 0$ and another at $k = N$. The phase term $e^{-k(N-1)i\,\pi/N}$ acts to twist the values of the transform $X(k)$ with a period of about 2. The Dirichlet kernel $D_N(k)$, shown in Figure 4.21 for $N = 100$, controls the magnitude of this transform; it has a peak, 2 units high, around $k - 0$. This is surrounded by 1 unit wide *side waveforms*, alternating in sign and gradually decreasing in magnitude as k increases or decreases away from zero. The phase term rotates by almost π radians each time the Dirichlet kernel changes sign, so that the product of the two stays roughly in the same complex half-plane for $k > 1$, and in opposite half-plane for $k < 1$. The phase rotates by almost 2π radians over the peak from $k = -1$ to $k = 1$. ∎

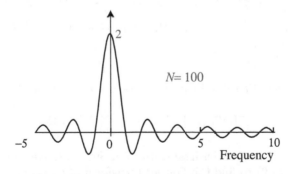

Figure 4.21 Dirichlet Kernel for $N = 100$.

4.6.2 Time and Phase Shifts. A time shift in a signal changes the phases of its sinusoidal components. First, we consider a time shift. Suposing that a signal $x(n)$ is complex-valued and repeats every N samples, we let $y(n)$ be the value of $x(n)$ delayed d samples, i.e., $y(n) = x(n - d)$, which, like $x(n)$, also repeats every N samples. The Fourier transform of $y(n)$ can be obtained using (4.43) as follows:

$$\begin{aligned} Y(k) &= \left[W^k\right]^0 y(0) + \left[W^k\right]^1 y(1) + \cdots + \left[W^k\right]^{N-1} y(N-1) \\ &= \left[W^k\right]^0 x(0-d) + \left[W^k\right]^1 x(1-d) + \cdots + \left[W^k\right]^{N-1} x(N-1-d) \\ &= \left[W^k\right]^d x(0) + \left[W^k\right]^{d+1} x(1) + \cdots + \left[W^k\right]^{d+N-1} x(N-1) \\ &= \left[W^k\right]^d \left(\{ \left[W^k\right]^0 x(0) + \left[W^k\right]^1 x(1) + \cdots + \left[W^k\right]^{N-1} x(N-1) \right\} \\ &= \left[W^k\right]^d X(k). \end{aligned}$$

Thus, we get the *time-shift formula* for discrete Fourier transforms:

$$\mathcal{F}\{x(n-d)\}(k) = e^{-dki\,\omega}\, \mathcal{F}\{x(n)\}(k). \tag{4.48}$$

Hence, the Fourier transform of $x(n-d)$ is a phase shift term times the Fourier transform of $x(n)$, where the phase is changed by $-dk\omega$ which is a linear function of the frequency k.

Next, we consider the phase shift. Let the starting signal $x(n)$ be changed by multiplying it by a complex exponential z^n with angular frequency α, i.e.,

$$y(n) = z^n x(n), \quad \text{where} \quad z = e^{i\,\alpha}.$$

The Fourier transform is

$$\begin{aligned} Y(nk) &= \left[W^k\right]^0 y(0) + \left[W^k\right]^1 y(1) + \cdots + \left[W^k\right]^{N-1} y(N-1) \\ &= \left[W^k\right]^0 x(0) + \left[W^k\right]^1 zx(1) + \cdots + \left[W^k\right]^{N-1} z^{N-1}x(N-1) \\ &= \left(zW^k\right)^0 x(0) + (zW^k)^1 x(1) + \cdots + (zW^k)^{N-1}x(N-1) \\ &= \left(k - \frac{\alpha}{\omega}\right) X(k). \end{aligned}$$

Thus, we have the *Phase Shift Formula* for discrete Fourier transform:

$$\mathcal{F}\{e^{i\,\alpha}x(n)\}(k) = \left(k - \frac{\alpha}{\omega}\right) \mathcal{F}\{x(n)\}. \tag{4.49}$$

Example 4.13. (Fourier Transform of a Sinusoid) We can use the phase shift formula (4.49) to find the Fourier transform of any complex sinusoid z^n with frequency α simply by setting $x(n) = 1$ and using the Fourier transform for DC:

$$\mathcal{F}\{z^n\}(k) = \left(k - \frac{\alpha}{\omega}\right) \mathcal{F}\{1\} = e^{i\,\Phi(k)}\, D_N\left(k - \frac{\alpha}{\omega}\right),$$

where $\Phi(k)$ is the phase defined by $\Phi(k) = -\pi\left(k - \frac{\alpha}{\omega}\right)\frac{N-1}{N}$, and D_N is the Dirichlet kernel. The Fourier transforms of the complex sinusoids with

$N = 100$ are shown in Figure 4.22 (a) with frequency 2ω, and (b) with frequency $1.5\,\omega$.

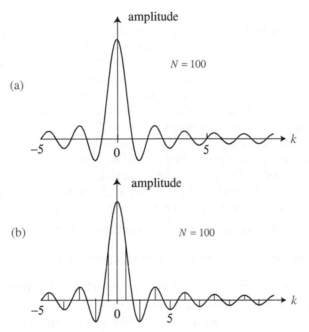

Figure 4.22 Fourier Transforms of Complex Sinusoids z^n with $N = 100$.

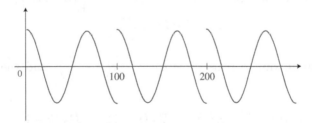

Figure 4.23 A Complex Sinusoid with Frequency $\alpha = 1.5\,\omega$.

   If the frequency α of the sinusoid is an integer multiple of the fundamental frequency ω, the Dirichlet kernel is shifted to left or right by an integer. In this case the zero crossings of the Dirichlet kernel line up with integer values of k, so that only one partial is nonzero. This is presented in Figure 4.22(a). However, Figure 4.22(b) presents the case when α is midway between integers (in this figure $\alpha = 1.5\,\omega = 3\pi/N$). The partials have amplitudes decaying almost as $1/k$ in both directions, measured from the actual frequency α. Although we started with a single sinusoid, the fact that the energy has been spread over many partials is very interesting. As shown in Figure 4.23, the signal repeats with a period N, which does not match with the frequency of the sinusoid. Thus, there is a discontinuity at the beginning of each period, and energy is spread over a wide range of frequencies. ∎

4.6.3 Nonperiodic Signals. In practice most signals are not periodic, and even a periodic signal has an unknown period. We will present discrete Fourier analysis of signals without the assumption that they have a fixed period N. In the analysis thus far we took N samples of the signal and made it periodic. In Figure 4.22(a) we obtained the Fourier transform for a pure sinusoid. However, we would like to obtain results where the response to a pure sinusoid is better localized around the corresponding value of k. Thus, given a signal $x(n)$, not necessarily periodic, defined at points from 0 to $N - 1$ in a window of N samples, we envelop the signal using a *window function*, and thereafter apply the Fourier analysis. Given a window function $w(n)$, we get the *windowed Fourier transform* $\mathcal{F}\{w(n)x(n)\}(k)$. Although there are different kinds of window functions, we will consider the simplest one, known as the *Hann window* which is defined as $w(n) = \dfrac{1}{2} - \dfrac{1}{2}\cos(2n\pi/N)$. Since the Hann window can be written as a sum of three complex exponentials: $w(n) = \dfrac{1}{2} - \dfrac{1}{4}u^n - \dfrac{1}{4}u^{-n}$, where $u = e^{-2\pi i/N}$, we can analyze the effect of multiplying a signal by the Hann window before taking the Fourier Transform. First, calculate the windowed Fourier transform of a sinusoid z^n with angular frequency α as

$$\mathcal{F}\{w(n)z^n\}(k) = \mathcal{F}\left\{\frac{1}{2}z^n - \frac{1}{4}(uz)^n - \frac{1}{4}\left(u^{-1}z\right)^n\right\}(k)$$
$$\approx e^{k\Phi i}M\left(k - \frac{\alpha}{\omega}\right),$$

where the approximate phase is $\Phi(k) = -\pi\left(k - \dfrac{\alpha}{\omega}\right)$, and the magnitude function is

$$M(k) = \left[\frac{1}{2}D_N(k) + \frac{1}{4}D_N(k+1) + \frac{1}{4}D_N(k-1)\right].$$

The graph of the magnitude function $M(k)$ of the Fourier Transform of the Hann window, expressed as sum of three shifted and magnified copies of the Dirichlet kernel D_N with $N = 100$, is presented in Figure 4.24.

The side waveforms (called *sidelobes*) reach their maximum amplitude near the midpoints. The amplitudes of the sidelobes in Figure 4.24 can be approximately calculated using the formula: $D_N(k) \approx \dfrac{N\sin(k\pi)}{k\pi}$. For $k = 3/2, 5/2, \ldots$ we get the sidelobe amplitudes, relative to the peak height N, as

$$\frac{2}{3\pi} \approx -13\text{dB}, \quad \frac{2}{5\pi} \approx -18\text{dB}, \quad \frac{2}{7\pi} \approx -21\text{dB}, \quad \frac{2}{9\pi} \approx -23\text{dB}, \ldots .$$

Sidelobes drop off progressively more slowly so that the tenth one is only attenuated to about -30 dB and the 32nd to about -40 dB. The Hann window sidelobes are attenuated by

$$\frac{2}{5\pi} - \frac{1}{2}\left[\frac{2}{3\pi} + \frac{2}{7\pi}\right] \approx -32.3 \text{ dB}.$$

and -40, -49, -54, and -59 dB for the next four sidelobes. Thus, applying a Hann window before taking the Fourier Transform allows us to better isolate sinusoidal components. If a signal has many sinusoidal components, the sidelobes produced by each will interfere with the main lobes of all others. Reducing the amplitude of the sidelobes reduces this type of interference.

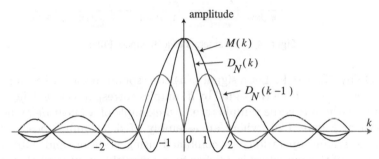

Figure 4.24 Magnitude $M(k)$ of the Fourier Transform of the Hann Window.

4.7 Theory of Filters

In the time domain, the convolution is simply the multiplication of Fourier transforms in the frequency domain. Thus, for a linear time-invariant (LTI) system the Fourier transforms of the input and output signals are related to each other by

$$Y(f) = H(f) X(f), \qquad (4.50)$$

where the transform $H(f)$ is, in general, a complex quantity. Thus, we can write $H(f) = |H(f)| e^{i \arg\{H(f)\}}$, where $|H(f)|$ is called the *amplitude response* and $\arg\{H(f)\}$ the *phase response* of the system. The *magnitude response* is generally expressed in decibels (dB) by $|H(f)|_{\text{dB}} = 20 \log_{10} |H(f)|$. In real systems, the signal $h(t)$ is a real-valued function, and so $H(f)$ has a conjugate symmetry in the frequency domain, i.e., $H(f) = \overline{H(-f)}$. If the input and output signals are expressed in terms of power spectrum density, then the *input-output relation* becomes

$$S_y(f) = |H(f)|^2 S_x(f). \qquad (4.51)$$

Thus, an LTI system acts like a *filter*. There are three types of filters: lowpass, passband, and highpass. They are characterized by stopband, passband, and half-power (3 dB) bandwidth, respectively. The parameters of a bandpass filter are shown in Figure 4.25. A bandpass signal has frequency content concentrated in a band of frequencies above zero frequency.

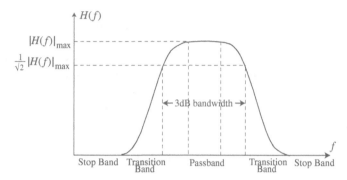

Figure 4.25 Parameters of a Bandpass Filter.

4.7.1 Delay Network. Let a signal $x(n)$ correspond to successive samples in time. Then time shifting the signal by d samples corresponds to a delay of d/R time units, where R is the sample rate. An example of a *linear delay network* is shown in Figure 4.26, where an assembly of delay units, possibly with amplitude scaling operations, is combined using addition and subtraction. The output of this operation is a linear function of the input such that adding two signals at the input is the same as processing each one separately and then adding the results. The original frequencies remain the same and no new frequency is created in the output so long as the network remains invariant. Hence, the gain and delay times do not change with time.

Figure 4.26 Delay Network.

An example of a linear delay network is shown in Figure 4.27 where the output is the sum of the input and its time shifted copy, like an assembly of delay units. Linear delay networks also create no new frequencies in the output: they keep their input frequencies as long as the network is time-invariant, i.e., the gain and delay times do not change with time.

An impulse is a pulse that lasts only one sample. If the output of the network for an impulse is known, it can be used to find the output for any other signal since any signal $x(n)$ is the sum of impulses, i.e., of amplitudes. Impulses are used to find the time-domain behavior of complicated networks. However, any network can be analyzed in the frequency domain by considering a complex-valued test signal $x(n) = z^n$, where $|z| = 1$ and $\arg\{z\} = \omega$. Let hz^n be the output of another complex sinusoid with the same frequency for some complex number h, yet to be determined. Since the output can be written as the sum of the input and its delayed copy, i.e., $z^n + z^{-d}z^n = (1 + z^{-d}) z^n$, we find that $h = 1 + z^{-d}$. The frequency-domain behavior can be analyzed by studying how h varies as a function of the angular frequency ω. That is, its magnitude and argument should tell us the relative magnitude and amplitude of the output sinusoid. As shown in Figure 4.28, the complex number z encodes the frequency of the input. The delay line output is the input multiplied by z^{-d}, where the total (complex) gain is h, and $|h|$ and $\arg\{h\}$ are determined by symmetrizing the sum, and rotating it by $d/2$ times the angular frequency of the input.

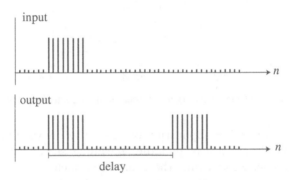

Figure 4.27 Time Domain View of a Delay Network.

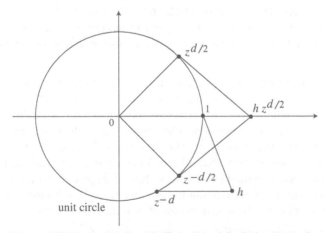

Figure 4.28 Frequency-Domain Behavior of the Delay Network.

Since the arguments of $z^{d/2}$ and $z^{-d/2}$ are one-half of those of z^d and z^{-d}, respectively, we have $h = z^{-d/2} \left(z^{d/2} + z^{-d/2} \right)$, where the first term has a phase shift of $-d\omega/2$. For the second term, note that $z^{d/2} + z^{-d/2} = \left\{\cos(\omega d/2) + i\,\sin(\omega d/2)\right\} + \left\{\cos(\omega d/2) - i\,\sin(\omega d/2)\right\} = 2\cos(\omega d/2)$, which is real. However, $|h| = 2|\cos(\omega d/2)|$. The quantity $|h|$ is called the *gain* of the delay network at the angular frequency ω, and is presented in Figure 4.29. The frequency-dependent gain of a delay network is the gain as a function of frequency and is called the network's *frequency response*. Since the network has a larger gain at some frequencies than at others, it may be regarded as a *filter* which is used to separate certain components of a sound from others. Since the shape of this kind of gain is expressed as a function of ω, such a delay network is called a (nonrecirculating) *comb filter*.

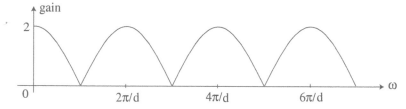

Figure 4.29 Delay Network as a Function of Angular Frequency ω.

There is an interesting application of this delay network in acoustics. If we have a periodic (or nearly periodic) incoming signal with frequency ω radians per sample, we can tune the comb filter such that the peaks in the gain are aligned at even harmonics and the odd ones fall where the gain is zero. Except for a factor of 2, the amplitudes of the remaining harmonics still follow the spectral envelope of the original sound. Thus, we have a tool for raising the pitch of an incoming sound by an octave without changing the spectral envelope. This *octave doubler* is the reverse of the *octave divider* (see Chapter 10).

The time and frequency domains offer complementary ways of looking at the same delay network. When the delays inside the network are smaller than a human ear's ability to resolve events in time (less than 20 milliseconds), the time domain picture becomes less relevant to our understanding of the delay network, and we turn mostly to the frequency-domain picture. However, when the delays are greater than about 50 ms, the peaks and valleys of plots showing gain versus frequency (such as that of Figure 4.29) crowd so closely together that the frequency-domain view becomes less important. Both are nonetheless valid over the entire range of possible delay times.

4.7.2 Classification of Filters. Some of the frequently used filters are described below.

(i) Lowpass and Highpass Filters. Figure 4.30 shows the frequency response of a lowpass filter. The highpass filter is simply a vertical reflection of the lowpass filter. The horizontal axis represents frequency and the vertical axis the gain.

(ii) Bandpass and Stopband Filters. A bandpass filter admits frequencies within a given band and rejects those that are larger or smaller. The frequency response to a bandpass filter is shown in Figure 4.31 in which the horizontal axis is the frequency and the vertical axis the gain. A stopband filter does the reverse, i.e., it rejects the frequencies within the band and lets through those outside it. In the case of Figure 4.31, the stopband filter would have a contiguous stopband surrounded by two passbands.

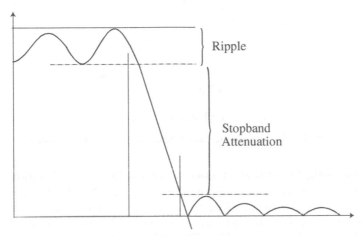

Figure 4.30 Lowpass Filter.

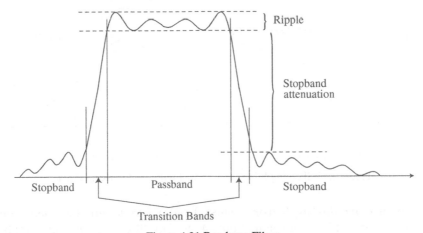

Figure 4.31 Bandpass Filter.

A bandpass filter is often described in a simpler terminology as shown in Figure 4.32, which has only two parameters: *center frequency* and *bandwidth*.

The passband is the region where the filter has at least half the power gain at the peak, i.e., the gain is within 3 dB of its maximum value. The center frequency is the point of maximum gain which is at about the midpoint of the passband. The bandwith is the width (in frequency units) of the passband.

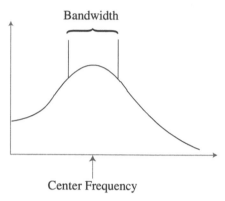

Figure 4.32 Bandpass Filter with Center Frequency and Bandwidth.

(iii) Equalizing Filters. In practice, we do not let signals of certain frequencies pass and stop others. It is always desirable to make adjustments that can be controlled, such as attenuating (boosting) a signal over a frequency range by a desired gain. For this purpose two types of filters are used: a *shelving filter* (Figure 4.33), and a *peaking filter* (Figure 4.34).

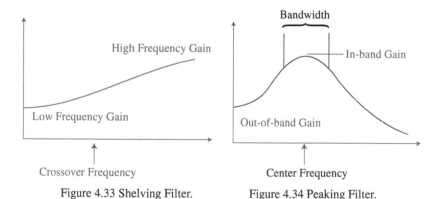

Figure 4.33 Shelving Filter.   Figure 4.34 Peaking Filter.

(iv) Nonrecirculating Filter. (FIRST FORM) The design of an elementary nonrecirculating filter is given in Figure 4.35. This filter has only one complex-valued parameter Q that controls the complex gain of the delayed signal from the original one.

To find the frequency response we feed the delay network a sequence of complex sinusoids $\{1, z, z^2, \dots\}$ with frequency $\omega = \arg\{z\}$. Then the nth sample of the input is z^n and that of the output is $H(z)z^n = \left(1 - Qz^{-1}\right) z^n$, where $H(z) = 1 - Qz^{-1}$ is the transfer function, and Q is a complex number with $|Q| = r$ and $\arg\{Q\} = \alpha$, i.e., $Q = r e^{i\alpha} = r(\cos\alpha + i \sin\alpha)$. The gain of the filter is the distance from the point Q to the point z in the complex plane, since $|z||1 - Qz^{-1}| = |Q - z|$. Graphically, Qz^{-1} is the complex number obtained from Q rotated clockwise by the frequency $\arg\{z\} = \omega$ of the incoming sinusoid. Also, the value $|1 - Qz^{-1}|$ is the distance from Qz^{-1} to 1 in the complex plane, which is the distance from Q to z. This is presented in Figure 4.36.

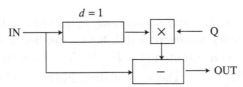

Figure 4.35 First Form of the Elementary Nonrecirculating Filter.

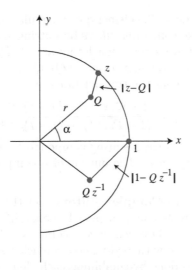

Figure 4.36 Frequency Response.

As the frequency of the input goes from 0 to 2π, the point z traverses the unit circle counter-clockwise. At the point where $\omega = \alpha$, the distance is minimum, equal to $1 - r$. The maximum occurs at the opposite point of the circle.

SECOND FORM: (Figure 4.37.) Sometimes a variant of the elementary nonrecirculating filter is needed. In the second form, instead of multiplying the delay output by Q, the direct signal is multiplied by its complex conjugate

$\overline{Q} = r\, e^{-i\,\alpha}$. Then the transfer function of this type of filter is $H(z) = \overline{Q} - z^{-1}$. Since $\left|\overline{Q} - z^{-1}\right| = \left|Q - \overline{z^{-1}}\right| = |Q - z|$ (as $\bar{z} = z^{-1}$). This filter has the same frequency response as in the first form, but its phase responses are different.

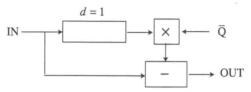

Figure 4.37 Second Form of Elementary Nonrecirculating Filter.

FEEDBACK GAIN: Let P denote a complex-valued *feedback gain* in either form of an elementary nonrecirculating filter. Feeding a sinusoid with nth sample z^n we obtain an output of $\left(1 - Pz^{-1}\right)^{-1} z^n$. Then the transfer function is $H(z) = \left(1 - Pz^{-1}\right)^{-1}$. The recirculating filter is stable when $|P| < 1$; however, if $|P| > 1$, the output grows exponentially as the delayed sample recirculates.

(v) Compound Filters. The elementary recirculating and nonrecirculating filters are used to make *compound filters* by putting several of them in series. If the parameters of the recirculating filters are $P_1, P_2, \ldots$ and those of the nonrecirculating filters of the first form are $Q_1, Q_2, \ldots$, then putting them all in series, in any order, gives the transfer function

$$H(z) = \frac{\left(1 - Q_1 z^{-1}\right) \cdots \left(1 - Q_j z^{-1}\right)}{\left(1 - P_1 z^{-1}\right) \cdots \left(1 - P_k z^{-1}\right)}. \tag{4.52}$$

The frequency response of the resulting compound filter is the product of the those of the elementary ones. The output of a compound filter is complex-valued.

(vi) Real Output from Complex Filters. Let the signals be real-valued, and suppose we need real-valued output. In general, a compound filter with a transfer function given above will give a complex-valued output. However, we can construct a filter with real-valued coefficients which will give real-valued outputs. This is done by combining each elementary compound filter, with coefficients P and Q, with another one with coefficients $\bar{P}$ and $\bar{Q}$. For example, put two nonrecirculating filters with coefficients Q and $\bar{Q}$ in series; this gives a transfer function $H(z) = \left(1 - Qz^{-1}\right) \cdot \left(1 - \bar{Q}z^{-1}\right)$, such that $H(\bar{z}) = \overline{H(z)}$. Then for real-valued sinusoids we get $AH(z)z^n + \overline{AH(z)z^n} = AH(z)z^n + \bar{A}\,\overline{H(z)}\,\overline{z^n}$, which is another real sinusoid. Thus, we always get a real-valued output from two conjugate compound filters. This result can be extended to a finite number of coefficients P_i and Q_i.

We can avoid double computing, one for each pair of recirculating filters if the input, as well as the output, is real. Let the input be a real sinusoid of the form $Az^n + \bar{A}z^{-n}$. Apply a single recirculating filter with coefficient P. Let $a(n)$ denote the real part of the output, which is

$$a(n) = \Re\left\{ \left(1 - Pz^{-1}\right)^{-1} Az^n + (1 - Pz)^{-1}\,\bar{A}\,z^{-n} \right\}$$

$$= \Re\left\{ \frac{1 - \Re\{P\}\,z^{-1}}{\left(1 - Pz^{-1}\right)\left(1 - \bar{P}z^{-1}\right)}\, Az^n + \frac{1 - \Re\{P\}\,z^{-1}}{\left(1 - Pz^{-1}\right)\left(1 - \bar{P}\bar{z}^{-1}\right)}\, \overline{Az^n} \right\}. \tag{4.53}$$

Comparing the input to the output we notice that the effect of passing a real signal through a complex one-pole filter, then taking the real part, is equivalent to passing the signal through a two-pole, one-zero filter with the transfer function

$$H_{\text{real}}(z) = \frac{1 - \Re\{P\}\,z^{-1}}{\left(1 - Pz^{-1}\right)\left(1 - \bar{P}z^{-1}\right)}.$$

If we take the imaginary part, which is a real signal, we obtain the transfer function

$$H_{\text{imag}}(z) = \frac{\Im\{P\}\,z^{-1}}{\left(1 - Pz^{-1}\right)\left(1 - \bar{P}z^{-1}\right)}.$$

Thus, taking either the real or imaginary part of a one-pole filter output gives filters with two conjugate poles. The two parts can be combined to synthesize filters with other possible numerators. That is, with one complex recirculating filter we can synthesize a filter that acts on real signals with two complex-conjugate poles and one real zero. This technique is known as *partial fractions* and it may be repeated for any number of stages as long as we compute an appropriate combination of real and imaginary parts of the output at each stage to form the real input of the next stage.

(vii) Designing Filters. Borrowing from complex analysis, each P_i is marked as '•' (calling it a *pole*), and each Q_i is marked with an 'o' (calling it a *zero*). A plot showing the poles and zeros is called a *pole-zero plot*. The frequency response tends to rise or dip as z is close to a pole or a zero. The effect of a pole or zero becomes more prominent and more localized the closer it is to the unit circle, and it is maximal if z lies on the unit circle. For a stable filter. i.e., the one whose response function goes to zero as $n \to \infty$, poles must lie within the unit circle. On the other hand, zeros may lie anywhere on the complex plane. However, any zero Q_i outside the unit circle may be replaced by one within it, at a point $1/\bar{Q}_i$, so as to give a constant multiple of the same frequency response. Thus, it is desirable that we keep all zeros inside the unit circle.

(viii) Different Types of Filters. As seen above, the total frequency response of a series of elementary recirculating and nonrecirculating filters is the product of all the distances from the point z to each Q_i divided by the product of the distances from z to each P_i.

(ix) One-Pole Lowpass Filter. The one-pole lowpass filter has a single pole at a positive real number p (see Figure 4.38(a))[2]. This is a recirculating comb filter with delay length $d = 1$. The maximum gain occurs at a frequency of zero, corresponding to the point on the circle closest to the point p. There the gain is $1/(1 - p)$. Assuming that p is close to 1, if we move a distance of $1 - p$ units up or down from the real axis, the distance increases by a factor of about $\sqrt{2}$. Thus, we expect the half-power point to occur at an angular frequency of about $1 - p$, provided p is real.

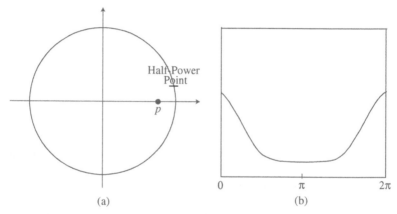

(a)   (b)

Figure 4.38 One-Pole Lowpass Filter.
(a) Pole-Zero Diagram; (b) Frequency Response.

An alternate method is as follows: let the half-power point lie at a given angular frequency ω. Set $p = 1 - \omega$. This approximation is valid if $\omega < \pi/2$, which is done in practice. The one-pole lowpass filter is normalized by multiplying it by the constant factor $1 - p$ so as to obtain a gain of 1 at zero frequency. Note that a nonzero frequency will get a gain of less than 1.

The frequency response is presented in Figure 4.38(b). The audible frequencies occur in the middle part of the graph. The right side of the graph lies above the Nyquist frequency π.

This type of filter is often used to smoothe out noisy signals. For example, if a physical controller is used and only changes of the order of about 0.1 second are needed, this filter with the half-power point of 20 or 30 cycles per second will smooth the noise.

(x) One-Pole One-Zero Highpass Filter. If the signal carries an un-

---

[2] In this and the following figures a pole is marked by a black circle, and a zero by an open circle.

wanted constant offset which is a zero-frequency component, it specifies electrical power that is sent to the speakers. This will reduce the level of loudness. This situation for a constant power signal component is known as DC (direct current). A method to remove this component is to use a one-pole lowpass filter to extract it and then subtract the result from the signal. The resulting transfer function is

$$H(z) = 1 - \frac{1-p}{1-pz^{-1}} = p\frac{1-z^{-1}}{1-pz^{-1}}, \qquad (4.54)$$

where the factor $1-p$ in the numerator of the lowpass transfer function is the normalization factor that yields the gain of 1 at zero frequency. This type of filter is presented in Figure 4.39, where (a) shows the pole-zero plot, and (b) the frequency response which is plotted in terms of the Nyquist frequency π.

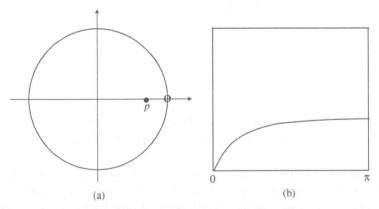

(a)   (b)

Figure 4.39 One-Pole One-Zero Highpass Filter.
(a) Pole-Zero Diagram; (b) Frequency Response.

(xi) Shelving Filter. A generalization of one-zero one-pole filter is as follows: place a zero at a point q on the real axis close to, but less than, 1. The pole at the point p is also situated on the real axis, either left or right of q, such that both p and q lie within the unit circle (see Figure 4.40, where part (a) shows the one-pole one-zero shelving filter, and part (b) the frequency response for the Nyquist frequency.)

At points of the circle far from p and q, the effect of the pole and the zero is almost inverse, assuming that the distances between them are almost the same. Then the filter passes these frequencies almost unaltered. But, in the neighborhood of p and q the filter has a gain larger or smaller than 1 depending on which of p or q is closer to the circle. Thus, this configuration acts like a low-frequency shelving filter. However, if p and q are placed close to -1, instead of 1, the configuration will give a high-frequency shelving filter.

To find the parameters of a shelving filter with a desired transition angular frequency ω and low-frequency gain g, first we choose an average distance d from the pole (as in Figure 4.40(a)) and then from zero to the boundary of the circle. For small values of d, the region of influence is about d radians. So, to get the desired transition frequency, we set $d = \omega$. Then place the pole at $p = (1-d)/g$ and the zero at $q = 1-d$. Thus, the gain at zero frequency is $g = \dfrac{1-d}{p}$, as desired.

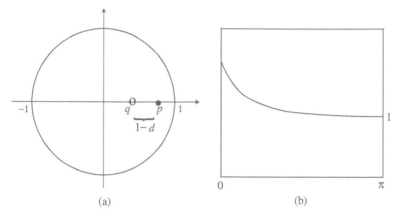

(a)   (b)

Figure 4.40 One-Pole One-Zero Shelving Filter.
(a) Pole-Zero Diagram; (b) Frequency Response.

(xii) **Bandpass Filter.** We will consider the type of filters which have complex-valued poles and zeros, i.e., they do not lie on the real axis. The lowpass, highpass, and shelving filters will then become bandpass, stopband, and peaking filters, respectively. Suppose we want a center frequency ω radians and a bandwidth b. Consider a lowpass filter with cutoff frequency b. Then, for small values of b, its pole is located nearly at $p = 1-b$. Now rotate this value by ω radians in the complex plane, i.e., multiply it by $e^{i\omega} = \cos\omega + i \sin\omega$. The new pole is at $P_1 = (1-b)\,e^{i\omega}$. To obtain a real-valued output this pole is paired with another pole $P_2 = \overline{P_1} = (1-b)\,e^{-i\omega}$. The resulting pole-zero plot is shown in Figure 4.41, in which the peak is almost, but not exactly, at the desired center frequency ω, and the frequency response drops by 3 dB about b radians above and below it. This filter is normalized to get a gain of nearly 1. To do this, we multiply together the inputs (or outputs) of the distances of the two poles to the peak on the circle, i.e., find $b \cdot (b + 2\omega)$. In some cases it is desirable to add a zero at both the points ± 1 so that gain drops to zero at angular frequencies 0 and π.

(xiii) **Peaking and Stopband Filter.** A peaking filter is obtained from a shelving filter by rotating the pole and the zero and by providing a conjugate pole and zero. Let the desired center frequency be ω, and the radii of the pole and zero be p and q, respectively. Then we place the upper pole and zero at

$P_1 = p\,e^{i\,\omega}$ and $Q_1 = q\,e^{i\,\omega}$, respectively (see Figure 4.42).

(xiv) Butterworth Filters. A filter with one real pole and one real zero can be configured as a shelving filter, or as a highpass filter (when the zero is located at the point 1), or as a lowpass filter (when the zero is at -1). The transition regions in the frequency response for these filters are wide. So it is desirable to get a sharper filter, either shelving, lowpass or highpass, whose two bands are flatter and separated by a narrower transition region. A technique from the analog filters transforms real, one-pole, one-zero filters to corresponding *Butterworth Filters*.

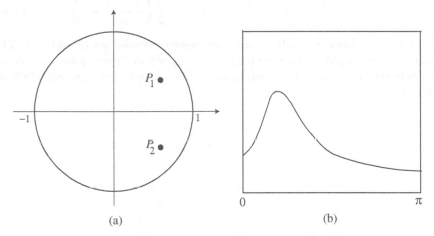

(a)   (b)

Figure 4.41 Two-Pole One-Zero Bandpass Filter.
(a) Pole-Zero Diagram; (b) Frequency Response.

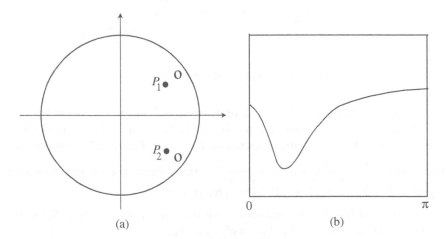

(a)   (b)

Figure 4.42 Peaking Filter.
(a) Pole-Zero Diagram; (b) Frequency Response.

For example, to make a Butterworth filter from a highpass, lowpass, or shelving filter, suppose that either the poles or the zeros are given by the expression $\dfrac{1 - r^2}{(1 + r)^2}$, where $r = 0, 1, \ldots, \infty$ is a parameter, and $r = 0$ corresponds to the point 1 and $r = \infty$ to the point -1. Then replace the pole or zero by n points given by

$$\frac{(1 - r^2) - 2ir\sin\alpha}{1 + 2r\cos\alpha + r^2}, \tag{4.55}$$

where

$$\alpha = \frac{\pi}{2}\left(\frac{1}{n} - 1\right), \frac{\pi}{2}\left(\frac{3}{n} - 1\right), \ldots, \frac{\pi}{2}\left(\frac{2n - 1}{n} - 1\right). \tag{4.56}$$

That is, α assumes n equally spaced arguments between $-\pi/2$ and $\pi/2$. The points are arranged in the complex plane as shown in Figure 4.43: they lie on a circle through the original real-valued point, which cuts the unit circle at right angles.

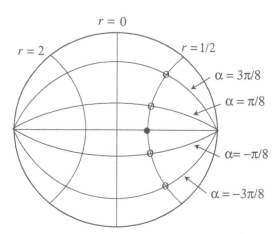

Figure 4.43 A Butterworth Filter with $r = 1/2$.

A good estimate for the cutoff or transition frequency defined by these circular collections of poles or zeros is simply the point where the circle intersects the unit circle, corresponding to $\alpha = \pi/2$. From (4.55) we find that this point is $\dfrac{(1 - r^2) - 2ir}{1 + r^2} = \dfrac{(1 - ir)^2}{1 + r^2}$, which yields the angular frequency equal to $b = 2\arctan(r)$, which is true for $\alpha = \pi/2$.

Figure 4.44 (a) shows a pole-zero plot and frequency response for a Butterworth lowpass filter with three poles and three zeros; part (b) shows the frequency response of the lowpass filter and three other filters obtained by choosing different values of b, and hence, of r, for the zeros, while leaving the poles stationary. As the zeros progress from $b = \pi$ to $b = 0$, the filter that starts as a lowpass filter becomes a shelving filter and then a highpass one.

Figure 4.44 (b) shows the frequency responses for four filters with the same pole configuration, with different locations of zeros, but leaving the poles fixed. The lowpass filter results from setting $b = \pi$ for the zeros, the two shelving filters correspond to $b = 0.3\pi$ and $b = 0.2\pi$, and finally the highpass filter is obtained by setting $b = 0$. The highpass filter is normalized for unit gain at the Nyquist frequency π, and the others for unit gain at DC.

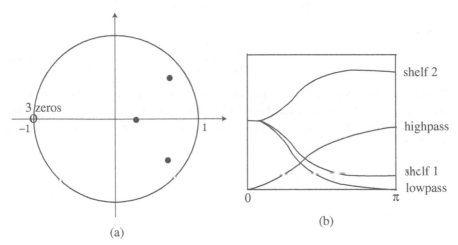

(a)   (b)

Figure 4.44 Butterworth Lowpass Filter with Three Poles and Three Zeros.
(a) Pole-Zero Diagram; (b) Frequency Responses for Four Pole Filters.

(xv) Butterworth Bandpass Filter. First, using the transformation $R(z) = -z^2$ the Butterworth filter can be converted into a high quality bandpass filter with center frequency $\pi/2$. Next, a bilinear transformation of the form

$$w = S(z) = \frac{az + b}{bz + a},\qquad (4.57)$$

with a and b real and not both zero, such that $a^2 + b^2 = 1$, is applied to shift the center frequency to any desired value $\omega \in (0, \pi)$. Since $S(\pm 1) = \pm 1$, the upper and lower halves of the unit circle are transformed symmetrically: if z goes into w, then $\bar{z}$ goes into $\bar{w}$. The effect of the transformation $S(z)$ is to squash points of the unit circle toward 1 or -1. In particular, given a desired center frequency ω, we choose S such that $S\left(e^{i\omega}\right) = i$. If we have $R = -z^2$, and if H is the transfer function for a lowpass Butterworth filter, then the combined filter with the transfer function $(H \circ R \circ S)(z) = H\left(R(S(z))\right)$ will be a bandpass filter with center frequency ω. Solving for a and b we get

$$a = \cos\left(\frac{\pi}{4} - \frac{\omega}{2}\right),\quad b = \sin\left(\frac{\pi}{4} - \frac{\omega}{2}\right).$$

The new function $H\left(R(S(z))\right)$ has $2n$ poles and $2n$ zeros, where n is the degree of the Butterworth filter H. The locations of all the poles and zeros of the new filter are known and they can be computed using elementary filters. If z is a pole of the transfer function $J(z) = H\left(R(S(z))\right)$, that is, if $J(z) = \infty$, then $R\left(S(z)\right)$ must have a pole of H. The same holds for the zeros. To find a pole or zero of J, set $R\left(S(z)\right) = w$, where w is a pole or zero of H, and solve for z, to get

$$w = \left[\frac{az+b}{bz+a}\right]^2, \quad \text{then} \quad \frac{az+b}{bz+a} = \pm\sqrt{w},$$

which gives

$$z = \frac{\pm a\sqrt{-w} - b}{\mp b\sqrt{-w} + a}. \tag{4.58}$$

A pole-zero plot and frequency response of J are shown in Figure 4.45, with center frequency $\pi/4$, and the bandwidth depends on both center frequency and the bandwidth of the original Butterworth lowpass filter used in producing this filter.

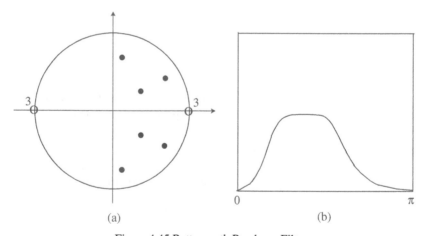

(a)   (b)

Figure 4.45 Butterworth Bandpass Filter.
(a) Pole-Zero Diagram; (b) Frequency Responses for Four Pole Filters.

4.7.3 Properties and Uses. Three important properties of filters are:

(i) INDEPENDENCE. Adding other kinds of filters should not alter the basic filtering process for which the filter was originally designed.

(ii) COMPLETENESS. The filter must capture every last aspect for which it was assembled.

(iii) COMBINABILITY. Different kinds of filters should be combinable.

The application of Fourier transforms starts a time-based signal and applies filters to measure and collect data on each one of the circular properties, namely, frequency, amplitude, and phase. Frequency defines the speed, amplitude the size, and phase angle the starting point of a signal. The measurement data of these three properties is required to apply one or more appropriate filters. This knowledge is comparable to the case of earthquake vibrations: a knowledge of their maximum speed and strength is necessary to design buildings that are capable of withstanding the strongest earthquakes, or to the case of separation of sound waves into bass and treble frequencies to boost the desirable features or avoid undesirable ones, and definitely remove random noise in music and audio communications. In the case of radio waves, filters can be used to listen to or view a particular channel.

5

Communication Systems

As noted in §4.1, a radio signal consists of a radio-frequency sinusoid (sine or cosine wave), called a *carrier wave*, that has undergone amplitude or frequency modulation. In AM the carrier amplitude is varied by the modulating signal, while in FM the carrier frequency is varied by the modulating signal. A bandwidth of as little as 5 kHz provides acceptable audible quality in an AM radio, and that of at least 10 kHz is required when the modulating signal broadcasts commercial music, whereas the range required for high-fidelity music must be wider in the range from 30 Hz (for a large organ pipe) to 15 kHz (for cymbals and trombone sounds).

5.1 Communication Channels

Any information in the form of a message is transmitted over a communication channel that may be regarded as a conduit through which the information is sent. There are different kinds of channels depending on the type of information: the channel is visual if the information is sent through smoke signals or semaphore flags; it is wired if the information is audio, as in telephone or telegraph (Morse code); and it is digital if the information is based on discrete (discontinuous) values. Another important concept is that of noise, which is interference that is always present during communications. Noise can be man-made or due to natural phenomena: the smoke or flag signals may get distorted by the presence of wind; several people speaking at once creates noise; atomic action and reaction in matter (above absolute zero) generates noise. Thus, noise is a fundamental fact of nature and it affects the information media (communication systems and data storage) very significantly.

Shannon's theorem states that if a discrete channel has capacity $\mathfrak{C}$ and a discrete source of entropy H per second, and if $H \leq \mathfrak{C}$, then there exists a coding system such that the output of the source can be transmitted over the channel with an arbitrarily small frequency of errors (Shannon [1948]). In this remarkable result, the word 'discrete' refers to information transmitted in symbols, which can be letters on a teletype, or dots and dashes of Morse code, or digital information. The notion of entropy per second implies that if our information source is transmitting data at a rate less than the communication channel can handle, we can add some extra bits to the data bits and push the error rate to an arbitrarily low level, with certain restrictions. These extra bits are called the *parity-check bits*. In general, the basic structure of a communication system is described in the following diagram:

$$\boxed{\text{Source}} \to \boxed{\text{Transmitter}} \to \boxed{\text{Noisy channel}} \to \boxed{\text{Receiver}} \to \boxed{\text{Sink}}$$

5.2 Noise

Noise is an undesirable and unwanted random signal. The theory of signals and random variables deals with noise encountered during signal transmission and reception. The purpose is to get rid of all noise, mostly by means of suitable filters and sometimes by improved system design. There are various types of noise which are described in the sequel, along with possible methods of their elimination.

There are various sources of audible noise, such as aircraft and jet noise, industrial noise, background noise, spectator sport and parade noise, traffic noise, and many others. It also means transmission and signal processing noise which is either audio or video in radio, television, digital photography, electronics, or atmospheric, and galactic (cosmic) signals. Then there is statistical noise, such as the Gaussian noise, white noise, pink noise, Brownian noise, genome noise, neuronal noise, phonon noise, synaptic noise, and meta noise.

The audio noise is encountered as a residual low level 'hiss or hum', while the video noise is 'snow' on video or television pictures. Some audio equipments generate ground noise at their ground terminals. Electronic noise is related to electronic circuitry. Radio noise is interference related to radio signals. Atmospheric noise is radio noise caused by lightning.

Noise also means cellular noise in biology which is a random variability among cells. Electrochemical noise is due to electrical fluctuations in electrolysis, corrosion, and others. There are statistical noises which are not discussed here. There are many technical measures of noise intensity, such as:

(i) Noise spectral density N_0 per unit of bandwidth measured in Watt/Hertz (equivalent to Watts-second), and used in communications, especially in signal-to-noise ratio calculations. If the noise is the white noise (i.e., with constant frequency), the total noise power is $N = BN_0$, where B is the bandwidth.

(ii) Noise figure is the ratio of the output noise power to attributable thermal noise, with thermal noise density given by $N_0 = kT$, where k is the Boltzmann constant in joules/kelvin, and T is the receiver system temperature in kelvins.

Noise in music is any departure from intention and pattern. It is an uncontrolled sound, or unmusical sound. In general terms, noise is any unwanted sound or signal. Sounds that are usually musical may become noise in a different context if they interfere with the reception of a message desired by the receiver (Attali [1985]). Although noise from tape hiss to bass drum is desirably prevented and removed in music, noise is also used very creatively in many ways, especially in almost all genres. Noise is generally employed by using all the frequencies associated with pitch and timbre, such as the white noise component of a drum roll on a snare drum, or the transients present in the prefix of the sounds of some organ pipes.

In physics and analog electronics, noise is an unwanted random addition to a signal. Signal noise becomes acoustic noise if the signal is converted into sound, and shows as 'snow' on a video image (television). High noise levels can block, distort, change or interfere with the meaning of a message in communication. In signal processing, noise is computed as an undesirable data without meaning. It means that the 'noise data' is not being used to transmit a signal, but is simply produced as an unwanted by-product of other activities. The amount of irrelevant information in a transmission exchange is known as the 'signal-to-noise ratio' (SNR). It is defined as the ratio of signal power to the noise power:

$$\text{SNR} = \frac{P_{\text{signal}}}{P_{\text{noise}}}, \tag{5.1}$$

where P is average power, and both signal and noise powers are measured at the same or equivalent point in the system. In the case when signal and noise are both measured across the same impedance, the SNR can be determined from

$$\text{SNR} = \frac{P_{\text{signal}}}{P_{\text{noise}}} = \left(\frac{A_{\text{signal}}}{A_{\text{noise}}}\right)^2, \tag{5.2}$$

where A is the root mean square (RMS) amplitude. Because of very wide dynamic range of many signals, SNR is often defined, using logarithmic decibel scale, by

$$\text{SNR}_{\text{dB}} = 10 \log_{10} \left(\frac{P_{\text{signal}}}{P_{\text{noise}}}\right) = P_{\text{signal, dB}} - P_{\text{noise, dB}}, \tag{5.3}$$

or, using amplitude ratio, by

$$\text{SNR}_{\text{dB}} = 10 \log_{10} \left(\frac{A_{\text{signal}}}{A_{\text{noise}}}\right)^2 = 20 \log_{10} \left(\frac{A_{\text{signal}}}{A_{\text{noise}}}\right). \tag{5.4}$$

The SNR usually gives an average signal-to-noise ratio, which is used for normalizing the noise level to 1 (0 dB) and then measuring how far the signal 'stands out'.

A ratio greater than $1:1$ means more signal than noise. The SNR, bandwidth, and channel capacity of a communication channel are related to one another by the Shannon-Hartley theorem. It states that the channel capacity $\mathfrak{C}$ (in bits/sec), which is the theoretical upper bound on the information rate (excluding error correcting codes) of data with arbitrary low bit error rate that can be sent with a given average signal power S through an analog communication channel subject to additive white Gaussian noise of power N, is given by

$$\mathfrak{C} = B \ln \left(1 + \frac{S}{N} \right), \tag{5.5}$$

where B is the bandwidth measured in Hz, S the average received signal power over the bandwidth measured in watts (or volts^2), N the average noise or interference power over the bandwidth, measured in watts (or volts^2), and S/N the signal-to-noise ratio (SNR) or carrier-to-noise ratio (CNR) of the communication signal to the Gaussian noise interference expressed as a linear power ratio. In the frequency-dependent (colored noise) case, where the additive noise is not white, or the S/N ratio is not constant with frequency over the bandwidth, the channel capacity is given by

$$\mathfrak{C} = \int_0^B \ln \left(1 + \frac{S(f)}{N(f)} \right) df, \tag{5.6}$$

where $S(f)$ and $N(f)$ are the signal power and noise power spectrum, respectively, and f is the frequency in Hz. This result applies only to Gaussian stationary process noise, and this process is stochastic.

5.3 Quantization

Quantization is the process of mapping a large set of input values to a smaller set, such as rounding values to some unit of precision. A device or algorithm that performs quantization is called a *quantizer*. The error introduced in this process is the round-off error.

The quantization process is used nearly in all digital signal processing since the representation of a signal in digital form involves rounding. It is a nonlinear process because it is a many-to-one mapping as the same output value is shared by multiple input values. The set of input values may be infinitely large, and possibly continuous and, hence, uncountable (such as the set of all real numbers $\mathbb{R}$ within a very limited range), while the set of output values may be finite or countably many.

There are two different classes of application where quantization is used:

(i) Simple rounding approximations, which include analog-to-digital conversion of a signal for digital signal processing, e.g., in using a sound card in a PC to hear an audio signal, and calculations performed in most digital filtering processes. This kind of application is aimed at retaining as much signal fidelity (i.e., accuracy) as possible while retaining the dynamic range of the signal within practical limits that may enter due to arithmetic overflow.

(ii) The so-called 'rate-distortion optimized quantization' is used in coding theory for 'lossy' data compression algorithms. It keeps distortion within limits of the bit rate supported by a communication channel or storage medium, thereby avoiding 'lossy' situations that cause attenuation or dissipation of electrical energy.

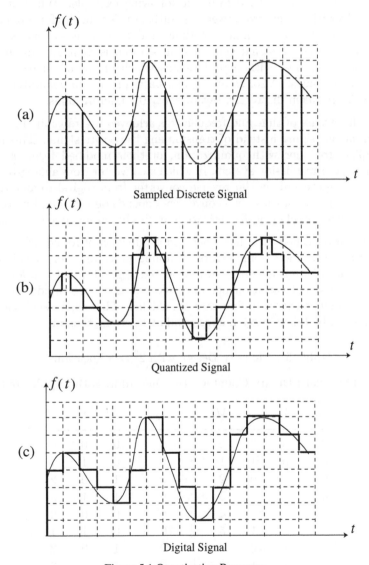

Figure 5.1 Quantization Processes.

A representation of quantization is given in Figure 5.1 where part (a) shows a sampled (discrete) signal with discrete time and continuous values; part (b) shows the quantized signal with continuous time but discrete values, and part (c) shows the digital signal (sampled and quantized) with discrete time and discrete values. Different discretization levels are shown in parts (b) and (c).

For example, if a signal at time $t = 0$ takes the value 0.276, then its quantized value may be taken as 0.3. A distinction between analog and digital signals is easy to understand from the following example. When an airline pilot speaks with the control tower, his (or her) voice (in the form of acoustic waves) is a continuous function of time, and thus, it is an analog signal. Similarly, the electric signal generated in a radio is analog. On the other hand, computers deal with digital signals which are assumed to take only finitely many values. Also, sampling is a basic process in computers, where the signals are taken discretely, say, at every 0.2 microseconds.

An efficient use of time and frequency is achieved by transmitting samples of a continuous signal instead of transmitting the entire signal. This process of sampling together with quantization, or their modified form, is always used in every transmission system, in which a converter from analog-to-digital is always reciprocated by the reverse converter from digital-to-analog. The reliability (fidelity) of these two converters depends on the sampling rate and on an adequate number of discretization levels chosen for quantization.

It is an established fact that whereas introduction of round-off (quantization) always introduces some errors, sampling poses another question as to which sampling rate is adequate to recover the original signal. An answer lies in the rate of signal variation per microsecond: the more rapidly the signal varies, the higher the sampling rate should be. That is, in such cases the samples must be taken every microsecond or fractions thereof.

Table 5.1 Memory Design with $2^3 = 8$ Locations.

(a) Standard Binary Counter   (b) One Address Bit A_2 shorted

A_2	A_1	A_0	M	A_2	A_1	A_0	M
0	0	0	0	0	0	0	0
0	0	1	1	0	0	1	1
0	1	0	2	0	1	0	2
0	1	1	3	0	1	1	3
1	0	0	4	**0**	**0**	**0**	**0**
1	0	1	5	**0**	**0**	**1**	**1**
1	1	0	6	**0**	**1**	**0**	**2**
1	1	1	7	**0**	**1**	**1**	**3**

An alternative procedure is as follows: given the samples, reconstruct the signal. Suppose two signals $x(t)$ and $y(t)$ are reconstructed with a given sample, and both are compatible as signals. Then which is the right one? In fact, there are infinitely many signals that are compatible with a given sample. This phenomenon is known as *aliasing*. In signal processing, aliasing refers to an effect that causes different signals to become indistinguishable (i.e., become aliases of one another) when sampled. For example, the signals $\sin(0.03\,\pi t)$ and $\sin(2.03\,\pi t)$ are aliases of each other. In computing, aliasing relates to a situation in which a data location in memory can be accessed through different symbolic names in the program. Thus, modifying the data through one name implicitly modifies the values associated to all aliased names, which may be unexpected in the program. Aliasing makes it difficult to understand, analyze and optimize programs. For example, C programming language never performs array bounds checking; it does not check whether a variable is within some bounds before its use.

Example 5.1. Let a memory design have 8 locations, thus requiring only 3 address bits since $2^3 = 8$. Three address bits, named $A_2\,A_1\,A_0$, are decoded to select unique memory locations (M), using a standard binary counter, as given in Table 5.1(a), in which each of the 8 unique combinations of address bits selects a different memory location M. Table 5.1(b) shows the effect on these combinations if one address bit, say A_2 is shorted to ground. In this case A_2 will always be zero, the first four memory locations M remain the same but repeat again as second four memory locations 4 through 7, which have now become inaccessible. The same result will happen if any of the other address bits A_1 or A_0 are aliased. ∎

A controlled aliasing, i.e., a specified aliasing behavior, is sometimes desirable. It is common in Fortran. The Perl programming language, in some constructs, specifies it, such as in 'foreach' loops. This helps modify certain data structure with less code.

From an operational point of view, the Shannon theorem, also known as the *Sampling Theorem* of Fourier analysis, answers the above question about selecting the right signal obtained from finitely many samples. It provides the assumption and the formula for the reconstruction of the correct signal. It is assumed that $x(t)$ does not have components beyond a certain signal frequency f_s, that is, its FT is

$$X(\omega) = 0 \quad \text{for } |\omega| \geq 2\pi f_s. \tag{5.7}$$

Then the signal is reconstructed by choosing the sample time $T = \frac{1}{2}\,f_s$, and applying the formula

$$x(t) = \sum_{k=-\infty}^{\infty} X(kT)\,\frac{\sin \pi\left(T^{-1}t - k\right)}{\pi\left(T^{-1}t - k\right)}. \tag{5.8}$$

The sample frequency $f_s = T^{-1}/2$ is known as the *Nyquist frequency*. If we choose another signal $y(t) = x(t) + \sin(2\pi f t)$, so that its Fourier transform

$Y(\omega) = Y(2\pi f) \neq 0$, this will violate the above assumption since $y(t)$ has frequency components beyond f_s.

Example 5.2. Consider a discrete signal processing system for a sinusoid with the sample rate f_s and Nyquist frequency $0.5f_s$. This is presented in Figure 5.2, where the black dot at $0.6f_s$ represents the amplitude and frequency of a sinusoid whose frequency is 60% of the sample rate f_s; the other three big dots are aliases of each other; the solid curve is an example of adjusting amplitude vs. frequency, and the dashed curves are the corresponding paths of the aliases. ∎

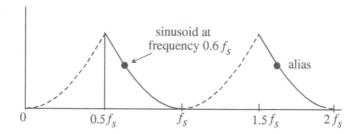

Figure 5.2 Aliases in a Sinusoid.

Note that the sample theorem requires $X(\omega) = 0$ outside some interval large enough to retain all essential information about the signal $x(t)$. To do this, filters are needed so that the filter's output can be sampled, transmitted, and reconstructed at the receiver. The Fourier analysis in this process is presented in Figure 5.3(a) which shows the signal $x(t)$ and its Fourier transform $X(\omega)$, whereas Figure 5.3(b) shows the sampled signal and its Fourier transform.

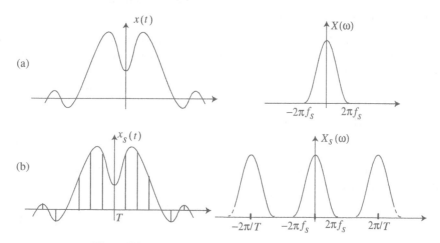

Figure 5.3 Sampled Signal and its Fourier Transform.

5.4 Digital Signals

In the case of digital signals, the number of bits used to represent the measurement determines the maximum possible SNR, since the minimum possible noise level is the error caused by quantization noise, that is by the quantization of the signal. This noise level is nonlinear and signal-dependent, and different computations are required for different types of signals. Although SNR can be used to determine noise levels in a digital system, it is more common to use the ratio E_b/N_0 which is energy per bit per noise power spectral density. The modulated error ratio (MER) measures the SNR in a digitally modulated signal. A fixed-point arithmetic is used to determine the dynamic range (DR) between quantization levels for n-bit integer. Assuming that the distribution of input signal values is uniform, the quantization noise is a uniformly distributed random signal with a peak amplitude of one quantization level, so that the amplitude ratio is $2^n : 1$. Then the dynamic range DR is given by

$$DR_{dB} = SNR_{dB} = 20 \log_{10} (2^n) \approx 6.02\, n. \qquad (5.9)$$

In technological terminology, this formula reads: '16-bit audio has a DR of 96 dB. Each additional quantization bit increases DR by about 6 dB.'

Let the quantization noise in a full-scale sinusoid approximate a sawtooth wave (see Figure 1.4(d)) with peak amplitude of one quantization level and uniform distribution. Then, in this case the SNR is given by

$$SNR_{dB} \approx 20 \log_{10} \left(2^n \sqrt{3/2} \right) \approx 6.02\, n + 1.761. \qquad (5.10)$$

On the other hand, the floating-point numbers provide a method to trade off SNR for an increase in DR. For n bit floating-point numbers, having $(n - m)$ bits with m bits in the exponent, we have

$$DR_{dB} = 6.02\,(2^m), \quad SNR_{dB} = 6.02\,(n - m). \qquad (5.11)$$

Note that the DR is much larger in this case than the fixed-point case, but SNR gets worse. Thus, the floating-point is preferred in situations where the DR is large or unpredictable. The fixed-point has simpler implementation and is preferably used in cases where $DR < 6.02\, m$.

Example 5.3. Let the channel capacity $\mathfrak{C}$ be measured in bits per channel use (modulated symbol). Consider an additive white Gaussian noise (AWGN) channel with one-dimensional input $y = x + n$, where x is a signal with average energy variance E_s and n is Gaussian with variance $N_0/2$. Then the capacity of AWGN with unconstrained input is

$$\mathfrak{C} = \frac{1}{2} \ln \left(1 + \frac{2E_s}{N_0} \right) = \frac{1}{2} \ln \left(1 + \frac{2RE_b}{N_0} \right),$$

where E_b is the energy per information bit, and R the transmission rate. This capacity is achieved by a Gaussian input x, and is shown in Figure 5.4. However, this is not a practical modulation. An antipodal BPSK modulation (see §E.4, Appendix E) yields a better result (see Kythe and Kythe [2012: 46–48]).

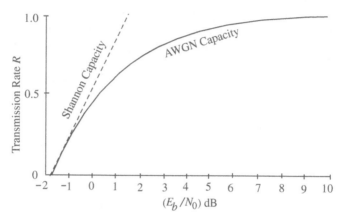

Figure 5.4 Additive White Gaussian Noise (AWGN) Channel.

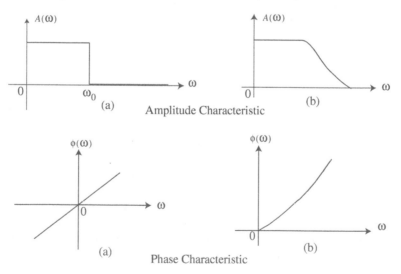

Figure 5.5 Amplitude and Phase Characteristics.

5.4.1 Amplitude and Phase Characteristic. In the case of a lowpass filter, let $x_i(t)$ and $x_0(t)$ denote the input and output functions, respectively, such that $x_i(t) \rightleftharpoons X_i(\omega)$ and $x_0(t) \rightleftharpoons X_0(\omega)$. These two Fourier transforms are related by $X_0(\omega) = e^{-i\omega t_0} A(\omega) X_i(\omega)$, where $A(\omega)$ is the amplitude characteristic function, also called the *rectangle function*, defined on the interval

$[0, \omega_0]$ such that $A(\omega) = \begin{cases} 1 & \text{for } \omega \in [0, \omega_0] \\ 0 & \text{for } \omega \notin [0, \omega_0] \end{cases}$. Similarly, the phase character-

istic $\phi(\omega)$ in a filter is assumed to be a linear function defined by $\phi(\omega) = t_0\omega$ (see Figure 5.5, where part (a) is the mathematical representation, and part (b) the actual one). Thus, in view of the translation property of Fourier transforms (see §3.1.2 (3)) the filtered signal will show a delay with respect to the original signal.

5.5 Interference

Interference is a physical phenomenon in which two waves superpose to form a resultant wave of lower or greater amplitude. It is an interaction of waves which are correlated or coherent with each other, either because they come from the same source or because they have the same or nearly the same frequency. All types of waves, such as light, radio, acoustic, and water waves, experience interference.

The principle of superposition of waves states that when two or more waves are incident upon the same point, the total displacement at that point is equal to the vector sum of the displacements of the individual waves. Mathematically, this means that if an input produces response x_1 and another input the response x_2, and so on, then the combined input will produce the response $x_1 + x_2 + \cdots$. A linear function satisfies the superposition principle (additivity), i.e., $F(x_1 + x_2 + \cdots) = F(x_1) + F(x_2) + \cdots$. The other property that a linear function satisfies is that of scalar multiplication (homogeneity of degree 1), defined by $F(\lambda x) = \lambda F(x)$, where λ is a scalar.

Linear systems are easy to analyze mathematically, using frequency domain linear transform methods such as Fourier and Laplace transforms. However, physical systems are in reality only approximately linear, and the superposition principle is only an approximation of the physical behavior. As an example, consider Fourier analysis, in which the stimulus is written as the superposition of many sinusoids. According to this principle, each of these sinusoids can be analyzed separately and the individual response (which is itself a sinusoid, with the same frequency as the stimulus but generally different amplitude and phase) can be computed. Then the response of the original stimulus is the sum (or integral) of all the individual sinusoidal responses.

Fourier analysis is particularly used for monochromatic waves. Such waves have a single frequency, which means that they propagate continuously in time. For example, in electromagnetic theory, ordinary light is described as a superposition of plane waves which have fixed frequency, polarization, and direction. The superposition principle holds only in the case of linear waves, not in the nonlinear cases.

Waves are usually described by variations in some parameter (known as the amplitude) through space and time, e.g., electromagnetic field in light waves, pressure in sound waves, or height in water waves. Thus, a wave is a function of the amplitude at each point. In any system with waves, the waveform at a given time is a function of the sources, which are external forces, if any, that affect the wave, and initial conditions of the system. In most cases the equation describing the wave is linear. If that is the case, the superposition principle can be applied, which means that the net amplitude caused by two or more waves traversing the same space is the sum of the amplitudes produced by individual waves separately. Thus, for example, when two waves travel towards each other in the same medium, they pass right through each other without any distortion on the other side. Similarly, two waves traveling in opposite directions in the same medium will combine linearly, and if they have the same wavelength, they will result in a standing wave. Thus, the *interference* between waves is based on the superposition principle. When two or more waves traverse the same space, the net amplitude at each point is the sum of the amplitudes of individual waves. Some interesting practical cases are as follows: (i) Noise-canceling headphones, in which the summed variation has a smaller amplitude than the component variations (this is called *destructive interference*), and (ii) line array, in which the summed variation has a bigger amplitude than any of the individual components (this is called *constructive interference*).

Since the frequency of light waves ($\sim 10^{14}$ Hz) is so high that it cannot be detected by present-day detectors, one can possibly observe only the intensity of an optical interference pattern. Since the intensity of the light at a given point is proportional to the square of the mean amplitude of the wave, the displacement $U(\mathbf{r})$ of two waves at a point $\mathbf{r}$ can be expressed as

$$U_{1,2}(\mathbf{r},t) = A_{1,2}(\mathbf{r}) \, e^{i \, \{\phi_{1,2}(\mathbf{r}) - \omega t\}}, \qquad (5.12)$$

where A is the magnitude of the displacement, ϕ the phase, and ω the radian frequency. Then the displacement of the summed waves is

$$U(\mathbf{r},t) = U_1(\mathbf{r},t) + U_2(\mathbf{r},t) = A_1(\mathbf{r}) \, e^{i \, \{\phi_1(\mathbf{r}) - \omega t\}} + A_2(\mathbf{r}) \, e^{i \, \{\phi_2(\mathbf{r}) - \omega t\}}, \quad (5.13)$$

and the intensity $I(\mathbf{r})$ of the light at $\mathbf{r}$ is given by

$$I(\mathbf{r}) = \int U(\mathbf{r},t) \, \overline{U(\mathbf{r},t)} \, dt \propto I_1(\mathbf{r}) + I_2(\mathbf{r}) + 2\sqrt{I_1(\mathbf{r}) \, I_2(\mathbf{r})} \, \cos\left(\phi_1(\mathbf{r}) - \phi_2(\mathbf{r})\right),$$
$$(5.14)$$

where $I_1 \propto A_1^2$ and $I_2 \propto A_2^2$. Hence, the interference maps the difference in phase between the two waves, where the maxima occur when the phase difference is a multiple of 2π. If the two light beams have the same intensity, i.e., if $I_1(\mathbf{r}) = I_2(\mathbf{r})$, then the maxima are four times as bright as each beam, and minima have zero intensity. To obtain interference, the two waves must have the same polarization, i.e., they oscillate with more than one orientation (e.g., the electromagnetic waves such as light, and gravitational waves, but not sound waves in gas), since waves of different polarization cannot cancel one another out or be added together.

5.5.1 Light Sources. The assumption that the light waves are monochromatic waves is, however, neither necessary nor practical. Two identical waves of finite duration with a fixed frequency over that duration produce a pattern of interference while they overlap. Similarly, two identical waves which consist of a narrow spectrum of frequency waves of finite duration will exhibit a series of fringe patterns of slightly different spacing, and again show a fringe pattern during the time they overlap, provided the spacing is much less than the average fringe spacing.

The conventionally used light sources emit waves of different frequencies and at different times from different source points. If the light is split into two waves and then recombined, each individual light wave may generate interference with its other wave pattern, but the individually generated fringe patterns will have different phases and spacing, and no overall fringe pattern will be observed. However, single-element light sources, such as mercury-vapor lamps have emission lines with very low frequency. When they are filtered (in space or in color) and then split into two waves, they can be superimposed to generate interference fringes (see Steel [1986]).

5.5.2 Laser Beam. A laser beam is generally a very close approximation to a monochromatic light source. It is a very straightforward technique to generate interference fringes. However, the ease with which such fringes can be observed sometimes causes problems when stray reflections are generated, creating spurious interference fringes that result in errors. A single laser beam is normally used in interferometry even in spite of the cases of occurrence of interference when two independent lasers with almost similar frequencies are used (Pfleegor and Mandel [1967]).

Interference fringes are generated when light from the source is divided into two waves and then re-combined by interferometers which are either amplitude-division or wavefront-division systems. In the former system a beam splitter is used to divide the light into two beams traveling in different directions which are then superposed to generate the interference pattern, whereas in the latter system the wave is divided in space. Interference is often observed in daily life. An example is the colors seen in a soap bubble, which arise due to interference of light reflecting off the thin front and back soap film.

5.5.3 Astronomical Interferometry. This subject developed in 1946 when astronomical radio interferometers, consisting of either arrays of parabolic dishes or two-dimensional arrays of omni-directional antennas, were constructed. All telescopes used in the array are connected together using coaxial cable, waveguide, optical filter, or other type of transmission line. A technique called *aperture synthesis* increases the resolution and eventually the total signal collected. This technique works by superposing (i.e., interfering) the signal waves from different telescopes using the fact that waves that coincide with the same phase add to each other while those with opposite phase cancel each other out. This creates a combined telescope that has the same resolution as a single antenna whose diameter is equal to the spacing of the antennas farthest apart in the array.

5.5.4 Acoustic Interferometry. An acoustic interferometer is an instrument for measuring the physical characteristics of sound waves in a fluid medium. It measures velocity, wavelength, absorption, or impedance. A vibrating crystal creates ultrasonic waves that are radiated in the medium. These waves strike a reflector placed parallel to the crystal. The reflected waves are then used to determine the measurements.

5.5.5 Quantum Interference. Let a system be in state ψ. Then its wave function is defined in the Dirac or bra-ket notation[1] by

$$\langle \phi | \psi \rangle, \tag{5.15}$$

where the left part $\langle \phi |$ is called the **bra**, and the right part $| \psi \rangle$ the **ket.** The expression (5.15) defines the probability amplitude for the state ψ to collapse into the state ϕ, where the collapse of a state implies that a wave function which is initially a superposition of several different possible eigenstates reduces to a single one after interaction with an observer, like an 'observer effect' which refers to the potential impact of observing a process output while the process is running. Let the quantum state of a physical system be described by a wave function as a vector

$$|\psi\rangle = \sum_i c_i |\phi_i\rangle. \tag{5.16}$$

The right side (kets) of Eq (5.16), i.e., $|\phi_1\rangle, |\phi_2\rangle, \ldots$, provides the list of different quantum 'alternatives' that are available; they form an orthonormal eigenvector basis such that $\langle \phi_i | \phi_j \rangle = \delta_{ij}$, where δ_{ij} is the Kronecker delta. The coefficients c_i are the (complex) probability amplitudes corresponding to each basis $|\phi_1\rangle, |\phi_2\rangle, \ldots$, such that $|c_i|^2 = c_i \overline{c_i}$ defines the probability of measuring the system to be in the state $|\phi_i\rangle$. If all wave functions are assumed to be normalized, then the total probability of measuring all possible states is unity, i.e., $\langle \psi | \psi \rangle = \sum_i |c_i^2| = 1$.

---

[1] This is a standard notation in quantum mechanics, introduced by Paul Dirac [1939], and is so called because the dot product (inner product on a complex vector space) of two states is denoted by $\langle$ **bra**|c|**ket** $\rangle$. It is also known as the *Dirac notation*.

5.6 Nonlinear Systems

Most real-life problems are definitely not linear, and neither are their governing equations. In nonlinear cases, the superposition principle does not hold, but may be made to hold approximately by making the amplitude of the wave smaller. For example, nonlinear optics analyzes the behavior of light waves through nonlinear media. Such media are produced by the dielectric polarization that responds nonlinearly to the electric field of the light. This nonlinearity is typically observed at intensities of the order 10^8 V/m such as electric fields provided by pulsed lasers. Over the Schwinger limit[2] the vacuum becomes nonlinear.

In a vacuum, the classical Maxwell's equations remain linear differential equations, and the superposition principle is applicable. These equations imply that the distribution of charges and currents are related to the electric and magnetic fields by a linear transformation. Details about Maxwell's electromagnetic equations and the sine-Gordon equation are given in Appendix D.

In music the theorist Schillinger [2005] defines rhythm as a very regularly recurring motion with symmetry and periodic movement. This movement is marked by a regulated sequence of strong and weak sinusoids, or of opposite or different conditions that have periodicity. This is a form of the so-called superposition principle that is used as a basis in his theory of rhythm.

5.7 SDR System

The software-defined radio (SDR) is a radio communication system, where the commonly used hardware components, such as filters, amplifiers, mixers, modulators/demodulators, detectors and others, are replaced by software on a PC or an embedded system that is controlled by a larger real-time computing system, with a sound card or analog-to-digital converter. The general processor in such a system performs signal processing rather than using a special hardware. This system then behaves like a radio that receives and transmits different radio protocols, which are mathematically waveforms, depending on software used. Software radios and cell phones are used in the military and civilian life and are showing a bright future in radio communications. SDRs with software defined antennas are the precursors of the cognitive radio. Such antennas 'lock' onto a directional signal. This feature enables the receivers to block or cut off interference from other directions so that even the fainter transmissions can be detected easily. In the cognitive radio system, each radio

---

[2] In quantum mechanics, the Schwinger limit is a scale above which the electromagnetic field becomes nonlinear. This limit is the maximum value of the electric field before nonlinearity for the vacuum: $E_s = \dfrac{m_e^2 c^3}{q_e \hbar} \approx 1.3 \times 10^8 V/m$, where m_e is the mass of the electron, c the speed of light in the vacuum, q_e the elementary charge, and $\hbar$ the reduced Plank constant.

measures the spectrum in use and transmits that information to other corresponding radios. This helps the transmitters avoid mutual interference by selecting unused frequencies.

A wireless mesh network increases its total capacity and reduces the power required at any one of the nodes, such that each node only transmits loudly enough for the message to jump over to the nearest node in that direction, thus reducing the near-far problem, or audibility problem[3], that is common in wireless communication.

The ideal receiver scheme is to attach an analog-to-digital converter to an antenna. Then the digital signal processor reads the converter, and its software transforms the data from the converter to any other form as needed. However, current limits on technology do not let us realize this ideal scheme. The main problem is the difficulty in converting between the digital and analog domains at a desirable high rate and high accuracy simultaneously, without reducing interference and electromagnetic resonance. Most receivers use a variable-frequency oscillator, mixer, and filter to tune a signal to baseband (i.e., intermediate frequency), where it is then sampled by the analog-to-digital converter. On the other hand, some applications bypass the baseband tuning and let the radio frequency signal be directly sampled by the analog-to-digital converter, after amplification.

Since analog-to-digital converters lack the dynamic range to pick up submicrovolt nanowatt-power radio signals, a low-noise amplifier precedes the conversion step. However, such a device suffers from the following drawback: If spurious signals are present, as it typically happens, they compete with the desired signals within the amplifier's dynamic range, and thus distort or completely block them. The solution usually is to put bandpass filters between the antenna and the amplifier, but they tend to reduce the radio's flexibility. These situations describe the drawbacks in the SDR system.

A solution to the drawbacks in the SDR system is the Joint Tactical Radio System (JTRS), which started as a U.S. military program. This system works at radio terminals that include hand-held, vehicular, airborne and dismounted radios, as well as base-stations (fixed or maritime). It is the SDR system based on an internally endorsed open software communication architecture (SCA), which uses CORBA (Common Object Request Broker Architecture) on POSIX (Portable Operating System Interface)[4] operating systems to coordinate various software modules.

The SDR software performs the following tasks: All demodulation, fil-

---

[3] This is a condition when a receiver captures a strong signal, thereby making it difficult for the receiver to detect a weaker signal. In some signal jamming techniques, the near-far problem is used to disrupt communication.

[4] Not to be confused with Unix, Linux, or Unix-like operating systems. POSIX defines the application programming interface (API) together with command line shells and utility interfaces, for software compatibility with variants of Unix and other operating systems.

tering of both radio frequency and audio frequency, and signal enhancement through equalization and binaural presentation. A most recent SDR project is the GNU Radio[5] with the Universal Software Radio Peripheral (USRP) that uses a USB 2.0 interface, an FPGA, and a high-speed set of analog-to-digital and digital-to-analog converters combined with reconfigurable free software. The sampling and synthesis of this system is a thousand times that of PC sound cards, which provides wideband operation.

The High-Performance Software Defined Radio (HPSDR) which is compatible to that of a conventional analogue HF radio, uses a 16-bit 135 MSPS analog-to-digital converter to obtain performance over the range 0 to 55 MHz. The receiver operates in the VHF and UHF ranges also, using either mixer image or alias responses.

The WebSDR system provides access via browser to multiple SDR receivers worldwide. It covers the complete software spectrum. It has recently analyzed Chirp Transmitter signals using the coupled system of receivers. A Chirp transmitter[6] is a short wave radio transmitter that sweeps the HF radio spectrum on a regular schedule.

5.8 Data Transmission

In communication systems the accuracy of data transmissions is very important. The technique used for this purpose is known as *checksum*, which is a function that computes an integer value from a string of bytes. Its purpose is to detect errors in transmission. There are different checksum functions with different speed or robustness. The most robust checksum functions are cryptographic hash functions that are typically very large (16 bytes or more). MD5 hash functions are typically used to validate Internet downloads.

The cyclic redundancy checks (CRC) are moderately robust functions and are easy to implement. They are described below. The checksum CRC32C (Castagnoli et al. [1993]) is extensively used in communications, and it is implemented on hardware in Intel CPUs as part of the SSE4.2 extensions. This checksum is easy to use because it uses certain built-in GNU functions with user-friendly instructions.

The CRC check is a type of function with input as a data of any length, and output as a value of certain space. A CRC can be used as a checksum to detect accidental alteration of data during transmission or storage. It is useful because (i) it is simple to implement in binary hardware; (ii) it is easy

---

[5] The applications for this radio are written using the Python, while the supplied performance-critical signal processing path is implemented in C++ using processor floating-point extensions, where available. For more information, see gnuradio.org.

[6] The name comes from the following fact while a specific frequency is being monitored: a chirp is heard (in CW or SSB mode) when the signal passes through. Such transmitters are used for over horizon radar systems and for probing ionospheric conditions.

to analyze mathematically; and (iii) it detects very efficiently common errors caused by noise. The computation method for a CRC resembles a long division operation, in which the quotient is discarded and the remainder becomes the result. The length of the remainder is always less than or equal to that of the divisor.

In deep space communications, Reed-Solomon (RS) codes have been used in several of NASA and ESA's planetary explorations. Initially it was realized that RS codes would not be suitable for deep space because deep space communication does not induce burst errors during transmission. However, once it was found that a combination of binary codes and RS codes can be used in a concatenated form, where the binary code is used as the 'inner code' while the RS code is used as an error correcting code, such concatenated codes become very powerful. A concatenated conventional binary code with an RS code was used in the Voyager expeditions to Uranus and Neptune, and the close-up photographs transmitted from those distant planets were received with the expected resolution. The Galileo mission to Jupiter is another example of the use of RS codes to successfully solve the erasures by a concatenated error correcting code.

In feedback systems, mobile data transmission and highly reliable military communication systems use RS codes with simultaneous error detection and correction. These systems distinguish between a decoder error and a decoder failure.

The spread-spectrum systems are of two types: frequency-hopping spread system (FH/SS) and direct-sequence spread system (DS/SS). The former system modulates information on to carrier that is systematically moved from frequency to frequency. This system is used in military communications, aimed at defeating partial-band jamming, and thus, providing protection against any partial-band disturbance. The latter system (DS/SS) creates a wideband signal by phase-shift-keying in RF carrier with the sum of the data sequence and a spreading sequence with pulse rate much higher than that of the data sequence. Combined with FH/SS system the DS/SS system provides excellent protection against partial-band disturbance, and therefore, this system is targeted for use in mobile radio applications. The RS codes are being used in the design of hopping sequences.

5.9 Space Exploration

Before we start with the developments achieved in space exploration, it is desirable to understand the role of some of the important related aspects. It will be assumed that the spacecraft design and construction is robust and able to withstand pressure. We shall also assume that there are smart devices on board to send and receive error-free signals (encoder and decoder). This involves reliable error-correcting codes (ECC). Finally, the physical problem of combustion and its solution must be addressed.

5.9.1 Error-Correcting Codes. The developments in error-correcting codes (ECC) are responsible in successful space exploration. During the early 1960s when the space race began as a result of the Sputnik crisis, the computers on-board unmanned spacecraft corrected any errors that occurred, and later the Hamming and Golay codes were used for better transmission of data, especially during the 1969 Moon landing. The Golay codes were then successfully used in the Jupiter fly-by of Voyager 1 and 2, launched by NASA Deep Space Missions during the summer of 1977 and the Saturn mission in 1979, 1980, and 1981 fly-bys from Cape Canaveral, Florida, to conduct closeup studies of Jupiter and Saturn, Saturn's Rings, and the larger moons of the two planets. These missions transmitted hundreds of color pictures of Jupiter and Saturn within a constrained telecommunication bandwidth. The Golay (24, 12, 8) code was used to transmit color images that required three times the amount of data. Although it was only a three-error correcting code, it could transmit data at a much faster rate than the Hadamard code that was used during the Mariner Jupiter-Saturn missions.

The new NASA standards for automatic link establishment (ALE) in high-frequency (HF) radio systems specified the use of the extended Golay code $\mathcal{G}_{24}$ for forward error correcting (FEC). Since this code is a (24,12) block code encoding 12 data bits to produce 24-bit codewords, it was systematic, i.e., the 12 data bits remained unchanged in the codeword, and the minimum Hamming distance between two codewords was eight, i.e., the number of bits by which any pair of codewords differed was eight.

Subsequent NASA missions used the Reed-Solomon (RS) codes together with the above codes for data transmission from the SkyLab and International Space Station and all other space missions to Mars (in 2001) and the Hubble Space Telescope. These codes provided error-free and high-speed data transmission with low overhead on the Internet and data storage. The LFSR encoder is an integral part of hardware used with this code. The encoding and decoding architecture is carried out in software or in special-purpose hardware.

The RS codes are based on the Galois fields arithmetic, and the RS encoder or decoder needs to carry out the arithmetic operations on finite field elements. These operations require special hardware or software functions to implement. The RS codeword is generated using a special polynomial. All valid codewords are exactly divisible by the generator polynomial of the general form of $\mathbf{g}(x) = g_0 + g_1 x + g_2 x^2 + \cdots + g_{2t-1} x^{2t-1} + x^{2t}$, and the codeword $\mathbf{c}(x)$ is constructed. For example, the generator polynomial for RS$(255, 223)$ code is $\mathbf{g}(x) = (x - \alpha^0)(x - \alpha^1)(x - \alpha^2)(x - \alpha^3)(x - \alpha^4)(x - \alpha^5)$. A general architecture for the RS decoder is shown in Figure 5.6. A detailed theory and implementation of

RS codes is available in Wicker and Bhargava [1999].

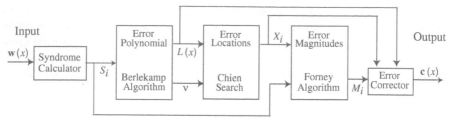

Figure 5.6 RS Decoder.

Here $\mathbf{w}(x)$: transmitted codeword; S_i: syndrome; $L(x)$: error locator polynomial; X_i: error locations; M_i: error magnitudes; $\mathbf{c}(x)$: transmitted codeword; and ν: number of errors. The received word $\mathbf{w}(x)$ is the original codeword $\mathbf{c}(x)$ plus errors. Thus, the RS decoder attempts to identify the location and magnitude of up to t errors (or $2t$ erasures) and correct the errors or erasures.

The success of error-correcting codes in these space missions led to the development of a new class of linear codes invented by Gallager [1962, 1963] and known as the low-density parity-check (LDPC) codes. Two types of communication channels, namely, the binary erasure channel (BEC) and the binary symmetric channel (BSC) were introduced, and the hard-decision and soft-decision algorithms presented as offshoots of the belief propagation for decoding of LDPC codes.

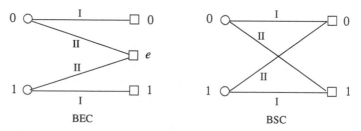

Figure 5.7 BEC and BSC.

where probability $1 - \mathfrak{p}$ is on edges I, and $\mathfrak{p}$ on edges II.

There are two types of models of communication channels used in coding theory and information theory: (i) Binary Erasure Channel (BEC), and (ii) Binary Symmetric Channel (BSC). They are presented graphically in Figure 5.7. In both channels the input alphabet is binary $\{0, 1\}$ and its elements are called *bits*. However, the output alphabet for the BEC is $\{0, 1, e\}$, where e is called the *erasure*, while for the BSC it is $\{0, 1\}$. The capacity of the BEC is $\mathfrak{C} = 1 - \mathfrak{p}$, where $\mathfrak{p}$ is the probability of error. This characterization suggests that on average the rate of successful transmission is $1 - \mathfrak{p}$, which is, therefore, an upper bound for this transmission channel. On the other hand, the capacity of the BSC channel is $1 - \mathcal{H}(\mathfrak{p})$, where $\mathcal{H}(\mathfrak{p})$ is the binary entropy function[7]. It appears at first sight that the BSC is simpler than BEC, but this is not true because the capacity of BSC is $1 + \mathfrak{p} \log_2 \mathfrak{p} + (1 - \mathfrak{p}) \log_2 (1 - \mathfrak{p})$.

The next aspect of development in ECC involved Erasure codes, Tornado codes, and Fountain codes. These codes, and especially the Tornado codes and the concept of a pre-code, provided the groundwork for the development of modern coding theory. A Fountain code, which is a rateless erasure code, solves the problem of transmitting a fixed set of data to multiple users over unreliable links. The encoder is a Fountain, a metaphor that provides an endless supply of water drops (encoded packets). To build a Fountain code, one uses Tornado erasure codes because they are fast, although they are sub-optimal, and reconstruction requires $(1 + \varepsilon)k$ packets. The server distributes a file containing n blocks of length m; it then encodes blocks into packets and distributes them into 2^n different packets, although infinitely many such packets can be generated. The number of blocks encoded into one packet is the *degree* of the packet. In the ideal case, the decoder needs to collect any n packets in order to decode the original content. This situation is metaphorically analogous to getting water drops from a (digital) water fountain. No retransmission of a specific packet is needed in the ideal case.

A Fountain code is *optimal* if the original k source symbols can be recovered from any k encoding symbols. There are Fountain codes known to recover the original k source symbols from any k' of the encoding symbols with high probability, where k' is just slightly larger than k. For any smaller k, say, less than 3,000, the overhead γ is defined by $k' = (1 + \gamma k)$, which amounts to a very insignificant value.

In practical communications, the data is transmitted in blocks. Suppose, a binary message of length n is transmitted in K blocks of m bits each; thus, $n = Km$. The transmitter continuously transmits blocks of m bits that are obtained by XOR-ing (i.e., by adding mod 2) subsets of the blocks, while the receiver generally collects a little bit more than the K blocks to decode and retrieve the original message.

The universal Fountain code family is composed of two classes: one consisting of the Tornado codes and the Fountain code, and the other consisting of the Luby transform (LT) codes and the Raptor codes (see Luby [2002]). The encoding and decoding of these modern codes and their performance are based on a strategy which involves introduction of a pre-code that detects and corrects erasures, by using an LDPC or a RS code, and then using the BP decoder for decoding the entire message at a rate close to Shannon's capacity $\mathfrak{C}$ with a very high probability of success.

---

[7] The binary entropy function is defined for a binary random variable 0 with probability $\mathfrak{p}$ and 1 with probability $1 - \mathfrak{p}$ as the function $\mathcal{H}(\mathfrak{p}) = -\mathfrak{p} \log \mathfrak{p} - (1 - \mathfrak{p}) \log(1 - \mathfrak{p})$. Note that $\mathcal{H}(0) = 0 = \mathcal{H}(1)$, with maximum at $\mathfrak{p} = \frac{1}{2}$.

The second class of Fountain codes is the Raptor code ('raptor' derived from *rapid tornado*), which was necessitated due to the fact that in many applications a universal Fountain code must be constructed that has a constant average weight of an output symbol and operates on a fast decoding algorithm. The basic idea behind Raptor codes is a pre-coding of the input symbols prior to the application of an appropriate LT code. A class of universal Raptor codes has been designed with linear time encoder and decoder for which the probability of decoding failure converges to zero polynomially fast as the number of input symbols increases. With proper bounds on the error probability, a finite length Raptor code can be constructed with low error probability. The decoding error probability of a Raptor code is estimated by the above finite-length analysis of the corresponding LT code and of the pre-code $\mathcal{C}$ using any code $\mathcal{C}$ for which the decoding error probability is already known. For example, $\mathcal{C}$ can be chosen from an ensemble of graphs $\mathbf{P}(\Lambda(x), n, r)$ defined in Kythe and Kythe [2012: 420]. Let $\mathfrak{p}_l^{\mathrm{LT}}$ denote the probability that the LT decoder fails after recovering l intermediate symbols, and let $\mathfrak{p}_j^{\mathcal{C}}$ denote the probability that the pre-code $\mathcal{C}$ fails to decode j randomly chosen erasures. Then the probability that the k input symbols cannot be recovered from the $k(1+\varepsilon)$ output symbols is $\sum_{l=0}^{n} \mathfrak{p}_l^{\mathrm{LT}} \mathfrak{p}_{n-j}^{\mathcal{C}}$. A presentation of different versions of Raptor codes is given in Figure 5.8, in which the pre-coding is accomplished in multiple stages: the first stage is Hamming, Golay, or RS coding, and the second a suitable LDPC coding. The choice of intermediate codes is always dependent on the stochastic process and achievement of desired error probabilities associated with different codes. Shokrollahi [2006] suggests irregular LDPC code as a suitable choice. A detailed account of these and other codes is available in Kythe and Kythe [2012].

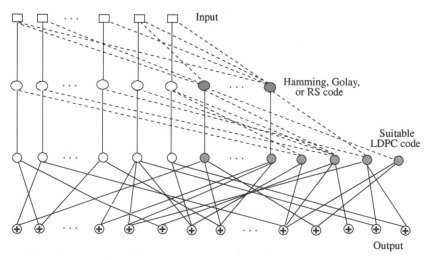

Figure 5.8 Different versions of Raptor codes.

5.9.2 Combustion. A signal $x(t)$ is better understood by its Fourier transform $X(f)$ defined by (3.15), in the sense that it provides an easier method to interpret the signal. It is particularly important that the FFT is used to analyze the experimental data obtained in many cases of signal processing. The phenomenon of combustion is an important aspect of engineering, physics, and applied chemistry. As trivial it may sound, burning hydrocarbons (fossil fuel) in automotive engines, natural gas or heating oil for heating dwellings, kerosene in heavy oil fuel in airplanes, and power stations are some of the well-known examples of fuel use. The energy consumption of industrialized countries of the world is about 75% in the form of natural gas and petroleum and its derivatives, and the remaining 25% in hydroelectric and nuclear power.

While these fuels together with air (oxidizer) are used in combustors in modern industries, there is another type of fuel, known as the solid propellants, which are used in rocket motors for space vehicles to generate thrust. The fuel and oxidizer are both solids: a polymer is used as fuel and binder, and a salt, e.g., ammonium perchlorate composite propellant (NH_4ClO_4), simply called APCP, is used as an oxidizer, in both manned and unmanned rocket vehicles. The APCP, which provides impulses of 180–260 seconds depending on the composition and operating pressure, is regularly implemented in the Space Shuttle Rocket Boosters, aircraft ejection seats, and specialty space exploration applications such as NASA's Mars Explorer Rover descent-stage retrorockets, in which the temperature of the combustion gases reaches $3,000°C$ (about $5,500°F$).

Technically, combustion is an exothermic reaction which is accompanied by development of heat and light at a rapid rate, so that temperature rises considerably. In general, combustion processes are oxidation reduction reactions, with the oxidizing agent known as the oxidant and the reducing agent as the fuel. Combustion is the phenomenon of burning or rapid oxidation. In a complete combustion reaction, a compound reacts with an oxidizing agent, such as oxygen or fluorine, and the products are compounds of each element in the fuel with the oxidizing element. As a chemical reaction, it can be expressed as $CH_4 + 2O_2 \longrightarrow CO_2 + 2H_2O$ + energy, where 'energy' is heat and light. Combustion in oxygen is a chain reaction, and that of hydrocarbons is initiated by hydrogen atom abstraction from the fuel to oxygen, to give a hydroperoxide radical which reacts further to cause hydroperoxides to break up and give hydroxyl radicals. Solid fuels undergo many pyrolysis reactions that give more easily oxidized, gaseous fuels. These reactions are endothermic and require constant energy input from the combustion reactions. The rate of combustion is the amount of a fuel that undergoes combustion over a period of time. It is expressed in grams per second (g/s) or kilograms per second (kg/s).

5.9.3 Combustion Instability. This phenomenon occurred in rocket propulsion during the launch of the Shuttle 'Challenger' disaster on January 28, 1986.

The minimum temperature of $-8°C$, as compared to $11°C$ during previous launches, made the rubber ring-seals of the booster case hard and brittle, and they failed to withstand the increasing pressure during the first few seconds of the booster ignition. The combustion gases reached the outer casing that was not designed for such high temperatures. It pushed against the rocket casing walls where a gap opened and, as a result of leaking gases, a combustion in the liquid propellant tank resulted in an explosion of the Shuttle a few seconds thereafter.

The combustion instability as a result of small changes in geometry of the size of propellants particles can cause a transition from stability to oscillatory behavior. It was discovered that many motors in the design of the Shuttle were operating close to the maximum stability thresholds, and severe spontaneous oscillatory motion started once these thresholds were crossed. Warranted by this disaster, this type of oscillatory behavior in a combustor was later studied at different frequencies ranging from 300 Hz to 15,000 Hz (see Feynman [1989]). The longitudinal, tangential, and radial pressure waves together with the combustion characteristic frequencies of the propellant depend on its gasification times. In the case of this disaster, experimental and theoretic (numerical) results on the self-sustained oscillatory burning of solid propellants were used to determine if these characteristic frequencies were close to one of the frequencies of the spacecraft engines at the operative conditions (Zanotti and Giuliani [1993]).

At NASA's White Sands Test Facility at the Lyndon B. Johnson Space Center, studies were conducted on combustion instability. Dynamic oscillations in spacecraft engines and related systems can often lead to disasters. Two of the more common phenomena are called *combustion instability*, and *pogo* which is a dynamic coupling of the combustion process with structure and feed system dynamics. Pogo causes rapid positive and negative acceleration along the thrust axis. Pogo-type instabilities can result in severe vibrations, interference with the guidance system, and possible destruction of the stage. FFT has been used to qualitatively determine the self-sustained oscillations at sub-atmospheric pressure by Zanotti et al. [1992].

5.10 Mars Project and Beyond

The Mars Exploration Program started in the late 20th century with probes sent from Earth to acquire knowledge about the Martian system, primarily its geology and possible habitation potential. Interplanetary journey by unmanned spacecraft is a very complicated engineering task. In January 2004, the NASA twin Mars Exploration Rovers named 'Spirit' (MER-A) and 'Opportunity' (MER-B) landed on the surface of Mars. Among the most valuable scientific returns has been the conclusive evidence that liquid water existed at some time in the past at both landing sites. MER-A ceased sending data after 2010.

On March 10, 2006, the NASA Mars Reconnaissance Orbiter (MRO) probe reached the orbit to conduct a two-year science survey, by mapping Martian terrain and weather, to find suitable landing sites for upcoming lander missions. The first image of a series of active avalanches near Mars's north pole were received on March 3, 2008.

The NASA Mars Odyssey orbiter entered Mars orbit in 2011. Odyssey's Gamma Ray Spectrometer detected a significant amount of hydrogen in the upper section of Regolith on Mars, a region which has layers of loose, heterogeneous material covering solid rock.

The Mars Express mission of the European Space Agency (ESA) reached Mars in 2003. It carried the Beagle 2 lander, but it was lost. In early 2004 the Mars Express Planetary Fourier Spectrometer team announced that the orbiter had detected methane in the Martian atmosphere. This spectrometer is an infrared spectrometer built by the Istituto Nazionale di Astrofisica (Italian National Institute for Astrophysics) with other agencies.

The Mars Science Laboratory mission was launched on November 26, 2011; it delivered the 'Curiosity' rover on the Martian surface on August 6, 2012. This rover is larger and more advanced than the previous rovers, and moves with a velocity of 295 feet per hour (90 meter/h), and has instruments such as a laser chemical sampler to detect and deduce the structure of rocks at a distance of 7 meters (23 feet).

NASA's Curiosity, a car-sized robotic rover exploring Gale Crater on Mars as part of NASA's Mars Science Project, reached Mars on August 6, 2012 and survived its harrowing and unprecedented Red Planet landing. In one year this one ton rover has achieved a lot, like discovering an ancient stream bed and gathering enough evidence for mission scientists to decide whether the planet could have supported microbial life billions of years ago.

The fastest spaceship was built for the Helios 2 probe that was launched in 1976 to monitor the Sun. It attained a top speed of 157,000 miles an hour. At this rate a starship heading to the nearest star Proxima Centuri would take more than 17,000 years to complete the 24-trillion-mile journey. According to modern thinking and speculation, different ways to propel a starship are suggested. Some of them are described below. In early 2012 the Defense Advanced Research Projects Agency (DARPA) has started the so-called 100 Year Starship Project. The general direction seems to be to harness nuclear fusion as an energy source that would help us reach the nearest star in a few decades provided the machinery could last that long. Some pseudo-designs of starship drives, as described in Folger [2013], are as follows:

(i) NUCLEAR PULSE. In this scheme, pioneered by Project Orion in the 1950s, thermonuclear bombs expelled by the spacecraft would explode against a 'pusher plate' propelling the ship forward (see Figure 5.9(a)).

(ii) Nuclear Fusion. Deuterium pellets injected into a reaction chamber 250 times a second would be heated by electron guns until the Deuterium nuclei fused and the pellets exploded. Powerful magnets would channel the blasts out of the back (see Figure 5.9(b)).

(iii) Antimatter. When protons collide with their antimatter twins, they annihilate, transforming into bursts of charged particles that travel at near light speed. It would be an ideal thrust for an interstellar rocket if magnets can be made to channel the particles in the right direction (see Figure 5.9(c)).

(iv) Beamed Solar Sail. A giant piece of reflective fabric pushed by solar photons, which are particles of light, could carry a probe to the edge of the solar system. To go farther would require the beam from an orbiting laser focused by an immense lens (see Figure 5.9(d)). A 3400-square-foot Mylar solar sail was tested in 2005 in a vacuum chamber at NASA's Plum Brook Station in Sandusky, Ohio. NASA plans to launch Sunjammer, a probe with a sail four times as big as a yearlong cruise toward the Sun in 2014.

(v) Warp Drive. According to GMA News of April 23, 2013, 'NASA is looking at a reimagined version of the controversial Alcubierre Drive to build a "faster-than-light" warp drive that can travel to the nearest star (Alpha Centuari) in just weeks.' Physicist Harold White who heads NASA's Warp Drive Project at Johnson Space Center without receiving DARPA or other external funding at this time is working on an advanced propulsion research project. His scientific talks and publications on space warp are given in the Bibliography.

The space-time metric engineering that forms the basis of the Alcubierre warp drive is based on the ADM formalism of general relativity[8]. The space-time is defined by a foliation of space-like hypersurfaces of constant coordinate time t. Then the general form of the metric is given by

$$ds^2 = -\left(\alpha^2 - \beta_i\beta^i\right)\,dt^2 + 2\beta_i\,dx^i\,dt + \gamma_{ij}\,dx^i\,dx^j, \qquad (5.17)$$

where a is the lapse function that determines the interval of proper time between adjacent hypersurfaces, β^i is the shift vector that relates the spatial coordinate systems on different hypersurfaces, and γ_{ij} is a positive-definite metric on each of the hypersurfaces.

---

[8] ADM formalism supposes that spacetime is foliated into a family of spacelike surfaces Σ_t, labeled by their time coordinate t, and with coordinates on each slice given by x^i. The dynamics of this theory is based on the metric tensor of three dimensional spatial slices $\gamma_{ij}(t, x^k)$ and their conjugate momenta $\pi^{ij}(t, x^k)$. In addition to the 12 variables γ_{ij} and π^{ij}, there are four Lagrange multipliers, called the lapse function, N_i. Using these variables a Hamitonian is defined, and thus, the equations of motion for general relativity in the form of Hamilton's equations are formulated. The acronym 'ADM' stands for the last names of the physicists Richard Arnowitt, Stanley Deser and Charles Misner, in honor of their research [1959].

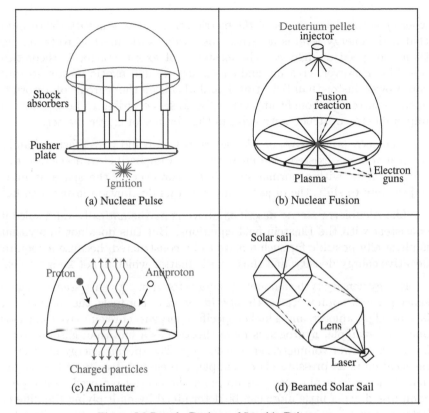

Figure 5.9 Pseudo-Designs of Starship Drives.

In a particular case, Alcubierre [1994] has taken $\alpha = 1$, $\beta^x = -v_s(t)f\left(r_s(t)\right)$, $\beta^y = \beta^z = 0$, $\gamma_{ij} = \delta_{ij}$, where δ_{ij} is the Kronecker delta, and

$$v_s(t) = \frac{dx_s(t)}{dt},$$

$$r_s(t) = \sqrt{\left(x - x_s(t)\right)^2 + y^2 + z^2},$$

$$f\left(r_s\right) = \frac{\tanh\left(\sigma(r_s + R)\right) - \tanh\left(\sigma(r_s - R)\right)}{2\tanh(\sigma R)}, \qquad (5.18)$$

where $\sigma > 0$ and $R > 0$ are arbitrary parameters. Then Alcubierre's specific form of the metric can be written as

$$ds^2 = \left(v_s(t)^2 f\left(r_s(t)\right)^2 - 1\right) dt^2 - 2v_s(t)f\left(r_s(t)\right)\ dx\,dt + dx^2 + dy^2 + dz^2. \quad (5.19)$$

Using the metric (5.19) it can be shown that the energy density measured by observers whose 4-velocity is normal to the hypersurfaces is given by

$$-\frac{c^4}{8\pi G}\frac{v_s^2(y^2 + z^2)}{4g^2 r_s^2}\left(\frac{df}{dr_s}\right)^2, \qquad (5.20)$$

where g is the determinant of the metric tensor. The quantity (5.20) implies that if the energy level is negative, one needs exotic matter[9] to travel faster than the speed of light c. The existence of exotic matter is theoretically possible. However, creating and sustaining enough exotic matter so that it can provide faster-than-light travel, and also keep the 'throat' of a wormhole open, is currently thought impossible, because, according to Low [1999], 'it is impossible to construct warp drive in the absence of exotic matter.'

Alcubierre warp drive, or Alcubierre metric, could travel faster than light if negative mass existed. We know that it is impossible to accelerate to the speed of light within normal spacetime. However, if the space around an object were to shift, the object would reach its destination faster than light.

The Alcubierre metric described above is mathematically valid since it is consistent with the Einstein field equations. But this does not imply that it is physically possible for such a drive to be constructed, because it requires a negative energy density, and thus, exotic matter which is not known to exist.

Energy density is the amount of energy stored in a given system or region of space per unit volume, called the *specific energy*. For fuels, the specific energy is a useful parameter, and a higher specific energy is more effective than a lower one. For example, hydrogen is more effective as a fuel than gasoline since it has a much lower volumetric energy density. The specific energy has the same physical units as pressure. For example, the energy density of the magnetic field behaves as a physical pressure, and the energy required to compress a compressed gas a little more can be determined by multiplying the difference between the gas pressure and the pressure outside by the change in volume. Thus, pressure is a measure of the volumetric enthalpy of a system. In other words, the enthalpy is the sum of the internal energy of a body or system and the product of its volume multiplied by the pressure.

Since $E = mc^2$, where $m = \rho V$, ρ the mass per unit volume, V the volume of the mass, and c the speed of light, the greatest source of energy is mass itself. This energy can be released only by the following processes: by nuclear fission (0.1%); nuclear fusion (1%), or the annihilation of some or all of matter by antimatter or by chemical reactions such as combustion (100%)[10].

---

[9] Exotic matter is matter with negative energy density, or more precisely, where the energy density tensor trace invariant is negative. While there are no known instances of exotic matter, there is not much we know about the full matter content of the universe. There is a large amount of dark energy driving the expanded acceleration of the universe. It might be due to a cosmological constant, it might be due to quintessence, or to negative energy, or hypothetical particles which have 'exotic' physical properties that would violate laws of physics, such as a particle having a negative mass, or matter that is poorly understood such as dark matter, or hypothetical particles or states of matter, which have not yet been encountered.

[10] In particle physics, antimatter is material composed of antiparticles, which have the same mass as particles of ordinary matter but have opposite charge and quantum spin. Antiparticles bind with each other to form antimatter in the same way as normal particles

The relationship between energy density of electric and magnetic fields is defined as follows. Let the electric and magnetic fields be denoted by $\mathbf{E}$ and $\mathbf{B}$, respectively. Then, in vacuum, the volumetric energy density U (in SI units) is given by

$$U = \frac{\varepsilon_0}{2}\mathbf{E}^2 + \frac{1}{2\mu_0}\mathbf{B}^2. \tag{5.21}$$

The units of U are joules per cubic meter. The magnetic energy density behaves like an additional pressure that adds gas pressure to a plasma, as in hydromagnetics. In the case of normal (linear) substances, the energy density U is given by

$$U = \frac{1}{2}\left(\mathbf{E}\cdot\mathbf{D} + \mathbf{H}\cdot\mathbf{B}\right), \tag{5.22}$$

where $\mathbf{D}$ is the electrical displacement field, and $\mathbf{H}$ the magnetic intensity.

Certain important problems arise with the metric (5.19), because all known warp-drive spacetime theories violate various energy conditions, such as the null energy condition, the weak energy condition, the dominant energy condition, and the strong energy condition. These conditions are needed in proofs of various theorems in general relativity. Certain experimentally verified quantum phenomena when described in the context of quantum field theories lead to stress-energy tensors that violate the energy conditions, such as negative mass-energy, and thus Alcubierre-type warp drives might be physically realized by using such quantum effects. Van der Broeck [2000] has pointed out some potential issues with such drives, namely, that 'the energy requirements for them may be absurdly gigantic as well as negative.' For example, they require energy equivalent of 10^{64} kg to transport a small spaceship across the Milky Way. But this is 'an order of magnitude greater than the estimated mass of the universe.' However, counter-arguments to such objections are also available, e.g., in Krisnikov [2003]. The physicist H. White [2011] announced that modifying the geometry of exotic matter could reduce the mass-energy requirements for a macroscopic spaceship from the equivalent of the planet Jupiter to that of the Voyager 1 spacecraft ($\sim$ 700 kg) or less, and state their intent to perform small-scale experiments in constructing warp fields.

5.10.1 SETI. The first article about the search for extraterrestrial intelligence (SETI) was written by Cocconi and Morrison [1959], suggesting detection by radio waves. This was followed by Bracewell [1960] who wrote, while speculating about existence of such civilizations, that "Their signals would have the appearance of echos having delays of seconds or minutes, such as were

---

do to form matter. Any mixing of antiparticles and particles can lead to annihilation of both, thus giving rise to high-energy photons (gamma rays) or other particle-antiparticle pairs. This process ends in release of energy proportional to the mass as is evident from the mass-energy equivalence equation $E = mc^2$.

reported thirty years ago by Størmer and van der Pol and never explained." The first project, known as Project Ozma, was carried out in 1960 at the National Radio Astronomical Observatory in Green Bank, WV, by Frank D. Drake. Since then a number of SETI experiments have been conducted as reported in Sagan and Drake [1975] and Puthoff [1996]. Currently the BETA (Billion Channel Extraterrestrial Assay) project at the Harvard-Smithsonian Observatory that started in October 1995 covers the band from 1400 to 1720 MHz, using a radiotelescope and a supercomputer. This is as if 250 million radios were all tuned to adjacent channels and received simultaneously. The frequency for such radio transmissions is most likely about 1420 MHz, which is the emission line of hydrogen, the most common element in the universe.

In 2012 Voyager I, which was launched in 1977, slipped into interstellar space beyond the Milky Way. Currently it is at a distance of 17 light years, traveling at about 61,115 km/h. Interviews with the NASA Jet Propulsion Laboratory in Pasadena, CA, compared this antiquated spacecraft's technology to an eight-track player and computers 240,000 times less powered than a 3-G iPhone. Yet, the communication signal powered by a radioactive battery similar in voltage to a refrigerator light continues to emit a signal back to Earth.

Just in case an extraterrestrial species encounters Voyager I, a diverse collection of music plays in an effort to introduce a mixture of our cultures, with selections such as Bach's Brandenberg Concerto # 2, Japanese flutes, dance rhythms from Stravinsky's Rite of Spring, Ragas from India, pan pipes from Peru, Night Chants from Navajo, and Melancholy Blues from Louis Armstrong. This represents a beautiful showcase of humanity. Yet, despite advances in radioastronomy, we have so far had no success.

6

Global Positioning System

The Global Positioning System (GPS) is a worldwide, satellite-based navigation system, funded by the U.S. Department of Defense (DoD). It was originally designed for military use, but later the system was made available for civilian use. We will provide some details about this technology.

6.1 GPS Structure

There are 32 GPS satellites located above the Earth (originally 24), each sending out signals with the following information: the satellite i ($i = 1, 2, \ldots, 32$), location of the GPS receiver $\mathbf{r}$, and the time t the information was sent. Each satellite sends out not only its location but also orbital data about the locations of other satellites in the form of ephemeris and almanac data that are stored by the GPS receiver for later calculations. To determine its own location $\mathbf{r}$ on Earth, the receiver computes the time between the signal sent by the satellite i and the time it was received. From this time difference the distance between the receiver and the satellite is calculated using the speed of light c. By taking data from other satellites the current location of the receiver can be computed by trilateration, which requires at least four satellites to determine the location $\mathbf{r}$. The method using three satellites is called a *dilateration* (2D-position fix) which is a two-dimensional location determination because the receiver is assumed to be located on Earth's surface regarded as a plane. However, using four or more satellites, the location of the receiver in a three-dimensional space can be absolutely fixed with the three spatial coordinates (*trilateration*).

The GPS receiver uses the messages so received to determine the transit time of each message and computes the distance to each satellite using the speed of light. Each of these distances and satellite locations define a sphere. The receiver is on the surface of each of these spheres when the distances and the satellite locations are used to compute and confirm that each location is correct. These distances and satellite locations are used to compute the position of the receiver using the navigation equations, which are defined as follows.

Let the spatial components of the satellite position and the time sent be denoted by (x_i, y_i, z_i, t_i), where $i = 1, 2, \ldots, n$ $(n \geq 4)$ denotes the ith satellite. When the time of message reception indicated by the on-board clock is T_i, let the true reception time be t_i, and let b denote the bias (or clock delay) of receiver's clock[1]. Then the message's transit time is $T_i - t_i + b$. Assuming the message travels at the speed of light $c(= 299792458\,\text{m/s} = 0.299792458\,\text{m/ns})$, the distance traveled is $c\,(T_i - t_i + b)$. Since the distance from receiver to satellite's position is known, the receiver is on the surface of the sphere centered at the satellite's position with radius equal to this distance. Thus, the receiver is at or near the intersection of the surfaces of the spheres if it receives signals from more than one satellite. In the ideal case of no errors, the receiver is at the intersection of the surfaces of the spheres. We will keep track of this information that we receive from the satellite i in the form of a vector $\mathbf{s}_i = [x_i, y_i, z_i, \rho_i]^T$, where T denotes the transpose of a vector (matrix). Thus, the receiver has four unknowns (x, y, z, b). The equations of the sphere's surface are given by

$$(x - x_i)^2 + (y - y_i)^2 + (z - z_i)^2 = c^2\,(T_i - t_i + b)^2, \quad i = 1, 2, \ldots, n, \quad (6.1)$$

or, in terms of pseudoranges $\rho_i = c\,(T_i - t_i)$, as

$$\rho_i = \sqrt{(x - x_i)^2 + (y - y_i)^2 + (z - z_i)^2} - cb, \quad i = 1, 2, \ldots, n. \quad (6.2)$$

These equations can be solved by algebraic or numerical methods.

6.1.1 Bancroft's Method. This method involves an algebraic approach to solve Eqs (6.1) for the case of four or more satellites (see Bancroft [1985]). It provides one of the two solutions for the four unknowns. When there are two solutions, only one of them will be a near-earth sensible solution. Since the clocks on all of the satellites are in sync with one another, the clock of the receiver will be out of sync with each satellite by the same amount. Let Δt be the error in the receiver clock. Then the error in the pseudorange is the bias $\tilde{b} = c\,\Delta t$, and each satellite's pseudorange ρ_i is off by the same distance $\tilde{b}$. If (x, y, z) is the actual position of the receiver, we put the unknown receiver data as the vector $\mathbf{u} = [x \ y \ z \ \tilde{b}]^T$. Thus, Eq (6.1) becomes

$$\sqrt{(x - x_i)^2 + (y - y_i)^2 + (z - z_i)^2} + \tilde{b} = \rho_i \quad \text{for each } i, \quad (6.3)$$

or

$$(x - x_i)^2 + (y - y_i)^2 + (z - z_i)^2 = \left(\rho_i + \tilde{b}\right)^2. \quad (6.4)$$

This equation is then solved for the four unknowns x, y, z, and $\tilde{b}$. If there are four satellites $(i = 4)$, then there will be four equations in four unknowns.

---

[1] The bias (clock error) b is the amount by which the receiver's clock is off.

But since this equation is nonlinear, a least-square solution is determined as follows: first, multiply out Eq (6.4), to get

$$x_i^2 - 2x_ix + x^2 + y_i^2 - 2y_iy + y^2 + z_i^2 - 2z_iz + z^2 = \rho_i^2 + 2\rho_i\tilde{b} + \tilde{b}^2,$$

which on rearranging the terms gives

$$\left(x_i^2 + y_i^2 + z_i^2 - \rho_i^2\right) - 2\left(x_ix + y_iy + z_iz - \rho_i\tilde{b}\right) + \left(x^2 + y^2 + z^2 - \tilde{b}^2\right) = 0. \tag{6.5}$$

Let $\langle \mathbf{s}_i, \mathbf{s}_i \rangle = x_i^2 - 2x_ix + x^2 + y_i^2 - 2y_iy + y^2 + z_i^2 - 2z_iz + z^2$ and $\langle \mathbf{u}, \mathbf{u} \rangle = x^2 + y^2 + z^2 - \tilde{b}^2$ define the Lorentzian inner product[2] (see Strang and Borre [1997]). Then Eq (6.5) can be written as

$$\frac{1}{2}\langle \mathbf{s}_i, \mathbf{s}_i \rangle - \langle \mathbf{s}_i, \mathbf{u} \rangle + \frac{1}{2}\langle \mathbf{u}, \mathbf{u} \rangle = 0, \tag{6.6}$$

which holds for each satellite i. Next, to analyze these equations for each satellite we introduce the notation:

$$\mathbf{B} = \begin{bmatrix} x_1 & y_1 & z_1 & -\rho_1 \\ x_2 & y_2 & z_2 & -\rho_2 \\ x_3 & y_3 & z_3 & -\rho_3 \\ x_4 & y_4 & z_4 & -\rho_4 \\ \vdots & \vdots & \vdots & \vdots \end{bmatrix}, \quad \mathbf{a} = \frac{1}{2}\begin{Bmatrix} \langle \mathbf{s}_1, \mathbf{s}_1 \rangle \\ \langle \mathbf{s}_2, \mathbf{s}_2 \rangle \\ \langle \mathbf{s}_3, \mathbf{s}_3 \rangle \\ \langle \mathbf{s}_4, \mathbf{s}_4 \rangle \\ \vdots \end{Bmatrix}, \quad \mathbf{e} = \begin{Bmatrix} 1 \\ 1 \\ 1 \\ 1 \\ \vdots \end{Bmatrix}, \quad \lambda = \frac{1}{2}\langle \mathbf{u}, \mathbf{u} \rangle.$$

$$\tag{6.7}$$

Note that in the case of n satellites, $\mathbf{B}$ is $n \times 4$ matrix, $\mathbf{a}$ and $\mathbf{e}$ are $n \times 1$ vectors, and λ is a scalar. Then Eq (6.7) becomes

$$\mathbf{a} - \mathbf{Bu} + \lambda\mathbf{e} = \mathbf{0}, \quad \text{or} \quad \mathbf{Bu} = \mathbf{a} + \lambda\mathbf{e}. \tag{6.8}$$

Thus, in the case of more than four satellites, a normal equation of the form $\mathbf{B}^T\mathbf{Bu} = \mathbf{B}^T(\mathbf{a} + \lambda\mathbf{e})$ has a least-square solution of the form

$$\mathbf{u}^* = \mathbf{B}^+(\mathbf{a} + \lambda\mathbf{e}), \tag{6.9}$$

where $\mathbf{B}^+ = \left(\mathbf{B}^T\mathbf{B}\right)^{-1}\mathbf{B}^T$. However, since Eq (6.8) involves λ, the solution $\mathbf{u}^*$ in (6.9) is to be modified accordingly. To do this, we substitute $\mathbf{u}^*$ into

---

[2] The Lorentzian inner product on $\mathbb{R}^4$ is defined as $-dx_0^2 + dx_1^2 + dx_2^2 + dx_3^2$, or for the vectors $\mathbf{v}$ and $\mathbf{w}$ as $\langle \mathbf{v}, \mathbf{w} \rangle = -v_0w_0 + v_1w_1 + v_2w_2 + v_3w_3$. This product is used in special relativity to measure distances, independent of the reference frame, where the variables x_1, x_2 and x_3 can be regarded as space variables and x_0 as the time variable, often replaced by t. In special relativity $x_0 = ct$, where c is the speed of light, with the convention that the units must be chosen so that $c = 1$ to simplify formulas. The vector $\mathbf{v}$ has the property that when it is positive, it is a space-like vector; when it is zero, it is a null vector, or light-like vector; and when it is negative, it is a time-like vector.

the definitions (6.7) of the scalar λ and use linearity of the Lorentzian inner product. This gives

$$\lambda = \frac{1}{2}\left\langle \mathbf{B}^+\left(\mathbf{a}+\lambda\mathbf{e}\right), \mathbf{B}^+\left(\mathbf{a}+\lambda\mathbf{e}\right)\right\rangle$$
$$= \frac{1}{2}\left\langle \mathbf{B}^+\mathbf{a}, \mathbf{B}^+\mathbf{a}\right\rangle + \lambda\left\langle \mathbf{B}^+\mathbf{a}, \mathbf{B}^+\mathbf{e}\right\rangle + \frac{1}{2}\lambda^2\left\langle \mathbf{B}^+\mathbf{e}, \mathbf{B}^+\mathbf{e}\right\rangle,$$

or

$$\lambda^2\left\langle \mathbf{B}^+\mathbf{e}, \mathbf{B}^+\mathbf{e}\right\rangle + 2\lambda\left(\left\langle \mathbf{B}^+\mathbf{a}, \mathbf{B}^+\mathbf{e}\right\rangle - 1\right) + \left\langle \mathbf{B}^+\mathbf{a}, \mathbf{B}^+\mathbf{a}\right\rangle = 0. \qquad (6.10)$$

This is obviously a quadratic equation in λ, which has coefficients $\left\langle \mathbf{B}^+\mathbf{e}, \mathbf{B}^+\mathbf{e}\right\rangle$, $2\left(\left\langle \mathbf{B}^+\mathbf{a}, \mathbf{B}^+\mathbf{e}\right\rangle - 1\right)$, and $\left\langle \mathbf{B}^+\mathbf{a}, \mathbf{B}^+\mathbf{a}\right\rangle$. It can be computed for two possible values of λ using the quadratic equation formula. We will get two solutions of Eq (6.10), say λ_1 and λ_2. Then we solve for the two solutions $\mathbf{u}_1^*$ and $\mathbf{u}_2^*$ in Eq (6.9). However, only one of these solutions will make sense, and it will be on the surface of Earth which has a radius of about 6378 km; the other solution will not, and will, therefore, be discarded.

6.1.2 Trilateration Method. The above description does not yet tell us how to determine the location (position); rather, it only computes a distance based on the speed of light. For example, if different people on fixed locations happen to determine the time span between lightning and thunder, this would allow the determination of the position where the lightning struck the ground, using both the speed of light and sound. We will now describe how the position determination by GPS works.

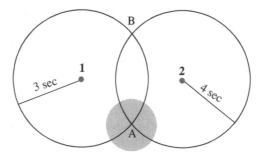

Figure 6.1 Dilateration Process.

For the sake of geometrical simplicity, let us assume that Earth is a two-dimensional disk. This gives us a two-dimensional model, which can be easily modified to a three-dimensional sphere. In Figure 6.1, suppose, for the sake of argument, that the time required by a signal to travel from the first of the two satellites (say, the satellite on the left) to the receiver at position A is determined to be 3 sec.[3] This means that the receiver is located somewhere on the circumference of the left circle of radius 3 sec and center at the left satellite **1**. Now, if we take another satellite **2** of radius, say, 4 sec (right circle), the two circles intersect at two points A and B, and the receiver must be located at one of these two points. This process is called *dilateration*. To achieve the *trilateration* process we need a third satellite to provide a third dimension. We also assume that the receiver is located close to Earth and not deep in space. The circle about the third satellite will intersect the two circles in only one of the points A and B. In Figure 6.1, it is taken as the point A, and the grey region about the point A indicates the realistic position of the receiver so that any position outside this region is discarded.

However, the problem with the third satellite lies in the determination of exact runtime of signals. As the satellite transmits a signal, it is always accompanied by the time data on each transmitted data package. These time data are always precise with zero bias as they are atomic clocks, and all satellites have the same type of clocks. But the problem lies in our GPS receiver's clocks since they have the usual quartz clocks with bias (error), and not the expensive and accurate atomic clocks. So going back to the example of Figure 6.1, suppose that the clock in the GPS receiver is 0.5 seconds early compared to the satellite clocks. Then the runtime of the signal will be 0.5 sec longer than it actually is. It leads us to believe that we are at point C instead of point A (see Figure 6.2). Because of the clock bias, the circles intersecting in point C are called *pseudoranges*. The clock bias is very significant in GPS. Even a very small bias may throw the location miles away from the accurate position. For example, a clock bias of $1/100 = 0.01$ sec would lead to an error in location by about 3000 km. To achieve an accuracy of 10 m of the position, the runtime of the signal must not exceed 3^{-8} sec. Since the receiver clock is not atomic, this problem is solved as follows.

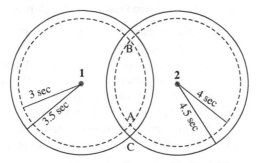

Figure 6.2 Dilateration Process with Clock Bias.

---

[3] This time in reality is too high if we take into account the speed of light (299,792,458 m/s). In fact, the actual time span for the signal from a satellite to a receiver on Earth is about 0.7 sec.

If a third satellite is employed for computing the position, we obtain another intersection point A as the actual position of the receiver, as explained in Figure 6.3, in which, because of the receiver clock bias of 0.5 sec, we have three intersection points C. By shifting the receiver clock until the three intersection points C merge to A, the clock bias is corrected and the receiver clock is synchronized with the atomic clocks in the satellites. This means, that the receiver clock, even in spite of its bias, behaves like an atomic clock, and the pseudoranges now correspond to actual distances. The position thus determined is accurate. The GPS receiver can also determine the speed and direction of its movement, known as *ground track* or *ground speed*, which can be determined using the Doppler effect[4] that occurs due to the movement of the receiver while receiving signals.

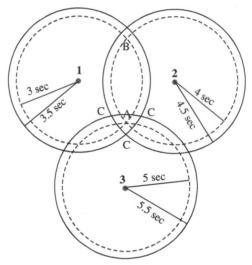

Figure 6.3 Trilateration Process with Clock Bias.

We will now consider spheres instead of circles. In the case of two intersections of three spherical surfaces, the point nearest the surface of the sphere corresponding to the fourth satellite is chosen. Let d denote the signed distance from the current estimate of receiver position to the sphere around the fourth satellite. Let the function d(correction) denote the clock bias. The problem is to determine this bias such that d(correction) = 0. This is the well-known problem of finding zeroes of one-dimensional nonlinear function of a scalar variable, which is solved using iterative numerical methods, such

---

[4] This phenomenon explains the change of frequency of a wave or any other periodic event for an observer moving relative to its source. It is commonly heard when a vehicle sounding a siren (like a police car) approaches, passes, and recedes from an observer. The received frequency is higher (compared to the emitted frequency) during the approach; it is the same when the siren is at the moment of passing, and it is lower during the recession. See Appendix J for more details.

as root-finding, as, e.g., in Press et al. [2007].

The Earth's geometry, including the analysis of the user's location, altitude and sidereal day, is available in Appendix H; for Kepler's laws see §9.1.

6.1.3 GPS Segments. According to Pike [2009], GPS consists of three major segments: space segment, control segment, and user segment. The U.S. Air Force develops, maintains, and operates the space and control segments. According to the website *GPS.gov* [June 26, 2010], the GPS satellites and the GPS receiver each uses these signals to compute its three-dimensional location (in terms of latitude, longitude, and altitude) and the current time.

The space segment consists of 32 GPS satellites, or Space Vehicles (SVs) in GPS terminology, located in medium Earth orbit (MEO), also called interme- diate circular orbit (ICO), which is the region of space around Earth bounded by the altitudes of 2000 km (1243 miles) and 35,780 km (22,236 miles), the most common altitude being 20,200 km (12,552 miles) for GPS. The original 24 SVs were divided eight each in three approximately circular orbits, but this was later modified to six orbital planes with four SVs each. The circular orbits are centered on Earth, but they do not rotate with Earth. Instead, they are fixed with respect to the distant stars. According to the NAVSTAR Joint Program Office [December 15, 2006], the six orbital planes are inclined at about 55° relative to Earth's equator, and are separated by 60° right ascen- sion of ascending node, which is the angle along the equator from a reference point to the orbit's intersection. The orbit period is one-half of a sidereal day (11 hrs and 58 mins)[5], and the orbits are arranged such that at least six SVs are always within sight from almost everywhere on Earth's surface. Thus, four SVs are unevenly (i.e., not exactly 90°) apart within each orbit. In general, the angular difference among SVs is 30°, 105°, 120°, and 105° apart, all adding to 360°. According to Agnew and Larson [2007], while orbiting at an altitude of 20,280 km (12,600 miles), with orbital radius of about 26,555 km (16,500 miles), each SV makes two complete orbits each sidereal day. As of December 2012, there are 32 SVs in the GPS constellation. The additional SVs improve precision of the GPS receiver computations. With this system the constellation has a changed nonuniform arrangement which has improved accuracy and reliability.

The control segment consists of a master control station (MCS), an al- ternate master control station, and four of dedicated and shared ground an- tennas and six dedicated motor stations. These antennas are a specialized type of radio antennas, and consist of two ground electrodes buried in the earth, separated by very large distances, and linked by overhead transmission lines to a power plant transmitter located between them. They radiate ex- tremely low frequency electromagnetic waves in the range of 3 Hz to 3 kHz (see

---

[5] For details of a sidereal and solar day, see Appendix H.

Jones [2012]). The MCS can also access U.S. Air Force Satellite Control Network (AFSCN) ground antennas and National Geospatial-Intelligence Agency (NGA) monitor stations. According to the United States Coast Guard *General GPS News 9-9-05*, the monitoring stations are located in Hawaii, Kwajalein, Ascension Island, Diego Garcia, Colorado Springs, Colorado and Cape Canaveral along with shared NGA monitor stations in England, Argentina, Ecuador, Bahrain, Australia and Washington, DC. The 2nd Space Operations Squadron regularly updates the ground antennas. These updates synchronize the atomic clocks on board the SVs to within a few nanoseconds of each other, and adjust the ephemeris (orbital position data) of each SV's internal orbital model.

The ephemeris is the orbital position data obtained from the navigation message. It is used to calculate the precise position of the satellite at the start of the message. It is important that the data receiver should be very fast to acquire the ephemeris data, especially if noise is present in the environment.

Ephemeris data are as follows: 6378137 ± 2 m is the semi-major axis of the earth; 6356752.3142 m is the semi-minor axis of the earth; $c = 2.99792458 \times 10^8$ meter/sec is the speed of light; 0.0818191908426 is the eccentricity of the earth; 0.00335281066474 is the ellipticity of the earth; 6368 km is average Earth radius; $\mu = 3.986005 \times 10^{14}$ meters^3/sec^2 is the Earth's universal gravitational parameter; and $7.2921151467 \times 10^{-5}$ rad/sec is the World Geodetic System (WGS)-84 value of Earth's rotation rate.

The almanac consists of coarse orbit and status information for each satellite. These data are used to relate GPS derived time to the universal coordinated time (UCT). The description of all frames of a satellite is given in Table 6.1.

Table 6.1. GPS Message Frames.

Subframes	Words	Description
1	1–2	TLM and HOW
	3–10	Satellite clock,
		GPS time relationship
2–3	1–2	TLM and HOW
	3–10	Ephemeris
4–5	1–2	TLM and HOW
	3–10	Almanac component
		(network synopsis, error correction)

In this table, 'TLM' and 'HOW' mean 'telemetry' and 'handover word', respectively. Each subframe contains a part of the almanac, and each satellite transmits the complete almanac in 12.5 minutes. There are many uses of the almanac, namely, (i) to assist in the acquisition of satellites at power-up by letting the receiver generate a list of available satellites based on its location and time, where ephemeris from each satellite is required to compute the position fix of the available satellite; (ii) to relate the time derived from the GPS, called *GPS time*, to the international time standard of UTC; and (iii) to allow a single-frequency receiver to correct the ionospheric error using a global ionospheric model.

The setup for the TLM and HOW words is as follows: each subframe has five pages. The first two words of all the subframes are the TLM and HOW. Each word contains 30 bits, so the message is transmitted from bit 1 through bit 30. These two words are shown in Figure 6.4, in which the TLM starts with an 8-bit preamble, followed by 16 reserved bits and 6 parity bits. The bit pattern of the preamble, shown in this figure, is used to match the navigation data to detect the start of a subframe.

The HOW word is divided into four parts:

(i) Bits 1–17 are the truncated time of the week (TOW) count; this provides the time of the week in units of 6 seconds. The TOW is the truncated least significant bit (LSB)[6] of the Z count described below.

(ii) The next two bits (18 and 19) are flag bits. They indicate the following features: For satellite configuration 001 (block II satellite)[7] the bit 18 is an alert bit and bit 19 is anti-spoofing bit, e.g., when bit 18=1, the accuracy of the satellite user range is worse than indicated in the subframe 1 and the user is using this satellite at his (her) own risk. Moreover, bit 19=1 means that the anti-spoofing mode is on. In other words, bit 18 indicates a synchronization flag for satellite configuration 000, but an anti-spoofing flag for satellite configuration 001; and bit 19 indicates a momentum flag for satellite configuration 000, but an alert flag for satellite configuration 001.

(iii) The next three bits (20, 21, and 22) indicate the subframe ID and their values are 001, 010, 011, 100, and 101, to identify one of the five subframes 1, 2, 3, 4, and 5, respectively.

---

[6] These terms take their meaning from the least significant bit (LSB) and the most significant bit (MSB), defined in a field containing more than one bit that has a single value. They are similar to the most (least) significant digit of a decimal integer. The LSB (sometimes called the *right-most bit*, by convention in positional notation) is the bit position in a binary integer giving the units value, i.e., determining whether the number is even or odd. A similar definition is valid for MSB, which is sometimes called the leftmost bit. The LSBs change rapidly if the number changes even slightly. For example, adding $(1)_{10} = (00000001)_2$ to $(3)_{10} = (00000011)_2$ yields $(4)_{10} = (00000100)_2$, where the (rightmost) three of the LSBs have changed from 011 to 100, whereas the three MSBs (which are leftmost) remain unchanged (000 to 000).

[7] Satellites are procured in blocks. Most block I satellites are of experimental nature and all satellites in orbit are from block II.

(iv) The last 8 bits (23 through 30) are used for parity bits, such that bits 23 and 24 are used to preserve parity check with zeros in bits 29 and 30.

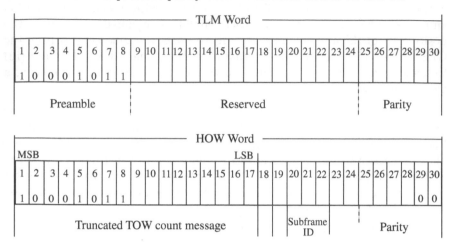

Figure 6.4 TLM and HOW Words.

The GPS time differs from UTC because the former is a continuous time scale, while the latter is corrected periodically with an integer number of loop seconds. The GPS time is always maintained within one μs of the UTC (measured module one sec). In each satellite, an internally derived 1.5-second epoch, called the Z *count*, makes a unit for precise counting and communication time. The Z count has 29 bits made up of two parts: the 19 LSBs define TOW and the 10 MSBs the week number. However, in actual data transmission by satellites, only 27 Z count bits are used. The 10-bit week number is the third word of subframe 1. The 17-bit TOW is in the HOW in every subframe. The two LSBs, which are left over out of 29 bits, are implied through multiplication of the truncated Z count. Since the TOW count has a time unit of 1.5 secs and covers one week (i.e., 604,800 seconds), the TOW count is from 0 to 403,199 because $604800/1.5 = 403200$.

The satellite Z count refers to the counting of time in UTC. The GPS zero time started at midnight of January 5/morning of January 6, 1980. The largest time unit used is one week which is equal to 604,800 seconds ($7 \times 24 \times 3600$).

The epoch counts approximately at midnight Saturday night/Sunday morning, and the midnight is defined as 0000 hours on the UTC scale which is the Greenwich meridian (Zulu time). But over the years the zero-epoch has occurred a few seconds from 0000 hours on the UTC scale. Since the 17-bit truncated TOW count covers an entire week and the time unit is 6 sec (1.5×4), which equals one subframe time, this truncated TOW is from 0 to 100799 (as $604800/6 = 100800$). For this timeline see Figure 6.5, in which the

Z count and TOW count are compared; note that the Z count is at the end and the start of a week.

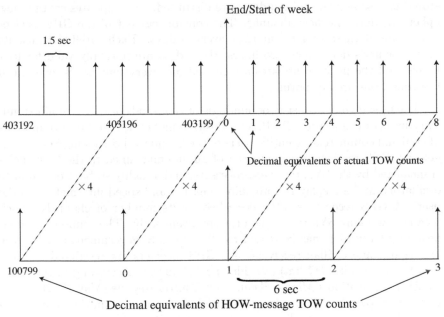

Figure 6.5 Z Count and TOW Count.

Almanacs are approximate orbital data parameters for all SVs. The ten-parameter almanacs describe SV orbits over extended periods of time (useful for months in some cases) and a set for all SVs is sent by each SV over a period of 12.5 minutes (at least). Signal acquisition time on receiver start-up can be significantly aided by the availability of current almanacs. The approximate orbital data is used to preset the receiver with the approximate position and carrier Doppler frequency (the frequency shift caused by the rate of change in range to the moving SV) of each SV in the constellation. Each complete SV data set includes an ionospheric model that is used in the receiver to approximate the phase delay through the ionosphere at any location and time. Each SV sends the amount to which GPS time is offset from Universal Coordinated Time (UCT). This correction can be used by the receiver to set UCT to within 100 ns. Other system parameters and flags are sent which characterize details of the system.

Clock data parameters describe the SV clock and its relationship to GPS time. Ephemeris data parameters describe SV orbits for short sections of the satellite orbits. Normally, a receiver gathers new ephemeris data each hour, but can use old data for up to four hours without much error. The ephemeris parameters are used with an algorithm that computes the SV position for any time within the period of the orbit described by the ephemeris parameter set.

The satellite data updates occur ever 24 hours, with up to 60 days of data stored in case regular daily updates are disrupted. Such updates contain new ephemeris data and new almanac. The control segment of the GPS receiver takes care of these updates at least every 6 days. Each satellite transmits new ephemeris data every two hours; these data are normally valid for four hours, and the position updates are carried out every four hours, or more if the conditions are not normal.

The user segment consists of hundreds of thousands of U.S. and allied military users of the secure GPS Precise Positioning Service, and tens of millions of civil and commercial scientific users of the Standard Positioning Service. In general, GPS receivers are composed of an antenna tuned to the frequencies transmitted by the SVs, receiver-processors, and a highly stable clock (quartz oscillator), and a display for providing location and speed information to the user. A GPS receiver is often described by its number of channels, which means how many SVs it can monitor simultaneously. The number of channels, originally four, has increased to 20. The U.S. government controls the export of some civilian receivers. All GPS receivers are restricted to function below 11 miles (17 km) altitude and 515 m/sec. According to the Arms Control Association's *Missile Technology Control Regime* (May 17, 2006), all receivers above these limits are designated as weapons (munitions) for which State Department export licenses are required. This rule also applies to civilian units that only receive the L1 frequency and the C/A (Coarse/Acquisition) code (see §6.3).

The hardware and software components of a typical GPS receiver are described in Table 6.2.

We will see in the sequel that Fourier transform and FFT play an important role in the P-code and C/A code phase search acquisition.

Table 6.2. Components of a GPS Receiver.

Hardware	Software	
Antenna	User Location	Subframe Identification
RF chain	Satellite Location	Tracking
ADC	Ephemeris and Pseudorange	Acquisition

6.1.4 Navigation. Precise positioning is possible using GPS receivers at reference locations providing corrections and relative positioning data for remote receivers. The surveying, geodetic control, and plate tectonic studies are carried out there. Time and frequency dissemination, based on the precise clocks on board the SVs and controlled by the monitor stations, is another use for GPS. Astronomical observatories, telecommunications facilities, and laboratory standards can be set to precise time signals or controlled to accurate frequencies by special purpose GPS receivers. Research projects have used GPS signals to measure atmospheric parameters.

The GPS Navigation Message consists of time-tagged data bits marking the time of transmission of each subframe at the time they are transmitted by the SV. A data bit frame consists of 1500 bits divided into five 300-bit subframes. A data frame is transmitted every thirty seconds. Three six-second subframes contain orbital and clock data. SV clock corrections are sent in subframe one and precise SV orbital data sets (ephemeris data parameters) for the transmitting SV are sent in subframes two and three. Subframes four and five are used to transmit different pages of system data. An entire set of 25 frames (125 subframes) makes up the complete Navigation Message that is sent over a 12.5 minute period. Data frames (1500 bits) are sent every thirty seconds. Each frame consists of five subframes. Data bit subframes (300 bits transmitted over six seconds) contain parity bits that allow for data checking and limited error correction.

6.1.5 Accuracy. The SPS predictable accuracy values are as follows: 100 meter horizontal accuracy; 156 meter vertical accuracy, and 340 nanoseconds time accuracy. These GPS accuracy figures are from the 1999 Federal Radio Navigation Plan. They are 95% accurate, and express the value of two standard deviations of radial error from the actual antenna position to an ensemble of position estimates made under specified satellite elevation angle (5°) and PDOP ($< 6°$) conditions. For horizontal accuracy, the figure 95% is the equivalent of 2drms (two-distance root-mean-squared), or twice the radial error standard deviation. For vertical and time errors, 95% is the value of two-standard deviations of vertical error or time error. Receiver manufacturers may use other accuracy measures.

Root-mean-square (RMS) error is the value of one standard deviation (68%) of the error in one, two or three dimensions. Circular Error Probable (CEP) is the value of the radius of a circle, centered at the actual position that contains 50% of the position estimates. Spherical Error Probable (SEP) is the spherical equivalent of CEP, that is the radius of a sphere, centered at the actual position, that contains 50% of the three dimension position estimates. As opposed to root-mean-square and other types of errors, CEP and SEP are not affected by large blundering errors making them an overly optimistic accuracy measure.

Some receiver specification sheets list horizontal accuracy in RMS or CEP and without Selective Availability, making those receivers appear more accurate than those specified by more responsible vendors using more conservative error measures.

6.1.6 Applications. Originally GPS was a military project, but now it is a dual-use technology for military and civilian use. Many civilian applications are divided into the following major areas: (i) Astronomy and celestial mechanics, amateur astronomy using small telescopes, for example finding extra-solar planets; (ii) Cartography, both military and civilian; (iii) Cellular telephony; (iv) Clock synchronization; (v) Disaster relief/emergency services; (vi) Fleet tracking; (vii) Geofencing, including vehicle tracking systems, person and pet tracking systems; (viii) Geotagging that applies location coordinates to digital objects such as photographs for purposes such as creating map overlays; (ix) GPS aircraft tracking; (x) GPS tours, in which the location determines what content to display, e.g., information about an approaching point of interest; (xi) Navigation to obtain data of digitally precise velocity and orientation measurements; (xii) Phasar measurements, which provide highly accurate time stamping of power system measurements, enabling to compute phasars; (xiii) Recreation, e.g., geocaching, geodashing, GPS drawing and waymaking; (xiv) Robotics which is used in self-navigating robots; (xv) Surveying to make maps and determine property boundaries; (xvi) Tectonics to enable direct fault motion measurements in earthquakes; and (xvii) Telemetrics where the GPS technology is integrated with computers and mobile communications technology in automobile navigation systems.

Carrier-phase tracking of GPS signals has resulted in a revolution in land surveying. A line of sight along the ground is no longer necessary for precise positioning. Positions can be measured up to 30 kms from reference point without intermediate points. This use of GPS requires specially equipped carrier tracking receivers. The L1 and/or L2 carrier signals are used in carrier phase surveying. L1 carrier cycles have a wavelength of 19 centimeters. If tracks are measured, these carrier signals can provide ranging measurements with relative accuracies of millimeters under special circumstances.

Tracking carrier phase signals provide no time of transmission information. The carrier signals, while modulated with time tagged binary codes, carry no time-tags that distinguish one cycle from another. The measurements used in carrier phase tracking are differences in carrier phase cycles and fractions of cycles over time. At least two receivers track carrier signals at the same time. Ionospheric delay differences at the two receivers must be small enough to insure that carrier phase cycles are properly accounted for. This usually requires that the two receivers be within about 30 kms of each other.

Carrier phase is tracked at both receivers and the changes in tracked phase are recorded over time in both receivers. All carrier-phase tracking is differential, requiring both reference and remote receiver tracking carrier phases at the same time. Unless the reference and remote receivers use L1-L2 differences to measure the ionospheric delay, they must be close enough to insure that the ionospheric delay difference is less than a carrier wavelength. Using

L1-L2 ionospheric measurements and long measurement averaging periods, relative positions of fixed sites can be determined over baselines of hundreds of kilometers.

Phase difference changes in the two receivers are reduced using software to find differences in three position dimensions between the reference station and the remote receiver. High accuracy range difference measurements with sub-centimeter accuracy are possible. Problems result from the difficulty of tracking carrier signals in noise or while the receiver moves.

6.2 CDMA Principle

GPS is a code division multiple access (CDMA) system, which is based on the *spread-spectrum system* (SS) technique. It was developed for military anti-jamming (spoofing) applications. Its wide bandwidth is necessary to support high bit rates and to combat fading in multi-path radio channels. Many users share the same RF carrier, and each user is assigned a unique random code (spreading sequence number) which is different from other codes and orthogonal to other codes. The sender XORs (mod 2 addition) the signal with this random number. The receiver can 'tune' into this signal if it knows the pseudorandom number. Tuning is done via a correlation function. It is an interference limited system, but quality degenerates as the number of users on a channel (carrier) increases. By spreading, codes keep the channels apart so that the same carrier can be used in the next cell with frequency re-use 1. The details are given in the sequel.

The advantages are as follows: all terminals can use the same frequency, so less planning is required. It has a very large code space (2^{32}) compared to the frequency space. The interference, i.e., the white noise, is not coded. Also, forward error correction and encryption can be easily integrated.

The disadvantages are significant not only to the medium but also to the reception of signals if there is one. All signals should have the same strength at a receiver.

CDMA principle (Uplink) is graphically presented in Figure 6.6, and the coding tree diagram is shown in Figure 6.7.

The binary logical elements 1 and 0 are represented electronically by voltage $+1$ and -1, respectively, and they can be graphically represented by a black square for 1 and a white square for -1. Such black-and-white squares produce Hadamard graphs of the signal frequencies (SFs) of order 2^m, $m = 0, 1, 2, \ldots$. These graphs for $m = 0$ and 1 corresponding to the bits for SF=1, and 2, as shown in Figure 6.7, are as follows:

$$\text{SF} = 1: \blacksquare \ ; \qquad \text{SF} = 2: \ \blacksquare \ .$$

For more information on Hadamard graphs, see Kythe and Kythe [2012: 187].

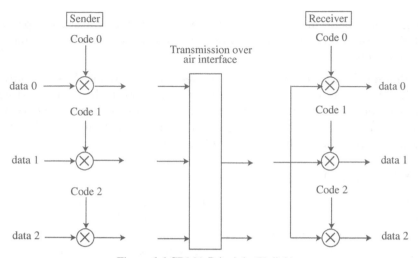

Figure 6.6 CDMA Principle (Uplink).

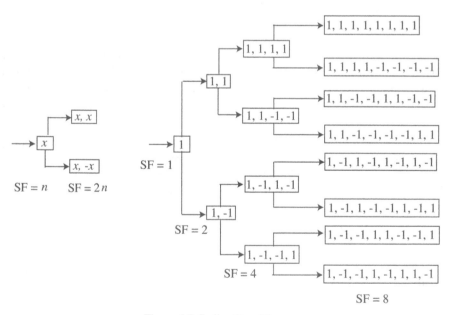

Figure 6.7 Coding Tree Diagram.

Example 6.1. Theoretically, CDMA operates as follows: Suppose there are two senders A and B who transmit two different signals: A sends $A_d = 1$, and code sequence $A_c = 1010011$, so A's signal is $A_s = A_d \cdot A_c = \{+1, -1, +1, -1, -1, +1, +1\}$. B sends $B_d = -1$, and code sequence $B_c = 01101101$, so B's signal is $B_s = B_d \cdot B_c = \{+1, -1, -1, +1, -1, +1, -1\}$, where we have used the Table F.2 (Appendix F) for multiplication of voltages. If we neglect interference, both signals superimpose in space as $A_s +$

$B_s = \{+2, -2, 0, 0, -2, +2, 0\}$. If the receiver wants to receive the signal from sender A, it will apply the sequence A_c chipwise, i.e., take the inner product of $(A_s + B_s) \cdot A_c$, and obtain $A_r = \{+2, -2, 0, 0, -2, +2, 0\} \cdot \{+1, -1, +1, -1, -1, +1, +1\} = 2 + 2 + 0 + 0 + 2 + 2 + 0 = 8$; thus, this result being greater than 0, the original bit was 1. Similarly, B receives $B_r \cdot B_c = \{+2, -2, 0, 0, -2, +2, 0\} \cdot \{+1, -1, -1, +1, -1, +1, -1\} = -2 - 2 + 0 + 0 - 2 - 2 + 0 = -8$, which being negative gives the original bit 0. Suppose another sender C sends a wrong sequence $C_c = 1100110$: then $C_r = \{+2, -2, 0, 0, -2, +2, 0\} \cdot C_c = 0$, which being 0, leads to the no-decision situation. ∎

6.2.1 CDMA Signals. Let a signal s be a sinusoid of the form

$$s = A \sin(2\pi f t + \phi), \tag{6.11}$$

where a is the amplitude, f the frequency, and ϕ the initial phase. These three parameters can be modulated to carry information. If they are modulated, they are respectively known as amplitude modulation, frequency modulation, and phase modulation which is described in Appendix E. The GPS is a phase-modulated signal with $\phi = 0, \pi$, and is called a *binary-phase shift keying* (BSPK) described in Appendix E. The rate of phase change is known as the chip rate, and the spectrum has the shape of the function $\text{sinc}(x) = \sin x / x$ such that the spectrum is proportional to the chip rate. For example, for the chip rate of 1 MHz, the main lobe of the spectrum has a null-to-null width of 2 MHz. Because of this feature, this type of signal is also called a *spread-spectrum system* (SS).

In general, a CDMA signal is a SS system. All signals in this system use the same center frequency, and they are modulated by a set of orthogonal (or near-orthogonal) codes. An individual signal is acquired only if its code is used to correlate with the received signal. Because of the fact that CDMA signals use the same carrier frequency, it is likely that the signals will interfere with one another, particularly when strong and weak signals are mixed together. To avoid this interference, all signals should have approximately the same power levels at the receiver. In the case when a cross-correlation peak of a strong signal is stronger than the desired peak of a signal, the receiver might get wrong information.

6.2.2 Differential GPS (DGPS) Techniques. The idea behind all differential positioning is to correct bias errors at one location with measured bias errors at a known position. A reference receiver, or base station, computes corrections for each satellite signal. Because individual pseudo-ranges must be corrected prior to the formation of a navigation solution, DGPS implementations require software in the reference receiver that can track all SVs in view and form individual pseudo-range corrections for each SV. These corrections are passed to the remote, or rover, receiver which must be capable of applying these individual pseudo-range corrections to each SV used in the navigation solution. Applying a simple position correction from the reference receiver to the remote receiver has limited effect at useful ranges because both receivers would have to be using the same set of SVs in their navigation solutions and have identical GDOP terms (not possible at different locations) to be identically affected by bias errors.

In differential code GPS (navigation), differential corrections may be used in real-time or later, with post-processing techniques. Real-time corrections can be transmitted by radio link. The U.S. Coast Guard maintains a network of differential monitors and transmits DGPS corrections over radio beacons covering much of the U.S. coastline. DGPS corrections are often transmitted in a standard format specified by the Radio Technical Commission Marine (RTCM). Corrections can be recorded for post-processing. Many public and private agencies record DGPS corrections for distribution by electronic means. Private DGPS services use leased FM sub-carrier broadcasts, satellite links, or private radio-beacons for real-time applications. To remove Selective Availability (and other bias errors), differential corrections should be computed at the reference station and applied at the remote receiver at an update rate that is less than the correlation time of SA. Suggested DGPS update rates are usually less than twenty seconds.

DGPS removes common-mode errors, those errors common to both the reference and remote receivers (not multipath or receiver noise). Errors are more often common when receivers are close together (less than 100 km). Differential position accuracies of 1-10 meters are possible with DGPS based on C/A code SPS signals.

All carrier-phase tracking is differential, requiring both a reference and remote receiver tracking carrier phases at the same time. In order to correctly estimate the number of carrier wavelengths at the reference and remote receivers, they must be close enough to insure that the ionospheric delay difference is less than a carrier wavelength. This usually means that carrier-phase GPS measurements must be taken with a remote and reference station within about 30 kilometers of each other. Special software is required to process carrier-phase differential measurements. Newer techniques such as Real-Time-Kinematic (RTK) processing allow for centimeter relative positioning with a moving remote receiver.

6.2.3 GPS Error Sources. GPS errors are a combination of noise, bias, and blunders. Noise errors are the combined effect of PRN code noise (around 1 meter) and noise within the receiver noise (around 1 meter). Bias errors result from Selective Availability and other factors. SA is the intentional degradation of the SPS signals by a time varying bias. SA is controlled by the DoD to limit accuracy for non-U.S. military and government users. The potential accuracy of the C/A code of around 30 meters is changed to 100 meters (two standard deviations). The SA bias on each satellite signal is different, and so the resulting position solution is a function of the combined SA bias from each SV used in the navigation solution. Because SA is a changing bias with low-frequency terms in excess of a few hours, position solutions or individual SV pseudo-ranges cannot be effectively averaged over periods shorter than a few hours. Differential corrections must be updated at a rate less than the correlation time of SA (and other bias errors). Other bias error sources are:

(i) SV clock errors uncorrected by Control Segment can result in one meter errors. Ephemeris data error is 1 meter.

(ii) Tropospheric delays: 1 meter. The troposphere is the lower part (ground level from 8 to 13 km) of the atmosphere that experiences the changes in temperature, pressure, and humidity associated with weather changes. Complex models of tropospheric delay require estimates or measurements of these parameters.

(iii) Unmodeled ionosphere delays: 10 meters. The ionosphere is the layer of the atmosphere from 50 to 500 km that consists of ionized air. The transmitted model can only remove about half of the possible 70 ns of delay leaving a ten-meter unmodeled residual.

(iv) Multipath: 0.5 meters. Multipath is caused by reflected signals from surfaces near the receiver that can either interfere with or be mistaken for the signal that follows the straight line path from the satellite. Multipath is difficult to detect and sometimes hard to avoid. Blunders can result in errors of hundreds of kilometers.

(v) Control segment mistakes due to computer or human error can cause errors from one meter to hundreds of kilometers.

(vi) User mistakes, including incorrect geodetic datum selection, can cause errors from 1 to hundreds of meters.

(vii) Receiver errors from software or hardware failures can cause blunder errors of any size.

(viii) Noise and bias errors combine, resulting in typical ranging errors of around 15 meters for each satellite used in the position solution.

GPS ranging errors are magnified by the range vector differences between the receiver and the SVs. The volume of the shape described by the unit-vectors from the receiver to the SVs used in a position fix is inversely proportional to Geometric Dilution of Precision (GDOP). Poor GDOP, a large value representing a small unit vector-volume, results when angles from receiver to the set of SVs used are similar. Good GDOP, a small value representing a large unit-vector-volume, results when angles from receiver to SVs are different. GDOP is computed from the geometric relationships between the receiver position and the positions of the satellites the receiver is using for navigation. For planning purposes GDOP is often computed from Almanacs and an estimated position. Estimated GDOP does not take into account obstacles that block the line-of-sight from the position to the satellites. Estimated GDOP may not be realizable in the field. GDOP terms are usually computed using parameters from the navigation solution process.

6.3 C/A Code Architecture

There are two types of signals: the coarse (or clear)/acquisition (C/A) and the precision (P) codes.

The GPS signal consists of two radio frequency components: link 1 (L1), and link 2 (L2). The center frequency of L1 is at 1557.42 MHz and of L2 is at 1227.6 MHz. These frequencies are consistent with a 10.23 MHz clock, and related to the clock frequency as

$$L1 = 1575.42 \text{ MHz} = 154 \times 10.23 \text{ MHz},$$

$$L1 = 1227.60 \text{ MHz} = 120 \times 10.23 \text{ MHz}.$$

These frequencies are very accurate in reference to an atomic frequency standard. When the clock frequency is generated, it is slightly lower than 10.23 MHz because of the relativistic effect. The reference frequency is off by -4.567×10^{-3} Hz, which corresponds to a fraction of $-4.567 \times 10^{-10} (= -4.567 \times 10^{-3} \times 10^{-7})$. Thus, the reference frequency used by an SV is $10.23 \times 10^6 - 4.567 \times 10^{-3} = 10.229999995433$ MHz rather than 10.23 HMz. Besides, the SV and GPS receiver produce a Doppler effect[8], and the motion of a SV at L1 creates a Doppler frequency shift of approximately ± 5 kHz.

Currently the L1 frequency contains the C/A and P(Y) signals, while the L2 contains only the P(Y) signal. The C/A and P(Y) signals in the L1 frequency are in quadrant phase (90°) of each other and they can be written as

$$s_{L1} = A_p P(t) D(t) \cos(2\pi f_1 t + \phi) + A_c C(t) D(t) \sin(2\pi f_1 t + \phi), \qquad (6.12)$$

where s_{L1} is the signal at L1 frequency, A_p and A_c denote the amplitude of the P code and C/A code, respectively, $D(t) = \pm 1$ represents the data code, f_1 is the L1 frequency, ϕ is the initial phase, and $C(t) = \pm 1$ represents the phase of the C/A code. Also, the P(Y), C/A, and the carrier frequencies are all phase-locked together. The power levels of GPS signals are as follows:

Frequency	P code	C/A code
L1	−133 dBm	−130 dBm
L2	−136 dBm	−136 dBm[†]

[†] Currently not in L2.

---

[8] For details about the Doppler effect, see Appendix J.

These power levels are very weak and the spectrum is spread, and therefore they cannot be observed directly from a spectrum analyzer. Any amount of amplification of the signal will fail to reveal the spectrum of the C/A code since it will result in noise that is stronger than the signal. Although the minimum beam width of the transmitting antenna to cover Earth is 13.87°, this width is kept wider at 21.3°, as shown in Figure 6.8.

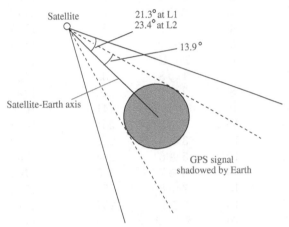

Figure 6.8 Width of GPS Signal Main Beam.

6.3.1 Gold Codes. According to Gold [1967] and Spilker [1978, 1996], the C/A code belongs to the family of pseudorandom noise (PRN) codes known as *Gold codes*. The signals are generated from the product of two 1,023-bit PRN sequences G1 and G2, both of which are generated by a maximum-length linear shift register of 10 stages and are driven by a 1,023 MHz clock (see Figure 6.9). Since the basic operating principles of these two generators are similar, we will discuss only G2. A maximum-length sequence (MLS) generator can be produced from a shift register with proper feedback. If the shift register has m bits, the length of the sequence generated is $2^m - 1$. Both shift generators in G1 and G2 have 10 bits, and thus, the sequence length is $1023(= 2^{10} - 1)$. The feedback circuit is accomplished through XOR (modulo-2) adders. The G1, G2 maximum-lengths are shown in Figure 6.9. The code phase assignments are provided in Table 6.3.

The C/A code generator is shown in Figure 6.10, which has another XOR-adder to generate the C/A code, using the outputs from G1 and G2 as inputs. The initial values of the two registers G1 and G2 are all 1s and they must be located in Register 1. The satellite identification is determined by the two output positions of the G2 generator.

Table 6.3 Code Phase Assignments.

Satellite ID #	PRN Signal #	Code Selection	Code Delay Chips	First 10 Chips C/A Octal
1	1	$2 \oplus 6$	5	1440
2	2	$3 \oplus 7$	5	1620
3	3	$4 \oplus 8$	7	1710
4	4	$5 \oplus 9$	8	1744
5	5	$1 \oplus 9$	17	1133
6	6	$2 \oplus 10$	18	1455
7	7	$1 \oplus 8$	139	1131
8	8	$2 \oplus 9$	140	1454
9	9	$3 \oplus 10$	141	1626
10	10	$2 \oplus 3$	251	1504
11	11	$3 \oplus 4$	252	1642
12	12	$5 \oplus 6$	254	1750
13	13	$6 \oplus 7$	255	1764
14	14	$7 \oplus 8$	256	1772
15	15	$8 \oplus 9$	257	1775
16	16	$9 \oplus 10$	258	1776
17	17	$1 \oplus 4$	469	1156
18	18	$2 \oplus 5$	470	1467
19	19	$3 \oplus 6$	471	1633
20	20	$4 \oplus 7$	472	1715
21	21	$5 \oplus 8$	473	1746
22	22	$6 \oplus 9$	474	1063
23	23	$1 \oplus 3$	509	1063
24	24	$4 \oplus 6$	512	1706
25	25	$5 \oplus 7$	513	1743
26	26	$6 \oplus 8$	514	1791
27	27	$7 \oplus 9$	515	1770
28	28	$8 \oplus 10$	516	1774
29	29	$1 \oplus 6$	859	1127
30	30	$2 \oplus 7$	860	1453
31	31	$3 \oplus 8$	861	1625
32	32	$4 \oplus 9$	862	1712
**	33	$5 \oplus 10$	863	1745
**	34*	$4 \oplus 10$	863	1713
**	35	$1 \oplus 7$	947	1134
**	36	$2 \oplus 8$	948	1456
**	37*	$4 \oplus 10$	950	1713

* 34 and 37 have the same C/A code.
** GPS satellites do not transmit these codes; they are reserved for other uses.

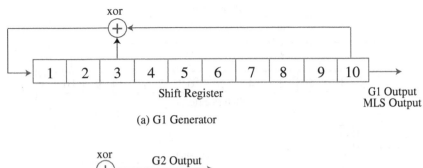

(a) G1 Generator

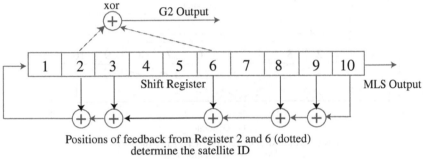

Positions of feedback from Register 2 and 6 (dotted)
determine the satellite ID

(b) G2 Generator

Figure 6.9 G1, G2 Maximum-Length Sequence Generators.

A list of the code phase assignments for the 32 outputs are given in Table 6.3, in which the first column is the satellite number. The second column gives the PRN signal number for satellites 1 through 7. Note that the PRN numbers for signals # 34 and # 37 are the same. The column gives the code phase selections to be used to form the output of the G2 generator. The fourth column gives the code delay measured in chips; this delay is the difference between the MLS output and the G2 output. The last column gives the first 10 bits of the C/A code, in an octal format, generated for each satellite, to check whether the generated code is correct or not.

The two codes, defined for the satellite i, are

$$C_i(t) = \text{G1}(t)\,\text{G2}(t + n_i T_c),\tag{6.13}$$

$$P_i(t) = \text{X1}(t)\,\text{X2}(t + (i-1)T_p),\tag{6.14}$$

where n_i is an integer from 0 to 1023 assigned to satellite i for C/A code, $i = 1, 2, \ldots, 37$ for P code, T_c is the C/A code chip width $= 1/1.023 \times 10^6$ secs, and T_p is the P code chip width $= 1/10.23 \times 10^6$ secs.

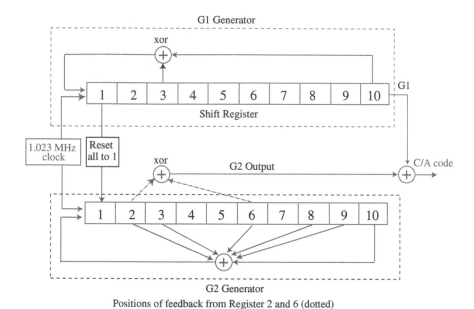

Figure 6.10 C/A Code Generator.

6.3.2 C/A Code and Data Format. According to Spilker [1978, 1996] and Dixon [1976], the C/A code is a bi-phase modulated signal with a chip rate of 1.023 MHz, giving it the null-to-null bandwidth of the main lobe of the spectrum as 2.046 MHz. Each chip is about 977.5 ns (1/1.023 MHz) long.

The transmitting bandwidth of the GPS satellite in the L1 frequency is about 20 MHz so that it can accommodate the P code signal. Thus, the transmitted C/A code contains the main lobe and several sidelobes. The total code period contains 1,023 chips which lasts 1 ms. Hence, the C/A code is 1 ms long, and it repeats itself every millisecond.

A very limited data record of 1 ms is needed to define the beginning of a C/A code. In the absence of a Doppler effect on the received signal, one millimeter of data contains all the 1023 chips. Different C/A codes are used for different satellites.

Example 6.2. Consider Satellite # 21. In view of Table 6.3, to generate the C/A code, the tabs 5 and 8 must be selected for the G2 output. Then the G2 output sequence is delayed 473 chips from the MLS output. The last column is 1746, which means $(1746)_8 = (1111110011)_2$. If the 10 bits generated for this satellite do not match this binary number, the code is incorrect. ∎

6.3.3 GPS Signal Waveform: L1 Channel. This waveform is shown in Figure 6.11. The data for GPS signal received power is as follows:

L1 Frequency (1575.42 MHz): (i) For C/A code : -160 dBW (10^{-16} watts); and (ii) for P code : -163 dBW (5×10^{-17} watts).

L2 Frequency (1227.6 MHz): P code : -166 dBW (25×10^{-17} watts).

In all these three cases, the actual levels are 5–6 dB higher, and the above data includes 2 dB atmospheric loss.

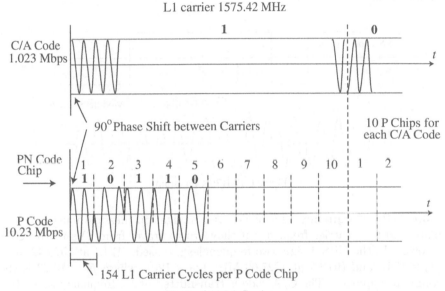

Figure 6.11 L1 Carrier.

6.3.4 DSSS. Digital communications use electronic pulses of amplitude +1 and −1 to form a binary code, as in §6.2. These pulses are then modulated to yield a clear signal using, in most cases, direct sequence spread spectrum (DSSS). This technique is used in GPS, and radio communication technologies such as CDMA (mostly used by Verizon Wireless).

Each pulse modulated by DSSS is called a chip. The chip rate of a device is the number of chips or pulses it received per second. It is measured in megachips (millions of chips per second). For example, the chip rate of GPS L1 C/A code is 1.024 MHz, or 1024 megachips per second.

The chip rate of a communication device contains the information signal, the useful data being transmitted, and the noise signal used to uniquely codify the information. In this sense, the chip rate amounts to several times the data transfer rate.

6.3.5 BPSK. As mentioned earlier, each GPS satellite transmits a unique navigation system centered on two L-band frequencies of the electromagnetic spectrum: L1 at 1575.42 MHz and L2 at 1227.60 MHz. The PRN codes and navigation message modulate L1 and/or L2 carrier using the binary phase shift keying (BPSK) modulation technique [Kaplan, 1990], in which the binary codes are directly multiplied with the carrier. This results in a 180° phase shift of the carrier each time the code changes its state.

A BPSK diagram is presented in Figure 6.12 [Braasch and Van Dierendonck, 1999].

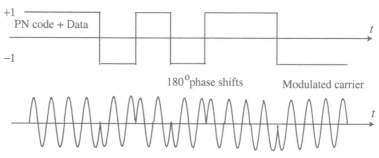

Figure 6.12 BPSK Diagram.

6.3.6 GPS Frequency. The codes and navigation message must be modulated on to a carrier frequency if they are to travel from SVs to the GPS receiver. In the GPS design, two frequencies are used: (i) L1 at 1575.42 MHz ($f_0 \times 154$), and (ii) L2 at 1227.60 MHz ($f_0 \times 120$), where $f_0 = 10.23$ is the common frequency. The C/A code is transmitted at L1 frequency as a 1,023 MHz signal using BPSK modulation, whereas the P(Y) code is transmitted on both L1 and L2 frequencies as a 10.23 MHz signal using the same BPSK modulation, with the difference that the P(Y)-code carrier is in quadrature phase with the C/A carrier, i.e., it is 90% out of phase. The advantage in this setup is that, besides redundancy and increased resistance to jamming, the two-frequency transmission from each of the SVs enables us to measure directly, and hence remove, the ionospheric delay error for the SV in use. Such an error is very critical and must be reduced to zero or to a negligibly small quantity so that both the SV and the receiver perform more accurately. A receiver with these features is called a *dual frequency receiver*.

All SV signals are modulated onto the same L1 carrier frequency, and they must be separated after demodulation. This is achieved by assigning each SV a unique binary sequence known as a Gold code, and signals are decoded after demodulation using modulo 2 addition (XOR) of the Gold codes corresponding to SVs i_1 through i_k, where k is the number of channels in the GPS receiver and i_1 through i_k are the PRN identifiers of the SVs. Each SV's PRN is unique, in the range from 1 through 32. As a result of these modulo

2 addition (XOR), the 50 bit/s navigation messages from SV i_1 through i_k are obtained. The Gold codes used in the GPS are a sequence of 1,023 bits with a period of 1 millisecond. These Gold codes are mutually orthogonal, so it is unlikely that one SV signal will be confused as from another. Also, all Gold codes behave like auto-correcting codes. There are a total of 1025 different Gold codes, but only 32 of them are used in GPS.

Initially, in the case of both C/A and P codes, the Galois field arithmetic was performed on $F(2^m)$, $m = 10$. There are two primitive polynomials of interest in this case. They are $g_1(x) = 1 + x^4 + x^{10}$, and $g_2(x) = 1 + x^2 + x^3 + x^6 + x^8 + x^9 + x^{10}$. Unlike tables available in Kythe and Kythe [2012: 163] for 2^3 elements, the addition and multiplication tables of $2^{10} = 1024$ elements, are only theoretically possible. The 1022 powers of the root α for either primitive polynomial will be the set $\{1 = \alpha^0, \alpha = \alpha^1, \alpha^2, \ldots, \alpha^{1022}\}$, followed by $\alpha^{1023} = 1$, and the set thereafter repeats. For example, in the case of the polynomial $p_1(x)$ the relation $\alpha^{10} = \alpha^3 + 1$ will yield the set $\{0, 1, \alpha, \alpha^2, \ldots, \alpha^9, \alpha^{10} = \alpha^3 + 1, \alpha^{11} = \alpha^4 + \alpha \ldots, \alpha^{20} = (\alpha^3 + 1)^2 = \alpha^6 + 1, \ldots, \alpha^{1022} = \alpha^3 + 1\}$, after which the set repeats. However, instead of going through such a monumental task, shift registers are used. The initial state of shift registers is the all one state. Addition is done mod 2 (XOR). The tabs of the lower shift register are different for different SVs. The sequence has 1023 chips[9]. The period of a sequence is 1 ms. Sequence of 0s and 1s are mapped to $+1$ and -1, i.e., $0 \rightarrow 1$, $1 \rightarrow -1$. The LFSRs are used to compute the values of the zeros α^i, $i = 1, \ldots, 1022$, of these irreducible primitive polynomials. A short introduction to the Galois field and the zeroes of irreducible primitive polynomials of small orders is presented in Appendix G.

The general rule of modulo 2 arithmetic is: when the two inputs are the same, the output is 0, otherwise it is 1. The position of the feedback circuit determines the output pattern of the sequence. For example, the feedback of G1 (polynomial $g_1(x)$) is from bits 3 and 10, while that of G2 (polynomial $g_2(x)$) is from bits 2, 3, 6, 8, 9, 10.

Once the receiver has obtained the ephemeris data, it knows, by their PRNs, which of the SVs to listen to. But if this information is not in memory, the receiver goes into a search mode and searches through the PRNs until it is locked to one of the SVs. To obtain a lock there must be an unobstructed line of sight from the receiver to the SV. The receiver can then obtain the ephemeris data and decide which SV it should listen for. As it detects each SV's signal, the receiver identifies the correct SV by its distinct C/A code.

---

[9] Electronic pulses of amplitude $+1$ and -1 are used in digital communications to form a binary code. These pulses are then modulated to obtain a clear signal using a variety of techniques, one of which mostly used is direct sequence spread spectrum (DSSS). Each pulse modulated by DSSS is called a *chip*. The chip rate of a device is the number of chips or pulses emitted per second. It is measured in megachips, or millions of chips per second. For example, the chip rate of a GPS L1 C/A code is 1.024 MHz or 1.024 megachips per second.

A graphic flowchart for demodulation and decoding GPS SV signals using Coarse/Acquisition (C/A) Gold code is presented in Figure 6.16 (on page 216).

As mentioned erlier, while receiving successive PRN sequences, the receiver will encounter a sudden change in the phase of the 1,023 bit received PRN signal. This situation will indicate the start of a data bit of the navigation message, and enable the receiver to begin reading the 20 millisecond bits of the navigation message. The telemetry word (TLM) at the beginning of each subframe of a navigation frame enables the receiver to detect the beginning of a subframe and determine the receiver's clock time at which the navigation subframe began. The handover word (HOW) then enables the receiver to determine which specific subframe is being transmitted. After a subframe is read and interpreted, the time at which the next subframe was sent is computed using the clock correction data and the HOW, both of which are known to the receiver. Next, the ephemeris data from the navigation message is used to precisely compute the position of the SV at the start of the message. The GPS frequencies are listed in Table 6.4.

Table 6.4 GPS Frequencies.

Band	Freq(MHz)	Phase	Original Use	Modern Use
L1	1575.42	In-Phase(I)	P(Y) code	
		Quad.-Phase(Q)	C/A code	C/A, L1C, and M code
L2	1227.60	In-Phase(I)	P(Y) code	
		Quad.-Phase(Q)	(a)	L2C code and M code
L3	1381.05		(b)	
L4	1379.913		(c)	(d)
L5	1176.45	In-Phase(I)	(c)	(e)
		Quad.-Phase(Q)		(e)

LEGEND: Freq = Frequency; P(Y) code = Encrypted Precision P(Y) code; Quad.-Phase = Quadrature-Phase; C/A = Coarse-acquisition code; L1C = L1 Civilian code; L2C = L2 Civilian code; M code = Military code; (a) = unmodulated carrier; (b) = used by Nuclear Detonation (NUDET) Detection System Payload (NDS); signals nuclear detonations/high-energy infrared events; used to enforce nuclear test ban treaties; (c) = no transmission; (d) under study for additional ionospheric correction; (e) = safety of life (SoL) data signal. Note that in the Frequency column, $10.23 \times 154 = 1575.42; 10.23 \times 120 = 1227.60; 10.23 \times 135 = 1381.05; 10.23 \times 1214/9 = 1379.913; 10.23 \times 115 = 1176.45$, where $f_0 = 10.23$ (MHz) is the common frequency.

6.3.7 Navigation. The L1 C/A navigation message is 1800 bits and is transmitted at 100 bits. It contains 9 bits of time information, 600 bits of ephemeris data, and 274 bits of packetized data payload. All SVs always broadcast at the two frequencies: 1575.42 MHz for L1 signal and 1227.6 MHz for L2 signal.

The SV network uses a CDMA spread-spectrum technique in which the low bit-rate message data is encoded with a high-rate pseudo-random (PRN) sequence which varies from SV to SV. This demands that the GPS receiver must know the PRN codes for each SV to reconstruct the correct message data. For example, the C/A code transmits data at 1.023 million chips per sec, whereas the P code at 10.23 million chips per sec. The L1 carrier is modulated by both C/A and P codes, while the L2 carrier only by the P code.

The P code can be encrypted as P(Y) code, and both C/A and P(Y) codes transmit precise time to the receiver. Each in-phase and quadrature-phase signal $x(t)$ is defined by

$$
\begin{aligned}
x(t) &= \sqrt{P_I}\, x_I(t) \cos(\omega t + \phi_0) - \sqrt{P_Q}\, x_Q(t) \sin(\omega t + \phi_0) \\
&= \sqrt{P_I}\, x_I(t) \cos(\omega t + \phi_0) + \sqrt{P_Q}\, x_Q(t) \cos(\omega t + \phi_0) \\
&= \left(\sqrt{P_I}\, x_I(t) + \sqrt{P_Q}\, x_Q(t) \right) \cos(\omega t + \phi_0),
\end{aligned}
$$

where P_I and P_Q denote signal power, and x_I and x_Q the codes with/without data $(= \pm 1)$.

6.3.8 GPS Signal Structure. The parameters used in GPS structure are shown in Table 6.5. The mathematical representation of signals is as follows: The signals L1 and L2 are defined as

$$
s_{L1} = A P(t) D(t) \cos\left(2\pi f_1 t + \phi_{01}\right) + \sqrt{2}\, A C(t) D(t) \sin\left(2\pi f_1 t + \phi_{01}\right),
\tag{6.15}
$$

$$
s_{L2} = \frac{A}{\sqrt{2}}\, P(t) D(t) \cos\left(2\pi f_2 t + \phi_{02}\right),
\tag{6.16}
$$

where A is the L1 P signal amplitude, $P(t)$ and $C(t) = \pm 1$ are P and C/A code RPN sequences, $D(t) = \pm 1$ the message data bit sequence, f_1 and f_2 are L1 and L2 carrier frequencies, respectively, and ϕ_{01} and ϕ_{02} are the ambiguous L1 and L2 carrier phases, respectively.

Table 6.5. GPS Parameters.

Parameter	C/A Signal	P Signal
Code clock rate R_c	1.023 mchips/sec	10.23 mchips/sec
Code length	1023 chips (1 ms)	$\sim$ 6 trillion chips (1 week)
Data[†] rate	50 bits/sec	50 bits/sec
Transmission Frequency	L1=1575.42 MHz =154 R_c	L1=1575.42 MHz =154 R_c 2=1227.6 MHz =120 R_C

[†] Data includes: (i) Telemetry; (ii) synchronization information (Preamble, time); (iii) satellite clock and ephemeris parameters; (iv) almanacs; and (v) ionospheric delay and UTC time models.

The first 10 bits of the generated C/A code can be compared with the result listed in Table 6.6, which provides the cross-correlations of the Gold code.

Table 6.6 Cross-Correlation of Gold Code.

Code Period	Number of Shift Register Stages	Normalized Cross- Correlation Level	Probability of Level
$p = 2^m - 1$	n = even	$-\dfrac{2^{(m+2)/2}+1}{p}$	0.125
		$-\dfrac{1}{p}$	0.75
		$-\dfrac{2^{(m+2)/2}-1}{p}$	0.125

Example 6.3. Consider the C/A code n = even = 10. Then $p = 1023$. Using the relations in Table 6.6, the cross-correlation values are: -65/1023 (occurs 12.5%); -1/1023 (occurs 75%); and 63/1023 (occurs 12.5%). The autocorrelation of the C/A codes of satellite 19 shows that the maximum of the autocorrelation peak is 1023, which equals the C/A code length. The cross-correlation of satellite 31 (arbitrarily chosen) has three values: 63, -1, and 65, calculated using Table 6.6. These values match exactly with those for satellite 19. The difference between the maximum of the autocorrelation to the cross-correlation determines the processing gain of the signal. The outputs

from the C/A code generator must be 1 and -1, rather than 1 and 0. ∎

6.3.9 C/A Code Acquisition. For acquisition of GPS C/A code signals, we will discuss the following methods: conventional and FFT, and delay and multiplication. The conventional and the FFT methods generate the same results, since FFT is a reduced computational version of the conventional method. The delay and multiplication methods can operate faster than the FFT, but with inferior performance. This will, therefore, involve a tradeoff between these two approaches. Thus, if the signal is strong, the desirable approach is the fast low-sensitivity acquisition method; if the signal is weak, the low-sensitivity approach will miss it but the conventional method will find it; and if the signal is very weak, the long data length acquisition method should be used. A desirable architecture to achieve fast acquisition is clearly a proper combination of these methods.

The following steps are required for an acquisition program:

STEP 1: Use a fast process to search for satellites visible to the GPS receiver.

STEP 2: Acquire the signal through hardware in the time domain, in which the acquisition is continuously performed on the input data. In case a software receiver is used, this acquisition is performed on a block of data. Some receivers are capable of this acquisition on many satellites in parallel.

STEP 3: Once the desired signal is found, pass the information to a tracking program. In case the receiver is operating in real time, the tracking program will work on data collected in STEP 2.

There is a time elapse between the data used for acquisition and the data being tracked. If the acquisition is slow, the receiver may not be able to track the signal, because of the time elapse and the acquired data being out-of-date. However, if the software receiver does not operate in real time, the acquisition time is not critical as the tracking program is able to process the stored data. Thus, the receiver should work in real time.

The C/A code is 1 ms long, which implies that at least 1 ms of data must be acquired. But in order to avoid data transition in the data, one should take two consecutive data sets to perform acquisition. In this way the data length can be increased to a maximum of 10 ms. If two consecutive 10 ms of data are taken to perform acquisition, there will be no data transition. In practice, the probability that a data record larger than 10 ms will not contain a data transition is found to be good.

Another important factor is the carrier frequency separation in the acquisition. Since the Doppler frequency range is ±10 kHz, the frequency step to cover it should be 20 kHz. The frequency depends on the length of the data used in the acquisition. Thus, when the input signal and the locally generated complex signal are off by 1 cycle, there is no correlation. When two signals are off by less than 1 cycle, there is partial correlation. By convention, the maximum frequency separation should be at 0.5 cycle in 1 ms. If the data record is 1 ms, a signal of 1 kHz will change 1 cycle in 1 ms. To maintain the maximum frequency separation at 0.5 cycle in 1 ms, the frequency step should be 1 kHz. For example, if the data record is 10 ms, a searching frequency step of 100 Hz is necessary. In other words, the frequency separation is the inverse of the data length, which is the same as a conventional FFT result.

6.3.10 Serial Search Algorithm. This feature is used to assess the available satellite and coarse values of the carrier frequency and code phase of the satellite signals. The code phase of each satellite is the time alignment of the PRN code in the data block currently in use, and it is used to generate a local PRN code which is in perfect alignment with the incoming code. The PRN codes have high correlation only when the phase lag is zero.

Another feature is the frequency of the carrier signal which corresponds to the intermediate frequency when in down-conversion. This intermediate frequency is known from the L1 carrier frequency (1575.42 MHz) and from the mixers in the down-converter, although its value may vary from its expected value. One cause is the Doppler effect that results in higher or lower frequency, and in the worst case it may vary ± 10 kHz. In any case the frequency of the signal must be known so as to generate a local carrier signal.

In any integrated circuit, the P code acquisition is performed in an ordinary GPS receiver. However, in a software receiver, its performance is software-oriented. A serial search acquisition is a common method for acquisition of CDMA, and its block diagram for the serial search algorithm is presented in Figure 6.13. The algorithm uses the multiplication (mod 2) of locally generated PRN code sequences related to the specific satellite. The code so generated has a code phase anywhere from 0 to 1022 chips. The input signal is initially multiplied ($\otimes$) by the PRN code generator; this locally generates a PRN sequence; thereafter the signal is multiplied by a locally generated carrier signal. This generates the in-phase signal I, while multiplication by a 90° phase-shifted version of the locally generated carrier signal generates the quadrature signal Q. Both the I and Q signals are then integrated over 1 ms, corresponding to the length of the C/A code, and finally squared, and added, indicated by $\int$, $(\ )^2$, and Σ, respectively, in Figure 6.13.

This serial search algorithm goes through two different sweeps: a frequency sweep over all possible carrier frequencies of intermediate frequency ± 10 kHz in steps of 500 Hz, and a code phase sweep over all 1023 different code phases, from 0 to 1022, yielding a sum total of a very large number of combinations, which are

$$\underbrace{1023}_{\text{code phases}} \underbrace{\left(2 \frac{10000}{500} + 1 \right)}_{\text{frequencies}} = 1023 \cdot 41 = 41943 \text{ combinations.} \qquad (6.17)$$

Therefore this serial search algorithm is not very robust, although its implementation is straightforward, as can be seen from Figure 6.13.

Legend: ⊗: multiplication mod 2 (using Table F.2, Appendix F).

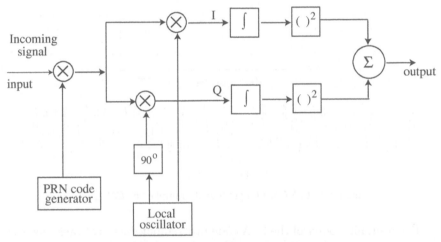

Figure 6.13 Serial Search Algorithm.

6.3.11 Time Domain Correlation. The acquisition process is based on the idea of despreading the input signal and finding the carrier frequency. If the C/A code with the correct phase is multiplied on the input signal, the signal will become a continuous wave or continuous waveform (CW) signal, as shown in Figure 6.14, where, for the sake of illustration only, plot (a) is the input signal (a radio frequency (RF) signal) phase coded by a C/A code (it does not represent the signal transmitted by a satellite); plot (b) represents the C/A code that has values ±1; and the plot (c) is a CW signal that represents the multiplication result of (a) and (b). The frequency of a CW signal can be determined from the FFT operation.

The desired frequency is the highest-frequency component crossing the threshold. For example, if the signal is digitized at 5 MHz, 1 ms of data will contain 5000 data points. A 5000-point FFT will generate 5000 frequency components. However, only the first 2500 of these frequency components contain useful information, whereas the last 2500 frequency components are the complex conjugates of the first 2500 components. Since the frequency resolution is 1 kHz, the total frequency range covered by the FFT is 2.5 MHz (half of the sampling frequency). Since the suggested frequency range is only 20 kHz, and not 2.5 MHz, computation of only 21 frequency components separated by 1 kHz are needed, using DFT, which saves computation time. The decision depends on the speed of the two operations.

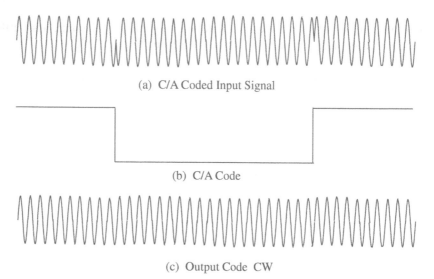

(a) C/A Coded Input Signal

(b) C/A Code

(c) Output Code CW

Figure 6.14 C/A Coded Input Signal Multiplied by C/A Code.

The beginning point of the C/A code in the input data is always unknown, and therefore must be determined. To find it, a locally generated C/A code must be digitized into 5000 points; and then multiplied with the input data point-by-point. The FFT or DFT is applied on this product to find the frequency. To search for 1 ms of data, the input data and the locally generated complex signals, one must move 5000 times against each other. If FFT is used, it will require 5000 operations, where each operation will consist of 5000-point multiplication and a 5000-point FFT. As mentioned above, the outputs are 5000 frames of data, each containing 2500 frequency components (recall that only 2500 frequency components provide useful data and the remaining 2500 components provide redundant information). There are then a total of 1.25×10^7 (5000×2500) outputs in the frequency domain available. Thus, the maximum amplitude among all these outputs is taken as the desired result if it also crosses the threshold. But the search for the maximum value among these large data is time consuming. Since only 21 frequencies of the FFT outputs which cover the desired 20 kHz are required, the total outputs can be reduced to $5000 \times 21 = 105000$. This will then provide the beginning point of the C/A code with a time resolution of 200 ns (1/5 MHz) and the frequency resolution of 1 kHz. If 10 ms of data are used, then the operation time increases from 1 ms to 10 ms. Although the time resolution for the beginning of the C/A code remains at 200 ns, the frequency resolution increases to 10^5 ns (100 Hz).

The GPS carrier modulation signals are presented in Figure 6.15.

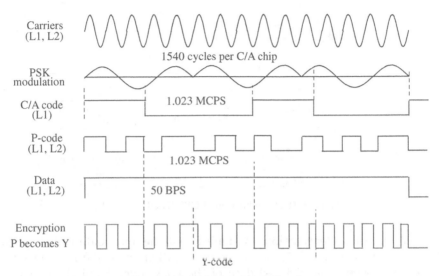

Figure 6.15 GPS Modulation Signals.

To prevent unauthorized users from interfering with the military signal, the P code is encrypted by modulating it with the W code which is a special encryption sequence. This generates the Y code which is transmitted by the satellite only after the spoofing module is set to the 'on' state. The encrypted signal is called the P(Y) code. The details of the W code are secret. However, it is known that it is applied to the P code at about 500 kHz (Litton et al. [1996]). This rate is slower than that of the P code by a factor of about 20.

All satellite signals are modulated onto the same frequency. Therefore it is necessary to separate the signals after demodulation. This is accomplished as follows: Assign each satellite $S_1, S_2, \ldots, S_k$ a unique binary sequence, which is simply the Gold code assigned to each satellite; then decode each signal after demodulation using mod 2 addition (XOR) of the Gold code for each satellite which are singled out by their PRN identifiers. This process is unique for each satellite and ranges from $k = 1, \ldots, 32$.

Finally, the results of these mod 2 additions provide the 50 bits/s navigation messages from each of the 32 satellites. The demodulating and decoding process of GPS satellite signals from satellites $S_1, S_2, \ldots, S_k$, using the C/A code, is presented as a flowchart in Figure 6.16, in which the details are provided for the satellites S_1 and S_2, but the demodulating and decoding processes repeat for the remaining satellites.

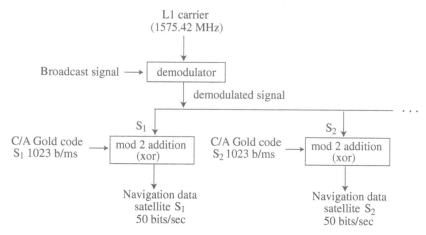

Figure 6.16 Demodulation of GPS Satellite Signals.

6.3.12 Domain Correlation. Assuming that the input data is digitized at 5 MHz, it will require a 5000-point multiplication every 200 ns in order to generate a 5000-point digitized data of the C/A code. Similarly, frequency analysis using a 5000-point FFT is performed on the products every 200 ns. A schematic scheme of such a setup is shown in Figure 6.17. This process generates 1.25×10^7 outputs in the frequency domain, as calculated above. However, the sorting process is much simplified because of the limitation of ±10 kHz imposed on the outputs within the frequency range.

An alternate approach to frequency domain implementation is to use DFT. Since the locally generated code consists of a C/A code and an RF, where the RF is complex-valued of the form $e^{i\omega t}$, the local code is the product of the complex RF and the C/A code, and hence is complex-valued. Let us assume that the L1 frequency (1575.42 MHz) is converted to 21.25 MHz and digitized at 5 MHz: then the output frequency of L1 is 1.25 MHz. Since there are 21 frequency components in the frequency acquisition of range 1250 ± 10 kHz in 1 kHz steps, let the local code l_{sj} be represented as

$$l_{sj} = C_s\, e^{2\pi i f_j t},$$

where the index s denotes the number of satellites and index $j = 0, 1, \ldots, 21$; C_s is the C/A code of the satellite s, and $f_j = 1250 - 10, 1250 - 9, 1250 - 8, \ldots, 1250+9, 1250+10$ kHz. The local signal is further digitized at 5 MHz to generate 5000 data points. Thus, the 21 data sets represent the 21 frequencies separated by 1 kHz.

If an input signal transmits through a linear and time-invariant system, the output can be determined either in the time domain using convolution, or in the frequency domain using Fourier transform. In the former case, let $x(t)$ and $y(t)$ denote the input signal and its output (through convolution), respectively. Let the impulse response to the system be $g(t)$. Then, in view of (3.11),

$$y(t) = \int_{-\infty}^{\infty} x(t-u)g(u)\, du = \int_{-\infty}^{\infty} x(u)g(t-u)\, du. \qquad (6.18)$$

In the latter case (frequency domain) the response of $y(t)$ is, using Fourier transform, given by

$$
\begin{aligned}
Y(f) &= \int_{-\infty}^{\infty}\int_{-\infty}^{\infty} x(u)g(t-u)\, du\, e^{2\pi i ft}\, dt \\
&= \int_{-\infty}^{\infty} x(u)\left(\int_{-\infty}^{\infty} g(t-u)\, e^{2\pi i ft}\right) dt \\
&= \int_{-\infty}^{\infty} x(u)\left(\int_{-\infty}^{\infty} g(u)\, e^{-2\pi i, f\tau}\, d\tau\right)e^{-2\pi i fu}\, du, \quad \text{setting } t-u = \tau \\
&= G(f)\int_{-\infty}^{\infty} x(u)\, e^{-2\pi i fu}\, du = G(f)X(f). \qquad (6.19)
\end{aligned}
$$

To find the output in the time domain, we use the inverse Fourier transform and get

$$y(t) = x(t) \star g(t) = \mathcal{F}^{-1}\left[X(f)G(f)\right]. \qquad (6.20)$$

Similarly, in the frequency domain where a convolution is equivalent to multiplication in the time domain, we have the duality of convolution

$$x(t) \star g(t) \rightleftharpoons X(f)G(f), \quad X(f) \star G(f) \rightleftharpoons x(t)g(t). \qquad (6.21)$$

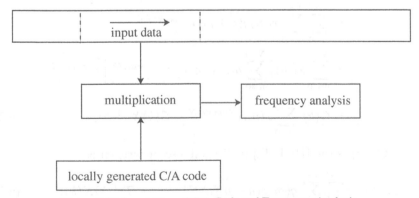

Figure 6.17 Acquisition of C/A Code and Frequency Analysis.

In discrete time, if $x(n)$ is an input signal, the response $y(n)$ is represented as

$$y(n) = \sum_{m=0}^{N-1} x(m)g(n-m),$$ (6.22)

where $g(n-m)$ is the system response. The time shift in $g(n-m)$ is circular because of the periodicity of the discrete operation. Taking DFT of (6.22) we get

$$Y(k) = \sum_{n=0}^{N-1}\sum_{m=0}^{N-1} x(m)g(n-m)\,e^{-2\pi i\,kn/N}$$

$$= \sum_{m=0}^{N-1} x(m)\left[\sum_{n=0}^{N-1} g(n-m)\,e^{-2\pi i\,(n-m)k/N}\right]e^{-2\pi i\,mk/N}$$

$$= G(k)\sum_{m=0}^{N-1} x(m)\,e^{-2\pi i\,mk/n} = X(k)G(k).$$ (6.23)

The results (6.22) and (6.23) are called as *periodic convolution* or *circular convolution*. However, this approach does not produce the expected result of a linear convolution. For example, if $x(n)$ and $y(n)$ both have N data points (samples), the output must consist of $2N - 1$ points (samples). But, by Eq (6.23) and because of the periodicity, it is easy to see that the output has only N points. That is why the acquisition algorithm does not use convolution. Instead it uses correlation $z(n)$, which is defined between $x(n)$ and $g(n)$ as

$$z(n) = \sum_{m=0}^{N-1} x(m)g(n+m).$$ (6.24)

This equation differs from (6.22) in the $\pm$ sign of the term $g(n \pm m)$. Recall that $g(n)$ is another signal, and not the pulse response of a linear system. The DFT on $z(n)$ yields

$$Z(k) = \sum_{n=0}^{N-1}\sum_{m=0}^{N-1} x(m)g(n+m)\,e^{-2\pi i\,kn/N}$$

$$= \sum_{m=0}^{N-1} x(m)\left[\sum_{n=0}^{N-1} g(n+m)\,e^{-2\pi i\,(n+m)/N}\right]e^{2\pi i\,mk/N}$$

$$= G(k)\sum_{m=0}^{N-1} x(m)\,e^{2\pi i\,mk/N} = G(k)X^{-1}(k),$$ (6.25)

where $X^{-1}(k)$ is the IDFT. Eq (6.25) can also be written as

$$Z(k) = \sum_{n=0}^{N-1}\sum_{m=0}^{N-1} g(m)x(n+m)\,e^{-2\pi i\,kn/N} = G^{-1}(k)X(k).$$ (6.26)

Since $\overline{x(n)} = x(n)$ for real $x(n)$, the magnitude of $Z(k)$ is given by

$$|Z(k)| = |\overline{G(k)}\,X(k)| = |G(k)\,\overline{X(k)}|. \tag{6.27}$$

This result provides a periodic (circular) correlation, and is hence known as *circular correlation*. It is used to determine the correlation of the input signal and the locally generated signal.

The following steps are needed to perform the acquisition on the input data:

STEP 1: Carry out the FFT on the 1 ms of input data $x(n)$ and convert the input into frequency domain as $X(k)$ for $k = 0, 1, \ldots, 4999$ for each 1 ms of data.

STEP 2: Determine the conjugate $\overline{X(k)}$ which become the outputs.

STEP 3: Use Eq (6.15) to generate 21 local codes l_{sj}, $j = 1, 2, \ldots, 21$. The frequencies f_j of the local codes are separated by 1 kHz.

STEP 4: Transform l_{sj}, using FFT, to the frequency domains L_{sj}.

STEP 5: Compute $\overline{X(k)} \times L_{sj}(k)$ point-by-point, and call the product $R_{sj}(k)$, $k = 1, 2, \ldots, 21$.

STEP 6: Take IFFT of $R_{sj}(k)$; this yields the result in time domain as $r_{sj}(n)$. Find all values of $|r_{sj}(n)|$ (a total of $21 \times 5000 = 105{,}000$ values).

STEP 7: The maximum of $|r_{sj}(n)|$ in the nth location and jth frequency is the beginning point of the C/A code in 200 ns resolution in the input data and the carrier frequency in 1 kHz resolution.

The above process is schematically illustrated in Figure 6.18.

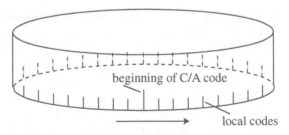

Figure 6.18 Acquisition with Periodic Correlation.

6.3.13 Delayed Signal. If the frequency information is eliminated in the input signal, the C/A code itself will determine its own initial point. Once the C/A code is found, the frequency can be determined using either FFT or DFT. Let $s(t)$ be a complex input signal defined by

$$s(t) = C_s(t)\,e^{2\pi i\,ft}, \tag{6.28}$$

where $C_s(t)$ denotes the C/A code of satellite s. In view of the translation property (§3.1.2(3)), the delayed version of this signal is

$$s(t - u) = C_s(t - u)\, e^{2\pi i\,(t-u)}, \tag{6.29}$$

where u is the delay time. Then the product

$$s(t)\,\overline{s(t - u)} = C_s(t)\overline{C_s(t - u)}\, e^{2\pi i\,ft}\, e^{-2\pi i\,f(t-u)} \equiv C_n\, e^{2\pi i\,fu}, \tag{6.30}$$

where

$$C_n(t) = C_s(t)\overline{C_s(t - u)}, \tag{6.31}$$

can be regarded as a 'new code', which is the product of a Gold code and its delayed version, and belongs to the family of Gold codes. The beginning point of this new code is the same as that of the C/A code. Note that Eq (6.30) is independent of frequency since the term $e^{2\pi i\,fu}$ is constant because both f and u are constant. However, the input signal of the new code is complex. To determine its initial point, we will use the Hilbert transform (see Appendix B). Following Lin and Tsui [1998], let $s(t)$ be a real signal $s(t)$ with its delayed version $s(t - u)$, defined respectively by

$$s(t) = C_s(t)\, \sin(2\pi ft), \quad s(t - u) = C_s(t - u)\, \sin\left(2\pi f(t - u)\right) \tag{6.32}$$

Then their product is given by

$$s(t)s(t - u) = C_s(t)C_s(t - u)\, \sin(2\pi ft)\, \sin\left(2\pi f(t - u)\right)$$
$$= \frac{C_n(t)}{2}\left\{ \cos(2\pi fu) - \cos\left(2\pi f(2t - u)\right) \right\}. \tag{6.33}$$

This equation has two terms: a DC term and a high-frequency term. The high-frequency term is usually filtered out. For this purpose, the quantity $|\cos(2\pi fu)|$ must be close to 1, which is theoretically very difficult to achieve as the frequency f is unknown. However, since the frequency is known to be within 1250 ± 10 kHz, a delay time can be chosen such that the frequency stays within these bounds. For example, take $2\pi \times 1250 \times 10^3 u = \pi$, which gives $u = 0.4 \times 10^{-6}$ s $= 400$ ns. Since the input data is digitized at 5 MHz, the sampling time is 200 ns (1/5 MHz). Thus, if the input signal is delayed by two samples, the delay time $u = 400$ ns. Then, $|\cos(2\pi fu)| = |\cos(\pi)| = 1$. If the frequency is off by 10 kHz, then the value of $|\cos(2\pi fu)| = |\cos\left(2\pi \times 1260 \times 10^3 \times 0.4 \times 10^{-6}\right)| = 0.9997 \sim 1$. Again, if the delay time $u = 1.6$ us, and frequency is off by ± 10 kHz, then $|\cos(2\pi fu)| = 0.995$. Hence, we find that the quantity $|\cos(2\pi fu)|$ decreases faster, the longer the delay time is for a frequency off the center value of 1250 kHz; so a very long delay time may not make this quantity close to unity. The main difficulty lies in the fact that when two signals with noise are multiplied together, the noise always increases, and it is not possible to search for just 1 ms of data to acquire a satellite. More data is needed for this purpose.

6.4 P Code Architecture

The Precision (P) code is also a PRN. However, each SV's P code PRN is 6.1871×10^{12} bits long (approximately equal to 720.213 gigabytes) and only repeats once a week as it is transmitted at 10.23 Mbits/s. The extreme length of the P code increases its correlation gain and eliminates any range of ambiguity within the Solar system. This code is very long and complex, so that it was believed that a receiver could not directly acquire and synchronize with this signal alone. The C/A PRNs are unique for each SV, but the P code PRN is, in fact, a small segment of a master P code which is approximately 2.35×10^{14} bits long (≈ 26.716 terabytes), and each SV repeatedly transmits its assigned segment of the master code.

According to Spilker [1978, 1996], the P code is bi-phase modulated at 10.23 MHz; thus, the width of its main lobe of the spectrum is 20.46 MHz from null-to-null. The chip length is about 97.8 ns (1/10.23 MHz). The P code is generated from two pseudorandom noise (PRN) codes with the same chip rate. One PRN sequence has 15,345,000 chips, and the difference is 37 chips, such that the two numbers 15,345,000 and 15,345,037 are relatively prime. Thus, the code length generated by these two codes is $15,345,037 \times 1.5 = 23,017,555.5$ seconds, which is a little over 38 weeks. In practice, however, the length of P code is 1 week as it is reset every week. Thus, this 38 week-long code can be divided into 37 different P codes and each satellite can use a different portion of the code. There are a total of 32 satellite identification numbers (1 – 32), and the remaining 5 code signals (33 – 37) are reserved for other uses such as ground transmission (Table 6.3). The navigation data rate carried by the P code through phase modulation is at 50 Hz. For acquisition on the signal, the time of the week must be very accurately known. This time is determined from the C/A code signal, which is defined below.

The GPS P code has higher chip rate, better accuracy and anti-jamming property than the C/A code. The P code generator with the main clock memory of 10.23 MHz has four 12-stage linear feedback shift registers (LFSRs)[10]: X1A, X1B, X2A and X2B, of which X1A and X2A each is shorted to 4092 chips while X1B and X2B to 4093 chips. The X_1 sequence is generated by the XOR (mod 2 sum) of the outputs of X1A and X1B, whereas the X_2 sequence is generated by the XOR of the outputs of X2A and X2B and then delayed by a selected finite number of chips i, $(i = 1, \ldots, 37)$, thus generating the X2 sequence. Each P_i is the XOR of X_1 and X_{2i}. The X1B LFSR is halted once its short cycles are counted to 3749. After X1A short cycles are counted 3750, the X1 epoch is generated and it resumes the X1B LFSR. Similarly, the X2B

---

[10] Galois field and the details of the LFSR encoder are available in Appendix G.

LFSR is halted once its short cycles are counted to 3749. But, after X2A short cycles are counted to 3750, the X2B LFSR to resume still needs an additional 37 clock cycles. Only then X2 epoch is generated, which accumulates 37 clock cycle delays for each epoch as compared with the X1 epoch.

At the beginning of each weekly period, the X1A, X1B, X2A and X2B shift registers are initialized to generate the first chip of the GPS week. The processing of the shift registers with respect to the X1A register continues until the last X1A period of the GPS week interval. During this period the registers X1B, X2A and X2B are held when they reach the last state of their respective cycles until the X1A cycle is completed. At this point, all four shift registers are again initialized and generate the first chip of the next GPS week.

For the P code, the Galois field arithmetic is performed using primitive polynomials of degree 12 with the X1A, X2A, X1B and X2B LFSRs, which are:

$$\begin{aligned}
&\text{X1A}: \ 1 + x^6 + x^8 + x^{11} + x^{12}, \\
&\text{X1B}: \ 1 + x + x^2 + x^5 + x^8 + x^9 + x^{10} + x^{11} + x^{12}, \\
&\text{X2A}: \ 1 + x + x^3 + x^4 + x^5 + x^7 + x^8 + x^9 + x^{10} + x^{11} + x^{12}, \\
&\text{X2B}: \ 1 + x^2 + x^3 + x^4 + x^8 + x^9 + x^{12}. \quad\quad (6.34)
\end{aligned}$$

The block diagram of these four LFSRs is shown in Figure 6.19, in which the registers are connected according to the exponents in each polynomial. These LFSRs perform the computational task required of the Galois arithmetic.

The P code design is verified using the following method: apply the input vectors to generate the first chip, then observe the 4091st, 4092nd or 4093rd output vectors and compare them with those in Table 6.7.

The verification of each LFSR design satisfies the data in Table 6.7. Other verification techniques involve halting and then resuming X1B, X2A and X2B registers at each epoch, and the chip generation at the end of the GPS week. There are two cases:

CASE 1: Verification of the code halting after one cycle of X1A, X1B, X2A and X2B. The P code generator initialization for this case is given in Table 6.8.

CASE 2: Verification of the P-code halting at the end of the GPS week. The P code generator initialization is given in Table 6.9.

The P code reset timing at the end of a GPS week is provided in Table 6.10.

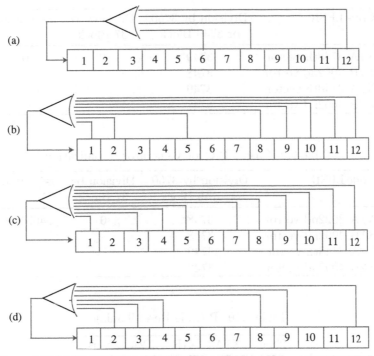

Figure 6.19 (a) X1A LFSR, (b) X1B LSFR, (c) X2A LFSR, and (d) X2B LFSR.

Table 6.7. P Code Vector States.

Code Chip #	Vector State (HEX)	Vector State for first chip following Epoch (HEX)
X1A 4091	100010010010 (892)	001001001000 (248)
4092	000100100100 (124)	
X1B 4092	100101010101 (955)	010101010100 (554)
4093	001010101010 (2AA)	
X2A 4091	111001001001 (E49)	100100100101 (925)
4092	110010010010 (C92)	
X2B 4092	000101010101 (155)	010101010100 (554)
4093	001010101010 (2AA)	

Table 6.8. Case 1 P Code Generator Initialization.

Code LFSR	Division by 3750 or 3749 Block	Division by 37 Block	z-Counter
X1A 3748th vector	3749	0	0
X1B 4092nd vector	3748		
X2A 3748th vector	3749		
X2B 4092nd vector	3748		

Table 6.9. Case 2 P Code Generator Initialization.

Code LFSR	Division by 3750 or 3749 Block	Division by 37 Block	z-Counter
X1A 3022nd vector	3748	0	403199
X1B 4092nd vector	3748		
X2A 3748th vector	3749		
X2B 4092nd vector	3748		

Table 6.10. P Code Reset Timing.

Time	X1A	X1B	X2A	X2B
↓	1	345	1070	967
	...	...	...	...
↓	3023	3367	4092	3989
	...	...	...	...
↓	3127	3471	4092	4093
	...	...	...	...
↓	3749	4093	4092	4093
	...	...	...	...
↓	4092	4093	4092	4093

The direct GPS P code acquisition algorithm is based on the Nallatech platform [Nallatech, 2002] and the Xilinx 1024-point FFT/IFFT core [Xilinx, 2000], which uses a Cooley-Tukey radix-4 decimation-in-frequency (DIF) FFT [Cooley and Tukey, 1965] to compute the DFT of a complex sequence. Details of the scaling strategies and their implications are available in Knight and Kaiser [1979] and Rabiner and Gold [1975]. Also, different features of Xilinx Virtex-E field programmable gate array (FPGA) are studied using the direct average method [Xilinx, 2002]. The preprocessed data sequences of averaging and FFT are generated by the local binary P code. The FFT is taken on the averaged local reference block data. Then the GPS signal FFT results are multiplied by the complex conjugate of the local reference FFT results. This is followed by IFFT which yields the complex correlation results. Then the square of the correlation amplitude is computed. The peak of this squared correlation amplitude and the peak location are searched in each block. This process continues until 10 ms data sequences are processed. The final output data in this process are the maximum and the second highest peak and the peak location values. This process is schematically presented in Figure 6.20.

The P code generator is started to generate unary P code in $(0, 1)$ format in each block cycle. Then it is converted to $(-1, 1)$ format by the binary converter. Then, every 20 chips of P code are averaged into one chip by the 'average' unit, and the averaged result is saved in a RAM. After 1 ms P code generation, 10220 chips are averaged by 20 and saved in RAM. The last 10 chips at the end of every 1 ms are averaged by 10 and saved in RAM. Hence, 10230 chips in 1 ms are averaged into 512 elements. The other 512 elements are zeros which are preloaded into RAM. Thus, 1024 averaged elements are now ready for further FFT processing. Finally, the RAM address is set to zero again, and the process is repeated. The direct P code acquisition architecture consists of eight units: (i) Local reference generation unit; (ii) local reference FFT processor; (iii) complex conjugate multiplication processor; (iv) IFFT processor; (v) correlation amplitude square unit; (vi) correlation peak unit; (vii) correlation peak location unit; and (viii) maximum selection unit. The entire design provides a sequential architecture: two FFT/IFFT cores are used, one for the forward FFT and the other for the inverse FFT. Complete architecture details of this entire process are available in Pang [2003: 85]. There may exist other possible architectures using more parallel P code generators, which require more FFT/IFFT cores. The upper limit for such an architecture is determined by the hardware capability available on FPGA chips.

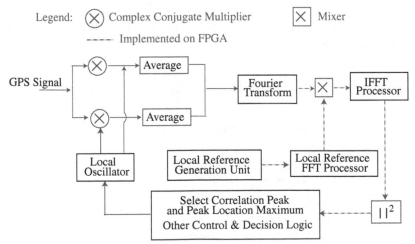

Figure 6.20 Direct GPS P Code Acquisition Processing Scheme.

A process called *spoofing* is used to prevent unauthorized use or potential interference with the military signal. As mentioned in §6.3.11, this is done by encrypting the P code by modulating the P code with a W-code, which is a special encryption sequence, to generate a Y-code. Recall that the SVs have been transmitting the Y-code since the anti-spoofing module was set to the 'on' state. The encrypted signal so obtained is called as the P(Y) code. The details of the W-code are secret, but it is known that it is applied to the P code at approximately 500 kHz, a rate slower than the P code itself by a factor of about 20. This has enabled companies to develop semi-codeless approaches for tracking P(Y) signal, without any knowledge of the W code. The GPS P code has higher chipping rate, better accuracy and anti-jamming (spoofing) property than the C/A code.

6.4.1 P Code Acquisition. As seen from Eq (6.17), there are 1023 search steps in the code phase domain compared to only 41 search steps in the frequency domain. This suggests that the acquisition should be parallelized in the code phase domain performing only 41 steps. The method, called the *parallel code phase search*, is as follows: since any acquisition must perform a correlation between an incoming signal and a PRN code sequence, it is easier to make a circular cross-correlation between the input and the PRN code without shifted code phase, as was done in §6.3.10 in the serial search acquisition method. For this purpose, a method of circular correlation through discrete Fourier transforms, is described in Oppenheim and Schäfer [2009: 746] and Tsui [2000: 140]. Accordingly, the discrete Fourier transforms of sequences $\{x(n)\}$ and $\{y(n)\}$, each of finite length, are computed as (see §3.4)

$$X(k) = \sum_{n=0}^{N-1} x(n)\, e^{-2\pi i\, kn/N}, \quad Y(k) = \sum_{n=0}^{N-1} y(n)\, e^{-2\pi i\, kn/N}. \tag{6.35}$$

Between these two sequences of length N and periodic repetition, the circular cross-correlation is given by

$$z(n) = \frac{1}{N} \sum_{j=0}^{N-1} x(j)y(j+n) = \frac{1}{N} \sum_{j=0}^{N-1} x(-j)y(j-n). \tag{6.36}$$

Then the discrete N-point Fourier transform of $z(n)$ is

$$Z(k) = \frac{1}{N} \sum_{n=0}^{N-1} \sum_{j=0}^{N-1} x(-j)y(j-n)\, e^{-2\pi i\, k/N}$$

$$= \sum_{j=0}^{N-1} x(j)\, e^{2\pi i\, kj/N} \sum_{n=0}^{N-1} y(j+n)\, e^{-2\pi i\, k(j+n)/N}$$

$$= \overline{X(k)}\, Y(k), \tag{6.37}$$

where the overline denotes the complex conjugate of $X(k)$. After the frequency domain representation of the cross-correlation is found, the time domain representation can be calculated using IDFT. The block diagram of this parallel code phase algorithm is presented in Figure 6.21, in which the incoming signal is multiplied (mod 2) by a locally generated carrier signal. This generates the I signal, and multiplication by a 90° phase-shifted version of the locally generated signal generates the Q signal. The I and Q signals are then combined to form a complex input signal $x(n) = I(n) + i\,Q(n)$ to the DFT function. Next, the so generated PRN code is transformed into the frequency domain and this yields a complex conjugate result. Finally, the DFT of the input is multiplied (mod 2) with the DFT of the PRN code. The result is transformed into the time domain using an IDFT. The absolute value of the IDFT squared is the correlation between the input and the PRN code. If a peak is present in this correlation, the index of this peak represents the phase of the PRN code of the incoming signal.

As compared to the serial search acquisition method of §6.3 and Figure 6.13, the present parallel code phase search acquisition cuts down the search to 41 different carrier frequencies. The DFT of the generated PRN code must be performed only once for each acquisition. Since one DFT and one IDFT are performed for each of the 41 frequencies, this method depends on the accuracy and speed of these two transforms. The accuracy is, however, the same as the method of §6.4, but the PRN code phase is more accurate than for the method of §6.3 since it provides a correlation value for each sampled code phase. For example, for a sampling frequency of 10 MHz, a sampled PRN code has 10000 samples, which means that the accuracy of the code phase has 10000 different values instead of 1023.

Legend: ⊗: multiplication mod 2 (using Table F.2, Appendix F).

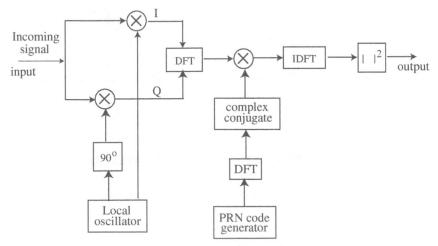

Figure 6.21 Parallel Code Phase Search.

6.4.2 P Code Spectral Density. Some of the PRN code spectral properties are as follows:

(i) P code spectral density is defined for a signal i as

$$S_{spi}(f) = P_{pi} T_c \frac{\sin^2(\pi f T_c)}{(\pi f T_c)^2} = P_{pi} T_c \operatorname{sinc}^2(\pi f T_c), \quad -\infty < f < \infty, \quad (6.38)$$

where P_{pi} is the P code carrier power, and T_c is the P code width. This property is bandlimited to protect radio astronomers by preventing a receiver from achieving full correlation. This results in correlation loss due to filtering.

(ii) C/A code spectral density, which has a short 1 ms repeating code, and line spectrum with components c_{ij}, $-\infty < j < \infty$, that are 1 kHz apart, such that

$$\sum_{j=-\infty}^{\infty} c_{ij} = P_{ci}, \qquad (6.39)$$

where P_{ci} is the C/A code carrier power, defined above. The envelope of line spectrum is defined as

$$S_{ci}(f) = 1000 P_{ci} T_c \frac{\sin^2(\pi f T_c)}{(\pi f T_c)^2} = 1000 P_{ci} T_c \operatorname{sinc}^2(\pi f T_c). \qquad (6.40)$$

These power spectral densities are presented in Figure 6.22 for L1 and L2 signals.

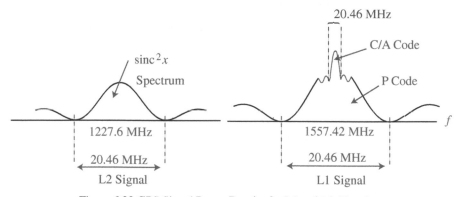

Figure 6.22 GPS Signal Power Density for L1 and L2 Signals.

The layout of the C/A coder with initial G2 state is schematically presented in Figure 6.23.

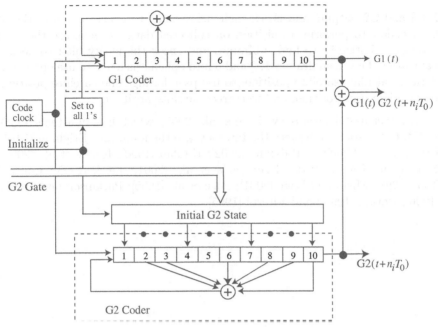

Figure 6.23 C/A Coder with Initial G2 State.

6.5 Computational Aspects

The generation of Gold codes is very simple. Using two preferred m-sequence generators of degree m, with a fixed non-zero seed in the first generator, $2m$ Gold codes are obtained by changing the seed of the second generator from 0 to $2m - 1$. Another Gold sequence can be obtained by setting all zeros to the first generator, which is the second m-sequence itself. In total, $2m + 1$ Gold codes are available. The software for Gold code generators (gold module) has been written intuitively using the above argument. The file *gold.h* must be included to use gold module; *test_gold.c* shows how to use gold module, where *test_gold* requires these three arguments: the first one is degree, and second and third are generator polynomials, and *lgold* is a 64-bit version of gold module (these files are available in Tsui [2000]).

Gold codes have three-valued autocorrelation and cross-correlation function with values $\{-1, -t(m), t(m) - 2\}$, where

$$t(m) = \begin{cases} 2^{(m+1)/2} + 1 & \text{for odd } m, \\ 2^{(m+2)/2} + 1 & \text{for even } m. \end{cases}$$

There are different computer programs in C and MATLAB available in the literature and on the Internet to compute various features of the GPS technology. For example, Tsui [2000] contains 17 useful MATLAB programs, to compute pseudorange, satellite positions, and user position; to generate

MLS and G2 outputs and check their delay time; to generate one of the 32 C/A codes; to perform acquisition on collected data; to generate the C/A code and digitize it; to find subframes and check the parity bits; to decode navigation data in subframes 1, 2 and 3 into ephemeris data; to use ephemeris data to calculate satellite position; to use pseudorange and satellite positions to calculate user position; and to correct satellite position.

Another useful work is by Borre et al. [2007], which has an attached CD-ROM. Other sources, besides the Internet, are the following: Brigham [1974]; Brown et al. [2000]; El-Rabbany [2002]; Kaplan [1990]; Kelley et al. [2002]; Knight and Kaiser [1979]; Krumvieda et al. [2001]; Leick [1995]; Lin and Tsui [1998]; Misra and Enge [2001]; Pace et al. [1995]; Parkinson and Spilker [1996]; Prasad [1996]; and Viterbi [1995].

7

Fourier Optics

We will study the scalar diffraction theory, acoustic-wave and radio-wave propagation, and physical optics. Different mathematical models of diffraction which lead to the Fourier transform in two dimensions are discussed. Such transforms are known as the *optical transforms*, which are presented for rectangular and circular apertures, and Talbot images are described in detail. X-rays, their related electromagnetic spectrum, and the phenomenon of synchrotron radiation are described. Some important aspects of the nature of light are also presented.

7.1 Physical Optics

Any understanding of the nature of light had been a puzzle in antiquity since it did not fit the model of the universe made up of the four elements: air, fire, water, and earth. The Greek philosopher Empedocles (c. 490 – c. 435 BC) thought that light was 'a stream of small particles, emitted from a visible body, that were to enter the eyes and return to the originating body, traveling with finite speed.' The Greek astronomer Ptolemy (c. 150 BC) wrote about the angle of incidence and of refraction, and measured it for both air-glass and water-glass interfaces. Then, after a long time the Dutch Willebrord Snell (1580–1626) established experimentally the law of refraction in 1621, and described it as a minute deviation from rectilinear propagation, caused by small apertures and exhibited by the presence of light in geometrical shadows. The English physicist Robert Hooke (1635–1703), well known for Hooke's law in the theory of elasticity, advanced his wave model to explain diffraction in 1672, and found that 'light consisted of short vibrations exceedingly quick, propagating every way in straight lines.' Hooke's work was advanced further by the Dutch physicist Christian Huygens (1629–1695), giving rise to the phenomenon named after him. He established, as we will see in the sequel, that every point of a wavefront of light may be regarded as a new source of spherical waves. Thereafter the nature of light was discussed by Isaac Newton (1642–1727) but he opposed the wave theory of light because in his time that meant only the 'longitudinal waves' that vibrate in the direction of propagation like those of a coiled spring. The English physician Thomas Young (1773–1829), while reviving the wave theory, explained that the different spectral colors correspond to different wavelengths.

When a wave encounters a boundary between two media with different refractive indices, the ray directions are changed abruptly as they pass through the interface. The angles of incidence θ_1 and of refraction θ_2 are related by Snell's law: $n_1 \sin\theta_1 = n_2 \sin\theta_2$, where n_1 and n_2 are the refractive indices of the first and the second medium, respectively (Figure 7.1(a)). Snell's law can be easily derived using Fermat's principle of least time, which states that light travels through a path that takes the least time to reach its destination.

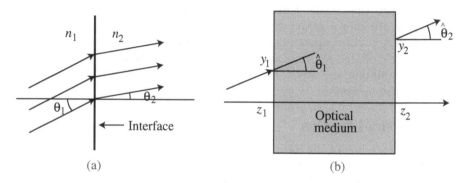

(a)   (b)

Figure 7.1 Snell's Law.

Under the assumption that the ray is traveling close to the optical axis and at small angles so that $\sin\theta \approx \theta$ and $\cos\theta \approx 1$, Snell's law is further simplified to $n_1\hat{\theta}_1 = n_2\hat{\theta}_2$, where $\hat{\theta}$ within the optical medium is called a *reduced angle*. This gives the paraxial version of Snell's law, which states that the reduced angle remains constant as light passes through a sharp interface between media of different refractive indices, i.e., the reduced angles $\hat{\theta}_1 = \hat{\theta}_2$ (Figure 7.1(b)).

The matrix formulation of the properties of rays in optical systems under paraxial conditions can be established from Figure 7.1(b), where the rays traveling in paths are contained in a single z-plane with axial coordinate z_1 at the left and z_2 at the right. A ray with transverse coordinate y_1 enters the optical system at an angle $\hat{\theta}_1$ and the same ray, now with transverse coordinate y_2 leaves the system at an angle $\hat{\theta}_2$. Under paraxial conditions the quantities $(y_1, \hat{\theta}_1)$ and $(y_2, \hat{\theta}_2)$ are related linearly as

$$y_2 = Ay_1 + B\hat{\theta}_1, \quad \hat{\theta}_2 = Cy_1 + D\hat{\theta}_1,$$

or in matrix form as

$$\left\{ \begin{matrix} y_2 \\ \hat{\theta}_2 \end{matrix} \right\} = \begin{bmatrix} A & B \\ C & D \end{bmatrix} \left\{ \begin{matrix} y_1 \\ \hat{\theta}_1 \end{matrix} \right\}, \tag{7.1}$$

where the reduced angles are used instead of the actual refractive angles. The square matrix on the right side of (7.1) is called the *ray-transfer matrix* or the ABCD matrix. This matrix is interpreted in terms of the local frequencies discussed in §3.3.1. In the (y, z)-plane under paraxial conditions, the reduced ray angle $\hat{\theta}$ is related, with respect to the z-axis, to the local frequency f by

$$ f = \frac{\theta}{\lambda} = \frac{\hat{\theta}}{\lambda_0}, $$

where λ is the wavelength of the optical medium and λ_0 that of the free space. Thus, the ray-transfer matrix (ABCD) can be regarded as establishing a transformation between the distribution of local frequency at the input and the corresponding distribution at the output.

7.1.1. Abbe Sine Condition. The German engineer Ernst Karl Abbe (1840–1905), known for the Abbe refrectometer and the Abbe number, established (Abbe [1873]) that in order for a lens or other optical system to produce sharp images of the off-axis as well as on-axis objects, the condition $\dfrac{\sin \theta'}{\sin \phi'} = \dfrac{\sin \theta}{\sin \phi}$ must be satisfied, where θ and ϕ are the angles, relative to the optic axis, of any two rays as they leave the object, and θ' and ϕ' the angles of the same rays where they reach the image plane (e.g., the film of the camera). This condition, known as the *Abbe sine condition*, applies to a paraxial ray (e.g., a ray parallel to the optical axis) or a marginal ray (e.g., a ray with the largest angle admitted by the aperture). In short, this condition states that the sine of the output angle must be proportional to the sine of the input angle.

In terms of the Fourier optics, consider an object in the plane of an optical system, which has a transmittance function at a point (x_0, y_0) of the form

$$ T(x_0, y_0) = \int_{-\infty}^{\infty} \int_{-\infty}^{\infty} T(f_x, f_y) \, e^{i\,(f_x x + f_y y)} \, df_x \, df_y. \tag{7.2} $$

where f_x, f_y are the spatial frequencies in the Fourier transform (7.2). Now, for the sake of simplicity, assume that the system is free of image distortion. Then the image plane coordinates (x_k, y_k) are linearly related to the coordinates (x_0, y_0) by $x_k = M x_0$ and $y_k = M y_0$, where M is the proportionality constant, or system magnification. Then Eq (7.2) can be expressed as

$$ T(x_0, y_0) = \int_{-\infty}^{\infty} \int_{-\infty}^{\infty} T(f_x, f_y) \, e^{i\,[(f_x/M)(Mx_0) + (f_y/M)(My_0)]} \, df_x \, df_y $$

$$ = \int_{-\infty}^{\infty} \int_{-\infty}^{\infty} T(f_x, f_y) \, e^{i\,[(f_x/M)x_k + (f_y/M)y_k]} \, df_x \, df_y. $$

This result provides another form of the Abbe sine condition as

$$f_x^k = \frac{f_x}{M}, \quad f_y^k = \frac{f_y}{M}, \tag{7.3}$$

which relates the object plane wavenumber spectrum to the image plane wavenumber spectrum. In view of the wavenumber condition (7.3), the transmittance function (7.2) becomes

$$T(x_0, y_0) = M^2 \int_{-\infty}^{\infty} \int_{-\infty}^{\infty} T\left(Mf_x^k, Mf_y^k\right) e^{i\left(f_x^k x_k + f_y^k y_k\right)} df_x^k \, df_y^k.$$

This form of the Abbe sine condition simply reflects Heisenberg's uncertainty principle (Appendix L) for Fourier transform pairs, namely, that as the spatial extent of any function is expanded (by the magnification factor M), the spectral extent contracts by the same factor, M, so that the space-bandwidth product remains constant.

7.2 Scalar Diffraction Theory

We will study scalar diffraction theory, acoustic-wave and radio-wave propagation, and physical optics.

Refraction is defined as the bending of light rays that occurs when they pass through a region in which there is a gradient of the local velocity of propagation of the wave, for example, when a light wave encounters a sharp boundary between two regions with different refractive indices. The wave propagation velocity in the first medium with refractive index n_1 is given by $v_1 = c/n_1$, where c is the velocity of light in vacuum. The propagation velocity in the second medium is $v_2 = c/n_2$. Figure 7.1 shows that the light rays are bent at the interface. The angles of incidence and refraction are related by Snell's law. This law is the foundation of geometric optics.

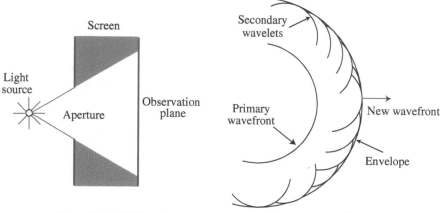

Figure 7.2 (a) Diffraction of Light.   (b) Huygens Principle.

Light rays are also bent upon reflection, which occurs at a metallic or dielectric interface. But in this case the angle of reflection is always equal to the angle of incidence.

Sommerfeld [1896] has defined diffraction as 'any deviation of light rays from rectilinear paths which cannot be interpreted as reflection or refraction.' Diffraction must not be confused with the penumbra effect which does not involve any bending of light. Figure 7.2 presents an arrangement for observing diffraction of light. Huygens in 1678 determined intuitively that if each point on the wavefront of a disturbance was regarded as a new source of a 'secondary' spherical disturbance, then the wavefront at a later time could be determined by constructing the 'envelope' of the secondary wavelets. This phenomenon is presented in Figure 7.2 (b).

In 1804 Thomas Young, a British physician, made significant progress in the wave theory of light by introducing the concept of *interference*. Then in 1818 Augustin Jean Fresnel combined the ideas of Huygens and Young, by making certain assumptions about the amplitudes and phases of Huygens's secondary sources. By allowing various wavelets to mutually interfere, he then calculated accurately the distribution of light in diffraction patterns. Laplace and Biot opposed Fresnel's theory, claiming that it was flawed. However, a prize for this work was awarded in 1818 to Fresnel (1788–1827). Then on the basis of Fresnel's theory, Poisson predicted the existence of a bright spot in the center of the diffraction pattern of a small circular disk. The chairman of the prize committee, A. Arago, performed the experiment and found the predicted spot. Since then this bright spot has been known as *Poisson's spot*. By 1882 the ideas of Huygens and Fresnel were firmly established mathematically by Gustav Kirchhoff, who proved that the amplitudes and phases of the secondary wavelets, as calculated by Fresnel, were the first logical consequences of the wave theory of light.

The next advance was the Rayleigh-Sommerfeld diffraction theory, where light is treated as a scalar phenomenon, free from the vector approach imposed by Maxwell's equations (Appendix D). If the polarization $\mathbf{P}$ of the medium is zero, then Eq (D.6) becomes

$$\nabla^2 \mathbf{E} - \frac{n^2}{c^2} \frac{\partial^2 \mathbf{E}}{\partial t^2} = 0, \tag{7.4}$$

where n is the refractive index of the medium, defined by $n = \sqrt{\epsilon/\epsilon_0}$, where ϵ_0 is the permittivity in the vacuum, ϵ the electric inductive capacity, and c is the velocity of propagation in vacuum. The magnetic field $\mathbf{B}$ satisfies an equation identical to (7.4). Since these two vector fields satisfy an identical vector equation, they also satisfy a scalar equation of the form (7.4) where the vector fields $\mathbf{E}$ and $\mathbf{B}$ are replaced by the scalar quantities E_x, E_y, E_z and B_x, B_y, B_z. In summary, Eq (7.4) can be replaced by a single scalar wave equation

$$\nabla^2 u(P,t) - \frac{n^2}{c^2}\frac{\partial^2 u(P,t)}{\partial t^2} = 0, \tag{7.5}$$

where $u(P,t)$ represents the scalar form of the vector $\mathbf{E}$ or $\mathbf{B}$, and u depends on both position P and time t.

7.2.1. Helmholtz Equation. Let the light disturbance at a point P and time t be denoted by a scalar function $u(P,t)$. For a monochromatic wave, we have

$$u(P,t) = A(P)\cos\left(2\pi f t + \phi(P)\right),$$

where $A(P)$ and $\phi(P)$ are the amplitude and phase, respectively, of the wave at the location P, and f is the optical frequency. This function $u(P,t)$ can also be expressed as

$$u(P,t) = \Re\{U(P)\,e^{-2\pi i\,ft}\}, \tag{7.6}$$

where $U(P)$ is a complex function of position (called a *phasor*), defined by $U(P) = A(P)\,e^{-i\,\phi(P)}$. If the real-valued disturbance $u(P,t)$ represents an optical wave, then it satisfies the scalar wave equation (7.5) at each source-free point. If we substitute (7.6) into (7.5), the complex function $U(P)$ satisfies the Helmholtz equation

$$\left(\nabla^2 + k^2\right)U = 0, \tag{7.7}$$

which is independent of time, where k is the wave number defined by $k = 2\pi/\lambda$ and $\lambda = c/f$ is the wavelength in the dielectric medium. Eq (7.7) holds for the complex amplitude of any monochromatic optical wave propagating in vacuum $(n = 0)$ or in a homogeneous dielectric medium $(n > 1)$.

7.2.2. Helmholtz-Kirchhoff Integral Theorem. The diffraction problem, as formulated by Kirchhoff, is based on an integral theorem obtained earlier by Helmholtz in acoustics. It provides a solution of the homogeneous wave equation at any point in terms of the values of the solution and its first derivative on an arbitrary closed surface surrounding that point. Following Goodman [1985: 42ff], let the point under observation be denoted by P_0, and let S denote an arbitrary closed surface containing the point P_0 (see Figure 7.3).

Since we want to express the optical disturbance at the point P_0 in terms of its values on the surface S, we apply Green's theorem and choose a unit-amplitude spherical wave expanding around the point P_0. Then the value of the Kirchhoff's Green's function at an arbitrary point P_1 on S is given by

$$G\left(P_1\right) = \frac{1}{r_{01}}\,e^{i\,kr_{01}}, \tag{7.8}$$

where r_{01} is the length of the vector $\mathbf{r}_{01}$ from the point P_0 to P_1. In particular, the Green's function G, which represents a disturbance generated by an expanding spherical wave, satisfies the Helmholtz equation (7.7), i.e.,

$$\left(\nabla^2 + k^2\right) G = 0. \tag{7.9}$$

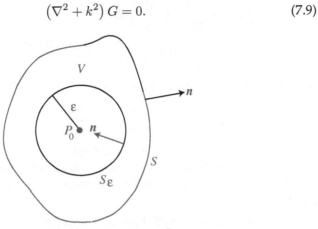

Figure 7.3 Integration Surface S.

Then the two Helmholtz equations (7.7) and (7.9), when substituted in the left side of Green's theorem[1], yield within a volume V (see Figure 7.3)

$$\iiint_V \left(U\nabla^2 G - G\nabla^2 U\right) dv = -\iiint_V \left(UGk^2 - GUk^2\right) dv = 0.$$

Thus, Green's theorem reduces to

$$\iint_S \left(U\frac{\partial G}{\partial n} - G\frac{\partial U}{\partial n}\right) ds = 0,$$

or

$$-\iint_{S_\varepsilon} \left(U\frac{\partial G}{\partial n} - G\frac{\partial U}{\partial n}\right) ds = \iint_S \left(U\frac{\partial G}{\partial n} - G\frac{\partial U}{\partial n}\right) ds. \tag{7.10}$$

Since (7.8) holds for an arbitrary point P_1 on S, and since

$$\frac{\partial G(P_1)}{\partial n} = \cos\left(\mathbf{n}, \mathbf{r}_{01}\right)\left(ik - \frac{1}{r_{01}}\right)\frac{e^{ikr_{01}}}{r_{01}}, \tag{7.11}$$

where $\cos\left(\mathbf{n}, \mathbf{r}_{01}\right)$ is the cosine of the angle between the outward normal $\mathbf{n}$ and the vector $\mathbf{r}_{01}$ joining the points P_0 to P_1. In particular, when P_1 is on S_ε, we have $\cos\left(\mathbf{n}, \mathbf{r}_{01}\right) = -1$, and these equation become

$$G(P_1) = \frac{e^{ik\varepsilon}}{\varepsilon}, \quad \text{and} \quad \frac{\partial G(P_1)}{\partial n} = \frac{e^{ik\varepsilon}}{\varepsilon}\left(\frac{1}{\varepsilon} - ik\right). \tag{7.12}$$

---

[1] Green's theorems can be found in any textbook on partial differential equations; also in Kythe [2011].

Then, in the limit as $\varepsilon \to 0$, we get

$$\lim_{\varepsilon \to 0} \iint_{S_\varepsilon} \left(U\frac{\partial G}{\partial n} - G\frac{\partial U}{\partial n} \right) ds$$

$$= \lim_{\varepsilon \to 0} 4\pi\varepsilon^2 \left[U(P_0) \frac{e^{i k \varepsilon}}{\varepsilon} \left(\frac{1}{\varepsilon} - i k \right) - \frac{\partial U(P_0)}{\partial n} \frac{e^{i k \varepsilon}}{\varepsilon} \right] = 4\pi U(P_0).$$

Substituting this result into (7.10) gives

$$U(P_0) = \frac{1}{4\pi} \iint_S \left\{ \frac{\partial U}{\partial n} \left(\frac{e^{i k r_{01}}}{r_{01}} \right) - U\frac{\partial}{\partial n} \left(\frac{e^{i k r_{01}}}{r_{01}} \right) \right\} ds. \qquad (7.13)$$

This result is the *Helmholtz-Kirchhoff integral theorem*. It is an important part of the scalar theory of diffraction, where any point P_0 can be expressed in terms of the 'boundary values' of the wave on any closed surface about that point.

The Helmholtz-Kirchhoff theorem is applied to the problem of finding the field at the point P_0, where it is important to choose a surface of integration that allows a successful calculation. A closed surface S is chosen with two parts, one a plane surface S_1, situated directly behind the diffracting screen, and joined and closed by a large spherical cap S_2, of radius R and centered at the point P_0. The total closed surface S is the sum of S_1 and S_2, as presented in Figure 7.4.

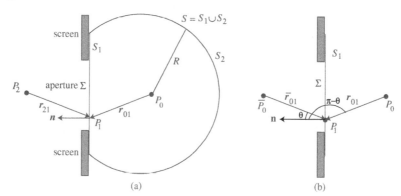

Figure 7.4 (a) Diffraction by a Screen. (b) Green's Function at P_0 and $\bar{P}_0$.

Then, in view of (7.13) we have

$$U(P_0) = \frac{1}{4\pi} \iint_{S_1 \cup S_2} \left(G\frac{\partial U}{\partial n} - U\frac{\partial G}{\partial n} \right) ds. \qquad (7.14)$$

Since $G = \dfrac{e^{i k R}}{R}$ on S_2, Eq (7.11) gives for large R:

$$\frac{\partial G}{\partial n} = \left(i k - \frac{1}{R} \right) \frac{e^{i k R}}{R} \approx i k G.$$

Thus, the integral in (7.14) reduces to

$$\iint_S \left(G\frac{\partial U}{\partial n} - ikGU\right) ds = \int_\Omega G\left(\frac{\partial U}{\partial n} - ikU\right) R^2\, d\omega,$$

where Ω is the solid angle subtended by S_2 at P_0. Since the quantity $|RG|$ is uniformly bounded on S_2, the integral over S_2 will vanish as R becomes arbitrarily large, provided the *Sommerfeld's radiation condition* (Sommerfeld [1896])

$$\lim_{R\to\infty} \left(\frac{\partial U}{\partial n} - ikU\right) = 0 \tag{7.15}$$

is satisfied uniformly in angle Ω, where U becomes zero at least as fast as a diverging spherical wave. Thus, the wave is an outgoing wave on S_2 and not an incoming wave, and since only outgoing waves fall on S_2, the integral over S_2 will be zero. Hence, we finally have

$$U\left(P_0\right) = \frac{1}{4\pi} \iint_{S_1} \left(G\frac{\partial U}{\partial n} - U\frac{\partial G}{\partial n}\right) ds. \tag{7.16}$$

The screen S_1 is opaque, except for an open aperture Σ. Thus, the major contribution to the integral (7.16) is from those points of S_1 that are located within the aperture Σ. We will use the following two assumptions made by Kirchhoff [1883] regarding both the field distribution U and its derivative $\frac{\partial U}{\partial n}$: (i) Across the aperture Σ both of these quantities are exactly the same as they would be in the absence of the screen; and (ii) over the portion S_1 that lies in the geometric shadow of the screen, both of them are identically zero. These conditions are known as the *Kirchhoff boundary conditions*. In view of these two conditions, Eq (7.16) reduces to

$$U\left(P_0\right) = \frac{1}{4\pi} \iint_{\Sigma} \left(G\frac{\partial U}{\partial n} - U\frac{\partial G}{\partial n}\right) ds. \tag{7.17}$$

Note that although the Kirchhoff conditions have simplified the equation, the screen definitely perturbs the field on the aperture Σ. But if the dimensions of the aperture are large compared with a wavelength, the effect of the field extending behind the screen for a distance of several wavelengths can be neglected.

7.2.3. Fresnel-Kirchhoff Diffraction. Note that the distance r_{01} in Figure 7.4 is generally equal to many optical wavelengths. Therefore, since $k \gg 1/r_{01}$, Eq (7.11) simplifies to

$$\frac{\partial G\left(P_1\right)}{\partial n} \approx ik\cos\left(\mathbf{n}, \mathbf{r}_{01}\right) \frac{e^{ikr_{01}}}{r_{01}}.$$

Substituting this result and the expression (7.12) for G into (7.17), we get

$$U(P_0) = \frac{1}{4\pi} \iint_\Sigma \frac{e^{i\,kr_{01}}}{r_{01}} \left[\frac{\partial U}{\partial n} - i\,kU\cos(\mathbf{n}, \mathbf{r}_{01})\right] ds. \qquad (7.18)$$

Next, suppose that the aperture Σ is illuminated by a single spherical wave $U(P_1) = \dfrac{A\,e^{i\,kr_{21}}}{r_{21}}$ that arises from a point P_2 which is at a distance r_{21} from P_1 (see Figure 7.4). If r_{21} is equal to many optical wavelengths, then Eq (7.18) can be reduced to

$$U(P_0) = \frac{A}{2i\,\lambda} \iint_\Sigma \frac{e^{i\,k(r_{01}+r_{21})}}{r_{01}r_{21}} \left[\cos(\mathbf{n}, \mathbf{r}_{01}) - \cos(\mathbf{n}, \mathbf{r}_{21})\right] ds. \qquad (7.19)$$

This is known as the *Fresnel-Kirchhoff diffraction formula*. Note that Eq (7.19) is symmetrical with respect to the point source P_2 and the observation point P_0. This means that a point source at P_0 will produce the same effect that a point source of equal intensity placed at P_2 will produce at P_0. This result is known as the *Helmholtz reciprocity theorem*.

7.2.4. Rayleigh-Sommerfeld Diffraction. The problem with Kirchhoff's theory is that one has to impose the 'boundary conditions' on both the field distribution U and its normal derivative $\dfrac{\partial U}{\partial n}$. According to a well-known theorem from potential theory, if both U and $\dfrac{\partial U}{\partial n}$ vanish along any finite segment of a path, then the distribution U must vanish over the entire plane. This is also valid in the case of the three-dimensional wave equation, i.e., if a solution of such an equation is zero on any finite surface element, then it is zero throughout the entire space. In the case of Kirchhoff's theory, the two boundary conditions imply that the field distribution is zero everywhere behind the aperture, which creates a contradiction. In fact, the Fresnel-Kirchhoff diffraction formula (7.19) fails to reproduce these boundary conditions as the point P_0 gets close to the aperture or the screen. However, even in spite of these serious drawbacks, the Kirchhoff theory has provided accurate results for practical problems.

Sommerfeld [1896] was able to remove the inconsistency in the Kirchhoff theory by eliminating the requirement of imposing the two above-mentioned boundary conditions. The so-called Rayleigh-Sommerfeld diffraction theory uses Green's functions and starts with Eq (7.16), and requires, in view of the scalar theory, that the following conditions must be satisfied: (i) Both U and G must satisfy the homogeneous scalar wave equation (7.5); and (ii) Sommerfeld's radiation condition (7.15) must be satisfied.

The existence of the Green's function G with the required properties was proved by Sommerfeld by considering that the function G is generated not only by a point source at P_0 but also by a second point source at $\bar{P}_0$ which is the mirror image of P_0 on the other side of the screen (see Figure 7.4(b)). We will find two Rayleigh-Sommerfeld solutions $U_{1,2}(P_0)$ of the diffraction problem governed by the Helmholtz equations (7.7) and (7.9).

(i) Assume that the source at the point $\bar{P}_0$ has the same wavelength λ as the source at P_0, and suppose that the two sources oscillate with a phase difference of 180°. Then let the Green's function at the point P_1 be given by

$$G_1(P_1) = \frac{e^{i k r_{01}}}{r_{01}} - \frac{e^{i k \bar{r}_{01}}}{\bar{r}_{01}}, \qquad (7.20)$$

which is zero on the aperture Σ. This yields the first *Rayleigh-Sommerfeld solution*

$$U_1(P_0) = -\frac{1}{4\pi} \iint_\Sigma U \frac{\partial G_1}{\partial n} \, ds, \qquad (7.21)$$

where

$$\frac{\partial G_1}{\partial n}(P_1) = \cos\left(\mathbf{n}, \mathbf{r}_{01}\right)\left(i k - \frac{1}{r_{01}}\right)\frac{e^{i k r_{01}}}{r_{01}} - \cos\left(\mathbf{n}, \bar{\mathbf{r}}_{01}\right)\left(i k - \frac{1}{\bar{r}_{01}}\right)\frac{e^{i k \bar{r}_{01}}}{\bar{r}_{01}},$$

and $\bar{r}_{01}$ is the distance from P_0 to P_1.

(ii) Since, for P_1 on S_1 we have $r_{01} = \bar{r}_{01}$ and $\cos\left(\mathbf{n}, \mathbf{r}_{01}\right) = -\cos\left(\mathbf{n}, \bar{\mathbf{r}}_{01}\right)$ (note that $\cos(\pi - \theta) = -\cos\theta$ in Figure 7.4(b)), we find that on the surface S_1

$$\frac{\partial G_1}{\partial n}(P_1) = 2\cos\left(\mathbf{n}, \mathbf{r}_{01}\right)\left(i k - \frac{1}{r_{01}}\right)\frac{e^{i k r_{01}}}{r_{01}}.$$

If $r_{01} \gg \lambda$, the term $(ik - 1/r_{01})$ can be neglected, and we get

$$\frac{\partial G_1}{\partial n}(P_1) = 2\cos\left(\mathbf{n}, \mathbf{r}_{01}\right)\frac{e^{i k r_{01}}}{r_{01}}, \qquad (7.22)$$

which is twice the value of the normal derivative of the Green's function in the Kirchhoff theory, i.e.,

$$\frac{\partial G_1}{\partial n}(P_1) = 2\frac{\partial G}{\partial n}(P_1).$$

Using this result, the Rayleigh-Sommerfeld solution (7.21) reduces to

$$U_1(P_0) = -\frac{1}{2\pi} \iint_\Sigma U \frac{\partial G}{\partial n} \, ds, \qquad (7.23)$$

Now, if the two point sources at P_0 and $\bar{P}_0$ oscillate in phase (i.e., zero phase difference), the Green's function becomes

$$G_2(P_1) = \frac{e^{i k r_{01}}}{r_{01}} + \frac{e^{i k \bar{r}_{01}}}{\bar{r}_{01}}, \qquad (7.24)$$

(compare this with Eq (7.20)). Since the normal derivative of this function is zero across the screen and aperture, we obtain the *second Rayleigh-Sommerfeld solution*:

$$U_2\left(P_0\right) = \frac{1}{4\pi} \iint_\Sigma \frac{\partial U}{\partial N} G_2 \, ds, \tag{7.25}$$

However, since G_2 is twice the Kirchhoff Green's function G on Σ and under the condition $r_{01} \gg \lambda$, we have

$$U_2\left(P_0\right) = \frac{1}{2\pi} \iint_\Sigma \frac{\partial U}{\partial n} G \, ds. \tag{7.26}$$

Then substituting the Green's function G_1 for G in Eq (7.16), using Eq (7.22), and assuming that $r_{01} \gg \lambda$, we get

$$U_1\left(P_0\right) = \frac{1}{i\lambda} \iint_{S_1} U\left(P_1\right) \frac{e^{i k r_{01}}}{r_{01}} \cos\left(\mathbf{n}, \mathbf{r}_{01}\right) ds. \tag{7.27}$$

If the Kirchhoff boundary conditions are applied to U, we obtain from (7.27)

$$U_1\left(P_0\right) = \frac{1}{i\lambda} \iint_\Sigma U\left(P_1\right) \frac{e^{i k r_{01}}}{r_{01}} \cos\left(\mathbf{n}, \mathbf{r}_{01}\right) ds. \tag{7.28}$$

Thus, the inconsistency in the Kirchhoff theory is removed since no boundary condition need be applied to the normal derivative $\dfrac{\partial U}{\partial n}$. Further, for the second solution we find that

$$U_2\left(P_0\right) = \frac{1}{2\pi} \iint_\Sigma \frac{\partial U\left(P_1\right)}{\partial n} \frac{i k r_{01}}{r_{01}}. \tag{7.29}$$

Having determined the second Rayleigh-Sommerfeld solution, we consider the special case of illumination with a diverging spherical wave. In this case Eqs (7.28) and (7.29) can be compared directly with Eq (7.19). Note that the illumination of the aperture in all cases is a spherical wave diverging from a point source at P_2 (see Figure 7.4(a)), which is given by $U\left(P_1\right) = A\, e^{i k r_{21}}/r_{21}$. If we use G_1, defined by Eq (7.20), we obtain the *first Rayleigh-Sommerfeld diffraction formula*:

$$U_1\left(P_0\right) = \frac{A}{i\lambda} \iint_\Sigma \frac{e^{i\lambda\left(r_{01}+r_{21}\right)}}{r_{01}+r_{21}} \cos\left(\mathbf{n}, \mathbf{r}_{01}\right) ds. \tag{7.30}$$

Similarly, using G_2 and assuming that $r_{21} \gg \lambda$, we obtain the *second Rayleigh-Sommerfeld diffraction formula*:

$$U_2\left(P_0\right) = -\frac{A}{i\lambda} \iint_\Sigma \frac{e^{i k\left(r_{01}+r_{21}\right)}}{r_{01}+r_{21}} \cos\left(\mathbf{n}, \mathbf{r}_{01}\right) ds, \tag{7.31}$$

where the angle between $\mathbf{n}$ and $\mathbf{r}_{21}$ is greater than $\pi/2$.

Let G_K denote the Green's function for the Kirchhoff theory (the subscript K for Kirchhoff). Note that we have already used G_1 and G_2, defined by (7.20) and (7.24), respectively, for the Green's functions of the two Rayleigh-Sommerfeld formulas. On the surface Σ of the aperture we have $G_2 = 2G_K$ and $\dfrac{\partial G_1}{\partial n} = 2\dfrac{\partial G_K}{\partial n}$. Hence, the two formulations, namely, Kirchhoff's and Rayleigh-Sommerfeld's, can be summarized as follows: for the Kirchhoff theory, Eq (7.17) is modified as

$$U(P_0) = \frac{1}{4\pi} \iint_\Sigma \left(G_K \frac{\partial U}{\partial n} - U \frac{\partial G_K}{\partial n} \right) ds. \qquad (7.32)$$

For the first Rayleigh-Sommerfeld solution, Eq (7.23) can be written as

$$U_1(P_0) = -\frac{1}{2\pi} \iint_\Sigma U \frac{\partial G_K}{\partial n} ds, \qquad (7.33)$$

while for the *second Rayleigh-Sommerfeld solution* Eq (7.26) can be written as

$$U_2(P_0) = \frac{1}{2\pi} \iint_\Sigma \frac{\partial U}{\partial n} G_K \, ds, \qquad (7.34)$$

A careful examination of these three solutions, namely, the Kirchhoff solution (7.19), and the two Rayleigh-Sommerfeld solutions (7.33) and (7.34), reveals that the Kirchhoff solution is the arithmetic mean of the two Rayleigh-Sommerfeld solutions.

Kottler [1965] made a comparison between the Fresnel-Kirchhoff diffraction solution (7.19) and the two Rayleigh-Sommerfeld solutions (7.30) and (7.31). Based on this comparison Goodman [1968: 51] has combined these three solutions into a three-in-one solution

$$U(P_0) = \frac{A}{i\lambda} \iint_\Sigma \frac{e^{i\,k(r_{01}+r_{21})}}{r_{01}r_{21}} \psi \, ds, \qquad (7.35)$$

where ψ is an *obliquity factor* ψ, which is the angular dependence introduced by the cosine terms and is defined by

$$\psi = \begin{cases} \frac{1}{2}\left[\cos(\mathbf{n},\mathbf{r}_{01}) - \cos(\mathbf{n},\mathbf{r}_{21})\right], & \text{for Kirchhoff solution (7.19)}, \\ \cos(\mathbf{n},\mathbf{r}_{01}), & \text{for First Rayleigh-Sommerfeld solution (7.23)}, \\ -\cos(\mathbf{n},\mathbf{r}_{21}), & \text{for Second Rayleigh-Sommerfeld solution (7.34)}. \end{cases}$$

In the special case when the point source is infinitely distant and produces normally incident plane wave illumination, the obliquity factor is given by

$$\psi = \begin{cases} \frac{1}{2}\left[1 + \cos\theta\right], & \text{for Kirchhoff solution}, \\ \cos\theta, & \text{for First Rayleigh-Sommerfeld solution}, \\ 1, & \text{for Second Rayleigh-Sommerfeld solution}, \end{cases}$$

where θ is the angle between $\mathbf{n}$ and $\mathbf{r}_{01}$. However, all these three diffraction solutions become identical when the angle θ is very small. The details of the Kirchhoff and Rayleigh-Sommerfeld theory are available in Sommerfeld [1896], Wolf and Marchand [1964], Goodman [1968], and Heurtley [1973].

7.2.5. Fresnel Diffraction. A photodetector is a semiconductor used to record the optical power P as it falls on its surface which is photosensitive. In such an optical region a photon generates an electron in the conduction band and a hole in the valence band. Under the effects of internal and applied forces, the hole and the electron move in opposite directions. This creates a photocurrent I as a result of the absorbed photon incident upon the photosensitive surface. Generally, the photocurrent I is linearly proportional to the incident power, i.e., $I = RP$, where R is the proportionality constant known as the *responsivity* of the photodetector and is defined as $R = \dfrac{\eta_{qe} q}{h\nu}$, where η_{qe} is the quantum efficiency of the photodetector, which is the average number of electron-hole pairs released by the absorption of a photon, this number being at most equal to 1 in the absence of internal gain; q is the electronic charge ($= 1.602 \times 10^{-19}$ coulombs); h is the Planck constant $= 6.626196 \times 10^{-34}$ joule-second; and ν is the optical frequency.

According to Saleh and Teich [1991: §5.3, 5.4], let us assume that the optical medium is isotropic, and the wave is monochromatic, behaving locally like a transverse electromagnetic plane wave. Then the electric and magnetic fields in the direction of the vector $\mathbf{k}$ can be represented as

$$\mathbf{E} = \Re\left\{\mathbf{E}_0\, e^{-i\left(2\pi\nu e^{\mathbf{k}\cdot\mathbf{r}}\right)}\right\}, \quad \mathbf{B} = \Re\left\{\mathbf{B}_0\, e^{-i\left(2\pi\nu e^{\mathbf{k}\cdot\mathbf{r}}\right)}\right\}. \tag{7.36}$$

The power flows in the direction of the vector $\mathbf{k}$, and power density p is given by

$$p = \frac{\mathbf{E}_0 \cdot \overline{\mathbf{E}}_0}{2\eta} = \frac{E_{0x}^2 + E_{0y}^2 + E_{0z}^2}{2\eta},$$

where $\mathbf{E}_0$ and $\mathbf{H}_0$ are complex-valued constants, $\overline{\mathbf{E}}_0$ being the complex conjugate, and η is the *characteristic impedance* of the medium, given by $\eta = \sqrt{\mu/\epsilon} = 37,752$ in vacuum. The total power incident on a surface of area A is equal to the integral of the power density over A, thus,

$$P = \iint_A p\, \frac{\mathbf{k}\cdot\mathbf{n}}{|\mathbf{k}|}\, dx\, dy,$$

where $\mathbf{n}$ is the outward unit normal vector to the surface of the photodetector, and $\mathbf{k}/|\mathbf{k}|$ is the unit vector in the direction of the power flow.

7.2.6. Huygens-Fresnel Principle. The diffraction geometry is presented in Figure 7.5. Using rectangular coordinates, let the diffraction aperture lie in the (ξ, η)-plane, and be illuminated in the positive z-direction (Figure 7.5), where the (x, y)-plane is parallel and orthogonal to the (ξ, η)-plane at a distance z, and the z-axis cuts both planes at their origins. In view of Eq (7.28), the Huygens-Fresnel principle is defined as

$$U(P_0) = \frac{1}{i\lambda} \iint_{\Sigma} U(P_1) \frac{e^{i\,kr_{01}}}{r_{01}} \cos\theta\, ds, \qquad (7.37)$$

where θ is the angle between the outward normal $\mathbf{n}$ and the vector $\mathbf{r}_{01}$ from the point P_0 to the point P_1 and $\cos\theta = z/r_{01}$. Thus, assuming that $r_{01} \gg \lambda$, i.e., the observation distance is many wavelengths from the aperture, the *Huygens-Fresnel principle* (7.37) can be restated as

$$U(x, y) = \frac{z}{i\,\lambda} \iint_{\Sigma} U(\xi, \eta) \frac{e^{i\,kr_{01}}}{r_{01}^2}\, d\xi\, d\eta, \qquad (7.38)$$

where

$$r_{01} = \sqrt{z^2 + (x - \xi)^2 + (y - \eta)^2}. \qquad (7.39)$$

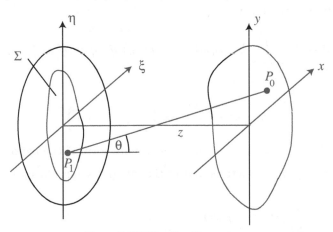

Figure 7.5 Diffraction Geometry.

Other approximations are as follows.

7.2.7. Fresnel Approximation. First, we approximate the distance r_{01} between P_0 and P_1, using the binomial series. Let b be a number, $b < 1$. Then $\sqrt{1 + b} = 1 = \frac{1}{2}b - \frac{1}{8}b^2 + \cdots$. Using this expansion, we find from (7.39) that

$$r_{01} = \sqrt{z^2 + (x - \xi)^2 + (y - \eta)^2}$$

$$= z\sqrt{1 + \left(\frac{x-\xi}{z}\right)^2 + \left(\frac{y-\eta}{z}\right)^2}$$

$$\approx z\left[1 + \frac{1}{2}\left(\frac{x-\xi}{z}\right)^2 + \frac{1}{2}\left(\frac{y-\eta}{z}\right)^2\right] \quad \text{retaining the first two terms.}$$

$$\tag{7.40}$$

We will now consider different cases of the expansion (7.40).

CASE 1. Note that for the term r_{01}^2 appearing in the denominator of (7.38), the error due to dropping all terms in the above expansion assumes that z will be very small. But for r_{01} appearing in the exponent the errors become very prominent. Also note that k is a very large number, usually greater than 10^7 in the visible portion of the spectrum, and that very small phase changes by as little as a fraction of a radian can change value of the exponential significantly. Hence, the expression (7.38) is approximated by

$$U(x, y) = \frac{e^{ikz}}{i\lambda z} \int_{-\infty}^{\infty} \int_{-\infty}^{\infty} U(\xi, \eta) \exp\left\{\frac{ik\left[(x-\xi)^2 + (y-\eta)^2\right]}{2z}\right\} d\xi\, d\eta,$$

$$\tag{7.41}$$

where $U(x, y)$ is the observed field strength, $U(\xi, \eta)$ is the aperture distribution, and $e^{ik(\xi^2+\eta^2)/(2z)}$ is a quadratic phase function. Note that Eq (7.41) is a convolution of the form

$$U(x, y) = \int_{-\infty}^{\infty} \int_{-\infty}^{\infty} U(\xi, \eta) h(x - \xi, y - \eta)\, d\xi\, d\eta = (U \star h)(x, y), \tag{7.42}$$

where

$$h(x, y) = \frac{e^{ikz}}{i\lambda z} \exp\left\{\frac{ik}{2z}(x^2 + y^2)\right\}. \tag{7.43}$$

CASE 2. Eq (7.41) takes another approximate form if the exponential part $\exp\left\{\frac{ik}{2z}(x^2 + y^2)\right\}$ is factored out of the integral signs, thus giving

$$U(x, y) = \frac{e^{ikz}}{i\lambda z} e^{ik(x^2+y^2)/(2z)}$$

$$\times \int_{-\infty}^{\infty} \int_{-\infty}^{\infty} \left\{U(\xi, \eta) e^{ik(\xi^2+\eta^2)/(2z)}\right\} \exp\left\{-\frac{2\pi i}{\lambda z}(x\xi + y\eta)\right\} d\xi\, d\eta, \tag{7.44}$$

which, except for the multiplying factor outside the integral, is the Fourier transform of the product of the complex field just right to the aperture and a quadratic phase exponential (see §3.3).

Both of these two cases, defined by Eqs (7.41) and (7.44), are known as the *Fresnel diffraction integral* valid in the near-field of the aperture. The Fresnel approximation gives accurate results only if the higher-order terms in (7.40) are small, such that they do not change the value of the Fresnel integral (7.41) or (7.44). From Eq (7.43) the real and imaginary parts of the quadratic-phase exponential, excluding the unit-magnitude phasor e^{ikz}, are

$$\frac{1}{i\lambda z} e^{i\pi(x^2+y^2)/(\lambda z)} = \frac{1}{i\lambda z}\left[\cos\left\{\frac{\pi(x^2+y^2)}{\lambda z}\right\} + i\sin\left\{\frac{\pi(x^2+y^2)}{\lambda z}\right\}\right],$$

where $k = 2\pi/\lambda$. The one-dimensional quadratic phase functions $\cos(\pi x^2)$ and $\sin(\pi x^2)$ have each an area $1/\sqrt{2}$. They are plotted in Figure 7.6.

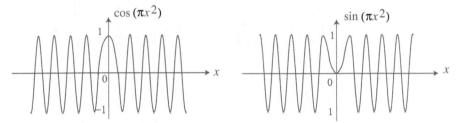

Figure 7.6 Quadratic-Phase Cosine and Sine Functions.

The accuracy of this approximation also depends on the above-mentioned exponent. As seen from Eq (7.43), this approximation replaces the spherical secondary wavelets with the parabolic wavefronts. A sufficient condition for accuracy is that the distance z satisfies the requirement:

$$z^3 \gg \frac{\pi}{4\lambda}\left[(x-\xi)^2 + (y-\eta)^2\right]^2_{\max}. \tag{7.45}$$

Thus, for example, for a circular aperture of diameter 1 cm, a circular region of size 1 cm, and a wavelength of 0.5 μm, the condition (7.45) requires that the distance z must be $\gg$ 25 cm for an accurate approximation.

Example 7.1. We will consider how to determine the signs of the phases of exponential expressions in spherical and plane waves. Let the observation point be in the (x, y)-plane which is orthogonal to the z-axis. Consider a spherical wave that is diverging from a point on the z-axis. Then the movement away from the origin will generate portions of the wavefront that were emitted earlier in time because the wave must propagate further to reach those points. This will imply that the phase must be positive and increase as the wave moves away from the origin. Hence, the exponential expressions $e^{ikr_{01}}$ and $e^{ik(x^2+y^2)/(2z)}$, $z > 0$, represent a diverging spherical wave and a quadratic-phase approximation, respectively, to this wave. Similarly, the exponential expressions $e^{-ikr_{01}}$ and $e^{-ik(x^2+y^2)/(2z)}$ will represent a converging spherical wave for $z > 0$. However, if $z < 0$, the nature of the spherical wave reverses itself.

Now, consider plane waves that travel at an angle θ with respect to the optical axis. Then the expression $e^{2\pi i \alpha y}$, $\alpha > 0$, represents a plane wave with a wave vector in the (y, z)-plane. Recall that a positive angle rotates counterclockwise with respect to the z-axis. Thus, if the movement is in the positive y direction, the argument of the above exponential expression increases in a positive sense. This means that we are moving to a portion of the wave that was emitted at an earlier time, implying that the wave vector makes a positive angle with the z-axis, as shown in Figure 7.7. ■

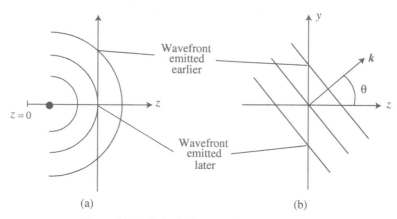

(a)   (b)

Figure 7.7 (a) Spherical Waves; (b) Plane Waves.

Example 7.2. (Talbot images) In the case of Fresnel approximations, the transfer function of the propagation through the free space is given by Goodman [1985: 72] as

$$H(f_x, f_y) = \begin{cases} \exp\left[2\pi i \dfrac{z}{\lambda} \sqrt{1 - (\lambda f_x)^2 - (\lambda f_y)^2}\right], & \text{if } \sqrt{f_x^2 + f_y^2} < \frac{1}{\lambda} \\ 0 & \text{otherwise.} \end{cases}$$

By taking the Fourier transform of the Fresnel diffraction impulse response function (7.43), we find a transfer function for Fresnel diffraction as

$$H(f_x, f_y) = \mathcal{F}\left\{\frac{e^{ikz}}{i\lambda z} e^{\pi i (x^2+y^2)/(\lambda z)}\right\} = e^{ikz} e^{-\pi i \lambda z (f_x^2 + f_y^2)}. \tag{7.46}$$

This is a quadratic phase distribution, where the term e^{ikz} on the right represents a constant phase delay for all plane waves that travel between the two parallel planes separated by a distance z. The second exponential term on the right represents different phase delays for the plane waves traveling in different direction (because of the minus sign). Now, we will find an approximate solution from Eq (7.46) by using the binomial expansion:

$$\sqrt{1 - (\lambda f_x)^2 - (\lambda f_y)^2} \approx 1 - \frac{(\lambda f_x)^2}{2} - \frac{(\lambda f_y)^2}{2}, \qquad (7.47)$$

which holds only if $\lambda f_x \ll 1$ and $\lambda f_y \ll 1$. These restrictions imply that the frequencies f_x and f_y are restricted to small angles. Thus, the Fresnel approximation remains accurate only for small angles of diffraction.

Consider a thin sinusoidal grating within the region of Fresnel diffraction. It is assumed that the structure is illuminated, from the light source, by a unit-amplitude, normally incident plane wave, so that the grating, modeled as a transmission, has amplitude transmittance $T_A(\xi, \eta) = \frac{1}{2}\left[1 + m \cos(2\pi\xi/L)\right]$ with period L, and the grating lines are parallel to the η-axis. We will compute the field and intensity on the (x, y)-plane which is at a distance z to the right of the grating. The geometry for the diffraction computation is shown in Figure 7.8.

As an example, consider a thin sinusoidal amplitude grating defined by the amplitude transmittance function t_A in a square aperture of width $2w$ by

$$t_A(\xi, \eta) = \left[\frac{1}{2} + \frac{m}{2} \cos(2\pi f_0 \xi)\right] \text{rect}\left(\frac{\xi}{2w}\right) \text{rect}\left(\frac{\eta}{2w}\right), \qquad (7.48)$$

The Fourier transform of the amplitude transmittance (7.47) gives

$$\mathcal{F}\{T_A(\xi, \eta)\} = \frac{1}{2}\,\delta\,(f_x, f_y) + \frac{m}{4}\delta\left(f_x - \frac{1}{L}, f_y\right) + \frac{m}{4}\delta\left(f_x + \frac{1}{L}, f_y\right).$$

The transfer function (7.48) is equal to 1 at the origin. Hence, when evaluated at the frequencies $(f_x, f_y) = (\pm\frac{1}{L}, 0)$, we find that

$$H\left(\pm\frac{1}{L}, 0\right) = e^{-\pi i\,\lambda z/L^2}.$$

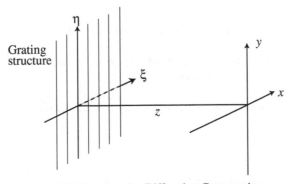

Figure 7.8 Geometry for Diffraction Computation.

Now, after propagating over the distance z behind the grating, the Fourier transform of the field becomes

$$\mathcal{F}\{U(x,y)\} = \frac{1}{2}\delta\left(f_x, f_y\right) + \frac{m}{4}e^{-\pi i \lambda z/L^2}\left\{\delta\left(f_x - \frac{1}{L}, f_y\right) + \delta\left(f_x + \frac{1}{L}, f_y\right)\right\},$$

which using inverse transform gives the field at a distance z as

$$U(x,y) = \frac{1}{2} + \frac{m}{4}e^{-\pi i \lambda z/L^2}\left\{e^{2\pi i x/L} + e^{-2\pi i x/L}\right\}$$

$$= \frac{1}{2}\left[1 + me^{-\pi i \lambda z/L^2}\cos\left(\frac{2\pi x}{L}\right)\right].$$

Thus, the intensity is given by

$$I(x,y) = \frac{1}{4}\left[1 + 2m\cos\left(\frac{\pi \lambda z}{L^2}\right)\cos\left(\frac{2\pi x}{L}\right) + m^2\cos\left(\frac{2\pi x}{L}\right)\right].$$

Now, we consider the following three cases.

CASE 1. If the distance z behind the grating is such that $\dfrac{\pi \lambda z}{L^2} = 2n\pi$, i.e., $z = \dfrac{2\pi L^2}{\lambda}$, where n is an integer, the intensity at this distance behind the grating becomes

$$I(x,y) = \frac{1}{4}\left[1 + m\cos\left(\frac{2\pi x}{L}\right)\right]^2.$$

This can be regarded as a *perfect image* of the grating, i.e., this is an exact image that would be observed behind the grating. Such images are called *Talbot images*, or *self images* (Talbot [1836]).

CASE 2. If the distance z is such that $\dfrac{\pi \lambda z}{L^2} = (2n+1)\pi$, i.e., $z = \dfrac{(2n+1)L^2}{\lambda}$, then

$$I(x,y) = \frac{1}{4}\left[1 - m\cos\left(\frac{2\pi x}{L}\right)\right]^2.$$

This distribution is also an image of the grating, only with a 180° spatial phase shift. This is also a Talbot image.

CASE 3. If the distance z satisfies $\dfrac{\pi \lambda z}{L^2} = \dfrac{(2n-1)\pi}{2}$, or $z = \dfrac{(n-\frac{1}{2})L^2}{\lambda}$, then $\cos\left(\dfrac{\pi \lambda z}{L^2}\right) = 0$, and

$$I(x,y) = \frac{1}{4}\left[1 + m^2\cos^2\left(\frac{2\pi x}{L}\right)\right] = \frac{1}{4}\left[\left(1 + \frac{m^2}{2}\right) + \frac{m^2}{2}\cos\left(\frac{4\pi x}{L}\right)\right].$$

This image has two times the frequency of the original grating and has reduced contrast. This type of image is called a *Talbot subimage*. If $m \ll 1$, then the periodic image vanishes at the subimage planes. ∎

7.2.8. Fraunhofer Diffraction. In the case of Fresnel diffraction, we know from Eq (7.44) that the observed field strength $U(x, y)$ can be calculated from a Fourier transform of the product of the aperture distribution $U(\xi, \eta)$ and a quadratic phase function $e^{i\,k(\xi^2+\eta^2)/(2z)}$. If we require that together with this Fresnel approximation the stronger *Fraunhofer approximation*

$$z \gg \frac{1}{2}\, k \left(\xi^2 + \eta^2\right)_{\max}$$

be satisfied, then the quadratic phase factor in the integral in Eq (7.44) is approximately equal to 1 over the entire aperture, and the observed field strength can be determined using a Fourier transform of the aperture distribution, up to a multiple phase factor in (x, y), as in the case of Eq (7.44). Thus, the observed field strength in the region of Fraunhofer diffraction (or, in the *far-field* in common usage) is given by

$$U(x, y) = \frac{e^{i\,kz}\, e^{i\,k(x^2+y^2)/(2z)}}{i\,\lambda z} \int_{-\infty}^{\infty}\int_{-\infty}^{\infty} U(\xi, \eta)\, e^{-2\pi i\,(x\xi+y\eta)/(\lambda z)}\, d\xi\, d\eta,$$

$$(7.49)$$

which is simply the Fourier transform of the aperture distribution evaluated at the circular frequencies f_x and f_y (see §3.1). Notice that there is no transfer function associated with the Fraunhofer diffraction.

This analysis is sufficient to develop the Fraunhofer diffraction. Some examples of specific diffraction patterns are given below. They will explain the concept of intensity.

Example 7.3. (Rectangular aperture) In this case the amplitude transmittance is defined as

$$T_A(\xi, \eta) = \text{rect}\left(\frac{\xi}{2w_x}\right)\, \text{rect}\left(\frac{\eta}{2w_y}\right),$$

where w_x and w_y are the half-widths of the aperture in the ξ and η directions, respectively. If the aperture is illuminated by a normal incident, monochromatic plane wave of unit amplitude, the field distribution across the aperture is equal to the transmittance function T_A. Using (7.49), the Fraunhofer diffraction is given by

$$U(x, y) = \frac{e^{i\,kz}\, e^{i\,k(x^2+y^2)/(2z)}}{i\,\lambda z}\, \mathcal{F}\{U(\xi, \eta)\}\Big|_{\substack{f_x=x/\lambda z, \\ f_y=y/\lambda z}},$$

where $\mathcal{F}\{U(\xi,\eta)\} = A \operatorname{sinc}(2w_x f_x) \operatorname{sinc}(2w_y f_y)$, A being the area of the aperture $(A = 4w_x w_y)$. Then, using Appendix A, we get

$$U(x,y) = \frac{e^{i\,kz}\,e^{i\,k(x^2+y^2)/(2z)}}{i\,\lambda z} A \operatorname{sinc}\left(\frac{2w_x x}{\lambda z}\right) \operatorname{sinc}\left(\frac{2w_y y}{\lambda z}\right),$$

$$I(x,y) = \frac{A^2}{\lambda^2 z^2} \operatorname{sinc}^2\left(\frac{2w_x x}{\lambda z}\right) \operatorname{sinc}^2\left(\frac{2w_y y}{\lambda z}\right),$$

where $I(x,y)$ is the normalized intensity. A cross-section of the Fraunhofer diffraction is shown in Figure 7.9. ∎

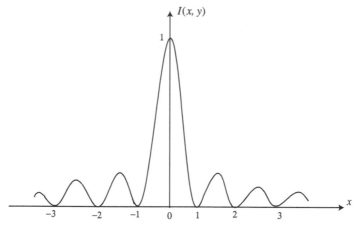

Figure 7.9 Fraunhofer Diffraction of a Rectangular Aperture.

Example 7.4. (Circular aperture) Let the radius of the aperture be w. If a radius coordinate in the plane of the aperture is ρ, then the amplitude transmittance is $T_A(\rho) = \operatorname{circ}\left(\frac{\rho}{w}\right)$ (see Taylor and Lipson [1964]). Since there exists a circular symmetry, the Fourier transform of Eq (7.49) can be written as a Fourier-Bessel transform (see Appendix K). Thus, if r is the radius coordinate in the observation plane, then

$$U(r) = \frac{e^{i\,kz}}{i\,kz} e^{i\,kzr^2/(2z)} \mathcal{B}\{U(\rho)\}\big|_{p=r/(\lambda z)},$$

where $\mathcal{B}$ denotes the Fourier-Bessel transform, $\rho = \sqrt{\xi^2 + \eta^2}$ is the radius in the aperture plane, $p = \sqrt{f_x^2 + f_y^2}$ is the radius in the spatial frequency domain (see §3.3.1), and for a unit-amplitude, normally incident plane-wave illumination

$$\mathcal{B}\left\{\operatorname{circ}\left(\frac{\rho}{w}\right)\right\} = A \frac{J_1(2\pi w\rho)}{\pi w\rho}, \quad A = \pi w^2.$$

Thus, the Fraunhofer amplitude distribution and the normalized intensity for the circular aperture are respectively given by

$$U(r) = e^{ikz}\, e^{ikr^2/(2z)}\, \frac{A}{i\lambda z}\, \frac{2J_1(kwr/z)}{kwr/z},$$

$$I(r) = \left(\frac{A}{\lambda z}\right)^2 \left[\frac{2J_1(kwr/z)}{kwr/z}\right]^2.$$

Note that $I(r)$ is an Airy-type distribution with successive maxima and minima and width of the central lobe (which is equal to $1.22\,\lambda z/w$), as shown in Figure 7.10, where the notation $\mathrm{Ai}(x) = \left[\dfrac{2J_1(\pi x)}{\pi x}\right]^2$ is used. The graph in Figure 7.10 is drawn for $x \geq 0$, and it can be reflected in the y-axis for $x < 0$. This complete graph is presented in Figure 7.11. ∎

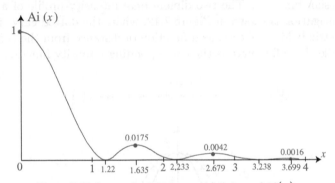

Figure 7.10 Successive Maxima and Minima of Ai(x).

Note that maxima, marked by black dots, are: $Ai(0) = 1$, $Ai(\pm 1.635) = 0.0175$, $Ai(\pm 2.679) = 0.0042$, $Ai(\pm 3.699) = 0.0016$, and minima are: $Ai(\pm 1.22) = 0$, $Ai(\pm 2.33) = 0$, $Ai(\pm 3.238) = 0$.

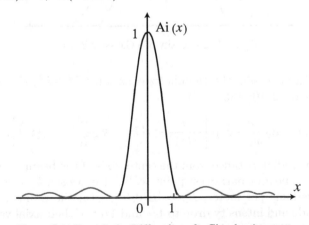

Figure 7.11 Fraunhofer Diffraction of a Circular Aperture.

7.3 Quasi-Optics

This topic deals with the propagation of electromagnetic radiation when the wavelength is about the same size as that of the optical components, like lenses, mirrors, and apertures. In particular, we study the propagation of Gaussian beams such that the beam width is comparable to the wavelength λ. A Gaussian beam is a beam of electromagnetic radiation such that its transverse electric field and intensity (irradiance) distributions are approximated by Gaussian functions. For example, many lasers emit beams that propagate a Gaussian pattern. In such a case the laser is said to be operating on the fundamental transverse mode (or TEM_{00} mode) of laser's optical resonator. When refracted by a diffraction-limited lens, a Gaussian beam is transformed into another Gaussian beam. The Gaussian beam is mathematically described as the solution of the Helmholtz equation in the paraxial form as a Gaussian function. The two-dimensional intensity profile of a Gaussian beam propagation is shown in Figure 7.12, where the dotted curve is the plot of the electric field amplitude as a function of distance from the center of the beam, while the solid curve is the corresponding intensity function $I(x, y)$.

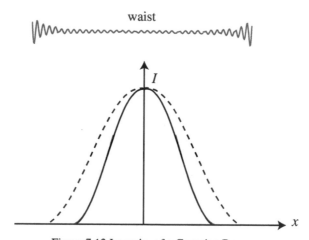

Figure 7.12 Intensity of a Gaussian Beam.

In the TEM_{00} mode, the Helmholtz equation $\left(\nabla^2 + k^2\right) E = 0$ yields the solution (Svelto [2010: 153–155])

$$E(r, z) = E_0 \frac{w_0}{w(z)} \exp\left[\frac{-r^2}{w(z)^2} - i\left(kz + k\frac{r^2}{2R(z)} - \zeta(z)\right)\right], \qquad (7.50)$$

where r is the radial distance from the center axis of the beam, z the axial distance from the beam's narrowest point (called the *waist*), $k = 2\pi/\lambda$ the wave number in radians per meter, $E_0 = |E(0,0)|$, $w(z)$ the radius at which the field amplitude and intensity drop to $1/e$ and $1/e^2$ of their axial value, respectively, $w_0 = w(0)$ is the waist size, $R(z)$ the radius of curvature of the beam's wavefronts, and $\zeta(z)$ the Gouy phase shift which is an extra contribution to the phase. The related time-averaged intensity (or irradiance) distribution is

$$I(r, z) = \frac{|E(r, z)|^2}{2\eta} = I_0 \left(\frac{w_0}{w(z)} \right)^2 \exp \left(\frac{-2r^2}{w^2(z)} \right), \qquad (7.51)$$

where $I_0 = I(0,0)$ is the intensity at the center of the beam at its waist, and η is a constant that defines the characteristic impedance of the medium through which the beam is propagating; thus, for free space $\eta = \eta_0 \approx 376.7\,\Omega$. Other references for quasioptics are Goodman [1985], Mandel and Wolf [1995], and Bandres and Gutierez-Vega [2004].

The name 'quasioptics' represents an intermediate state between conventional optics as described above, and electronics. It often describes the signals in the far-infrared or terahertz region of the electromagnetic spectrum. The far-infrared has wavelength from 1.5 μm to 1 mm that corresponds to a range of about 20 THz to 300 GHz. The visible light has wavelength ranging from 400 mm to 700 mm. This entire spectrum is presented in Figure 7.13.

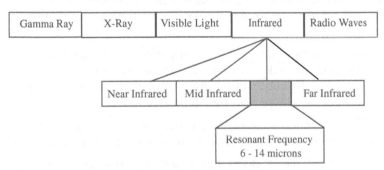

Figure 7.13 Electromagnetic Spectrum.

The phenomenon of diffraction of waves plays an important role not only in the case of light waves but in all kinds of waves, since diffraction is the result of all waves of given wavelength along all obstructing paths. In the case of light waves it affects the visibility of objects around us. For example, we cannot see the motion of a single molecule even under a microscope, because it is much smaller than a polystyrene bead or a grain of pollen. Such objects are seen through a microscope only because they block some of the light that illuminates them from below as we look down upon them. If an object is smaller than one half the wavelength of light, the light around it eliminates most of the shadow it casts and we do not see it. However, some objects which are fluorescent and emit light are visible and refraction no longer makes them invisible. For example, individual DNA molecules are visible only when they are tainted with fluorescent dyes, but they are not visible in normal microscope light because the width of the DNA helix is much smaller than the wavelength of light.

If we consider the contribution of an infinitely small neighborhood around a path of waves, we obtain what is known as a wavefront, and if we add all wavefronts (i.e., integrate over all paths from the source to the detector which is a point on the screen), we obtain a procedure to study the phenomenon of diffraction. A diffraction pattern can be determined by the phase and the amplitude of each wavefront, which enables us to find at each point the distance for each source on the incoming wavefront. If this distance for each source differs by an integer multiple of wavelength, then all wavefronts will be in phase, which will result in constructive interference. But if this distance is an integer plus one half of the wavelength, there will be a complete destructive interference. Usually such values of minima and maxima are determined to explain diffraction results.

The simplest type of diffraction occurs in a two-dimensional problem, like water waves since they propagate only on a surface. In the case of light waves we can neglect one dimension if the diffracting object extends in the direction of light over a distance far greater than its wavelength. However, in the case of light emanating through small circular holes we must consider the three-dimensional problem. Some qualitative observations about diffraction are as follows:

(i) The angular spacing in diffraction patterns is inversely proportional to the dimension of the object causing diffraction, i.e., the smaller the diffracting object, the wider the resulting diffraction pattern, and conversely. This pattern depends on the sines of the angles.

(ii) The diffraction angles are invariant under scaling, i.e., they depend only on the ratio of the wavelength to the size of the diffracting object.

(iii) If the diffracting object has a periodic structure, like a diffracting grating, the diffraction becomes sharper.

The problem of computing the pattern of a diffracted wave depends on the phase of each of the simple sources on the incoming wavefront. In Fraunhofer diffraction (or far-field diffraction) the point of observation is far from the diffracting object. It is mathematically simpler than Fresnel diffraction (or near-field diffraction), in which the point of observation is closer to the diffracting object. The structure of near-field and far-field in electromagnetic radiation is shown in Figure 7.14, where microwaves from an antenna radiate into infinite space. Note that these waves decrease in intensity by the inverse-square law.

The best way to study diffraction of light waves is presented by multi-slit arrangements which can be regarded as multiple simple wave sources if the slits are very narrow. A light wave going through a very narrow slit extends to infinity in one direction and this situation results in a two-dimensional problem. The simplest case is that of a two-slit arrangement, where the slits are a distance a apart (Figure 7.15). Then the maxima and minima in the amplitude are determined by the path difference to the first slit and to the other slit.

In the case of the *Fraunhofer approximation*, if the observer is far away from the slits, the difference in path lengths to the two slits as seen from the image is $\Delta s = a \sin \theta$. The maxima in the intensity occur if this path length difference Δs is an integer multiple of wavelength, i.e., if $a \sin\theta = m \lambda$, where m is the number of the order of each maximum, a the distance between the slits, and θ the angle at which constructive interference occurs.

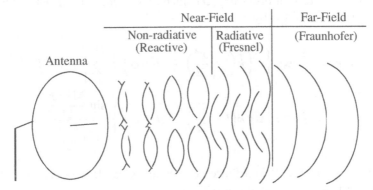

Figure 7.14 Near-Field and Far-Field Electromagnetic Radiation.

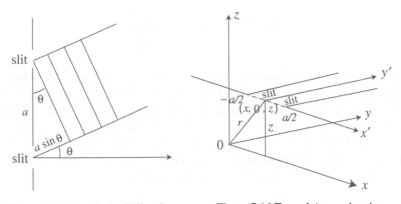

Figure 7.15 Fraunhofer Diffraction.   Figure 7.16 Fresnel Approximation.

A radial wave is represented by $E(r) = \dfrac{A}{r} \cos(k r - \omega t + \phi)$, where $k = 2\pi/\lambda$, ω the frequency and ϕ the phase of the wave at the slits. The function $E(r)$ is the real part of the complex wave $\Psi(r)$ which is defined as $\Psi(r) = \dfrac{A}{r} e^{i(kr - \omega t + \phi)}$. The amplitude of this complex wave is given by $|\Psi(r)| = A/r$. For M slits, the radial wave at a point x on the screen, which is at a distance

l away, is given by

$$E_M(r) = A\, e^{i\,(-\omega t + \phi)} \sum_{m=0}^{M-1} \frac{e^{i\,k\sqrt{l^2+(x-ma)^2}}}{\sqrt{l^2 + (x - ma)^2}}.$$

Since $\sqrt{l^2 + (x - ma)^2} = l \left[1 + \frac{(x - ma)^2}{l^2}\right]^{1/2} \approx l + \frac{x^2}{2l} - \frac{max}{l} + \frac{m^2 a^2}{2l}$, we can neglect the term of order $a^2/(2l)$ since $a^2 \ll l$ for the far-field. Then the complex wave is defined by

$$\Psi(r) = \frac{A}{l}\, \exp\left\{i\,\left(k\left(l + \frac{x^2}{2l}\right) - \omega t + \phi\right)\right\} \sum_{m=0}^{M-1} e^{-i\,kxma/l}$$

$$= \frac{A}{l}\, \exp\left\{i\,\left(k\left(l + \frac{x^2}{2l}\right) - \omega t + \phi\right)\right\} \frac{\sin \dfrac{Mkax}{2l}}{\sin \dfrac{kax}{2l}}, \qquad (7.52)$$

where the sum in (7.52) is a finite geometric series. If we set $A/l = I_0$, the intensity of this wave is given by

$$I(x) = |\Psi|^2 = I_0 \left(\frac{\sin \dfrac{Mkax}{2l}}{\sin \dfrac{kax}{2l}}\right).$$

In the case of the *Fresnel approximation*, let $\Psi(r)$ define a monochromatic complex plane wave of wavelength λ passing through a slit of width a centered at the point $(x, 0, z)$. The 3-D situation can be reduced to a 2-D problem if we consider a coordinate system (x', y') through the point $(x, 0, z)$ so that the slit is defined by $\{-a/2 < x' < a/2, -\infty < y' < \infty\}$ (see Figure 7.16). Notice that this slit is located on the (x, z)-plane and the light beam is diffracted along the y'-axis. This complex wave is defined by

$$\Psi(r) = \int_{\text{slit}} \frac{i}{r\lambda} \Psi_0\, e^{-i\,kr}\, d(\text{slit}), \qquad (7.53)$$

where Ψ_0 is a constant, and

$$r = \sqrt{(x - x')^2 + y'^2 + z^2} = z\sqrt{1 + \frac{(x - x')^2 + y'^2}{z^2}} \approx z + \frac{(x - x')^2 + y'^2}{2z}.$$

Since the factor $1/r$ in (7.53) is non-oscillatory, its contribution to the magnitude of the intensity is very small compared to the exponential factor. We will, therefore, approximate $1/r$ by $1/z$. Then (7.53) becomes

$$
\begin{aligned}
\Psi(x) &= \frac{i\,\Psi_0}{\lambda z} \int_{-a/2}^{a/2} \int_{-\infty}^{\infty} e^{-i\,k\left(z+\dfrac{(x-x')^2+y'^2}{2z}\right)} \, dx'\, dy' \\[2mm]
&= \frac{i\,\Psi_0}{\lambda z} e^{-i\,kz}\, e^{-i\,kx^2/(2z)} \left[\int_{-a/2}^{a/2} e^{i\,kxx'/z}\, e^{-i\,kx'^2/(2z)}\, dx' \right] \sqrt{\frac{2\pi z}{i\,k}} \quad (7.54) \\[2mm]
&= \Psi_0 \sqrt{\frac{i}{z\lambda}}\, e^{-i\,kz}\, e^{-i\,kx^2/(2z)} \left[\int_{-a/2}^{a/2} e^{i\,kxx'/z}\, e^{-i\,kx'^2/(2z)}\, dx' \right],
\end{aligned}
$$

where we have used the formula $\displaystyle\int_{-\infty}^{\infty} e^{-i\,ky'^2/(2z)}\, dy' = \sqrt{\dfrac{2\pi z}{i\,k}}$. Since the quantity kx'^2/z is small in the Fresnel approximation, we can approximate the factor $e^{-i\,kx'^2/(2z)} \approx 1$, and the terms $e^{-i\,kz}$ can also be neglected if we take z to be a very small positive quantity, say $z = \varepsilon$. Then (7.54) reduces to

$$
\begin{aligned}
\Psi(x) &= \Psi_0 \sqrt{\frac{i}{\varepsilon\lambda}}\, e^{-i\,kx^2/(2\varepsilon)} \int_{-a/2}^{a/2} e^{i\,kxx'/\varepsilon}\, dx' \\[2mm]
&= \Psi_0 \sqrt{\frac{i}{\varepsilon\lambda}}\, e^{-i\,kx^2/(2\varepsilon)}\, \mathrm{sinc}\left(\frac{kax}{2\varepsilon}\right) \\[2mm]
&= \Psi_0 \sqrt{\frac{i}{\varepsilon\lambda}}\, e^{-i\,kx^2/(2\varepsilon)}\, \mathrm{sinc}\left(\frac{\pi a x}{\varepsilon\lambda}\right), \quad \text{where } k = 2\pi/\lambda.
\end{aligned}
$$
$$(7.55)$$

Since $\lim_{x\to 0} \mathrm{sinc}(x) = 1$, the sinc term in the above expression is almost equal to 1 in this near-field approximation, and hence, can be neglected. This leads to

$$
\Psi(x) = \frac{2\lambda\,\Psi_0}{\pi} \sqrt{\frac{i}{\varepsilon\lambda}}\, e^{-i\,kx^2/(2\varepsilon)},
$$

which defines a diffracted wave, with intensity, from (7.55), given by $I = I_0\,\mathrm{sinc}^2\left(\dfrac{\pi a x}{\varepsilon\lambda}\right)$, and $I_0 = \dfrac{|\Psi_0|}{\varepsilon\lambda}$. The details can be found in Kythe [2011: 201–205].

7.4 Electromagnetic Spectrum

The electromagnetic spectrum in Figure 7.13 shows a very small part as the visible light. Its length is less than an octave ranging from 4×10^{-5} cm (violet) to 7×10^{-5} cm (red). On both sides of the visible light there is invisible radiation with longer and shorter wavelengths, to the right and to the left, respectively: radio waves and microwaves even thousands of times longer, and the gamma rays a billion times shorter. A detailed spectrum is presented in Figure 7.17.

The German physicist Wilhelm Conrad Röntgen (1845–1923) discovered x-rays in 1895, for which he won the Nobel Prize in 1901. In his experiments with cathode rays he noticed that light was emitted by an adjacent fluorescent screen even when the tube was completely wrapped in black paper. He even took the first x-ray photograph of his wife's hand. This discovery revolutionized the medical diagnosis in orthopedics and dentistry. He called it the x-radiation, although it was named Röntgen rays. Now it is known that x-rays are generated by energetic electrons as they accelerate or decelerate. The electrons in the x-ray tubes decelerate abruptly while hitting a metallic object, whereas in synchrotons they accelerate while bending their path.

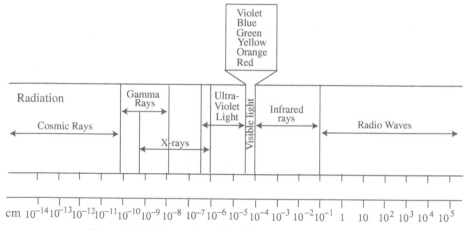

Figure 7.17 Wavelengths of the Electromagnetic Spectrum.

In 1897 the British physicist Joseph John Thomson (1856–1940), while working on cathode rays, discovered electrons, called 'corpuscules' by him. About fifteen years later, the German physicist Max von Laue postulated that the regular arrangement of atoms in crystals might provide a natural grating to generate interference if radiation of wavelength of about 10^{-8} cm were used. The experiment by Walter Friedrich, Paul Knipping and von Laue [1912] established the electromagnetic nature of x-rays. The 1903 Nobel Prize was shared by Henri Becquerel and Pierre and Marie Curie for their work on gamma rays. Based on Laue's experiment Ernest Rutherford discovered the electromagnetic nature of gamma rays, and showed that these rays penetrated metal castings to show their defects, but were very harmful for living beings. Both x-rays and gamma rays are present in radiation coming from outer space, though mostly prevented by Earth's atmosphere and Earth's magnetic field which acts like a filter.

In the electromagnetic spectrum, shown in Figure 7.17, the wavelength λ and the frequency f are related by formula: $\lambda f = c = 3 \times 10^{10}$ cm/sec. Since electron volt (eV) is used as the unit of energy, defined as the energy gained by one electron as it is accelerated in vacuum through a potential difference of one volt, its multiples are as follows: one kiloelectron volt (keV) = 10^3 eV; one megaelectron volt (MeV) = 10^6 eV; one gigaelectron volt (GeV) = 10^9 eV; and one teraelectron volt (TeV) = 10^{12} eV.

In Figure 7.18, there are two windows under atmospheric transparency, marked 1 and 2: the first is the *optical window* (visible light) with wavelengths from 3,000 Å to 300,000 Å, and the second is the *radio window* (VLF, LF, MF, HF, UHF, and VHF) covering radiation of wavelengths from 3 kHz to 300 MHz, through which electromagnetic radiation can reach Earth. Conversely, the radiation can also escape Earth through these windows to outer space. The radio window is much larger than the optical window.

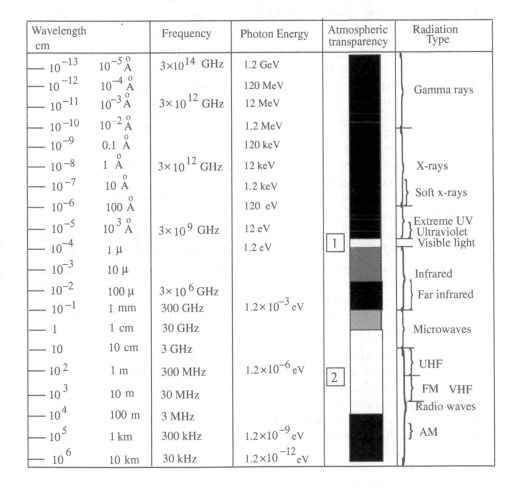

Wavelength cm		Frequency	Photon Energy	Atmospheric transparency	Radiation Type
10^{-13}	10^{-5} Å	3×10^{14} GHz	1.2 GeV		
10^{-12}	10^{-4} Å		120 MeV		Gamma rays
10^{-11}	10^{-3} Å	3×10^{12} GHz	12 MeV		
10^{-10}	10^{-2} Å		1.2 MeV		
10^{-9}	0.1 Å		120 keV		
10^{-8}	1 Å	3×10^{12} GHz	12 keV		X-rays
10^{-7}	10 Å		1.2 keV		Soft x-rays
10^{-6}	100 Å		120 eV		Extreme UV
10^{-5}	10^3 Å	3×10^9 GHz	12 eV		Ultraviolet
10^{-4}	1 μ		1.2 eV	1	Visible light
10^{-3}	10 μ				Infrared
10^{-2}	100 μ	3×10^6 GHz			Far infrared
10^{-1}	1 mm	300 GHz	1.2×10^{-3} eV		
1	1 cm	30 GHz			Microwaves
10	10 cm	3 GHz			
10^2	1 m	300 MHz	1.2×10^{-6} eV	2	UHF
10^3	10 m	30 MHz			FM VHF
10^4	100 m	3 MHz			Radio waves
10^5	1 km	300 kHz	1.2×10^{-9} eV		AM
10^6	10 km	30 kHz	1.2×10^{-12} eV		

Figure 7.18 Electromagnetic Spectrum on a Logarithmic Scale.

Wave theory failed to explain the ionizing power of x-rays and gamma rays and, in general, the interaction between light and matter. The experiment to demonstrate the photoelectric effect was conducted by Hertz in 1877. Using emissions of electrons by metal and many other materials, he found them to be photosensitive when struck by radiation of sufficiently high frequency (depending on the material). However, he could not explain the one characteristic of this phenomenon that the maximum kinetic energy of the emitted energy was independent of the intensity of light. Finally, Albert Einstein [1905] provided an explanation in his work on the theory of light. He examined the hypothesis put forth by Max Planck a few years earlier, and assumed that light was composed of quanta, or photons, whose energy is proportional to the frequency of light. Although Figure 7.17 provides the details of visible light, a complete chart of wavelength (λ), frequency, and the atmospheric transparency on a logarithmic scale is presented in Figure 7.18.

Cosmic rays originate outside the solar system. They consist of very high-energy particles and produce showers of secondary particles which penetrate and impact Earth's atmosphere and sometimes its surface. They are composed of high-energy protons and atomic nuclei, as known from the data from the Fermi space telescope in 2013 (see Ackerman et al. [2013]). According to their research it is now recognized that a major portion of primary cosmic rays originate from the supernovas of massive stars. However, there may be other sources which are not yet known. In modern scientific usage, cosmic rays are regarded as high-energy particles (with intrinsic mass), and photons which are quanta of electromagnetic radiation (without intrinsic mass), and are called 'gamma rays' and 'x-rays' depending on their frequency.

It is observed that the most energetic cosmic rays (UHECRs) reach an energy of 3×10^{20} eV, which is about 40 million times the energy of particles accelerated by the Large Hadron Collider operated by the European Organization for Nuclear Research. Most cosmic rays, however, do not possess such extreme energies. According to Nave [2013], the energy distribution of cosmic rays peaks at 3×10^{-12} J. The cosmic ray flux versus the particle energy is shown in Figure 7.19.

The visible ultraviolet (UV-Vis) refers to the case when light is visible and adjacent to near-UV and near-infrared ranges. The absorption and reflection of these UV rays change the color of the affected chemicals. In this region of the electromagnetic spectrum (Figure 7.18), molecules undergo electronic transitions, similar to the fluorescence spectroscopy, where the absorption measures transition from initial state to the excited state. UV-Vis is used in analytical chemistry for the quantitative determination of different analytes such as transition metal ions, highly conjugated organic compounds, and biological macromolecules. A UV-Vis spectrophotometer is sometimes used to detect high-performance liquid chromatography (HPLC). In such cases the presence of an analyte gives a response proportional to the concentration, and the maximum value (e.g., peak height) of the response is known as the 'response factor'. The wavelengths of the absorption follow the Woodward-Fieser-Rajagopalan rules, which are a set of empirical rules that attempt to predict the wavelength of the absorption maximum (λ_{max}) in an UV-Vis spectrum of a given compound, e.g., conjugated carbonyl compounds, conjugated dienes, and polyenes.

Microwaves range from as long as one meter to as short as one millimeter, i.e., with frequency ranging between 300 MHz (0.3 GHz) and 300 GHz. This broad range includes both UHF and VHF which are millimeter waves, and others with different boundaries. The prefix 'micro' in these waves does not mean that these waves have frequencies in the micrometer range. Instead, it indicates that they are waves with shorter wavelengths. The boundaries between far-infrared light, terahertz radiation, microwaves, and ultra-high-frequency radio waves (FM and AM) are not rigid as used in different studies and experiments.

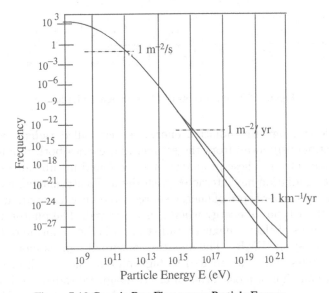

Figure 7.19 Cosmic Ray Flux versus Particle Energy.

Microwave technology is used extensively in point-to-point non-broadcast telecommunications. Microwaves are more suited for broadcast purposes since they are more easily focussed into narrow beams than radio waves, and thus re-use the frequency. The comparatively higher frequencies of microwaves allow for broad bandwidth and a high data transmission rate, with smaller antenna size, since the antenna size is inversely proportional to transmitted frequency. As such, microwaves are used in spacecraft communications, data transmission, TV, and telephone communications, and the GPS systems. The details of such communications are also discussed in §4.1, and Kythe and Kythe [2012: §3.2].

7.5 Electromagnetic Radiation

Electromagnetic radiation (EMR) is a form of energy emitted and absorbed by charged particles. This energy behaves like a wave as it travels through space. It has both electric and magnetic components, which are always in a fixed ratio of intensity with each other, oscillate in phase perpendicular to each other, and are orthogonal to the direction of energy and wave propagation. These properties are presented in Figure 7.20, where the solid curves are in the vertical $(\mathbf{k}, \mathbf{E})$-plane which corresponds to the electric field $\mathbf{E}$, while the dotted curves are in the horizontal $(\mathbf{k}, \mathbf{B})$-plane which corresponds to the magnetic field $\mathbf{B}$. Electromagnetic waves can be regarded as a self-propagating transverse oscillating wave of electric and magnetic fields. In vacuum, electromagnetic radiation propagates at the speed of light.

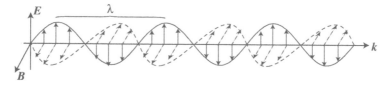

Figure 7.20 Components of Electromagnetic Radiation.

The electromagnetic radiation energy is also called *radiant energy* that moves continuously away from the source, and carries both momentum and angular momentum. These properties may be found in substances after electromagnetic radiation has interacted with them. Electromagnetic radiation is generated from other types of energy sources when it is created, and it is converted to other types of energy when it is destroyed. The photon is the basic unit of all forms of electromagnetic radiation; it is the quantum of the electromagnetic radiation (EMR). The quantum nature of light is more apparent at high frequencies which are related to high photon energy. Electrons are the vehicle for emission of most forms of electromagnetic radiation, because they have low mass and are thus easily accelerated by many devices. Fast moving electrons are greatly accelerated when they encounter a region of electromagnetic force. This leads to the production of high-frequency electromagnetic radiation encountered in nature.

The classification of electromagnetic radiation is based on the frequency of its waves. It is governed by the laws of electrodynamics (see Appendix D for Maxwell's equations of electrodynamics). Electric and magnetic fields obey the superposition principle. Since they are vector fields, all electric and magnetic field vectors add together by laws of vector addition.

Light, being an oscillatory wave, is not affected by traveling through static electric and magnetic fields in a linear medium like vacuum. But in nonlinear media, like crystals, there are interactions between light and static electric and magnetic fields, such as the Faraday effect and the Kerr effect.

Electromagnetic radiation exhibits both wave properties and particle properties simultaneously. This is known as the *wave-particle duality*. As a wave, it is transverse, so that the oscillations of the wave are perpendicular to the direction of the energy transfer. The electric and magnetic parts of the field remain in a fixed ratio and satisfy Maxwell's equations involving both $\mathbf{E}$ and $\mathbf{B}$ fields. As a particle, in 1900 Max Planck developed the theory of black-body radiation in which black bodies only emit light and other electromagnetic radiation as discrete bundles or packets of energy (called *quanta*). Later, Albert Einstein proposed that quanta of light might be regarded as real particles, and called the particle of light the *photon*. Thus, a photon has an energy E which is proportional to its frequency f; i.e., it is governed by the Planck-Einstein equation $E = hf = \dfrac{hc}{\lambda}$, where h is Planck's constant, λ the wavelength, and c the speed of light in vacuum. Similarly, the momentum p of a photon is also proportional to the frequency of light and inversely proportional to wavelength; i.e., $p = \dfrac{E}{c} = \dfrac{hf}{c} = \dfrac{h}{\lambda}$. Through experimental measurements, it was shown that the energy of individual ejected electrons was proportional to frequency of light, and not its intensity. Moreover, there is a certain minimum frequency, depending on the metal used in experiments, below which no current would flow regardless of intensity.

After a photon is absorbed by an atom, it excites the atom and elevates an electron to a higher energy level. Such an electron is generally the one that is farthest from the nucleus. When an electron in an excited atom or molecule descends to a lower energy level, it emits a photon of light equal to the difference in energy. The energy levels of electrons in atoms are discrete, such that each molecule emits and absorbs its own characteristic frequency. When the emission of the photon is imminent, the phenomenon of fluorescence occurs, which is a type of *photoluminescence*, for example, the visible light emitted from fluorescent paints, in response to ultraviolet (blacklight). When the emission of the photon is delayed, the phenomenon is called *phosphorescence*.

7.6 Adaptive Additive Algorithm

Adaptive Additive Algorithm (AA algorithm) is useful in Fourier optics, sound synthesis, stellar interferometry, optical tweezers, and diffractive optical elements (DOEs). This is an iterative algorithm that uses the Fourier transform to compute the spatial frequency phase of a propagating wave source with input as an observed amplitude (position domain) and an assumed starting amplitude. To determine the correct phase, the algorithm uses error between the desired and theoretical intensities.

The AA Algorithm follows the following ten steps:

STEP 1. Define input amplitude and random phase.

STEP 2. Apply forward Fourier transform.

STEP 3. Separate transformed amplitude and phase.

STEP 4. Compare transformed amplitude and intensity to desired output amplitude and intensity.

STEP 5. Check convergence.

STEP 6. Mix transformed amplitude with the desired output amplitude and combine with transformed phase.

STEP 7. Apply inverse Fourier transform

STEP 8. Separate new amplitude and new phase.

STEP 9. Combine new phase with the original input amplitude.

STEP 10. Loop back to forward Fourier transform, to Step 1 if error is too large.

Example 7.5. To reconstruct the spatial frequency phase (k-space) for a desired intensity in the image plane (x-space), assume that the amplitude and the starting phase of a wave in the k-space is A_0 and ϕ_n^k, respectively. Using the forward Fourier transform on the wave in the k-space to x-space, we get:

$$A_0\, e^{i\,\phi_n^k} \xrightarrow{\text{FFT}} A_n^f\, e^{i\,\phi_n^f}.$$

Then comparing the transformed intensity I_n^f with the desired intensity I_0^f, where $I_n^f = \left(A_n^f\right)^2$, and $\varepsilon = \sqrt{\left(I_n^f\right)^2 - \left(I_0\right)^2}$. Next, check if convergence conditions are satisfied. If these conditions are not met, then mix the transformed amplitude A_n^f with the desired amplitude A^f, to get $\bar{A}_n^f = aA^f + (1 - a)A_n^f$, where a is the mixing ratio ($0 \le a \le 1$), and $A^f = \sqrt{I_0}$. Then, combine the mixed amplitude with the x-space phase and apply the inverse Fourier transform, to get

$$\bar{A}^f\, e^{i\,\phi_n^f} \xrightarrow{\text{IFT}} \bar{A}_n^k\, e^{i\,\phi_n^k}.$$

Finally, separate $\bar{A}_n^k$ and ϕ_n^k, and combine A_0 with ϕ_n^k. Increase the loop by one, $n \to n + 1$ and repeat the algorithm.

Note that if $a = 0$, then $\bar{A}_n^k = A_0$. Also, if $a = 1$, the AA algorithm reduces to the Gergberg-Saxton algorithm. ∎

For details, see Soifer and Doskolovich [1997], Robel [2005], and Grier [2000], and also the following links: http://www.physics.nyu.edu/ dg86/cgh2b /node6. html; http://physics.nyu.edu/grierlab/cgh2b/node6.html; also see http://www-ccrma. stanford .edu/~ roebel1/addsyn/index.html

8

X-Ray Crystallography

Matter exists in solid, liquid or gaseous state. Most of the solid matter (about 95%) is in the form of crystals, and the rest is amorphous. Crystals have all their molecules arranged in completely ordered structures. The diffraction of visible light is used in crystallography to investigate and understand the three-dimensional structure of matter, by using radiation of very short wavelength and the Fourier transform to interpret the data.

8.1 Historical Notes

The discovery of x-rays and the electromagnetic spectrum have been discussed in §7.4. Further studies of x-rays exhibited an interesting property that they could be polarized but not refracted. The controversy arose whether the x-rays were particles or waves. If they were waves, their wavelength must be very small, of the order of at most 10^{-10} meters. Max van Laue theorized that x-rays could be diffracted if the slits were small enough. The molecular spacings in crystals were understood to be of the order of one tenth of a nanometer (= one angstrom, denoted by Å). Note that a nanometer (nm) is 10^{-9} meters; a micrometer (μm) or micron is 10^{-6} meters; a millimeter (mm) is 10^{-3} meters; and an angstrom (Å) is 10^{-10} meters. Von Laue performed an experiment in which x-rays were allowed into a lead box containing a crystal, with a sensitive film behind the crystal. After the film was developed, there was a large central point from the incident x-rays, and also many smaller points in a regular pattern. They could only be due to the diffraction of the incident beam and interference of many beams. Thus, using a crystal as a diffraction grating, von Laue established in 1912 that x-rays were not particles, and that they were light waves with very small wavelengths. A further advancement was made by the father and son Bragg in 1915 when they published their work on constructive interference of reflected x-rays and symmetries in crystals.

8.2 Bragg's Law

The Australian born Lawrence Bragg (1890–1971) and his father William H. Bragg (1862–1942) used von Laue's discovery to show that, in the case of monochromatic radiation, diffraction could be treated geometrically like reflection. They were awarded the Nobel Prize in Physics in 1915. Bragg's law can be stated as follows: the atoms of a crystal are arranged in a regular periodic pattern. They can be regarded as lying on a family of parallel and equidistant planes, where every plane of a fixed family reflects a small part of the incident radiation through the electrons of its atoms, while the larger part passes through and gets reflected over and over by many inner planes of the family. This pattern is shown in Figure 8.1, in which the reflected rays exhibit a phase difference as they go along several different paths. Among all possibilities there are two extreme cases depending on interference which is represented by the path difference ABC in Figure 8.1. They are

(i) *Constructive interference*: the path difference ABC is an integer multiple of the wavelength λ, in which case the rays reflected by two consecutive planes reinforce each other; and

(ii) *Destructive interference*: the path difference ABC is an integer multiplier of $\lambda/2$, in which case the reflected rays annihilate each other, as shown in Figure 8.2.

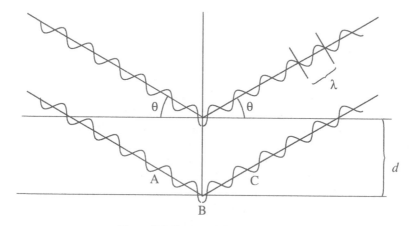

Figure 8.1 Constructive Interference.

Let d denote the distance between consecutive planes, and θ the diffraction angle (angle between the direction of the incident rays and the chosen direction of observation), then $AB = BC = d\sin\theta$, and Bragg's law for constructive interference is

$$n\lambda = 2d\sin\theta, \qquad\qquad (8.1)$$

where n is an integer, and λ the wavelength of the x-ray radiation. The condition $\lambda \leq 2d$ must be satisfied in order for diffraction to occur. Since the interplanar distance d can range from a few angstroms to tens of angstroms, the visible light as well as the ultraviolet light, as seen from Figures 7.17 and

7.18, will fail to generate the diffraction phenomenon in crystals.

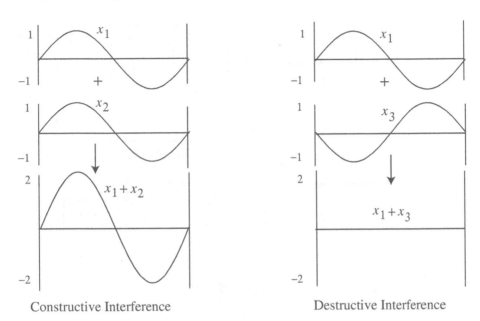

Constructive Interference   Destructive Interference

Figure 8.2 Interference.

One single radiation determines the direction of the reflecting planes, as Bragg's law relates the directions of maximum intensity of the reflected waves to the (unknown) interplanar distance d. The determination of both n and d is a crucial part of this law, which is conducted via experiments. Since $\theta \neq 0$ is the incident ray, we consider the consecutive reflections starting with the first reflection ($n = 1$), and then continue with $n = 2, 3, \ldots$, thus producing the second, third, and subsequent reflections. Along the direction of θ that satisfies (8.1), the reflected waves add to each other, and result in a wave that at a great distance (relative to the dimensions of the crystal) makes a mark on the photographic plate, i.e., it gives one of the peaks of intensity detected by a counter. Thus, all the geometrical settings of the crystal are determined by reversing the direction of the incident ray while keeping the crystal fixed, or by rotating the crystal while maintaining the direction of the x-ray fixed. These settings conform to Braggs's law and are used to measure the diffraction angle θ. All such data then determine the families of lattice planes and the distance d of the crystal using Bragg's law, which also determines the structure of crystal lattices and unit cells of crystalline compounds. Seven different types of such unit cells are presented in Figure 8.3. An accurate picture of the structure of crystals requires an x-radiation that is as close to monochromatic as possible. Filters, monochromators, specially tuned detectors, and software are used to refine the frequency of x-rays used in experiments to determine the crystalline structures.

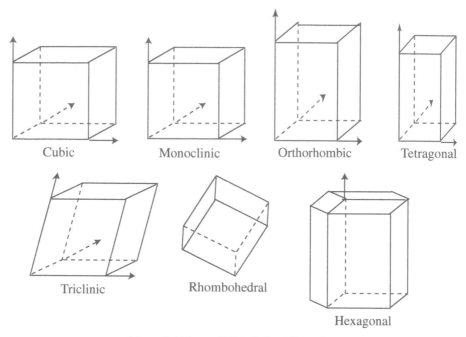

Figure 8.3 Types of Unit Cells of Crystals.

8.3 X-Ray Diffraction

In the x-ray region, synchrotron radiation provides a continuous band of wavelengths, as well as radiation of extremely high intensity, brilliance and collumination. There are five characteristics of any radiation: spectral range, photon flux, photo flux density, illumination (or brilliance), and polarization. Of these, photon flux is the total flux collected in an experiment; photon flux density is the flux per unit area, and the illumination is determined in terms of the flux per unit area. Collumination means beams of rays highly parallel (i.e., beams with extremely small divergence). When such beams pass through very small crystals of dimension of the order of 10^{-2} mm, x-ray diffraction occurs due to the spatial atomic structure of crystals. This feature becomes very important when the diffraction peaks are densely packed as in the case of macromolecular crystals. However, synchrotron radiation has good coherence but not complete coherence since all points of the source fail to emit radiation in phase. Synchrotron builders, like the storage rings in which the electrons move in a circular path, always try to improve coherence as well as intensity.

The 'first generation' synchrotrons were constructed in 1960s at University of Wisconsin, in Frascati (Italy), Hamburg (Germany), Novosibirsk (Siberia), and Tokyo. In 1970, the first storage ring (380 MeV SOR ring) was designed at the University of Tokyo. Subsequently others were constructed in the U.S., Russia, Europe, Brazil, and China. Then came the 'third generation' storage rings capable of enormous productivity and powerful undulators, which can be divided into two classes: rings of 30–60 meters in diameter designed for ultraviolet and soft x-rays; and rings 300–500 meters in diameter designed for hard x-rays. Both synchrotrons and storage rings consist of a metal vacuum chamber in the shape of a ring or a doughnut, with magnets distributed around it, and with a diameter up to a few hundred meters, thus making the circumference more than one kilometer; for example, ESRF (European Synchrotron Radiation Facility, 1994) has circumference of 844 meters, APS (Advanced Photon Source, 1996) of 1,104 meters, and SPring-8 (Super Photon ring-8 GeV, 1998) of 1,436 meters. The detailed information of all third generation sources is as follows: ALS 1.9 GeV Berkeley CA; APS 7 GeV Argonne IL; ESRf 6 GeV Grenoble, France; ELECTRA 2.0 GeV Trieste Italy; MAX II 1.5 GeV Lund Sweden; 1.7 GeV Berlin; PLS 2 GeV Pohang, S. Korea; SPring-8 8GeV Hyogo, Japan; and SRRC 1.5 GeV Taiwan.

8.3.1 Electron Distribution. The position of atomic nuclei is determined by the maximum values of electron distribution. Since the diffracted radiation cannot be focused, it can be slightly deviated using lenses as in an optical microscope, or using magnetic fields as in the electron microscope. A mathematical model for electron distribution depends on the directions of the diffracted rays as well as their intensities, which determine this distribution. In the rectangular coordinate system, let one of the atoms of the crystal be situated at the origin. The incident radiation of wavelength λ is independent of time t, and therefore, defined as

$$E(\mathbf{x}) = E_0 \, e^{i\,\mathbf{k}\cdot\mathbf{x}}, \tag{8.2}$$

where $\mathbf{x}$ is a point of the crystal and $\mathbf{k}$ is a vector in the direction of the incident radiation with length $|\mathbf{k}| = k = 2\pi/\lambda$. The electron placed at $\mathbf{x}$ scatters a part of its incident radiation as shown in Figure 8.4.

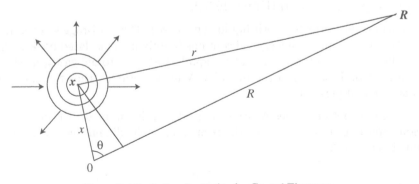

Figure 8.4 Radiation Scattering by Crystal Electrons.

Then the scattered radiation at a point $\mathbf{R}$ which is at a distance r from $\mathbf{x}$ is given by $\frac{1}{r} E_0\, e^{i\,\mathbf{k}\cdot\mathbf{x}}\, e^{i\,kr} = \frac{1}{r}\, e^{i\,(\mathbf{k}\cdot\mathbf{x}+kr)}$. Note that the term $\mathbf{k}\cdot\mathbf{x} + kr$ is a function of both the phase of the incident radiation and the distance r after the radiation scattering.

Let θ denote the (acute) angle between the vectors $\mathbf{x}$ and $\mathbf{R}$. If R is large enough in comparison with the dimensions of the crystal, then we can approximate $r \approx R - x\cos\theta$, and thus $kr \approx kR - kx\cos\theta$. Now, let $\mathbf{k}'$ be a vector parallel to the vector $\mathbf{R}$ such that $|\mathbf{k}'| = k$. Then $kx\cos\theta = \mathbf{k}'\cdot\mathbf{x} = |\mathbf{k}'||\mathbf{x}|\cos\theta$. Hence,

$$\mathbf{k}\cdot\mathbf{x} + kr \approx kx\cos\theta + kR - kx\cos\theta = (\mathbf{k}-\mathbf{k}')\cdot\mathbf{x} + kr.$$

Since the scattering is caused by the electrons, and not by the nuclei of the atoms in the crystal, we can safely assume that the amplitude of the scattered radiation is proportional to the electron density $n(\mathbf{x})$. Since we have assumed that $r \approx R$, the radiation at $\mathbf{R}$ is given by

$$\frac{1}{R}\, e^{i\,kR} \int_V n(\mathbf{x})\, e^{-i\,(\mathbf{k}'-\mathbf{k})\cdot\mathbf{x}}\, d\mathbf{x}, \tag{8.3}$$

where V denotes the volume of the crystal. Note that the vectors $\mathbf{k}$ and $\mathbf{k}'$ have the same length but different directions: $\mathbf{k}$ in the direction of the incident ray and $\mathbf{k}'$ in the direction of the scattered radiation. The direction of the vector $\mathbf{K} = \mathbf{k}' - \mathbf{k}$ measures the change in the direction of the incident ray such that $\max|\mathbf{K}| = 4\pi/\lambda$. If we drop the constant factor $\frac{1}{R}\, e^{i\,kR}$, we find from Eq (8.3) that the radiation at $\mathbf{R}$ is given by

$$\int_V n(\mathbf{x})\, e^{-i\,\mathbf{K}\cdot\mathbf{x}}\, d\mathbf{x} = N(\boldsymbol{\omega}), \tag{8.4}$$

which is the Fourier transform of the electron density $n(\mathbf{x})$. Thus we have the relationship $n(\mathbf{x}) \rightleftharpoons N(\boldsymbol{\omega})$ (Kittel [1971]).

This model works for both liquids and gases. Due to Bragg's reflections, only the Fourier image of a crystal contains finitely many isolated peaks, while this phenomenon is totally absent in liquids and monatomic gases, as shown in Figure 8.5. In a crystal the quantity $N(\boldsymbol{\omega})$ is periodic, thus the spectrum is discrete and periodic.

Note that not only does $N(\boldsymbol{\omega}) \to 0$ at infinity, it can be set to zero outside a solid sphere of radius equal to the reciprocal of the dimension which ranges from 1 Å to 1.7 Å.

8.4 Generation of X-Rays

X-rays are part of the spectrum of electromagnetic radiation that occupy the region between the ultraviolet and gamma rays (Figure 7.17). The wavelengths of radiation commonly used for x-ray diffraction lie between 0.7 and 2.3 Å. This is very close to the interplanar distances of most crystalline substances. The more penetrating radiation used for medical purposes has a smaller wavelength.

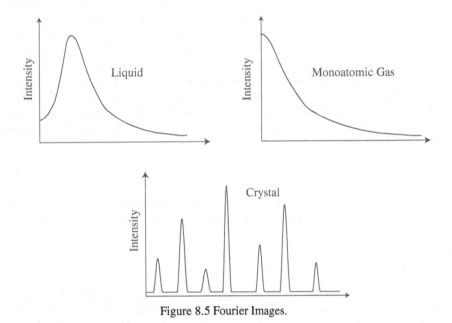

Figure 8.5 Fourier Images.

X-rays are composed of short-wavelength high-energy electromagnetic radiation. They possess the properties of both waves and particles, and can be described in terms of both photon energy (E) or wavelength (λ) and frequency f. Recall that λ is the distance between the peaks, and f is the number of peaks passing a point per second. The relationships between these quantities are given by the following equations:

$$f = c/\lambda, \quad \text{and} \quad E = hf, \tag{8.5}$$

where E is the energy of the electron flux (in KeV); h the Planck constant (4.135×10^{-15} eV), c the speed of light (3×10^{18} Å/s), and λ the wavelength (in Å). If we substitute the first of Eqs (8.5) into the second, we get

$$E = \frac{hc}{\lambda}, \tag{8.6}$$

which describes the energy of x-rays in terms of their wavelength. When the values of h and c are substituted into (8.6), it reduces to

$$E = \frac{12.398}{\lambda}, \tag{8.7}$$

which shows that there is an inverse relationship between the energy and wavelength of x-rays.

Since x-rays are produced whenever a crystalline substance is irradiated with a beam of high-energy charged particles (photons), interactions occur between the beam (i.e., electrons) and the crystal. This always results in a loss of energy. According to the law of conservation of energy, this energy loss results in the release of x-ray photons of energy (wavelength) equal to the energy loss defined by Eq (8.7). This process generates a broadband continuous radiation, also known as *white radiation*. The minimum wavelength at which this continuous radiation begins is related to the maximum accelerating potential V of the beam of electrons directed at the crystal, and is calculated from Eq (8.7) as

$$\lambda_{min} = \frac{hc}{V} = \frac{12.398}{V}. \tag{8.8}$$

For a successful analysis of crystals by x-ray diffraction, Bragg's diffraction condition must be met. This, in turn, requires high-intensity monochromatic radiation.

8.5 DNA

DNA (DioxyriboNucleic Acid) is composed of a sugar backbone — something like table sugar — that binds together nucleic acid bases that make another of its components, as shown in Figure 8.6. DNA is a molecule that encodes the generic instructions used in the development and functioning of all known living organisms and many viruses. The four building blocks of the DNA molecule are the four nucleic acid bases, known as Adenine, Cytosine, Guanine, and Thymine, also designated by the letters A, C, G, and T. Besides their names, the order of these nucleotide bases in the genome vary the generic markers to identify and classify all living creatures. The genes scattered around the genome are between 5,000 and 50,000 nucleotide bases in length. On average, the human genome contains around 30,000 genes.

How does DNA work? It is quite simple. The DNA of parents passes on to the children, and a genetic hand-off is repeated in every generation. This may be a simplistic reason why the offspring look more like their parents than other people. As this copying process goes on, known as *replications*, the entire DNA molecule serves as a template to create another version of itself. This process generates millions of tiny little copying enzymes that carry on the replications until all pieces are assembled at the end. Although these enzymes remain very careful in their work, sometimes mistakes occur, like substituting a C for a G in one instance, or more. Most of these mistakes are caught by the molecular 'proofreader', yet occasional mistakes creep into the final output. Such mistakes are called *mutations*, which occur at a very low rate every generation, about 5×10^{-8} per cent.

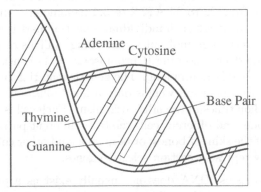

Figure 8.6 Structure of DNA's Double Helix.

Bragg's law opened up some crystallographic methods to determine the spatial structure of complex organic molecules. In 1951 Linus Pauling discovered the helical structure of an important protein (myoglobin). He began the huge task of constructing models for the DNA molecule, and neglected analytical methods of crystallography. The first experiment was to see the x-ray pictures of DNA. Linus Pauling showed that DNA possessed a three-chain helix with sugar-phosphate backbone in the center and bases on the outside. However, there were errors in the chemistry of the model. This was recognized by James Watson, and he and his assistant Maurice Wilkins were successful to obtain a new three-dimensional form of DNA. Francis Crick (1916–2004) while working at the Cavendish Laboratory in Cambridge (U.K.) discovered one of the salient points in the history of genetics, and was awarded the Nobel Prize in 1962. He and Watson discovered the correct structure of DNA (Watson and Crick [1953]). Their DNA pattern had two helices, the backbone on the outside and the bases inside (Figure 8.6). The two ribbons symbolize the two phosphate-sugar chains and the horizontal rungs on the pairs of bases holding the chain together. The DNA molecule is tightly packed inside the cells, but once it is extracted and pulled, it takes the shape of a fiber.

The DNA structure is not periodic in three dimensions; it is not a crystal, and therefore its spectrum is not discrete. The phase problem can be controlled by considering the phase on a continuum. The method used to find the DNA spectrum is called a 'trial and error method'; it depends on making an initial 'guess' for a hypothetical structure and computing the Fourier transform for such a structure. Then the spectrum thus obtained must be matched against the experimental data.

The DNA spirals are approximately one meter long. They are so thin and tightly packed to fit into a cell that has transverse dimension of only a few microns. The chemical structure of DNA is shown in Figure 8.7, in which the hydrogen bonds are represented as dotted lines. The structure of DNA of all species consists of two helical chains, each coiled around the same axis, and each with a pitch from 22 to 34 Å (2.2 to 3.4 nanometers) and a radius of 10 Å (1 nanometer). Although each individual repeating part is very small, DNA polymers can be very large molecules containing millions of nucleotides. For example, the largest chromosome (Chromosome # 1) consists of about 220 million base pairs and is 85 mm long. The genetic material that is transmitted from generation to generation weighs only a few thousandths of a gram, yet grows several billion times in the adult organism. Each cell is a complete copy of the genetic code, and different cells relate to different parts of the genome. In short, a cell of the skin, of bones, or of nerves, and other cells have different functions as they represent certain selective genes.

In living organisms, DNA does not usually exist as a pair of molecules that are held tightly together. These two spirals entwine like vines, in the shape of a double helix. The backbone of the DNA strand is made from alternating phosphate and sugar residues; the sugar is 2-deoxyribose, which is a pentose (five-carbon) sugar. These sugars are joined together by phosphate groups that form bonds between the third and fifth carbon atoms of adjacent sugar rings. The asymmetric bonds provide a strand of DNA a direction. In the double helix the direction in one strand is opposite to the direction in the other strand, i.e., the strands are antiparallel. The asymmetrical ends of DNA are called the 5′ (5 prime) and 3′ (3 prime) ends, with the 5′ end having a terminal phosphate group and the 3′ end a terminal hydroxyl group.

The DNA double helix is stabilized by two forces: hydrogen bonds between nucleotides and base-stacking interactions among aromatic nucleotides. The four bases in the DNA are A, C, G, and T, which are attached to the sugar/phosphate to form the complete nucleotide. All functions of DNA depend on interactions with protein, marked 'P' (inside a small circle) in Figure 8.7. These protein interactions can be nonspecific, or the protein can bind to a single DNA sequence. Proteins are macromolecules and have the shape of a long chain or multiple chains, composed of repeating blocks called *aminoacids*, in 21 types. The total number of aminoacids in a chain are in the hundreds. Two proteins can be generated from the same aminoacid, but they will be different, with different tasks, because of the different order of aminoacids that appear in the chain. The purpose of protein is to regulate the metabolic process in living organisms.

The protein diffraction spectrum is very complex, having spatial distribution of thousands of atoms. The complexity of protein's structure was solved by Perutz [1964], who indicated the positions of approximately ten thousand atoms of hydrogen, nitrogen, oxygen, and sulphur, in addition to the four atoms of iron. Yet he could not fix the location of about four thousand scattered hydrogen atoms. Their scattering occurred because one electron is rather weak. Each chain exhibits a heme group which is the structure that binds oxygen to the molecule and carries the four atoms of iron. The technique used is called the *heavy atoms method* that was developed by the crystallographers to solve the phase problem in relatively simple structures. A detailed account of Perutz's experiments is described in Prestini [2004: 208].

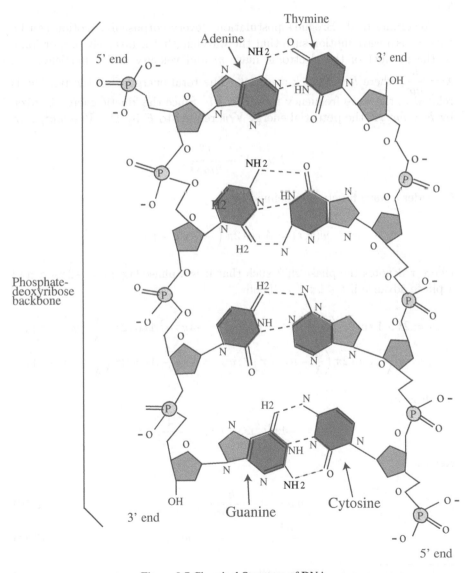

Figure 8.7 Chemical Structure of DNA.

8.6 Hydrogen Atom

The hydrogen atom is an atom of the chemical element hydrogen (protium, ^{1}H). This is an electrically neutral element (i.e., has no neutron) that contains a single positively charged proton and a single negatively charged electron bound to the nucleus by the Coulomb force. The diameter of a hydrogen atom is 1.1 Å, which is twice the Bohr model radius. The atomic hydrogen makes up about 75% of the elemental mass of the universe (Griffiths [1995, §4.2]).

According to de Broglie's postulation[1] 'every corpuscular motion can be treated as a wave motion such that the wavelength λ is inversely proportional to the product of the electronic mass m and velocity c of a particle', i.e., $\lambda = \dfrac{h}{mc}$, where h is Planck constant. The total energy E of the particle is related to the wave frequency ν by $E = h\nu$. Since the kinetic energy is given by $E = mc^2/2$, the potential energy V is related to E by $E = V + mc^2/2$, or

$$h\nu = V + \frac{h^2}{2m\lambda^2}.$$ (8.9)

Consider a wave function w defined by

$$w(x,t) = A \cos 2\pi \left(\frac{x}{\lambda} - \nu t - \tau\right),$$

where τ denotes the phase shift such that it is a phase lag if $t \geq -2\pi + \tau$ and a phase advance if $t \leq 2\pi + \tau$. Then

$$w_t = 2\pi \nu A \sin 2\pi \left(\frac{x}{\lambda} - \nu t - \tau\right), \quad w_{tt} = -4\pi^2 \nu^2 A \cos 2\pi \left(\frac{x}{\lambda} - \nu t - \tau\right),$$

$$w_x = -\frac{2\pi}{\lambda} A \sin 2\pi \left(\frac{x}{\lambda} - \nu t - \tau\right), \quad w_{xx} = -\frac{4\pi^2}{\lambda^2} A \cos 2\pi \left(\frac{x}{\lambda} - \nu t - \tau\right),$$

which gives

$$w_{tt} = -4\pi^2\nu^2 w, \quad w_{xx} = -\frac{4\pi^2}{\lambda^2} w.$$

Hence,

$$h\nu w = -\frac{h}{4\pi^2\nu} w_{tt},$$ (8.10)

$$\frac{h^2}{2m\lambda^2} w = -\frac{h^2}{8\pi^2 m} w_{xx}.$$ (8.11)

---

[1] As mentioned in Sagan [1989: 292], this postulation was made by de Broglie in 1924 and provided a basis for quantum mechanics.

Multiplying Eq (8.9) by w and substituting into Eqs (8.10) and (8.11), we get

$$\frac{\partial^2 w}{\partial x^2} - \frac{8\pi^2 m}{h^2} V w = \frac{2}{\lambda^2 \nu^2} \frac{\partial^2 w}{\partial t^2},$$

which is Schrödinger's wave equation in 1-D. In 3-D this equation becomes

$$\nabla^2 w - \frac{8\pi^2 m}{h^2} V w = \frac{2}{\lambda^2 \nu^2} \frac{\partial^2 w}{\partial t^2}. \tag{8.12}$$

Now, let $w = A \cos 2\pi \left(\frac{x}{\lambda} - \nu t - \tau\right) u(x, y, z)$. Then Eq (8.12) gives

$$\nabla^2 u + 8\pi^2 \left(\frac{1}{\lambda^2} - \frac{m}{h^2} V\right) u = 0.$$

Since $E = \frac{h^2}{m\lambda^2}$, or $\frac{1}{\lambda^2} = \frac{mE}{h^2}$, the above equation becomes

$$\nabla^2 u + \frac{8\pi^2 m}{h^2} (E - V) u = 0, \tag{8.13}$$

where $u(x, y, z) \equiv u(\mathbf{x}) \to 0$ as $\cos |\mathbf{x}| \to \infty$ such that $\iiint_{-\infty}^{\infty} u^2 \, dx \, dy \, dz$
$< +\infty$, and the norm $\|u\|$ is defined by $\iiint_{-\infty}^{\infty} u^2(\mathbf{x}) \, dx \, dy \, dz = 1$, (or
$\iiint_{-\infty}^{\infty} |u|^2 \, dx \, dy \, dz = 1$ in the case when u is complex). This condition
can be expressed in spherical coordinates as an iterated integral

$$\int_0^r r^2 R^2 \, dr \int_0^{2\pi} \Phi^2 \, d\phi \int_0^{\pi} \Theta^2 \sin\theta \, d\theta = 1. \tag{8.14}$$

The function u is sometimes called a *probability wave*; that is, it is a function
for which $\iiint_{-\infty}^{\infty} u^2 \, dx \, dy \, dz$ represents the probability that a particle is found
in the volume element $dx \, dy \, dz$.

The hydrogen atom consists of one nucleus of charge e and one electron of
charge $-e$ with electronic mass m. Hence, the potential energy is $V = -e^2/r$,
where r is the distance between nucleus and electron. Thus, Eq (8.13) becomes

$$\nabla^2 u + \frac{8\pi^2 m}{h^2} \left(h\nu + \frac{e^2}{r}\right) u = 0, \tag{8.15}$$

where we have chosen spherical coordinates such that the nucleus is at the
origin, i.e., the source point is $(r', \theta', \phi') = \mathbf{0}$. The spherical coordinate system

is defined by $x = r \sin\theta \cos\phi$, $y = r \sin\theta \sin\phi$, $z = r \cos\theta$. Using Bernoulli's separation method, with $u = R(r)\Theta(\theta)\Phi(\phi)$, where $r > 0$, $0 \leq \theta \leq \pi$, $0 \leq \phi < 2\pi$, Eq (8.15) becomes

$$\frac{R'' + 2R'/r}{R} + \frac{\cot\theta}{r^2}\frac{\Theta'}{\Theta} + \frac{1}{r^2}\frac{\Theta''}{\Theta} + \frac{1}{r^2 \sin^2\theta}\frac{\Phi''}{\Phi} + \frac{8\pi^2 m}{h^2}\left(h\nu + \frac{e^2}{r}\right) = 0,$$

or

$$\frac{r^2 R''}{R} + \frac{2r R'}{R} + r^2 \frac{8\pi^2 m}{h^2}\left(h\nu + \frac{e^2}{r}\right) = -\cot\theta\,\frac{\Theta'}{\Theta} - \frac{\Theta''}{\Theta} - \frac{1}{\sin^2\theta}\frac{\Phi''}{\Phi}. \quad (8.16)$$

Since both sides of this equation must be equal to a constant which we take to be κ, we get

$$r^2 R'' + 2r R' - \kappa R + \frac{8\pi^2 m}{h^2}r^2\left(h\nu + \frac{e^2}{r}\right)R = 0, \quad (8.17)$$

To separate equations in Θ and Φ, we take the constant to be $-\mu^2$. Then the right side of Eq (8.16) gives two equations:

$$\sin^2\theta\,\Theta'' + \sin\theta\cos\theta\,\Theta' + \kappa\sin^2\theta\,\Theta - \mu^2\Theta = 0, \quad (8.18)$$

$$\Phi'' + \mu^2\Phi = 0, \quad (8.19)$$

To solve Eq (8.18), let $t = \cos\theta$, $\Theta(\theta) = P(t)$, and $\kappa = n(n+1)$. Then this equation becomes

$$\frac{d}{dt}\left[(1-t^2)\frac{dP}{dt}\right] + \left[n(n+1) - \frac{\mu^2}{1-t^2}\right]P = 0. \quad (8.20)$$

This equation would be Legendre's equation if it were not for the term $\frac{\mu^2}{1-t^2}$. The solutions to Eq (8.20) are associated Legendre polynomials $\Theta = P_n^\mu(\cos\theta)$, where $P_n^\mu(t)$, and $\mu = 0, \pm 1, \pm 2, \ldots$.

The solution of Eq (8.19) is $\Phi = A\cos\mu\phi + B\sin\mu\phi$, where ϕ is a periodic function of period 2π, i.e., $\Phi(\phi) = \Phi(2\pi + \phi)$. Now, to solve Eq (8.17), first we divide it by r^2, and get

$$R'' + \frac{2}{r}R' + \left[a + \frac{b}{r} - \frac{n(n+1)}{r^2}\right]R = 0,$$

where $a = \frac{8\pi^2 m\nu}{h}$ and $b = \frac{ae^2}{h\nu}$. Introducing a new variable $\rho = 2r\sqrt{-a}$, and taking $K = \frac{b}{2\sqrt{-a}} = \frac{2\pi me^2}{h\sqrt{-2mh\nu}}$, the above equation becomes

$$\frac{d^2 R}{d\rho^2} + +\frac{2}{\rho}\frac{dR}{d\rho} - \left(\frac{1}{4} - \frac{K}{\rho} + \frac{n(n+1)}{\rho^2}\right)R = 0,$$

or

$$\frac{d}{d\rho}\left(\rho^2 R'\right) - \frac{\rho^2 - 4K\rho + 4n(n+1)}{4} R = 0. \qquad (8.21)$$

This equation is closely related to Laguerre's equation, and its solution is

$$R(\rho) = \rho^n\, e^{-\rho/2}\, L_{K+n}^{2n+1}(\rho), \qquad (8.22)$$

where $n \in \mathbb{Z}$, K is any number, and $L_k^\sigma(\rho)$ are the generalized Laguerre polynomials defined by $L_k^\sigma(\rho) = \dfrac{\rho^{-\sigma}\,e^\rho}{k!}\,\dfrac{d^k}{d\rho^k}\left(\rho^{n+\sigma}\,e^{-\rho}\right)$. If we impose the condition that K be an integer N, then we find that $E = -\dfrac{2\pi^2 m\,e^4}{h^2 N^2}$, $N = 1, 2, \ldots$, are the possible values of the kinetic energy such that these values represent a discrete energy spectrum with a limit point at $E = 0$. For $N = 1$ we get a *normal quantum state* (or *ground state*, as it is commonly called) of the hydrogen atom, which is the most stable state occupied by the spectrum. For $N > 1$, we have a manifold of higher quantum states, described as follows: Assuming that $m \le k$, we have $2n+1 \le N+n$, that is, $n \le N-1$, which means that for every N there are N possible values of n, namely $n = 0, 1, 2, \ldots, N-1$. Since $\mu \le n$, we will have $\mu = 0, 1, 2, \ldots, n$ for every n. Hence, for every N there are $\dfrac{N(N+1)}{2}$ independent quantum states. Hence, the solutions of Eq (8.16) are given by (see Kythe [2011: 189])

$$u_{Nn\mu}(r, \theta, \phi) = \frac{h^2 N}{4\pi^2 m\,e^2}\, r^2\, P_n^\mu(\cos\theta)\, \exp\left[-\frac{2\pi^2 m\,e^2}{h^2 N} r\right] \times$$

$$\times L_{N+n}^{2n+1}\left(\frac{4\pi^2 m\,e^2}{h^2 N} r\right)(A\cos\mu\phi + B\sin\mu\phi), \qquad (8.23)$$

where the constants A and B can be determined from the boundary conditions (to be provided), and N is called the *principal quantum number*, n the *azimuthal quantum number*, and μ the *magnetic quantum number*.

Thus, Bernoulli's separation method leads to three equations for the three spatial variables, and their solutions produce the above three quantum numbers associated with the hydrogen energy levels. The wave functions for certain spins of the hydrogen atom are listed below, where $a_0 = \dfrac{\hbar^2}{me^2} = 0.529$Å denotes the first Bohr radius which is the nuclear charge, and the subscripts $1s$, $2s$, $3s$ indicate the first, second, and third state of its radial density:

$$u_{1s} = \frac{1}{\sqrt{\pi}}\, a_0^{-3/2} r\, e^{-r/a_0}, \quad u_{2s} = \frac{1}{4\sqrt{2\pi}}\, a_0^{-3/2}\left(2 - a_0^{-1}\right) e^{-r/(2a_0)},$$

$$u_{3s} = \frac{1}{81\sqrt{3\pi}} a_0^{-3/2} \left(27 - 18ra_0^{-1} + 2r^2a_0^{-2}\right) e^{-r/(3a_0)}.$$

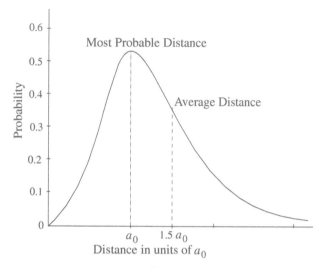

Figure 8.8 Plot of Probability of Radial Density.

The probability of finding the electrons in terms of the three dimensions, i.e., its radial density, is proportional to the square of the respective wave function. Figure 8.8 shows a plot of the radial density which indicates that the distance of the highest probability (most probable distance) is at $r = a_0$.

To prove this statement, we find that squaring the $1s$ radial wave we obtain the probability function $P(r)$ as

$$P(r) = \frac{1}{\pi} a_0^{-3} r^2 e^{-2r/a_0}. \tag{8.25}$$

Differentiating with respect to r and setting $P'(r) = 0$ gives

$$P'(r) = 0 = \frac{2}{\pi} r(r - a_0)a_0^{-2} e^{-2r/a_0},$$

which yields $r = 0$, or $r = a_0$. Since $r = 0$ is the absolute minimum (see Figure 8.8), we find that $P''(a_0) = -8\pi a_0^{-3} e^{-2}$, which is negative. Hence, $P(r)$ has the relative maximum at the distance $r = a_0$.

For the $2s$ wave function, we use the value of u_{2s} given above. Then the probability of finding the point where the electron has a node follows from setting the probability $P(r) = 0$, and it occurs is at $r = 2a_0$, which can be seen from the $2 - r/a_0$ part of the wave function. Similarly, for the $3s$ function, the only part of the wave function u_{3s}, and hence the square of this function, that can make $P(r) = 0$ is the term $\left(27 - 18ra_0^{-1} + 2r^2a_0^{-2}\right)$, or the polynomial

$27 - 18x + 2x^2 = 0$ where $x = r/a_0$, i.e., at $x = \dfrac{18 \pm \sqrt{18^2 - 4(2)(27)}}{4} \approx 7.1$ or 1.9. Thus, the $3s$ wave has 2 nodes, at $r = 1.9$ and $r = 7.1$. In general the probability has $N - 1$ nodes where N is the principal quantum number.

Example 8.1. To calculate the average probability distance for the hydrogen atom, we start with the normalized wave function u_{1s} and let r denote the spatial position coordinate of this distance. Then this distance, denoted by $|\mathbf{r}|$ is

$$|\mathbf{r}| = \int_0^\infty \int_0^\pi \int_0^{2\pi} \frac{1}{\sqrt{\pi}} a^{-3/2} e^{-r/a_0} \, m \, \frac{q}{\sqrt{\pi}} a_0^{-3/2} \, e^{-r/a_0} \, r^2 \sin\theta \, dr \, d\theta \, d\phi$$

$$= \frac{a_0^{-3}}{\pi} \int_0^\infty \int_0^\pi \int_0^{2\pi} r^3 e^{-2r/a_0} \sin\theta \, dr \, d\theta \, d\phi$$

$$= \frac{a_0^{-3}}{\pi} \int_0^\infty r^3 e^{-2r/a_0} \, dr \int_0^\pi \int_0^{2\pi} \sin\theta \, dr \, d\theta \, d\phi$$

$$= \frac{a_0^{-3}}{\pi} \frac{3! \, a_0^{-4}}{16} \, 4\pi = \tfrac{3}{2} a_0,$$

where we have used the formula $\displaystyle\int_0^\infty x^n e^{-bx} \, dx = \frac{n!}{b^{n+1}}.$ ∎

The geometry of the hydrogen atom is given in Figure 8.9.

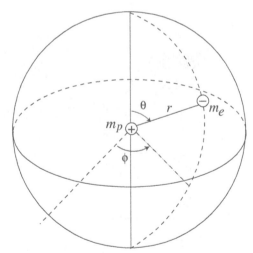

Figure 8.9 Hydrogen Atom.

8.6.1 Hydrogen Spectrum. Schrödinger used the model of a standing wave to represent the electron within the hydrogen atom and solved the resulting wave equation. The reasoning behind using the spherical coordinate system is based on the fact that the hydrogen atom has a spherically symmetric potential. The potential energy is due to a point charge.

The hydrogen atom, being an atom of the chemical element H, is an electronically neutral atom which contains a single positively charged proton and a single negatively charged electron bound to the nucleus by the Coulomb force. In fact, the $1/r$ Coulomb factor leads to the Laguerre functions of the first kind. The hydrogen atom is very significant in the study of quantum mechanics and quantum field theory of a simple two-body problem. In 1914 Niels Bohr made a number of simplifying assumptions and obtained the spectral frequencies of the hydrogen atom (see Griffiths [1995]). The Bohr model explained only the spectral properties of the hydrogen atom, and his results were confirmed by an analytical solution of the Schrödinger wave equation (8.15) using Bernoulli's separation method. These solutions use the fact that the Coulomb potential produced by the nucleus is isotropic, i.e., it is radially symmetric in space and depends only on the distance to the nucleus. They were useful in calculating hydrogen energy levels and frequencies of the hydrogen spectral lines. In this sense, the solutions of the Schrödinger equation are far more useful than the Bohr model because these solutions also contain the shape of the electron's wave function (orbital part) for various quantum mechanical states, which eventually explains the anisotropic character of atomic bonds. The hydrogen spectrum showing the Bohr model is presented in Figure 8.10.

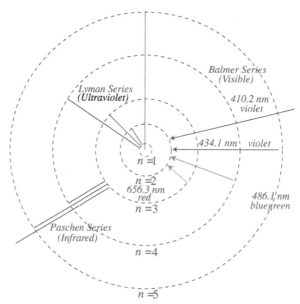

Figure 8.10 Hydrogen Spectrum.

Pauli [1926] solved the problem of the hydrogen atom by using a rotational symmetry in four dimensions (the so-called $O(4)$-symmetry) generated by the angular momentum and the Laplace-Runge-Lenz vector. The non-relativistic problem of the hydrogen atom was later solved by Duru and Kleinert [1979] using Feynman's path integral formulation of quantum mechanics.

The energy levels E of a hydrogen atom are defined by $E = E_0/n^2$, where $E_0 = 13.6\,\mathrm{eV}$ (1 eV $= 1.602 \times 10^{-19}$ Joules, or $= 3.29 \times 10^{15}$ sec^{-1}), and $n = 0, 1, 2, \ldots$. The energy is expressed as a negative number because it takes much energy to ionize (unbind) the electrons from the nucleus. A common saying is that 'an unbound electron has zero (binding) energy.' When an electron absorbs a photon, it gains the energy of the photon. Since an electron bound to an atom can only have certain energies, the electron can only absorb photons of certain energies. For example, an electron in the ground state E_1 has the energy of -13.6 eV. If the second level energy is $E_2 = -3.4$ eV at the second state, then it would take $E_2 - E_1 = -3.3 + 13.6 = 10.2$ eV to excite the electron from the ground state to the next excited state.

However, the excited state is not the most stable state of an atom. The energy of the emitted photon is given by *Rydberg formula* which defines the subtraction of two energy levels. Thus, $E_{\mathrm{photon}} = E_0\left(1/n_1^2 - 1/n_2^2\right)$, where E_0 is defined above and $n_1 < n_2$ so that, e.g, if $n_1 = 1, 2, \ldots$, then $n_2 = n_1 + 1, n_2 + 2, \ldots$. The restriction $n_1 < n_2$ guarantees that the energy of the photon is always positive. However, if this restriction is removed, the negative energy would imply that a photon (of positive energy) is absorbed. The Bohr model has been successful in explaining the Rydberg formula for the spectral emission lines of the hydrogen atom. This formula confirms that the energy levels of the hydrogen atom are inversely proportional to n^2.

In classical mechanics, an electron is held in a circular orbit of radius r by electrostatic attraction. The centripetal force is equal to the Coulomb force, i.e., $m_e v^2 = Z C_e e^2 / r$, where m_e is the mass and c is the charge of the electron, C_e is the Coulomb constant, and Z is the atomic number of the atom. Hence, the electron's speed v and its total energy E at any radius r are determined by

$$v = \sqrt{\frac{Z C_e e^2}{m_e r}}, \quad E = \frac{1}{2} m_e v^2 - \frac{Z C_e e^2}{r} = -\frac{Z C_e e^2}{2r}, \tag{8.26}$$

which shows that the total energy is negative and inversely proportional to r. That is, it takes energy to pull the orbiting electron away from the proton. Also, E becomes zero as $r \to \infty$, signifying a motionless electron infinitely far away from the proton. The total energy is one-half of the potential energy, which also holds for non-circular orbits.

In the quantum state, the angular momentum $L = m_e v r = n\hbar$, where $\hbar$ is the Planck's constant. Substituting it into Eq (8.26) for velocity yields $\sqrt{Z C_e e^2 m_e r} = n\hbar$. Thus, the allowed orbit radius at ant state n is $r_n = \dfrac{n^2 \hbar^2}{Z C_e e^2 m_e}$. The smallest possible value of r in a hydrogen atom is called the

Bohr radius and it is equal to $r_1 = \dfrac{\hbar^2}{C_e e^2 m_e} \approx 5.29 \times 10^{-11}$ m. Since the energy at the nth level in any atom is defined as

$$E = -\frac{ZC_e e^2}{2r_n} = -\frac{Z^2 \left(C_e e^2\right)^2 m_e}{2\hbar^2 n^2} \approx -\frac{13.6}{n^2} \text{ eV},$$

the electron in a hydrogen atom ($n = 1$, $Z = 1$) has the lowest energy level of about 13.6 eV less energy than a motionless electron infinitely far from its nucleus. The subsequent energy levels are: -3.4 eV at $n = 2$; -1.51 eV at $n = 3$, and so on. Other details and more information can be found in Griffiths [1995], Hecht [2000], and Kythe [2011: 274].

8.7 Units of Measurement

In any quantum theory of gravity, the units of measurement of length and time are incredibly small, simply because the strength of the gravitational force is measured by Newton's constant. On the other hand, physicists use different sets of units in which the speed of light c is set as unity[2]. Under this scale, one second is then equivalent to 299,792,458 meters. Moreover, Planck constant $\hbar = h/(2\pi)$ is set to unity, which establishes a relationship between seconds and ergs of energy. Under these units everything, including Newton's constant, can be reduced to centimeters. For example, Newton's constant of gravitation $G = 6.67384 \times 10^{-11}$ N m^2 kg^{-2}. This is different from g which is the acceleration due to gravity. Force of gravity is $F = (G \times m_1 \times m_2)/r^2$, where m_1 and m_2 are the masses of two objects, and r the distance between them. To have an idea how small G is, let us write it out: then it is simply $G = 0.00000000006673$ N m^2 kg^{-2}. Then the force F exerted between two masses of 1 kg each spaced 1 meter apart comes to $F = 0.00000000006673$ N. If, instead, we use the heaviest recorded mass 12,000 kg of an elephant, and consider two of them 1 meter apart from their center of mass, the value of $F \approx 0.01$ N. Compare it to the force exerted by the earth on an apple, which is 1 N. So even when the heaviest masses are set together, they exhibit practically no force of attraction, unless they are very massive, like planets. If we calculate the length associated with Newton's constant, it is precisely the Planck length, or 10^{-35} cm, or 10^{19} billion electron volts. All gravitational effects are, thus, measured in terms of this tiny distance. Even the size of the unseen space dimensions $n \geq 4$ is the Planck length (Kaku [1994: 336]).

In the case of time related structure, the biological activity of a macro-molecule is so fast that its lifespan ranges from a femtosecond (10^{-15} s) to a millisecond (10^{-3} s). Time-related experiments using x-rays use an intrinsic property of synchrotron spectrum, where the x-rays pulses of 10–50

---

[2] The speed of light is an important physical quantity because it is the speed at which all electromagnetic waves propagate. It is the limit on how fast matter, energy, and information can travel.

picoseconds (10^{-12} s) are generated. The x-ray diffraction data collected at the European Synchrotron Radiation Facility in Grenoble used the crystals of dimensions 0.4 mm × 0.3 mm × 0.07 mm, and used fast laser pulses. The smallest pulse obtained so far is an attosecond (10^{-18} s).

A nanosecond is one billionth of a second (10^{-9} s). The following proportion gives an idea about this time unit:

$$\frac{1 \text{ nanosecond}}{1 \text{ second}} = \frac{1 \text{ second}}{31.7 \text{ years}},$$

i.e., one nanosecond to one second is as one second to 31.7 years. Times of nanosecond magnitude are encountered in electromagnetic field, telecommunications, pulsed laser, and electronics. All small time units are defined as

$$\text{millisecond (ms)} = 10^{-3} \text{ s}$$
$$\text{microsecond } (\mu s) = 10^{-6} \text{ s}$$
$$\text{nanosecond (ns)} = 10^{-9} \text{ s}$$
$$\text{picosecond (ps)} = 10^{-12} \text{ s}$$
$$\text{femtosecond (fs)} = 10^{-15} \text{ s}$$
$$\text{attosecond (as)} = 10^{-18} \text{ s}$$

Thus, 1 min= 6×10^{10} ns. Other common measurements are:

(i) Average life of a molecule of positronium hydride is 0.5 ns.

(ii) Cycle time for radio frequency is 1.0 ns, or 1 GHz (1×10^{9} hertz). This time corresponds to a radio wavelength of 1 light-nanosecond or 0.3 m, calculated by multiplying 1 ns by the speed of light ($\approx 3 \times 10^{8}$ m/s) to determine the distance traveled.

(iii) Cycle time for a 1 GHz processor is 1 ns. The average frequency of most processors is about 1–3.5 GHz, thus their cycle time is shorter than 1 ns.

(iv) Cycle time for frequency 10 MHz (1×10^{7} Hz), radio wavelength 30 m (shortwave) is 100 ns.

(v) Cycle time for frequency 100 MHz (1×10^{8} Hz), radio wavelength 3 m (VHF, FM band) is 10 ns.

(vi) Cycle time of highest medium wave radio frequency, 3 MHz, is 333 ns.

(vii) Time taken by light to travel one foot in vacuum is about 1.017 ns.

(viii) Time taken for light to travel 1 meter in vacuum is about 3.33564095 ns; in air and in water it travels more slowly.

(ix) Time of fusion reaction in a hydrogen bomb is 20–40 ns.

(x) Planck constant $\hbar = 1.054 \times 10^{-34}$ joules-sec.

(xi) In the FPS units, the calculations are carried out as follows:

$$\left(299,792,458\,\text{m s}^{-1}\right) \times \left(1\,\text{s} \times 10^9\,\text{ns}\right) = 0.299792458 \text{ meters per ns.}$$

Since 1 meter = 39.37 inches, we get

$$\left(0.299792458\,\text{m ns}^{-1}\right) \times \left(39.37\,\text{in m}^{-1}\right) = 11.8 \text{ inches per nanosecond.}$$

Thus, since 1 foot = 12 inches, light travels about one foot per nanosecond.

8.7.1 Ultrashort Pulse. An ultrashort pulse of light is an electromagnetic waveform with time duration of the order of at most one femtosecond (10^{-15} s). These pulses have a broadband optical spectrum. They can be generated using mode-locked oscillators. The technique of chirp pulse amplification is used to amplify these pulses. An ultrashort pulse of light is presented in Figure 8.11. Its field amplitude is a sinusoid with a Gaussian envelope; its phase function is quadratic, which results in a chirp (i.e., an instantaneous frequency sweep, as explained in §3.3.1).

In Figure 8.11, the solid graph defines the electric field $e(t)$ that oscillates at an angular frequency ω_0 relative to the central wavelength of the pulse, and the dotted curve is the time-averaged intensity $I(t)$. Mathematically, $e(t)$ is defined as an analytical signal corresponding to the real field, such that

$$e(t) = \sqrt{I(t)}\, e^{i\omega_0 t}\, e^{i\,\phi(t)}, \tag{8.27}$$

where $\phi(t)$ is the phase function. The Fourier transform of (8.27) yields

$$\mathcal{F}\{e(t)\} = E\left(\omega - \omega_0\right), \tag{8.28}$$

where the shifting property (§3.1.2(1)) has been used. Similarly, in the time domain, intensity and a phase function is defined in the angular frequency domain by the Fourier transform

$$E(\omega) = \sqrt{S(\omega)}\, e^{i\,\Phi(\omega)}, \tag{8.29}$$

where $S(\omega) = \mathcal{F}\{I(t)\}$ is the spectral density (or spectrum) of the pulse, and $\Phi(\omega) = \mathcal{F}\{\phi(t)\}$ is the spectral phase. In the particular case when $\Phi(\Omega) = $ const, the pulse is called a *bandwidth-limited pulse*, or in the case when $\Phi(\omega)$ is a quadratic functions, it is called a *chirped pulse*. Such a chirp can be obtained as a pulse propagates through glass due to its dispersion, resulting in a timewise broadening of the pulse.

Because of Heisenberg's uncertainty principle (Appendix L), the product of the functions $I(t)$ and $S(\omega)$ has a lower bound that depends on the definition used for the time period and the shape of the pulse. The minimum time-bandwidth product, obtained by a Fourier transform-limited pulse (i.e., for a

constant spectral phase $\Phi(\omega)$), yields the shortest pulse, while high values of
the time-bandlimited product generate a more complex pulse.

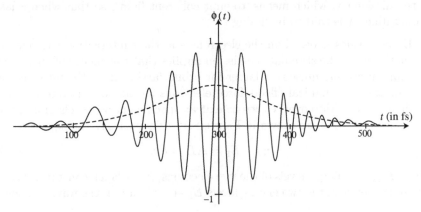

Figure 8.11 An Ultrashort Pulse of Light.

8.8 Laser

Laser is the acronym for 'light amplification by stimulated emission of radiation'. It is a device that emits light using a process of optical amplification based on stimulated emission of electromagnetic radiation. Laser emits light coherently, unlike other sources of light. It is spatially coherent, which keeps the laser beam collimated (i.e., keep light rays parallel) over long distances. Laser also has high temporal coherence, which allows it to have a very narrow spectrum, i.e., they only emit a single color of light, and also emit pulses of light which only last a femtosecond.

Coherence is the property that distinguishes lasers from other light sources. Spatial coherence means that the output is a narrow beam (also called a *pencil beam*) which is diffraction-limited. Thus, laser beams can achieve a very high irradiance and can be focused on very tiny spots at larger distances, or they can be launched into beams of very low divergence so that they can be concentrated. As mentioned above, besides spatial coherence, lasers possess temporal coherence, which implies a longitudinal single-frequency wave with its phase correlated over a large distance along the beam. Such a distance is called the 'coherence length'. This type of wave is in sharp contrast with a wave that is produced by a thermal or other incoherent light source, as the latter has an instantaneous amplitude and phase which vary with respect to position and time and have very short coherence length.

As we have seen in §7.4, light generally implies an electromagnetic radiation of any frequency, including visible light. So there are infrared laser, ultraviolet laser, x-ray laser, and so on. Historically, masers, which is the acronym for 'microwave amplification by stimulated emission of radiation', are microwave predecessor of laser. Masers operate at microwave and radio frequencies, and so initially laser was called an optical maser, especially at Bell Telephone Laboratories. The term laser has also added the word 'lase' in the dictionary, which means 'to emit coherent light', so that when a laser is operating, it is said to be 'lasing'.

Laser physics is based on the electrons and their interaction with electromagnetic field. The stimulated emission implies that the energy of an electron orbiting an atomic nucleus is larger for orbits farther from the nucleus. But the quantum mechanical effects force electrons to take on discrete positions on the orbitals. Moreover, electrons exit at specific energy levels of an atom. Two such electrons are shown in Figure 8.12. They satisfy the equation

$$E_2 - E_1 = \Delta E = h\nu, \tag{8.30}$$

where $E_{1,2}$ are energy levels of the two electrons, h is Planck constant, ΔE the difference between the two energy levels E_1 and E_2, and ν the wave frequency.

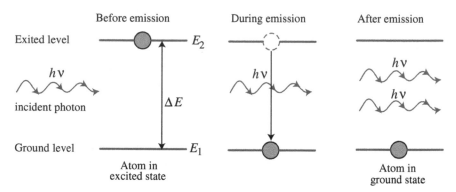

Figure 8.12 Electrons in Two Energy Levels of an Atom.

When an electron absorbs energy either from light (photons) or from heat (phonons), it receives the incident quantum of energy. However, transitions are only allowed between discrete energy levels, as is seen from Figure 8.12. This leads to emission lines and absorption lines. The excitation of an electron from a lower to a higher energy level is only temporary. An electron in an excited state may decay to a lower energy state if that is not occupied. In the case of such a decay of an electron, if it occurs without external influence, a photon is emitted, and this is called a *spontaneous emission*. The phase associated with the emitted photon is random. Thus, a material with many atoms in such an excited state generates radiation which is spectrally very limited: it is centered around one wavelength of light. The individual photons do not have any common phase relationship and thus emanate in random directions. This is known as *fluorescence* and *thermal emission*.

The process of stimulated emission can be explained as follows: the quantum mechanical state is affected by an external electromagnetic field at a given frequency. As the electron in an atom transits between two stationary states, which are not dipoles, it enters a transition state that does have a dipole field and acts like an electric dipole. This dipole oscillates at a characteristic frequency. As a response to the external electric field at this frequency, the probability that an atom will enter this transition state is greatly increased. Thus, the rate of transition between two stationary states increases beyond the spontaneous emission. Such an emission at the higher state is called absorption, and it destroys the incident photon. Hence, the transition from a higher to a lower energy state produces an additional photon, and this is the process of stimulated emission. The light emitted by stimulated emission is similar to the input signal in terms of wavelength, phase, and polarization. This provides laser light its coherence, and allows it to maintain its uniform polarization and monochromaticity established by the optical cavity design of a laser.

When traveling in free space or in a homogeneous medium excluding a waveguide (as in an optical fiber laser), a laser beam is approximately similar to a Gaussian beam in most cases. However, some high power lasers may be multimode with transverse modes which can be approximated by Hermite-Gaussian or Laguerre-Gaussian functions. As a result of diffraction theory, the Gaussian beam laser, which is the beam of a single transverse mode, eventually diverges at an angle that varies inversely with the beam diameter. This effect was noticed when a beam generated by a common helium-neon laser spread out to a size of about 500 km when shone on the Moon from the distance of the Earth. On the other hand, the laser generated by a semiconductor emerging through a crystal diverges up to 50°. But a proper lens system can transform such divergent lasers into a collimated beam originating from a laser diode. This property, however, cannot be achieved using standard light sources.

Lasers are classified into either continuous or pulsed mode, depending on whether the energy output is a continuous waveform (CW) over time or whether the output is in the pulse waveform on one or different time scale. A laser is also called a 'modulated' or 'pulsed' continuous wave when the modulation rate on time scale is much slower than the cavity lifetime and the time period over which energy can be stored in the lasing medium, as, for example, in most laser diodes used in communication systems.

The first application of laser was the barcode scanner introduced in supermarkets in 1974. The laser disk player came into market in 1978, followed by laser printers in 1982. Many important applications of lasers are as follows: (i) in medicine: bloodless laser surgery, laser healing, kidney stone and eye treatments, and dentistry; (ii) in industry: cutting, welding, material heat treatment, marking parts, noncontact measurement of parts; (iii) in military: marking targets, guiding munitions, missile defense, electro-optical countermeasures (EOCM), alternative to radar, and blinding troops; (iv) in law enforcement: laser fingerprint detection in forensic identification; (v) in commercial and product development: laser printers, optical disks, barcode scanners, thermometers, fluorescence microscopy; (vi) in laser lighting displays: laser light shows; and (vii) in cosmetic skin treatment: acne treatment, cellulite reduction, and hair removal.

Consider the family of Gaussian functions

$$g_\varepsilon(x) = \frac{1}{\sqrt{2\pi\varepsilon}}\, e^{-x^2/(2\varepsilon)}, \quad \varepsilon > 0, \tag{8.31}$$

which are normalized by the condition that $\int_{-\infty}^{\infty} g_\varepsilon(x)\, dx = 1$. Let $f_k(x) = g_{1/k}(x),\ k = 1, 2, \ldots$ be a weakly approximating sequence. Then $\varepsilon \to 0$ as $k \to \infty$, and the graphs of the approximating Gaussian functions $g_k(x)$ show higher and higher peaks and move more and more closeness toward $x = 0$, but the area under each one of them remains constant (see Figure 8.13).

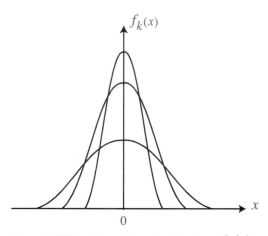

Figure 8.13 First Three Gaussian Functions $f_k(x)$.

The Gaussian beam is a transverse electromagnetic mode. The solution for its complex electric field amplitude is found by solving the paraxial Helmholtz equation, and is given by

$$E(r, z) = E_0 \frac{w_0}{w(z)} \exp\left\{ -\frac{r^2}{w(z)^2} - i\,kz - i\,k\frac{r^2}{2R(z)} + i\,\zeta(z) \right\}, \tag{8.32}$$

where r is the radial distance from the center axis of the beam, z is the axial distance from the beam's narrowest point (called the 'waist'); $k = 2\pi/\lambda$ is the wave number (in rads/m), $E_0 = |E(0,0)|$, $w(z)$ is the radius at which the field amplitude and intensity drop to $1/e$ and $1/e^2$ of their axial values, respectively, $w_0 = w(0)$ is the waist size, $R(z)$ is the radius of curvature of the beam's wavefronts, and $\zeta(z)$ is the Gouy phase shift[3], which is an extra

---

[3] The Gouy phase shift is the $n\pi/2$ axial phase shift that a converging light wave expe-

phase shift seen in the Gaussian beams. The time dependence factor $e^{i\omega t}$ has been suppressed in (8.32). The corresponding time-averaged intensity (or irradiance) distribution is given by

$$I(r,z) = \frac{|E(r,z)|^2}{2\eta} = I_0 \left(\frac{w_0}{w(z)} \right) \exp \left\{ -\frac{2r^2}{w^2(z)} \right\}, \qquad (8.33)$$

where $I_0 = I(0,0)$ is the intensity at the center of the beam at its waist, and the constant η is the characteristic impedance of the medium through which the beam is propagating, while in free space $\eta = \eta_0 \approx 376.7\,\Omega$.

Optical-resonator modes and optical-beam propagation are generally analyzed using the Hermite-Gaussian eigenfunctions as basis set. These functions consist of a Hermite polynomial $H_n \left(\sqrt{2}\, x/w(z) \right)$ of real argument times the complex Gaussian function $e^{-i k x^2/2q(z)}$, where $q(z)$ is a complex quantity. Siegman [1973] has provided a more elegant yet simple set of eigensolutions to the basic wave equation, which constitute a Hermite-Gaussian set $\hat{\psi}_n(x,z)$ of the form $H_n \left(\sqrt{c}\, x \right) e^{-cx^2}$, where the Hermite polynomial and the Gaussian function have the same complex argument $\sqrt{c}x \equiv \sqrt{i\,k/2q}\,x$. The new functions $\hat{\psi}_n$ are, however, not solutions of a Hermitian operator in x and, hence, form a biorthogonal set with a conjugate set of functions $\bar{\psi}_n \left(\sqrt{c}\, x \right)$. Although the new eigenfunctions $\hat{\psi}_n$ are not eigenfunctions of the conventional spherical-mirror optical resonators, since the wavefronts of $\hat{\psi}_n$ are not spherical for $n > 1$, they may still be used as a basis set for other optical resonators and beam-propagation problems.

A Laguerre-Gaussian laser mode has a well-defined orbital angular momentum equal to $l\hbar$ per photon, where l is the azimuthal mode index. This orbital angular momentum can be removed from the mode and converted into a mechanical torque. This change is achieved using astigmatic optical elements, which may be used to produce Laguerre-Gaussian modes from the more commonly occurring Hermite-Gaussian modes. Details of this analysis can be found in Allen et al. [1992].

8.8.1 Luminescence. Spontaneous emission of light from a light source such as an atom, molecule, nanocrystal or nucleus in an excited state that undergoes a transition from a ground state (i.e., a state with lower energy) and emits a photon is known as *luminescence*. It is a fundamental process that is important in many phenomena in nature and in many applications like the fluorescent tubes, older TV screens (cathode ray tubes), plasma display panels, lasers, and light emitting diodes.

Luminescence is an emission of light by a substance not resulting from heat; it is thus a form of cold body radiation. It can be caused by chemical

---

riences as it passes through its focus in propagating from $-\infty$ to ∞, where the dimension are: $n = 1$ for a line focus (cylindrical wave), and $n = 2$ for a point focus (spherical wave). This phase anomaly was first discovered by Gouy [1890]; see also Siegman [1986].

reactions, electrical energy, subatomic motions, or stress on a crystal. This distinguishes luminescence from incandescence which is light emitted by a substance as a result of heating. The rate of spontaneous emission, also known as the *radiative rate*, is described by Fermi's second golden rule, which is described as follows. Details can also be found in Loudon [2001].

In quantum physics, Fermi's golden rule is used to calculate the rate of transition from one energy eigenstate to another of a quantum system into a continuum of energy eigenstates. This transition is caused by perturbation and defined by the probability of transition per unit time. Suppose that a system is described by a Hamiltonian $\mathbf{H}$ as

$$\mathbf{H} = i\hbar \frac{\partial \psi}{\partial t}, \tag{8.34}$$

where $\mathbf{H}$ is of the form $\mathbf{H} = \mathbf{H}_0 + \mathbf{H}'$; here $\mathbf{H}_0$ is the unperturbed part of the Hamiltonian for which the eigenstates ψ_n are known, and $\mathbf{H}'$ is the time-dependent perturbation. The known eigenfunctions ψ_n satisfy the orthogonality conditions: $\mathbf{H}_0 \psi_n = E_n \psi_n$ such that $\langle \psi_a | \psi_b \rangle = \delta_{ab}$, where Dirac's bra-ket notation (§5.5.5) is used, and δ_{ab} is the Kronecker delta. The solution of Eq (8.34) can be expressed as a sum over the eigenstates of $\mathbf{H}_0$ with time-dependent coefficients

$$\psi(t) = \sum_n a_n(t) \psi_n \, e^{-i E_n t / \hbar}. \tag{8.35}$$

Then, after substituting (8.35) into (8.34) and using the above orthogonality conditions, we get

$$i\hbar \frac{da_k(t)}{dt} = \sum_n \mathbf{H}'_{kn} a_n(t) \, e^{i\omega_{kn}t}, \tag{8.36}$$

where $\mathbf{H}'_{kn} \equiv \langle \psi_k | \mathbf{H}'(t) | \psi_n \rangle$ and $\hbar \omega_{kn} \equiv E_k - E_n$. The part $\mathbf{H}'_{kn}$, known as the *transition amplitude*, connects the state n to the state k. Notice that Eq (8.36) is the Schrödinger equation in the coefficients $a_k(t)$; it defines a '2-level' system which can be solved explicitly. For example, if the system involves a continuum of states, Eq (8.36) is generally approximated by the method of perturbation expansion such that the $(p+1)$th order approximation is found from the pth order solution by

$$i\hbar \frac{da_k^{(p+1)}(t)}{dt} \approx \sum_n \mathbf{H}'_{kn} a_n^{(p)}(t) \, e^{i\omega_{kn}t}, \tag{8.37}$$

where the zeroth order approximation is given by $da_k^{(0)}(t)/dt = 0$, i.e., $a_k^{(0)}(t) =$ const. To compute the first order approximation, assume that the system is initially in the state m, in which case $a_n^{(0)}(t) = \delta_{mn}$. Then integrate Eq (8.37), which gives

$$i\hbar a_k^{(1)}(t) = \int_{-\infty}^{t} \mathbf{H}'_{km}(\tau)\, e^{i\omega_{mn}\tau}\, d\tau. \qquad (8.38)$$

If we assume that the perturbing force defined by $\mathbf{H}'$ starts at $t = 0$ and remains constant over the interval $0 \le \tau \le t$, then Eq (8.38) can be integrated to yield

$$i\hbar a_k^{(1)}(t) \approx 2\mathbf{H}'_{km}\, e^{i\,\omega_{km}(t)/2}\left(\frac{\sin \omega_{km}t/2}{\omega_{km}}\right). \qquad (8.39)$$

If the perturbation is terminated after the first-order term, we have $a_k(t) \approx a_k^{(1)}(t)$. Thus, the probability $\mathfrak{p}_k(t)$ that the system passes from the state m to the state k is given by

$$\mathfrak{p}_k(t) = |a_k(t)|^2 \approx \frac{4|\mathbf{H}'_{km}|^2}{\hbar^2}\frac{\sin \omega_{km}t/2}{\omega_{km}^2}. \qquad (8.40)$$

The mean rate w_k of this transition is $w_k = \mathfrak{p}_k(t)/t$, which peaks in $g(\omega, t) \equiv \frac{1}{t}\frac{\sin \omega t/2}{\omega^2}$ near $w = 0$ as shown in Figure 8.14.

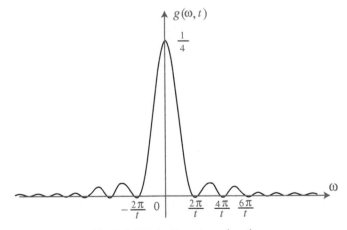

Figure 8.14 The Function $g(\omega, t)$.

Note that Eq (8.40) requires that the states to which transitions can occur must have $\omega_{km} \approx 0$, due to energy conservation. The function $g(\omega, t) \to \frac{\pi}{2}\delta(\omega)$ as $t \to \infty$. This demonstrates the Heisenberg uncertainty relation (Appendix L) between energy and time through the half-width of the peak $\Delta\omega$ and the time t of the perturbation such that $\Delta\omega\, t \sim \pi$. In general, there are some number of states dn within an interval $d\omega_{km}$ so that the number of all possible states is $dn = \rho(k)dE_k$, where $\rho(k) = dn/dE_k$ is the so-called *density of states per unit energy interval* near E_k (units in E_k^{-1}), and $d\omega_{km} = dE_k/\hbar$. Then the total transition rate to states near the state k is given by

$$W_k = \frac{1}{t} \sum_{l\,near\,k} \mathfrak{p}_l(t) = \frac{1}{t} \int \mathfrak{p}_l(t)\rho(l)\, dE_l \quad \text{replacing summation by integration}$$

$$= \int \rho(k)\frac{4|\mathbf{H}'_{km}|^2}{\hbar^2}\frac{1}{t}\left(\frac{\sin\omega_{km}t/2}{\omega_{km}}\right) dE_k$$

$$= \frac{4}{\hbar}|\mathbf{H}'_{km}|^2\rho(k)\int_{-\infty}^{\infty}\frac{1}{t}\frac{\sin\omega_{km}t/2}{\omega_{km}}\, d\omega. \tag{8.41}$$

Since the last integral in (8.41) has the value $\pi/2$ as determined from Figure 8.14, Fermi's *second golden rule* becomes

$$W_k = \frac{2\pi}{\hbar}|\mathbf{H}'_{km}|^2\rho(k). \tag{8.42}$$

Although named after Fermi, most of the work in this area was done by Dirac [1947] who formulated an equation identical to (8.42). Fermi himself called it "Golden Rule No. 2". Luminescence was introduced by Eilhard Wiedermann [1888].

9

Radioastronomy

We will discuss synchrotron radiation and its development from astronomy to radioastronomy. Subsequent discoveries in radio interferometry has led to cosmic microwave background radiation. The phenomena of quasars and pulsars, Big Bang, Olbers paradox, and black holes are presented.

9.1 Historical Notes

Many cosmological observations were made by ancient Babylonians and Greeks. In the second century AD the astronomer Claudius Ptolemy of Alexandria in Egypt wrote the *Almagest*, a work containing thirteen books, and the Greek Hipparchus catalogued about 850 stars. These astronomers predicted motion of the planets with good accuracy and provided their study to Earth's central position with regard to the solar system. Their calculations and observations remained in effect until the fifteenth century AD.

Then came the Polish astronomer Nicolaus Copernicus (1473–1563) who studied astronomy in Krakow, and then in Bologna, Italy. He was named a cannon of the chapter of Frauenburg in the late 1490s. He may have received holy orders from the Polish King Sigismund in 1537 who named him one of the four possible candidates to a vacant Episcopal seat. Whatever his clerical status, he was an astronomer of high renown in ecclesiastical circles. He was consulted by the Fifth Lateran Council (1512-1517) for the calendar reform. In 1531 he prepared an outline of his astronomy, which caught the attention of Pope Clement VII. Copernicus was persuaded by churchmen and academic colleagues to publish his work, which finally appeared in 1543 as *Six Books on the Revolution of the Celestial Orbits*. In these books Copernicus placed the Sun, rather than Earth, at the center of the system of perfectly spherical heavenly bodies, with circular orbits and constant orbital speed. This system was vigorously opposed by Protestants for its (alleged) opposition to Holy Scripture. However, the Catholic church placed no censure on it until the Galileo case.

Galileo Galilei (1564–1642) was a physicist and astronomer. He, in cooperation with the famous French mathematician abbé Marin Mersenne (1588–

1648), studied the effect of a change in mass and tension on the frequency of vibration of a string, and established that for a given length, an increase in mass or a decrease in tension produces lower notes, especially in the cases of stringed instruments like the piano. Galileo and Hooke established that every sound is characterized by a precise number of vibrations per second. Besides these and other achievements, Galileo designed his refracting telescope, discovered four moons orbiting Jupiter, and the phases of Venus. He believed in the Copernicus system, and started teaching it. But the Church authorities told him to teach the Copernicus theory as a hypothesis, and not as a truth. In 1624 Pope Urban VIII told Galileo that the Church had never declared Copernicanism to be heretical and that the Church would never do so. However, the publication of his work *Dialogo sopra i due massimi sistemi del mondo, tolemaico e copernicanto* (Dialogue concerning the Two Chief World Systems, Ptolemaic and Copernican) in 1632, written in Italian for very wide circulation, Galileo was suspected of heresy and was ordered to cease from publishing on Copernicanism. 'Certainly the condemnation of Galileo, even when understood in its proper context rather than in the exaggerated and sensational accounts so common in the media, proved to be an embarrassment to the Church, establishing the myth that the Church is hostile to science' (Woods [2005: 74]).

Johannes Kepler (1571–1630) was a royal astronomer at the court of Rudolph II in Bohemia. He published in 1609 his work *Astronomia nova* or *Commentarius de stella Martis* ('New Astronomy' or 'Commentary on the Planet Mars'). While working on his predecessor Tycho Brahe's theory on Mars and following the calculations provided him, he was not satisfied with the results which were based on the Copernicus system. Kepler eventually abandoned the Earth-centered circular model, and developed an independent theory published in 1619 as *Harmoniē Mundi* (The science of the Harmony of the World), in the form of three laws which are:

FIRST LAW: The orbit of each planet is an ellipse with the Sun at its focus.

SECOND LAW: The line joining the planet to the Sun sweeps out equal areas in equal times.

THIRD LAW: The square of the period of a planet is proportional to the cube of its mean distance from the Sun.

Subsequent developments in astronomy were made by William Herschel (1738–1822) who laid the foundation of modern astronomy.

9.2 Synchrotron Radiation

X-ray diffraction has been discussed in §8.3, where it is mentioned that synchrotron radiation provides a continuous band of wavelengths, as well as the radiation of extremely high intensity, brilliance and collimation. In the study of synchrotron radiation, computations are mostly carried by considering a relativistic electron turning around a circular orbit and emitting a radiation characteristic, as shown in Figure 9.1(a) and the geometry involved in synchrotron radiation emission in Figure 9.1(b).

The electrons in a storage ring travel continuously from several hours to a few days at a speed v very close to the speed of light c, for example at the ALS, Berkeley, CA. The synchrotron radiation emitted by a relativistic electron moving in a circle is shown in Figure 9.1. According to Schwinger [1951], the shape and intensity within the radiation cone is given by the differential equation

$$\frac{d^2\Phi}{d\theta\,d\psi} = \frac{3\alpha\gamma^2}{4\pi^2}\frac{\Delta\omega}{\omega}\frac{I}{e}(1+X^2)\left[K_{2/3}^2(\xi) + \frac{X^2}{1+X^2}\,K_{1/3}^2(\xi)\right], \qquad (9.1)$$

where Φ is the photon flux (i.e., number of photons per sec); θ and ψ the observation angle in the orbital and vertical plane, respectively; α the fine structure constant $(= 1.137)$; γ the electron energy$/m_e c^2$ (m_e = electron mass, c velocity of light $= 2.9979 \times 10^8$ m/s); ω the angular frequency of photons ($h\omega$ the photon energy); I the beam current; e the electron charge ($= 1.602 \times 10^{-19}$ coulomb or 0.5 MeV); E_c the critical photon energy $= 3hc\gamma/2\rho$, $\rho(m) = 3.3356 \times (E/\text{GeV})/(B/\text{T})$; E the electron beam energy, $X = \gamma\psi$ the normalized angle in the vertical plane; $\xi = y(1 + X^2)^{3/2}/2$; and $K_{1/3}$ and $K_{2/3}$ are modified Bessel functions of the second kind (see Appendix K). The two terms within the square brackets in (9.1) are the polarization terms.

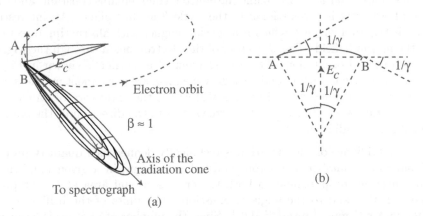

Figure 9.1 (a) Angular Distribution; (b) Geometry of Synchrotron Radiation.

The radiation is coherent, and the aperture of the cone (Figure 9.1(a)) is less than 0.0001 radians. It is defined as $1/\gamma = \sqrt{1-\beta^2}$, $0 < \beta < 1$. The angular radiation pattern in the orbital plane for different values of $\beta = v/c$

is shown in Figure 9.2.

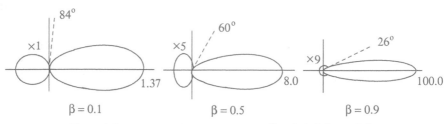

Figure 9.2 Angular Radiation Pattern for $\beta = 0.1, 0.5$ and 0.9.

In Figure 9.1(a), the length of the arc AB is $d = 2R/\gamma$, where R is the local radius of curvature. The angle of emission is $2/\gamma$ (Figure 9.1(b)). The small arc AB is approximated by the segment $\overline{AB}$. In time Δt, the front edge of the wave (pulse) travels a distance $D = c\Delta t$. During the same time Δt, the electron moving at speed v will travel in the same direction a distance d. The rear end of the pulse ends in a burst at a distance $D - d$ behind the front end. Since the series expansion of β is $\beta = \sqrt{1 - 1/\gamma^2} \approx 1 - \frac{1}{2}\gamma^{-2}$, $0 < \beta < 1$, we see that when $\beta \approx 1$, γ^{-2} is close to zero, and $1/\beta \approx 1 + \frac{1}{2}\gamma^{-2}$. Thus, the duration of the pulse is $\dfrac{D - d}{c} = \dfrac{2R}{c\gamma}\left(\dfrac{1}{\beta} - 1\right) \approx \dfrac{R}{c\gamma^3}$. This is an extremely short duration for the pulse which has a wide range of angular frequencies, with the critical frequency $\omega_c \approx c\gamma^3/R$. The electron keeps circulating in the synchrotron facility, and the burst will repeat every $2\pi R/c$ seconds.

The photon flux of the synchrotron radiation from a bending magnet has been studied. In the orbital plane the emitted cone remains constant, and thus the photon flux is proportional to the angle θ in that plane. An interesting case is that of a wiggler, which is a special magnet with alternating directions of the magnetic field. The trajectory of the electron beam through the wiggler is a sinusoidal oscillation. The arrangement of a wiggler is shown in Figure 9.3(a), in which the white and dark gray magnets are permanent magnets, and light gray is magnet steel. Figure 9.3(b) shows the trajectory of an electron beam within a wiggler, where the arrows denote the direction of the emitted synchrotron radiation.

The brilliance of the radiation is technically defined as a quantity proportional to the number of photons emitted per second in a given bandwidth, and inversely proportional to both the cross-sectional area at the origin of the radiation and to the angle of emission. The units of the brilliance are, therefore, $\text{s}^{-1} \text{ mm}^{-2} \text{ mrad}^{-2} \text{ Hz}^{-1}$. Since the number of photons is large and the other two quantities are small, synchrotron radiation has extremely high brilliance, millions of times higher than any other source in the same range. The undulators which are the third generation of storage rings, are used in experiments on the brilliance aspect of synchrotron radiation.

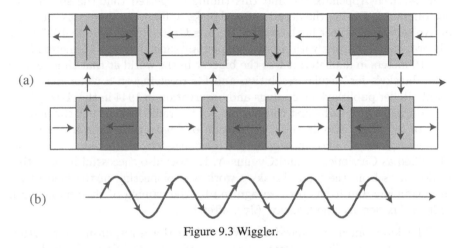

Figure 9.3 Wiggler.

9.3 Radioastronomy

Radioastronomy is a branch of astronomy that examines the celestial bodies at radio frequencies. Karl G. Jansky (1905–1950) was the first astronomer who detected in the 1930s radio radiation (waves) from an astronomical object in the Milky Way while working as a radio engineer at the Bell Telephone Laboratories in New Jersey. He constructed an array of rotating radio antennas designed to operate at a wavelength of 14.6 m (20 MHz). Such antennas are referred to as radio telescopes, employed either singly or with multiple linked telescopes using the techniques of radio interferometry. It was in the spring of 1933 that a cosmic 'hiss' was broadcast and millions of people heard it. A very delicate receiving set in New Jersey noted this noise for the first time from some source in the Milky Way, as reported by the *New Yorker* magazine. However, the hiss did not have any effect on radio communications, and the event was eventually forgotten. Subsequent studies identified a number of different sources of radio emission, like the stars and galaxies, as well as entirely new classes of objects such as radio galaxies, quasars, pulsars, and masers. The discovery of cosmic microwave background radiation later provided compelling evidence for the Big Bang.

Before Jansky observed the Milky Way, physicists speculated that radio waves could be observed from astronomical sources. In the 1860s, Maxwell's equations had shown that electromagnetic radiation is associated with electricity and magnetism, and could exist at any wavelength. Several attempts to detect radio emission from the Sun by experiments were performed by Nicola Tesla and Oliver Lodge, but they failed to detect any emission.

The astronomer Albert Melvin Skellett, who was a friend and teacher of

Jansky, pointed out that the signal (hiss) repeated the exact length of a sidereal day (see Appendix H), and this timing suggested that the source was astronomical. During the decade after Jansky's discovery there was only one active radioastronomer in the world. He was Grote Reber, who in 1937 built his own radiotelescope with a bowl-shaped antenna that was 15 meters high and 10 meters in diameter; it was the biggest in the world at that time, better than Jansky's for angular resolution and for receiving on a wider frequency band. Reber published his findings and observation in 1944 in the *Astrophysical Journal*, which had a first contour map of the radio brilliance in the densest part of the Milky Way in the constellation of Sagittarius. He discovered the intense regions of radio emission in a strong magnetic field, which were later identified as Cassiopeia A and Cygnus A. He was also successful in detecting radio waves from the Sun. Jansky's work was pioneering in the field of radioastronomy and it has been recognized by the naming of the fundamental unit of flux density, the 'Jansky' (Jy), after him.

The development of radar during World War II was responsible for further developments in radioastronomy. A system of radars, while tracking the enemy attacks on February 27, 1942, detected an intense noise that could not be eliminated. Although this noise was confirmed to be an exceptional activity of solar spots, the intensity of the signal was extremely high, a hundred times higher than that predicted by Planck's law for a body at $6000°$ K. The latest system of radars in the 1940s detected a number of false alarms at an altitude of about 100 km, which were finally confirmed the so-called Jansky's hiss.

The Mullard Radio Astronomy Group, founded near Cambridge University in 1950s by some members of the Telecommunications Research Establishment that had carried out war time research into radar, utilized in 1970s the computational power of computers, such as the Titan, to study the computationally intense Fourier transform inversions. They made use of aperture synthesis to create a 'One-Mile' and later a '5 km' effective aperture using the One-Mile and Ryle telescopes, respectively. They used the Cambridge interferometer to map the radio sky, producing the 2C and 3C surveys of radio sources (see the website http://www.phy.cam.ac.uk/history/years/radioast.php for more details).

9.4 Radio Interferometry

The two-antenna interferometer, developed by Martin Ryle (1918–1984), became the subsequent instrument of discovery in radioastronomy. Ryle was awarded the Nobel Prize in Physics in 1974 for this invention and astronomical research, jointly with Antony Hewish who had discovered in 1967 the first pulsar by analyzing the data obtained from Ryle's interferometer.

Radio interferometry deals with a class of techniques in which electromagnetic waves are superposed in order to extract information about the waves, as described in Bunch and Hellemans [2004: 695]. Interferometers are also used in industry to detect small displacements. However, their efficient and widespread use covers many fields like radioastronomy, fiber optics, oceanography, seismology, spectroscopy, quantum mechanics, nuclear and particle physics, plasma physics, remote sensing, biomolecular interactions, and mechanical stress-strain measurements. The Michelson interferometer, constructed in 1890 by Albert Abraham Michelson (1852–1931), was the first such device as reported in Michelson [1898]. It was successfully used to measure the diameter of some nearby stars such as Orionis, Arcturus, and Betelgeuse; the data is available in Michelson and Pease [1921]. The light path through a Michelson interferometer is shown in Figure 9.4.

If there is an extended source of diffraction, various parts of the image will overlap. This will destroy the finest details of the image. In the case of two stars with their images closer together than the size of the diffraction disk of either, a single spot is slightly elongated to show the image. The resolution of a radiotelescope as well as of a lens or the reflecting mirror of an optical telescope is defined as the smallest angle α between two stars which provide two separate images. It is computed using formula $\alpha = 2.1 \times 10^5 \dfrac{\lambda}{D}$ in seconds of an arc, where λ is the wavelength and D the aperture diameter of the paraboloid disk or lens to be measured in the same units (in cm). The resolution of the human eye is almost one minute of arc, with λ close to 0.5 microns and D equal to a few millimeters. Radio waves are much longer than light waves. For example, radio waves of 20 cm are about 400,000 times longer. This means that in order to resolve a given angle, a radio telescope must be 400,000 times larger than an optical telescope.

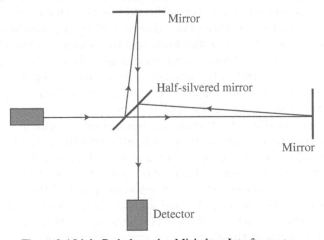

Figure 9.4 Light Path through a Michelson Interferometer.

Modern radio interferometers consist of widely separated radio telescopes which observe the same objects that are connected together using coaxial cable, waveguide, optical fiber, or other type of transmission line. This increases the total signal collected. It can be also used in a process called *aperture synthesis* to vastly increase resolution. The process involved in this system works by superposing (interfering) the signal waves from the different telescopes. This is based on the principle that waves that coincide with the same phase will add together while two waves with opposite phases will cancel each other out.

Our senses perceive infrared radiation as heat and light, but we are not aware of the related propagating waves because they have extremely high frequencies, some with thousands of millions of oscillations per second. Such frequencies never give our eyes any chance to perceive them. Radio waves have frequencies that start at a million oscillations per second and go higher. These radiations cannot be captured by the human eye or the photographic emulsion. Hence, other means are employed to detect them, like a dipole antenna pointing toward the sky that can detect radio waves, but not the source of their origin. The phenomena of diffraction and interference cause the image of a celestial point source, created with a paraboloidal antenna, to become a tiny spot sounded by faint concentric rings. These rings are clearly visible under good atmospheric conditions if viewed with an optical telescope, but are not so clear if the image is generated by a radio dish. The central spot of the radio image is called its *diffraction disk*.

The aperture in the paraboloid of a radio telescope provides the two-dimensional distribution of the radiation intensity of a celestial body as an image with a resolution that depends on the aperture dimensions. Since in these cases diffraction occurs at infinity (extremely distant objects), there is a Fourier transform relation between the electric field that exists on a remote plane, parallel to the aperture plane, determined by the two-dimensional intensity distribution of the celestial body and the electric field on the aperture plane of the paraboloid. This, in turn, establishes a Fourier transform relation between the electric field of the aperture surface and the electric field at the focal plane of the paraboloid generated after reflection from the paraboloidal surface. Thus, the field on the focal plane constitutes an image of the celestial body. Mathematically, this phenomenon can be best visualized by considering a one-dimensional aperture, at an angle θ, although the aperture is in reality two-dimensional. Let the origin be fixed at O and let one-dimensional aperture be represented by a segment described by the variable x (Figure 9.5). Let the unit be chosen similar to that of the wavelength λ of the observed radiation. The electric field at a point x and time t is then given by $e(x)\,e^{i\omega x}$, where $e(x)$ is the aperture function, which is complex-valued with an amplitude and spatial phase, and ω is the angular frequency under reception.

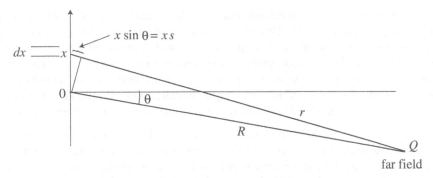

Figure 9.5 Electric Fields in the One-Dimensional Model.

At time $t = 0$ the infinitesimal element between x and $x + \Delta x$ induces an electric field at a distance R from the origin in the direction θ. This electric field is the far field which is proportional to $Ce(x)dx$, where C is a constant that accounts for the fact that the radiation power is distributed over the entire spherical surface of radius R, and thus C becomes closer to unity as the distance R increases, and so it can be set to 1 in the case of an object in the far field. Also, the wave cycles in the electric field at Q create a phase delay, which depends on the length r of the radiation path. Then the far field at Q due to the infinitesimal element dx is defined by $e(x)\,dx\,e^{-2\pi i r}$, and the electric field at Q is obtained by integrating this infinitesimal field over all such elements. The far field can be interpreted as follows: since Q is situated at an extremely large distance R, the approximation $r \approx R + x\sin\theta = R + xs$ holds, where $s = \sin\theta$ (Figure 9.5). Then $e^{-2\pi i r} = e^{-2\pi i R}e^{-2\pi i xs}$, and the far field $E(s)$ at Q can be approximated by

$$E(s) \approx \int_{-\infty}^{\infty} e(x)\,e^{-2\pi i xs}\,dx, \qquad (9.2)$$

where the constant $e^{-2\pi i R}$ is removed and ignored in (9.2) just like the constant C. Although the limits of integration are from $-\infty$ to ∞ in (9.2), the function $e(x) = 0$ is outside the aperture. However, the formula (9.2) clearly shows that the far field $E(s)$ at Q is the Fourier transform of $e(x)$ by definition (3.1) and it depends on the direction θ since $s = \sin\theta$. The square of the amplitude $|E(s)|^2$ of this field is called the *angular power spectrum*.

For example, consider the case of a point source defined by the Dirac δ-function. When the radio waves, which are originally spherical and have constant amplitude, reach the telescope far away, they can be regarded approximately as plane waves. Then the electric field in the paraboloid becomes uniform and is defined as the Fourier transform of the δ-function, i.e., $E(s) \approx 1$ (Appendix A). In fact, $E(s) = 1$ exactly, if the paraboloid were infinite.

9.5 Aperture Synthesis

Aperture synthesis, or synthesis imaging, is a kind of interferometry that mixes signals from a collection of telescopes and produces images that have the same angular resolution. It requires only two telescopes at a time, movable along rails, and longer observation times during which the image of the radio source becomes fixed. At each separation and orientation, the lobe-pattern of the interferometer produces an output which is one component of Fourier transform of the spatial distribution of the brilliance of the observed object.

Most aperture synthesis interferometers use the rotation of the Earth to increase the number of baseline orientations included in an observation. As shown in Figure 9.6, the Earth is represented as a sphere, and the baseline between telescopes A and B changes angle with time when viewed from the radio source as the Earth rotates. Thus, collecting data at different times provides measurements with different telescope separations.

Aperture synthesis is possible only if both the amplitude and the phase of the incoming signal are measured by each telescope. For radio frequencies it is achieved by electronics, while for optical light the electromagnetic field cannot be measured directly, but must be propagated by sensitive optics and interfered optically. For aperture synthetic technique, accurate optical delay and correction of the atmospheric wavefront aberration are required. This became possible in the 1990s. Since then, the imaging with aperture synthesis has become successful in radio astronomy, and since the 2000s in optical/infrared astronomy.

Let two equal apertures in an interferometer be set apart by a units measured in multiples of the wavelength λ, and let θ be the angle between the direction of observation and the line perpendicular to the axis of the interferometer. Let the two apertures be set at $-a/2$ and $a/2$, respectively, so that origin is located at their mid-point. Assume that the two apertures are equally illuminated. Then the total electric field is $b(x-a/2)+b(x+a/2)$, where $b(x)$ denotes the field of one of the apertures placed at the origin (Figure 9.6(b)). Then the sum of the signals on the focal planes is

$$E(s) = B(s)\left(e^{-2\pi i\, sa/2} + e^{2\pi i\, sa/2}\right) = 2B(s)\cos\psi/2, \qquad (9.3)$$

where $\psi = 2\pi as = 2\pi a\sin\theta$, $s = \sin\theta$, and $B(s)$ is the Fourier transform of $b(x)$.

The corresponding intensity distribution or *angular power spectrum* is given by

$$P(s) = |E(s)|^2 = 2|B(s)|^2\left(1+\cos\psi\right). \qquad (9.4)$$

Note that the factor $1+\cos\psi$, which depends on the direction of observation θ, is a periodic function in θ. Thus, since

$$I = \frac{\Delta\psi}{2\pi} = a\sin(\theta + \Delta\theta) - a\sin\theta = 2a\sin\frac{\Delta\theta}{2}\cos\frac{2\theta + \Delta\theta}{2} \approx a\cos\theta\,\Delta\theta,$$

the period, or separation between two maxima, is $1/(a\cos\theta)$, where $a\cos\theta$ is the projection of the baseline a on the direction perpendicular to the direction of observation and is equal to a if the vertical direction of observation $\theta = 0$. Also, $B(s)$ is constant, and the fringes (lobes) of the radio interferometer pattern in the case of a point source are described by $1 + \cos\psi$, as shown in Figure 9.7.

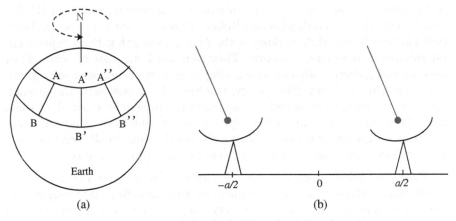

(a)   (b)

Figure 9.6 (a) Aperture Synthesis, (b) A Simple Interferometer.

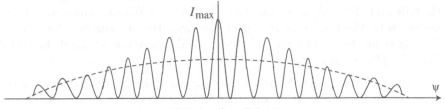

Figure 9.7 Lobes for a Point Source.

The *complex visibility* $V(a) = |V|\,e^{2\pi i\,a\Delta\theta}$ at the interferometer a is a complex-valued function, which is calculated from the data obtained by the interferometer lobes, where the amplitude V is given by

$$|V| = \frac{I_{\max} - I_{\min}}{I_{\max} + I_{\min}}, \qquad (9.5)$$

and where $I_{\max}$ and $I_{\min}$ are the maximum and minimum intensity at the lobes. Both $|V|$ and θ depend on 'frequency' $a\cos\theta \equiv a_p$ which is the projection of the baseline on the direction perpendicular to the direction of observation. Note that the visibility evaluated at a_p is the inverse Fourier transform of the brilliance, except for a multiplicative constant. Thus,

$$B(\theta) = \int_{-\infty}^{\infty} V(a_p) \, e^{-2\pi i \, a_p \theta} \, da_p = \int_{-\infty}^{\infty} |V(a_p)| \, e^{2\pi i \, a_p \Delta\theta} \, e^{-2\pi i \, a_p \theta} \, da_p.$$

(9.6)

After a sufficient number of calculations for the values of $|V|$ and $\Delta\theta$ using the interferometer, the value of the brilliance $B(\theta)$ is obtained using FFT.

9.6 Big Bang, Quasars, and Pulsars

In 1965 the phenomenon of cosmic microwave background radiation (CMB) was discovered by Arno Penzias and Robert Wilson at Bell Telephone Laboratories in New Jersey while working at the 7 cm wavelength with an experimental antenna in the shape of a horn. They were receiving a lot of noise. They also noticed galactic radiation emanating from certain unknown directions. They checked the Milky Way galaxy, the Sun, the atmosphere, the ground, and even the equipment to find the source of the noise, and ruled all of them out. They finally determined through careful measurements that the noise was from a supplementary emission, which was the same in all directions and appeared to come from the sky. They had no clue as to its source.

Based on Hubble's research, the physicist R. H. Dicke at Princeton University was of the opinion that the universe was expanding. He thought that initially the universe must have been very hot, and that the residual radiation from the distant past can still be detected. His calculations convinced him that this radiation must be in the millimeter or centimeter wavelengths. He constructed a suitable receiver on the roof of the Princeton biology building. He relied on the evidence of noise by Penzias and Wilson, which was determined to be the fossil relic of a primeval age. His attempt was focused on the Big Bang theory formulated before WWII. Dicke was awarded the Nobel Prize for physics in 1978.

The Big Bang theory describes the early cosmological development of the universe. It is a scientific theory that is most consistent with observations of the past and present states of the universe. According to this theory, the Big Bang occurred approximately 13.798 ± 0.037 billion years ago (Wollack [2010]). It says that the universe expanded from an extremely dense and hot state, and continues to expand even today. A general view is that space itself is expanding, carrying galaxies with it. Because of the extremely dense and hot initial state, it began expanding rapidly to cool down. After the initial expansion and sufficient cool-down, the energy began to convert into various subatomic particles, including protons, neutrons, and electrons. After a few first minutes of this Big Bang, thousands of years passed before the first electrically neutral atoms formed, and the majority of these atoms were hydrogen, along with helium and traces of lithium. Later, giant clouds of these primordial elements coalesced through gravity to form stars and galaxies, and heavier elements were synthesized either within stars or during supernova.

9.6.1 Quasars. A *quasar* (acronym for 'quasi-stellar radio source') is a very energetic and distant galactic nucleus. Quasars are extremely luminous. They were first identified as high redshift point sources (i.e., shift towards the infrared frequencies) of electromagnetic energy, including radio waves and visible light, similar to stars. The nature of these objects remained controversial as recently as the 1980s. However, now scientists agree that a quasar is a compact massive galaxy, which surrounds its supermassive black hole. Its size is 10 to 10,000 times the Schwarzschild radius of the black hole (see §9.7). A quasar is powered by an accretion disk around the black hole, and shows a very high redshift, which is due to the expansion of the universe between itself and the Earth. The redshift velocity can be measured by determining the wavelength of a known transition, such as hydrogen α-lines for distant quasars, and finding the fractional shift compared to a stationary reference. An extensive discussion on redshift is available in Harrison [1992]. The redshift velocity v_{rs} is defined by $v_{rs} = cz$, where c is the velocity of light, and the redshift z is calculated using the so-called Fizeau-Doppler formula

$$z = \frac{\lambda_0}{\lambda_e} - 1 = \sqrt{\frac{1 + v/c}{1 - v/c}} - 1 \approx \frac{v}{c}, \qquad (9.7)$$

where λ_0 and λ_e are the observed and emitted wavelengths, respectively. The redshift velocity v_{rs} is, however, not so simply related to the real velocity at larger velocities. So this terminology may become confusing if interpreted as a real velocity.

Hubble's law, also known as Lamaître's law, states that (i) all objects observed in deep space (intergalactic space) are found to have a Doppler shift observable relative to the velocity of the Earth, and to each other; and (ii) this Doppler-shift-measured velocity, of various galaxies receding from the Earth, is proportional to their distance from the Earth and all other interstellar bodies. It simply says that the space-time volume of the observable universe is expanding and this law is the result of direct physical observation of this process. The law is mathematically expressed as $v = H_0 D$, where H_0 is the constant of proportionality (called the Hubble constant), and D is the proper distance to a galaxy, and $v = dD/dt$ the velocity of the galaxy's movement. The constant H_0 has the SI unit s^{-1} but is usually expressed in (km/s)/Mpc, thus giving the speed in km/s of a galaxy 1 megaparsec (1 Mpc = 3.09×10^{19} km) away. Hubble's law is regarded as a fundamental relation between recessional velocity and distance, whereas the relation between recessional velocity and redshift depends on the cosmological model adopted for experiment and observation.

The recessional velocity is defined as follows: let $R(t)$ be called the scale factor of the universe, which increases as the universe expands in a manner that depends on the cosmological model selected. Then all measured distances $D(t)$ between co-moving points increase proportionally to $R(t)$, i.e., $\dfrac{D(t)}{D(t_0)} = \dfrac{R(t)}{R(t_0)}$, where t_0 is some reference time. If light is emitted from a galaxy at time t_e and received by us at t_0 due to the expansion phenomenon, then the redshift z is $z = \dfrac{R(t_0)}{R(t_e)} - 1$. Now, suppose that a galaxy at a distance D is expanding so that this distance will change with time at a rate $d_t D$, which can be called the rate of recession. Then the recession velocity v_r is given by

$$v_r = d_t D = \frac{d_t R}{R} D.$$

Now, define the Hubble constant $H \equiv \dfrac{d_t R}{R}$, and Hubble's law becomes

$$v_r = HD, \tag{9.8}$$

Then the redshift z can be determined, using a Taylor series expansion, approximately by

$$z = \frac{R(t_0)}{R(t_e)} - 1 \approx \frac{R(t_0)}{R(t_0)\,[1 + (t_e - t_0)H(t_0)]} - 1 \approx (t_e - t_0)H(t_0).$$

If the distance D is not too large, most of the serious problems with this model become small corrections. Recalling that the time interval is simply the distance divided by the speed of light, the redshift z can be approximated by

$$z \approx (t_0 - t_e)H(t_0) \approx \frac{D}{c} H(t_0), \quad \text{or} \quad cz \approx DH(t_0) = v_r. \tag{9.9}$$

This relation for v_r holds at low redshifts.

For distance D larger than the radius of the Hubble sphere r_{HS}, objects recede at a rate faster than the speed of light c, i.e., $r_{HS} = \dfrac{c}{H_0}$. Since H_0 is a constant only in space, and not in time, the radius r_{HS} may increase or decrease over various time intervals. The subscript zero in H_0 indicates the current value of this constant. Although the universe is expanding, the Hubble parameter is thought to be decreasing with time. This means that if we were to look at some fixed distance D and watch different galaxies pass that distance, later galaxies would pass the same distance with a smaller velocity than the earlier ones.

"Fingers of God" is an effect in observational cosmology that causes clusters of galaxies to be elongated in redshift space, with an axis of elongation pointed toward the observer (Jackson [1972]). It is caused by a Doppler shift

associated with the large velocities of galaxies in a cluster. These velocities
are caused by the gravity of the cluster, and change the observed redshift
of the galaxies in the cluster. The result is the deviation from the Hubble
law relationship between distance and redshift, defined by Eq (9.8), and this
leads to inaccurate measurements. A similar effect, known as the *Kaiser effect* (Kaiser [1987]), is also caused by large velocities leading to an additional
Doppler shift to the cosmological redshift. It also leads to a distortion; however, it is not caused by the random internal motions of the cluster, but rather
by the coherent motions as the galaxies fall inwards toward the cluster center
and form a galactic group. It normally does not lead to an elongation, but an
apparent flattening of the structure, known as the 'pancakes of God'.

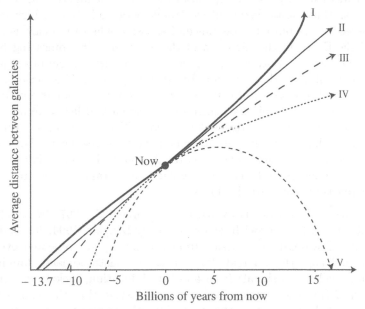

Figure 9.8 Age of the Universe.
Legend. I: Accelerating; II: $\Omega = 0$; III: $\Omega < 1$; IV: $\Omega = 1$; V: $\Omega > 1$.

The age and ultimate fate of the universe can be determined by measuring
the Hubble constant H_0 now and extrapolating with the observed value of
the deceleration factor $q = -(1 + \dot{H}/H^2)$, where $H = 1/t$, which is uniquely
characterized by values of density parameters (Ω_M for matter and Ω_λ for dark
energy)[1]. A *closed universe* satisfies $\Omega_m > 1$ and $\Omega_\lambda = 0$, but this situation

---

[1] In physical cosmology and radioastronomy, dark energy is a hypothetical form of
energy that permeates the space and tends to accelerate the expansion of the universe.
This energy is mostly accepted since 1990s to explain cosmological observations. According
to the Planck Mission Team, the total mass-energy of the universe contains 4.9% ordinary
matter, 26.8% dark matter and 68.3% dark energy. The existence of Type 1A supernova

will come to an end in a Big Crunch[2] and is considerably younger than its Hubble age. An 'open universe' with $\Omega_m \leq 1$ and $\Omega_\lambda = 0$ will expand forever and has an age that is closer to the Hubble age. For our current accelerating universe with $\Omega \neq 0$, the age of the universe is very close to the Hubble age[3]. The age of the universe and its ultimate fate is graphically presented in Figure 9.8.

The Hubble age is based on the following theorem: any two points which are moving away from the origin, each along straight lines and with speed proportional to distance from the origin, will be moving away from each other with a speed proportional to their distance apart.

9.6.2 Pulsars. A pulsar (i.e., pulsating star) is a highly magnetized, rotating star that emits beams of electromagnetic radiation, which can only be observed in its pulsed appearance of emission when the beam is pointing toward the Earth in the direction of the observer. This pulsating behavior is very precise and continues at regular, short rotational periods that range from about milliseconds to seconds for different pulsars. This periodic feature makes the pulsars very useful in making observations. For example, pulsar activity observed in a binary neutron star system had led to indirectly confirm the existence of gravitational radiation. Neutron stars are very dense and exhibit short periodic pulsating activity very regularly. For example, the first extrasolar planets were discovered around the pulsar known as PSR B1257+12. Their periodicity is so regular that certain pulsars are known to keep better time than atomic clocks.

The first pulsar was discovered on November 28, 1967, by Jocelyn Bell Burnell and Anthony Hewish, as described by Longair [1994: 99] and Ghosh [2007: 2]. The observed emission from this pulsar was repeated every 1.33 seconds and kept the sidereal time. These two astronomers, thinking that they have picked up signals from another civilization, nicknamed the emitted signal *LGM-1* for 'little green men', suggesting that the signals were of extraterrestrial origin. But this LGM hypothesis was rejected after a second pulsar, later named CP 1919, was discovered in a different part of the sky. In 1974 the astronomer Antony Hewish was awarded the Nobel Prize in

---

(one-A) by the High-S Supernova Team followed by the Supernova Cosmology Project suggested that the universe is accelerating. For this work the Nobel Prize was awarded in 2011 (Perlmutter et al. [1999]).

[2] The Big Crunch is one of the possible ultimate fate of the universe, in which metric expansion of space eventually reverses as the universe recollapses, ultimately ending as a black hole singularity, or causing a reformation of the universe starting with another Big Bang.

[3] In physical cosmology, according to the Lambda-CDM concordance model and based on the best measurements, the age of the universe since the Big Bang is 13.798 ± 0.037 billion years ($= 13.798 \pm 0.037 \times 10^9$ years, or $4.354 \pm 0.012 \times 10^{17}$ seconds). It has an uncertainty of 37 million years, which has been agreed upon by a number of scientific research projects, such as microwave background radiation measurements by the Planck satellite, the Wilkinson Microwave Anisotropy Probe, and others.

Physics. Although pulsars emit in radio wavelengths, they have been subsequently found to emit visible light, x-rays, and gamma ray wavelengths (see http://space.newscientist.com/article/).

Pulsars are formed when the core of a massive star is compressed during a supernova, which then collapses into a neutron star. The neutron star retains most of its angular momentum, and since its moment of inertia is sharply reduced, it rotates with very high speed, emitting beams of radiation along its magnetic axis which spins along with the rotation of the pulsar. The magnetic axis of the pulsar determines the direction of the electromagnetic beam. Recall that the magnetic axis and the rotating axis need not be the same. This misalignment of the two axes leads to the pulsed nature of the pulsar. The rotation, however, slows down over time as the electromagnetic source is emitted out, so that when this rotation slows down significantly, the radio pulsar mechanism terminates. This phenomenon is called the *death line* of the pulsar. The death line seems to occur after about 10–100 million years, and of all the neutron stars in their 13.7 billion years age of the universe, about 99% no longer pulsate. The periods of regular pulses range from 1.33 seconds to 9.437 seconds.

9.6.3 Olbers Paradox. Named after the German astronomer Heinrich Wilhelm Olbers (1758–1840), this paradox is also known as *dark night sky paradox*. The darkness of the night sky contradicts the assumption of an infinite and eternal static universe, because this darkness suggests a nonstatic universe. It is one of the pieces of evidence for the Big Bang theory. If the universe is static and populated by an infinity of stars, any sight from the Earth must catch the bright surface of the star, so that the night sky should be completely bright. But this contradicts the observation that even in spite of a myriad of bright stars, the night sky remains black.

The first account of the Olbers paradox is found in Edward Robert Harrison's book *Darkness at Night: A Riddle of the Universe* [1978]. According to Harrison, the first to mention this paradox was the British Thomas Digges (1546–1595) who expounded the Copernican system in English, and also postulated an infinite universe with infinitely many stars. Kepler also posed this paradox in 1610. The paradox attained a final form in the 18th century work of Halley and Cheseaux. Then Olbers described and postulated this paradox in 1823, but Olbers did not provide any answer to this paradox. Finally, it was Lord Kelvin who in 1901 settled this paradox in a little known paper as mentioned in Harrison [1978: 227-228].

The paradox is that a static, infinitely old universe with an infinite number of stars distributed in an infinitely large space would be bright rather than dark at night. To show this, we follow Harrison [1978] and argue as follows: divide the universe into a series of concentric shells, say, each shell being one light year thick. Then a certain number of stars will be in the shell

1,000,000,000 to 1,000,000,001 light years away. If the universe is homogeneous at a large scale, then there would be four times as many stars in a second shell between 2,000,000,000 and 2,000,000,001 light years away. But the second shell is twice as far away, so each star in it would appear four times dimmer than the one in the first shell. Thus the total light received from the second shell is the same as the total light received from the first shell. Thus each shell of a given thickness will produce the same net amount of light regardless of how far it is. Thus, the light from each shell adds to the total amount. The more shells there are, the more light there will be, and with infinitely many shells there would be a bright night sky.

The above argument was perceived by Kepler for an observable universe, or at least for a finite number of stars. In terms of the general relativity, it is still possible for the paradox to hold in a finite universe, though the sky would not be infinitely bright; every point in the sky would be like the surface of the star.

Under the assumption of a finite universe of finite age, only finitely many stars can be observed within a given volume of the surface (of the sky) visible from Earth. The density of stars within this finite volume is sufficiently low so that any line of sight from Earth is unlikely to reach a star. However, the Big Bang theory introduces a new paradox, which states that the sky was much brighter in the past, especially at the end of the reconstruction era, when it first became transparent. All points of the total sky at that era were brighter than the surface of the sun, due to the high temperature of the universe in that era, and most light rays terminated not in the star but in the relic of the Big Bang. This paradox is explained by the fact that the Big Bang theory also involves the expansion of space which can cause the energy of emitted light to be reduced due to redshift. This explains the relatively low light densities present in most of the sky despite the assumed bright nature of the Big Bang.

On the other hand, the steady-state cosmological model, that has no Big Bang, assumes that the universe is infinitely old and uniform in time as well as space. There are stars and quasars at arbitrarily great distances. The light from these distant stars and quasars will be redshifted accordingly by the Doppler effect and thermalization so that the total light flux from the sky remains finite. Thus, the observed radiation density can be independent of the finiteness of the universe. Mathematically, the total electromagnetic energy density (i.e., radiation energy density) in thermodynamic equilibrium using Planck's law is

$$\frac{U}{V} = \frac{8\pi^5 (kT)^4}{15(hc)^3}. \tag{9.10}$$

For example, for temperature 2.7° K, this ratio is 40 J/m^3 to 4.5×10^{-31} kg/m^3, and for visible temperature 6000° K it is 1 J/m^3 to 1.1×10^{-17} kg/m^3. But the total radiation emitted by a star is at most equal to the total nuclear binding energy of isotopes in the star. Thus, for the observable universe of about 4.5×10^{-34} kg/m^3 and given the known abundance of the chemical elements, the corresponding maximal radiation energy density is 9.2×10^{-28} kg/m^3, i.e., the temperature 3.2 K. This is close to the total energy density of the cosmic microwave background and the cosmic neutrino background combined.

On the contrary, the Big Bang hypothesis predicts the cosmic background radiation (CBR) should have the same energy density as the binding density of the primordial helium, which is much greater than the binding energy density of the nonprimordial elements. So it gives almost the same result. But the Big Bang theory would also predict a uniform distribution of CBR, while the stead-state model does not predict so.

Since the speed of light is constant regardless of the redshift, i.e., shift towards the infrared frequencies, the universe is still sharply constrained to finite sizes in space and time. Some models of an infinite universe are still viable, like the steady-state theory or static universe, and Olbers paradox cannot help sharply distinguish between them from some variants of the Big Bang theory. For more details on this paradox, see Wesson [1991].

9.7 Black Holes

The Schwarzschild radius, named after the German astronomer Karl Schwarzschild [1916], is the radius of a sphere such that if all the mass of an object is compressed within that sphere, the escape speed from the surface of the sphere would be equal to the speed of light. An object smaller than its Schwarzschild radius is a black hole. Once a stellar object or remnant collapses below this radius, light cannot escape and the object is no longer visible (Chaisson and McMillan [2008]). The Schwarzschild radius r_s of an object is proportional to its mass m with a proportionality constant involving the gravitational constant G and the speed of light c, i.e., $r_s = \dfrac{2Gm}{c^2}$, where the constant $2G/c^2 \approx 1.48 \times 10^{-27}$ m/kg, or 2.95 km/solar mass. For example, this radius for the Sun is 2.08×10^{15}, and for the Earth 8.87×10^{-3} m which is the size of a peanut.

The Schwarzschild radius of an object, being proportional to its mass, is also proportional to the cube root of its volume, and hence the mass. This means that as matter accumulates at normal density, its Schwarzschild radius increases faster than its radius. For example, at around 150 million times the mass of the Sun, this accumulation of matter will fall inside its own Schwarzschild radius and would thus be a supermassive black hole of about 150 million solar mass. Supermassive black holes up to 18 billion solar mass have been observed, as mentioned in WMAP at the website http://map.gsfc.nasa.gov/universe/uni_matter.html. The supermassive black hole in the center of our galaxy, of about 4.5 ± 0.4 million solar mass, is the most convincing evidence for the existence of black holes.

9.7.1 A Mathematical Theory. The problems of modern stellar dynamics and astrophysics are indeed very classical in nature. The progress made in recent years has been due to the use of the theory of quasilinear elliptic equations. Based on steady radially symmetric models in stellar systems, this theory has useful applications in certain cosmological problems, especially those connected with the black holes. The theory has been used to develop the fundamental solutions for the p-Laplacian, which are applicable to some very diverse problems such as fluid flows through certain types of porous media, the Lane-Emden equations for equilibrium configurations of spherically symmetric gaseous stellar objects, singular solutions for the Emden-Fowler and Einstein-Yang-Mills equations, and existence and nonexistence of black hole solutions.

Soon after the publication of Einstein's theory of general relativity, the Schwarzschild relativistic model [1916] implied the theoretical existence of black holes, unintentionally and unbeknownst. An investigation into the existence of black hole solutions of the Einstein-Yang-Mills equations is available in Smoller et al. [1993].

The early articles on the derivation of the stationary distribution of stars near a massive collapsed object, such as a black hole, located at the center of a globular cluster were published by Peebles [1972]. One of these articles entitled *Black Holes are Where You Find Them* discussed the question of the existence of black holes, and determined that 'there can be no conclusions until we find a black hole.' Since then a vast literature has appeared on the theory of black holes, e.g., by De Sabbata and Zang [1992], Israel [1992], Page [1992], Shapiro and Teukolsky [1992], Wald [1992], Sparkle and Gallagher [2000], Hooft [2009], Hawking [2009], and Kovacs et al. [2009]. The question of existence of a black hole was settled by the announcement on May 25, 1994 by the astronomers of the NASA Headquarters that the Hubble Space Telescope has found a supermassive black hole in the heart of the giant galaxy M87, more than 50 light-years away.

9.7.2 Black Hole Solutions. After every thermonuclear source of energy is exhausted, a star cools down and becomes a *cold* star. There are two basic aspects related to this phenomenon: (i) During the cooling process, a star collapses gravitationally and forms a superdense object called a compact star; and (ii) compact stars can be divided into three types: white dwarfs, neutron stars, and black holes. These phenomena have been subject of extensive research.

In the Newtonian theory, an ideally spherically symmetric star is defined by the equation

$$\frac{dP(r)}{dr} = -\frac{GM(r)}{r^2}\rho(r),$$ (9.11)

where P denotes momentum, m mass, and ρ density. Its relativistic equation is

$$\frac{dP(r)}{dr} = -\frac{[GM(r) + 4\pi G P r^3/c^2][\rho + P/c^2]}{r^2[1 - 2GM(r)/c^2 r]}.$$ (9.12)

A comparison of Eqs (9.11) and (9.12) shows that the transition from the Newtonian mechanics to general relativity is equivalent to the following changes:

$$GM(r) \longrightarrow GM(r) + 4\pi G P r^3/c^2,$$
$$\rho \longrightarrow \rho + P/c^2,$$
$$r^2 \longrightarrow r^2[1 - 2GM(r)/c^2 r].$$

Notice that Eq (9.12) reduces to (9.11) as $c \to \infty$. We shall confine our analysis to the case of black holes. In general, stars do differ from one another in their properties of mass, angular momentum, charge, magnetic momentum, and chemical properties. A collapsing star which has more mass than the critical mass reduces itself to a black hole. There is a proposition which states that stationary black holes are of only one type, viz., their metric is defined in the spherical coordinate system (r, θ, ϕ) by

$$ds^2 = \frac{\Delta}{\rho^2}\left[dt - \frac{1}{c}a\sin^2\theta\,d\phi\right]^2 - \frac{\sin^2\theta}{\rho^2}\left[\frac{1}{c^2}(r^2 + a^2)\,d\phi - a\,dt\right]^2$$
$$- \frac{\rho^2}{c^2\Delta^2}\,dr^2 - \frac{1}{c^2}\rho^2\,d\theta^2,$$ (9.13)

where $\Delta = r^2 - \frac{2GM}{c^2}r + a^2 + \frac{GQ^2}{c^4}$, $\rho = R^2 + a^2\cos^2\theta$, $a = \frac{L}{Mc}$, and L is the angular momentum, and Q is the charge of a black hole. The metric (9.13) is known as the Kerr-Newman metric in the spacetime $R^3 \times R^1$ for $M \neq 0$, $L \neq 0$, and $Q \neq 0$. Some special cases of the metric (9.13) are as follows:

(i) If $L = Q = 0$, but $M \neq 0$, then (9.13) becomes the Schwarzschild metric which is a solution of Einstein's field equations and expresses the spacetime curvature near a spherically symmetric body of mass M. As a spacetime metric it has the form

$$g_{ij} = \begin{pmatrix} -e(r) & 0 & 0 & 0 \\ 0 & c^{-2}e^{-1}(r) & 0 & 0 \\ 0 & 0 & c^{-2}r^2 & 0 \\ 0 & 0 & 0 & c^{-2}r^2\sin^2\theta \end{pmatrix},$$ (9.14)

for $i, j = 1, 2, 3, 4$, where $ds^2 = e(r)\,dt^2$, and $e(r) = 1 - 2GM(r)/c^2 r$. This metric is valid in the region exterior to the spherical body.

(ii) If $L = 0$, but $Q \neq 0$, and $M \neq 0$, then (9.13) becomes the *Reissner-Nordström metric*, which is defined by

$$ds^2 = \left(1 - \frac{2GM}{c^2 r} + \frac{GQ^2}{c^4 r^2}\right) dt^2$$

$$- \frac{1}{c^2}\left[\left(1 - \frac{2GM}{c^2 r} + \frac{GQ^2}{c^4 r^2}\right)^{-1} dr^2 + r^2\, d\theta^2 + r^2 \sin^2 \theta\, d\phi^2\right]. \quad (9.15)$$

(iii) If $Q = 0$, but $L \neq 0$, and $M \neq 0$, then (9.13) becomes the *Kerr metric*:

$$ds^2 = \left(1 - \frac{2GM}{c^2 \rho^2}\right) dt^2 - \frac{2}{c}\left(\frac{2GMr}{c^2 \rho^2}\right) a \sin^2 \theta\, dt\, d\phi$$

$$- \frac{1}{c^2}\frac{\rho^2}{\Delta'}\, dr^2 - \frac{1}{c^2}\, d\theta^2 - \frac{1}{c^2}\frac{\Lambda}{\rho^2} \sin^2 \theta\, d\phi^2, \quad (9.16)$$

where

$$\Delta' = r^2 + a^2 - \left(\frac{2GMr}{c^2}\right), \quad \Lambda = (r^2 + a^2)^2 - a^2 \sin^2 \theta.$$

The above proposition suggests that black holes can be specified by three physical parameters, M, L, and Q, and therefore, are simple objects. Thus, the collapsing process causes complex stars to reduce to such simple objects. This simplification has given rise to the adage that 'a black hole has no hair!'

Einstein's equations are so complicated that analytical methods as well as modern computer technology do not have the answer to the problem of collapse of a general odd-shaped (not spherically symmetric) star. But the lack of an exact solution has not prevented researchers to conjecture about the ultimate fate of such collapsing objects. Moreover, in the case of a body that deviates slightly from perfect spherical symmetry, one can assume very small perturbations in the surface, rotation and electromagnetic fields so that they do not produce any significant effect on the geometry of the spacetime exterior of the body. Then it is possible to *linearize* Einstein's equations.

The laws of black hole physics are similar to those of thermodynamics. The *first law* is simply a collection of well-known conservation laws in physics, namely, the laws of conservation of matter and energy, of momentum, of angular momentum, of electric charge, and so on. Since black holes are products of ordinary matter, they are subject to the same basic laws of physics, which govern matter in general. Thus, e.g., if a black hole has mass M and absorbs a quantity of matter of mass m and energy of quantity E, then the mass will increase to $M' = M + m = E/c^2$, which yields Einstein's relation $E = mc^2$. The *second law* of black hole physics is, therefore, related to the question, 'Can we proceed in a direction opposite to that in the above example, i.e., can we decrease the mass of the black hole by extracting energy out of it?' Schwarzschild's black hole clearly answers this question in the negative. It concludes that the black hole mass can only increase: it can never decrease, and, therefore, no energy extraction is possible. However, the Kerr-Newman black hole allows energy extraction, which can be explained by the well-known Penrose process (see Narlikar [1982]). But this process cannot operate when all angular momentum and electric charges have been extracted from the black hole (Schwarzschild's case, $L = Q = 0$). Hence, the second law states that the total surface area of all participating black holes cannot decrease in any physical process. Finally, the *third law* of black hole physics states that the surface gravity of a black hole cannot be made to vanish under any finite system of physical processes.

These three laws of black hole physics can be compared to those of thermodynamics. To wit, the first law of thermodynamics states that the conversion from heat to work is done subject to the law of conservation of energy. The second law of thermodynamics introduced the property called *entropy*. It states that an isolated system behaves in such a manner that its entropy can never decrease; at best, the entropy may remain constant. The third law of thermodynamics maintains that under no finite system of physical processes can the absolute zero of temperature be ever attained.

An analogy between the radiation property of thermodynamics and of black hole is very interesting. A hot body radiates if it is placed in a cooler environment. But a black hole, by its very definition, cannot be hot and therefore cannot radiate in the classical sense. However, in the quantum mechanical sense, where the quantum *vacuum* is full of virtual particles and antiparticles, the Hawking effect showed that a black hole would radiate as if it were a black body. Hawking concluded that the number of photons of energy E per unit energy band per unit volume emitted by a black hole is given by

$$N(E) = \frac{E}{\pi c^3 \hbar^3} e^{-E/\kappa\theta}, \qquad (9.17)$$

where κ is the Boltzmann constant, θ the temperature, and $\hbar$ the Planck's constant.

For a steady radially symmetric model for black hole solutions, Matukuma [1938] proposed the equation

$$\nabla^2 \phi + \frac{\phi^p}{1 + r^2} = 0 \quad \text{in } R^3, \qquad (9.18)$$

where $r = |x|$, and $\phi > 0$ denotes the gravitational potential of a globular cluster of stars of density $\rho = \phi^p/4\pi(1+r^2)$ and mass $\int_{R^3} \rho \, dx$. If a black hole is located at the center of the globular cluster, it must produce a gravitational potential behaving like $1/r$ near the origin. An interesting problem then is to study the existence and nonexistence of black hole solutions for this model. Other important problems include the stability of black hole solutions, and the existence and asymptotic behavior of naked singularities.

Li and Santanilla [1995] have considered the model

$$\nabla^2 v + A(r)\, v^p = 0 \quad \text{in } R^n\backslash\{0\},\ p > 1. \qquad (9.19)$$

subject to the conditions

$$v > 0 \quad \text{in } R^n\backslash\{0\}, \quad v \sim \frac{1}{r^{n-2}} \quad \text{near } x = 0 \text{ and at } \infty,$$

where A is a nonnegative and locally Hölder continuous on $(0, +\infty)$. They have found that the model (9.19) has infinitely many radial solutions with finite total mass provided that $\int_0^\infty r^{2-p} A(r)\, dr < +\infty$. This result can be extended to include nonradial $A = A(x)$ with the additional requirement that $\int_0^\infty r^{2-p} \left\{ \sup_{|x|=r} A(x) \right\} dr < +\infty$. Moreover, the result for (9.19) when applied to the model (9.18) implies that for $A(r) = 1/(1 + r^2)$, the problem (9.19) has infinitely many radial solutions with finite mass, provided that $1 < p < 3$.

Ni's model (Ni [1986]) considers the problem

$$\nabla^2 v + r^{-l} v^p = 0, \quad \text{in } B_{r_0}\backslash\{0\}, \qquad (9.20)$$

subject to the conditions $v > 0$ in $B_{r_0}\backslash\{0\}$, $v = 0$ on ∂B_{r_0}, and $v \to \infty$ as $r \to 0$, where $B_{r_0} = B(0, r_0) \subset R^n$, $n \geq 3$, is an open ball. This system has infinitely many radial solutions if $1 < p < (n + 2 - l)/(n - 2)$ and $l < 2$. If $v \in C^2(B_{r_0})$ with $1 < p < (n-l)/(n-2)$ and v has a nonremovable singularity at the origin, then there exist positive constants C_1 and C_2 such that near the origin we have

$$\frac{C_1}{r^{n-2}} \leq v(x) \leq \frac{C_2}{r^{n-2}}, \qquad (9.21)$$

which means that the singular solutions are not distributions at the origin. Li and Santanilla [1995] have generalized the model (9.20) to

$$\nabla^2 v + f(r, v) = 0 \quad \text{in } B_{r_0}\backslash\{0\}, \qquad (9.22)$$

subject to the conditions $v > 0$ in $B_{r_0}\backslash\{0\}$, and $v = 0$ on ∂B_{r_0}, where f is continuous on $(0, r_0] \times [0, \infty)$. This system has infinitely many radial solutions that behave like $1/r$ near the origin, if $1 < p < 3 - l$, and there are no such solutions if $p \geq 3 - l$. For $n = 3$, the model (9.22) includes the model (9.21). Details of this analysis are also available in Kythe [1996: 338 ff].

10

Acoustics, Poetry, and Music

We will study how sound waves propagate and are measured. Their relationship with the evolution of poetry, music, and computerized music is presented.

10.1 Introduction

Acoustics is the science of propagation of sound waves, which surround us and interact with our hearing in everyday activities, like talking, and hearing speech, poetry, and music. Our hearing organs are stimulated by vibrations propagating through an elastic medium like air. However, we cannot 'see' sound; we can only see the vibrations on an oscilloscope. Mathematically, harmonic analysis and repeated application of fast Fourier transform help clarify the structure and nature of sound.

Modulation of sound waves, which are combinations of sinusoids, is a useful tool not only in processing all kind of waves, but also in poetry and music. Historically, it was Pythagoras (c. 550 BC) and his school who made progress in analyzing sound and its quantities that turn it into poetry and music. The art and science of prosody, or versification, and music were part of mathematics. Prosody, which includes accent, breathing, and quantity, is the mode to determine the discipline by which successive syllables are arranged to form a verse as a series of rhythmical syllables, divided by pauses, and represented in script by a single line, as understood by the ancient Greeks (στίχος). They also regarded versification, a branch of music, as a kind of musical art (μουσοικη').

The Romans adopted all Greek forms of poetry and music, but for the satire which was purely Roman. It is, therefore, quite logical to take the classical Latin poetry as the basis for constructing a sinusoidal theory of prosody and music. Almost all forms of its prosodic and musical advancement laid the foundation for the development of Western music. There is an important difference between most Latin poetry and most English poetry: in English verse the rhythm is due mainly to a systematic arrangement of accented and unaccented syllables, whereas in Latin verse the rhythm is due mainly to a systematic arrangement of long and short syllables.

The musical notation dates back from a time when the alphabetical method of delineation was prevalent among the Greeks who were perhaps pioneers in realizing and evolving this system. We learn from Boethius, a distinguished Roman statesman, philosopher and theologian of the post-classical period (AD 524), that this Greek notation was not adopted by the Romans, although it is not certain whether he was the first to apply the fifteen letters of the Roman alphabet to the scale of sounds included within the two octaves, or whether he was only the first to make a record of this application. The reduction of the scale to the octave is ascribed, perhaps wrongly, to St. Gregory, as is the naming of the seven notes. Indications of a scheme of notation based not on the alphabet, but on the use of dashes, hooks, curves, dots and strokes are found to exist as early as the sixth century AD, while specimens illustrating this different method do not appear until the eighth century AD. The origin of these signs, known as neumēs ($\nu\epsilon\upsilon\mu\alpha\tau\alpha$), is the full-stop (puntus), the comma (virga), and the mound or undulating line (clīvus), where the first indicates a short sound, the second a long sound, and the third a group of the two notes. The musical intervals were suggested by the distance of these signs from the words of the text.

The Greeks produced musical sounds by plucking strings of the same material, thickness, and tension, but with different lengths. They discovered that the strings of length $1, 3/2, 2$ (notes C, G, C), when plucked simultaneously or in succession, produced a pleasant and harmonious sound. The Latin meter (or measure, $\mu\epsilon'\tau\rho o\nu$) is a system composed of 'feet', which are divided into seven types of periodic sinusoids of values of multiples of a quaver ($= 1/8$).

Abbé Martin Mersenne (1588–1648) was the first mathematician to publish qualitative analysis of a complex tone in terms of harmonics. The French mathematician and philosopher Pierre Gassendi was the first to measure the speed of sound in 1635 at 478 m/s, an incorrect result. The correct speed was measured under the Royal Academy of Sciences in Paris in 1750 to be 332 m/s at 0°C; subsequently this speed in water was determined to be 1,435 m/s at 8°C.

Every sound is characterized by a precise number of vibrations per second. This fact was announced by both Galileo and Hooke. But in the 17th century Leonard Euler, Daniel Bernoulli, Jean le Rond d'Alembert, and Joseph Louis Lagrange all investigated the vibrating string, and d'Alembert provided the solution of the wave equation. In 1753, Bernoulli found a method of decomposing every motion of a string as a sum of elementary sinusoidal motions. Then Fourier, in the beginning of the 18th century, developed the Fourier transform and harmonic analysis. The German physicist Georg Simon Ohm (1789–1854) applied the Fourier method to the study of sound. Later the physicist Hermann von Helmholtz (1821–1894) published in 1863 his work entitled *On the Sensation of Tone as a Physiological Basis for the Theory of Music*, reprinted much later as Helmholtz [1954]. In this work he provided a detailed account of the sensation of sound from the ear through the sensory nerves to the brain. Finally, in 1877 and 1878 John William Stuart (Lord Rayleigh) published the two volumes of his work *The Theory of Sound*. Then Alexander Woods published his work entitled *The Physics of Music* in 1913.

Although the frequency limits of audibility cover a range of about 20 Hz to 20,000 Hz, depending on the tone loudness and person's age, varying from person to person, the human ear with its super sensitivity can hear in the range between 500 to 4000 Hz and responds very well in this range to change of pressure of the order 10^{-10} of atmospheric pressure. However, when the change in pressure reaches about 10^{-4} of atmospheric pressure, hearing becomes painful. In between these two limits audible speech ranges from minimum audibility to very loud noise.

10.2 Harmonic Analysis

All musical sounds are compressional waves caused by vibrations. All sound waves are characterized by their speed, pitch, loudness and quality (or timbre). In hearing people talk, the frequencies of vowels and consonant vary. According to Nei Bauman (*www.hearinglosshelp.com*), 'the vowels are clustered around the frequencies between 300 and 750 Hz; some consonants like j, z, v are at very low frequency at about 250 Hz; others like k, t, f, s, th are at high frequency between 3,000 and 8,000 Hz. All these consonants are voiceless and are produced by air hissing from around the teeth. On the other hand, the vowels are loud using about 95% of the voice energy, so the consonants are left with only 5% of this energy.'

Sound is detected by a microphone, or an oscilloscope, or the human (animal) ear, as a plane wave. A simple sound, generally produced by a tuning fork struck close to the tip of the fork, as a single tone, agitates the molecules of the air in its vicinity, which are immediately subjected to repeated compressions followed by expansions. The motion of each molecule over time t is governed by

$$y = A \sin 2\pi f t, \qquad (10.1)$$

where f is the frequency, or number of oscillations (cycles) per second, A is the amplitude of the oscillation, i.e., A is the maximum displacement of the molecule, and the oscillations take place in the direction of propagation. Thus, this sound is a longitudinal compressional wave (Figure 1.12(b)).

The hearing starts around 1×10^{-12} Watts/meter^2. The ear drum is a wonderful receptor of sound waves. Even at and near the lowest perceptible sound intensities, the eardrum vibrates much less distance (i.e, with much less amplitude) than the diameter of a hydrogen atom, which is twice the Bohr radius ($a_0 = 5.29 \times 10^{-12}$ m = 5.29 pm). It is so amazing to learn that if the energy in a single 1-Watt night light were converted into acoustical energy and distributed into equal amount to every person in the world, it would still be audible to every person with normal hearing.

The intensity (loudness) of sound is proportional to the square of the amplitude, which is measured in 'decibels' (dB), the one-tenth part of a 'bel' (see §4.4.4). The unit 'bel' was chosen by electric engineers at the Bell Telephone Company in 1920s to honor the inventor of the telephone, Alexander Graham Bell (1847–1922), from his last name slightly shortened. Let I denote the sound intensity, and let I_0 correspond to the minimum audibility level for intact hearing. The value $I_0 = 0$ dB was fixed by international agreement. It corresponds to a pressure of 20 μPa (micropascal), where 1 atmospheric pressure is equal to 10^5 pascals. The sound intensity I is then compared to $I_0 > 0$, as the ratio I/I_0 which is a rational number (no units). Then $\log_{10}\{I/I_0\}$ is the measure of sound intensity I in 'bels'. This implies that we can hear sound loudness over 14 orders of magnitude, i.e., the jet noise at 140 dB on a runway has a loudness of 10^{14} dB greater than threshold which is 1×10^{-12} Watts/meter^2. Thus, our ears can measure loudness over a very large range. Table 10.1 provides the decibel levels for typical sounds.

Table 10.1 Decibel Levels for Typical Sounds

Source	dB	Intensity
Threshold	0	1×10^{-12}
Breathing	20	1×10^{-10}
Whispering	40	1×10^{-8}
Soft talking	60	1×10^{-6}
Loud talking	80	1×10^{-4}
Yelling	100	1×10^{-2}
Loud concert	120	1
Jet takeoff	140	100

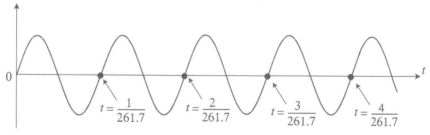

Figure 10.1 Waveform of a Single Tone.

In view of formula (10.1), if the single tone of a tuning fork is converted to an electric current and fed into an oscilloscope, the result is a graph of a single sinusoid with frequency f, shown in Figure 10.1. The value of $f = 261.7$ cycles per second, which is the *first harmonic*, is called the *fundamental*, since the frequency of the whole sound wave is 261.7 Hz. The higher harmonics, or *overtones*, are integral multiples of the fundamental (see §4.4.7 also). The waveform is then decomposed as the sum of harmonics with appropriate amplitudes. Thus, harmonic analysis of sound analyzes the sound spectrum.

10.3 Huygens Principle Revisited

Consider the Cauchy problem for the three-dimensional wave equation

$$u_{tt} = c^2 \left(u_{xx} + u_{yy} + u_{zz} \right), \quad -\infty < x, y, z < \infty, \ t > 0, \qquad (10.2)$$

subject to the initial conditions $u(x, y, z, 0) = 0$, $u_t(x, y, z, 0) = g(x, y, z)$, $-\infty < x, y, z < \infty$, where c is the wave speed. Assuming that g is absolutely integrable, and u and its first and second partial derivatives are absolutely integrable in x, y, z for each t, and using the Fourier transform, we find that

$$\frac{\partial^2 U}{\partial t^2} + c^2 (f_x^2 + f_y^2 + f_z^2) U = 0,$$

$$U(f_x, f_y, f_z, 0) = 0,$$

$$\frac{\partial U}{\partial t}(f_x, f_y, f_z, 0) = G(f_x, f_y, f_z).$$

The solution of this problem is

$$U(f_x, f_y, f_z, t) = G(f_x, f_y, f_z) \frac{\sin \rho \, ct}{\rho c}, \qquad (10.3)$$

where $\rho = \sqrt{f_x^2 + f_y^2 + f_z^2}$. Then the formal solution of the Cauchy problem can be written as

$$u(x, y, z, t) = \int\limits_{f_x^2 + f_y^2 + f_z^2 < t^2} G(f_x, f_y, f_z) \frac{\sin \rho \, ct}{\rho c} e^{-i (f_x x + f_y y + f_z z)} \, df_x \, df_y \, df_z,$$

$$(10.4)$$

Since $\sin(\rho ct) = \dfrac{1}{2i} \left(e^{i \rho ct} - e^{-i \rho ct} \right)$, the solution (10.4) becomes

$$u(x, y, z, t) = \frac{1}{2} \int\limits_{f_x^2 + f_y^2 + f_z^2 < t^2} \frac{G(f_x, f_y, f_z)}{\rho c} \times$$

$$\left\{ e^{-i \rho(ct - f_x x/\rho - f_y y/\rho - f_z z/\rho)} - e^{-i \rho(ct + f_x x/\rho + f_y y/\rho + f_z z/\rho)} \right\} df_x \, df_y \, df_z.$$

$$(10.5)$$

Notice that the function $e^{-i \rho(ct + f_x x/\rho + f_y y/\rho + f_z z/\rho)}$ is a solution of the wave equation obtained by Bernoulli's separation method. It also represents a plane wave propagating with speed c, and for fixed (x, y, z) it varies sinusoidally in time with frequency ρc in the direction $\left(-\dfrac{f_x}{\rho}, -\dfrac{f_y}{\rho}, -\dfrac{f_z}{\rho} \right)$. Thus, the solution u in (10.5) represents a plane wave in various directions and with various frequencies.

To obtain the solution of the Cauchy problem (10.2) in a sphere of radius r, we introduce the spherical coordinates by setting $f_x - x = r \sin \theta \cos \phi$, $f_y - y = r \sin \theta \sin \phi$, $f_z - z = r \cos \theta$. Then the solution becomes (Weinberger [1965: 335])

$$u(x, y, z, t) =$$
$$\frac{1}{2} \int_0^{2\pi} \int_0^{\pi} f(x + ct \sin \theta \cos \phi, y + ct \sin \theta \sin \phi, z + ct \cos \theta) \sin \theta \, d\phi \, d\theta.$$
$$(10.6)$$

Note that this solution does not depend on the behavior of the function f both outside as well as inside the above spherical domain. It depends only on the values of f on the surface of this sphere. This fact provides another insight into the Huygens principle: if 'a signal concentrated in the neighborhood of a point P at time zero is concentrated at time $t > 0$ near a sphere of radius ct centered at P, then a listener at a distance d from a musical instrument hears exactly what has been played at time $t - d/c$, rather than a mixture of all the notes played up to that time'.

10.4 Modulation

The concept and types of modulation, and waveshaping methodology is discussed in detail in Appendix E. This material is useful in almost all aspects of signal processing, including computerized music. In ordinary speech, modulation is an inflection of the tone or pitch of the voice, generally used to convey meaning. In music, it is a change from one musical key to another. It is generally the process of varying any one of the following three properties: amplitude (volume), frequency (pitch), or phase (timing) of a carrier signal using a modulating signal which typically contains information to be transmitted. An alternative method of modulating signal is the waveshaping index (§E.3), which determines the altitude of an input waveform. Musically, the waveshaping index affects the timbre. The waveshaping of an input signal is shown in Figure E.3, where the size of this index affects the output, such that a small index results in very little distortion of the output, but a larger index distorts the output relatively, which produces a much richer timbre in music. The effect of the index also depends on the output waveform where the input and output remain unchanged so long as the transfer function f remains in the interval $(-0.3, 0.3)$. However, outside this interval a more severe effect is related to larger amplitude, as shown in Figure E.4. This effect becomes prominent when an instrument is played through an overdriven amplifier.

The louder the input, the more distorted the output. Hence, waveshaping is sometimes called *distortion*.

The composition of two sinusoidal signals with different amplitudes, when sent separately, exhibit an effect of intermodulation (as explained in Example E.3). Composite periodic tones at constant intervals (i.e., periodic) can produce distortion at subharmonics. For example, if two periodic signals x and y are a musical fourth apart (i.e., periods in the ratio 4:3), then the sum of the two repeats at the lower rate are given by the common subharmonic. In fact, $x(t + p/3) = x(t)$ and $y(t + p/4) = y(t)$ yield $x(t + p) + y(t + p) = x(t) + y(t)$, and the composition $f \circ (x + y)$ is given by $f((x + y)(t + np))$, since p is the period, and n has the respective values. This phenomenon has been observed in electric guitars when setting the amplifier to 'overdrive' and playing the open B and high E strings together, resulting sometimes in the distortion sounding at the pitch of the low E string, two octaves below the high one.

The octave, which is a doubling in frequency, is an important concept in music. For example, 40 Hz is one octave higher than 20 Hz. The human ear is sensitive over a frequency range of about 10 octaves: within the range 20 Hz $\rightarrow$ 40 Hz $\rightarrow$ 80 Hz $\rightarrow$ 160 Hz $\rightarrow$ 320 Hz $\rightarrow$ 640 Hz $\rightarrow$ 1,280 Hz $\rightarrow$ 2,560 Hz $\rightarrow$ 5,120 Hz $\rightarrow$ 10,240 Hz $\rightarrow$ 20,480 Hz, the ear can distinguish between thousands of differences in frequency. As the *just noticeable difference* (JND) in frequency below 1,000 Hz is 1 Hz, it rises sharply beyond 1,000 Hz, such that at 2,000 Hz the JND is about 2 Hz and at 4,000 Hz the JND is about 10 Hz. The idea here about a JND of 1 Hz at a frequency f Hz is that if one were to listen alternately to tones of f Hz and $(f + 1)$ Hz, one would be able to distinguish clearly between these two tones.

10.4.1 Audio Signals. The multiplication of two audio signals is usually carried out by slowly varying signals, as in amplitude envelopes (§4.4.4). Assume that neither of the two audio signals is slowly varying. Let us first consider the product of two sinusoids $x_1 = \cos(\alpha_1 n + \phi_1)$ and $x_2 = \cos(\alpha_2 n + \phi_2)$:

$$x_1 x_2 = \cos(\alpha_1 n + \phi_1) \cos(\alpha_2 n + \phi_2)$$
$$= \frac{1}{2} \left[\cos\left((\alpha_1 + \alpha_2)n + (\phi_1 + \phi_2)\right) + \cos\left((|\alpha_1 - \alpha_2|)n + (\phi_1 - \phi_2)\right) \right],$$
$$(10.7)$$

The two components in (10.7) are called the *sidebands*.

The shifting of the component frequencies of a sound is called *ring modulation*. A very simple form is shown in Figure 10.2, in which an oscillator provides a carrier signal which is simply multiplied by the input (called the *modulating signal*).

In general, the term 'ring modulation' is used to mean multiplication of any two signals, although we have considered sinusoidal signals in the above definition. Eq (10.7) implies that in terms of two phases $\phi_{1,2}$ and the same amplitude a we have

$$2a\cos(\alpha n+\phi_1)\cos(\alpha n+\phi_2) = a\cos\left(2\alpha n + (\phi_1 + \phi_2)\right)+a\cos(\phi_1-\phi_2). \quad (10.8)$$

The second term on the right side in (10.8) has zero frequency; its amplitude depends on the relative phases of the two sinusoids, ranging from $+a$ to $-a$ as the phase difference varies from 0 to π radians. The sidebands obtained from multiplying two sinusoids of frequency $\alpha_{1,2}$ are shown in Figure 10.3 for the following cases: (a) $\alpha_1 > \alpha_2$; (b) $a_2 > \alpha_1$; (c) $\alpha_1 = \alpha_2$; and (d) $\alpha_1 = 0$. Note that in Figure 10.3(b) the lower sideband is not symmetric about α; in (c) the amplitude of the zero-frequency sideband depends on the relative phases of the two sinusoids; and in (d) there exists a carrier signal with zero frequency, amplitude a, and only one sideband of amplitude $a/2$.

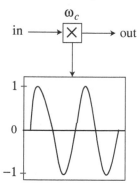

Figure 10.2 Ring Modulation.

In the case when a signal consists of more than one partial each, the signal of frequency α_1 can be replaced by a sum of finitely many sinusoids, such as $a_1\cos(\alpha_1 n) + \cdots + a_k\cos(\alpha_k n)$. If this signal is multiplied by another signal of frequency β, we obtain partials at frequencies $\alpha_1 + \beta$, $|\alpha_1 - \beta|$, ... , $\alpha_k + \beta$, $|\alpha_k - \beta|$. The resulting spectrum is the original spectrum together with its reflection about the vertical axis. This composite spectrum is then shifted right by the carrier frequency ω_c. Finally, if any component of the shifted spectrum is still left on the vertical axis, it is reflected back to make its frequency positive.

The multiplication of two audio signals, assumed to be slowly varying, is defined by the trigonometric identity (10.8), valid for $\phi_1 - \phi_2 > 0$. If $\phi_1 - \phi_2 < 0$, simply interchange the two sinusoids, so that $\phi_1 - \phi_2$ becomes positive. The formula (10.8) means that the multiplication of two sinusoids yields a sum of two terms, one involving the sum of the two frequencies and the other their difference. It also provides a method for ring modulation (shifting the component frequencies in waveshaping), as shown in Figure 10.2, in which an oscillator provides a carrier signal ω_c which is then multiplied by the input.

Note that ring modulation generally implies multiplication of any two signals. In the case of audio signals it always uses a carrier signal ω_c.

In the case of multiplying a complex signal which has components tuned in the ratio $1 : 2 : 3 : \ldots$ by a sinusoid, the resulting spectrum is shown in Figure 10.4, in which part (a) represents the spectrum of the original complex signal and its envelope at the original frequency ω; part (b) shows the modulation by a low frequency $\omega/3$, in which the carrier frequency (of the original sinusoid) is below the original frequency, and so the shifting is relatively small, and the spectrum re-folds, and part (c) shows modulation at a higher frequency $10\omega/3$, in which the unfolding effects are different: only the leftmost partial is reflected so that the frequency remains positive, and the spectrum envelope becomes widely displaced as compared to that in part (a).

10.5 Rhythm and Music

A necessary condition to compose a work of musical value is to devise a system of disciplined and linked movements. This system, called rhythm, consists of the reproduction of sound of human voices or musical instruments, and orchestras, and deals with the movement of the human body. Such expressions in metrical art are attained by an arrangement of syllables, which is known as rhythm. The science of meter consists in establishing laws which define the rhythmic forms of poetry and music. Thus, rhythm, not to be confused with rhyme which merely marks the ends of lines, means a harmonious movement. The classical Latin verse is rhythmic, and the rhythm occurs with a precise regularity.

In vocal music, rhythm is marked by the stress of voice, i.e., accent. The rhythmic accent is called the *ictus* (beat, blow). The accented part of the syllables is called the *thesis* ($\theta\epsilon'\sigma\iota\varsigma$), and the unaccented the *arsis* ($\alpha'\rho\sigma\iota\varsigma$)[1].

10.5.1 Pitch, Loudness, and Timbre. The process of hearing is complex. When speech or music reaches the ear in the form of vibrations, they create similar vibrations of the ear drum which are picked up in the inner ear by about 16,000 hair cells, the so-called *receptors*, and then finally in the brain by hundreds of millions of cortical neurons. These details of the hearing process have been recently provided by NMR imaging methods.

---

[1] *Thesis* and *arsis* are Greek terms, meaning putting down and raising of the foot, respectively, in military marching. The Roman grammarians, perhaps misunderstanding Greek, applied these terms to the lowering and raising of the voice, and thus reversed their significance, so that in the old books on Latin grammar, and even up to the first half of the 20th century, we find the incorrect statement that 'the syllable on which the ictus or stress of the voice falls is said to be in arsis.'

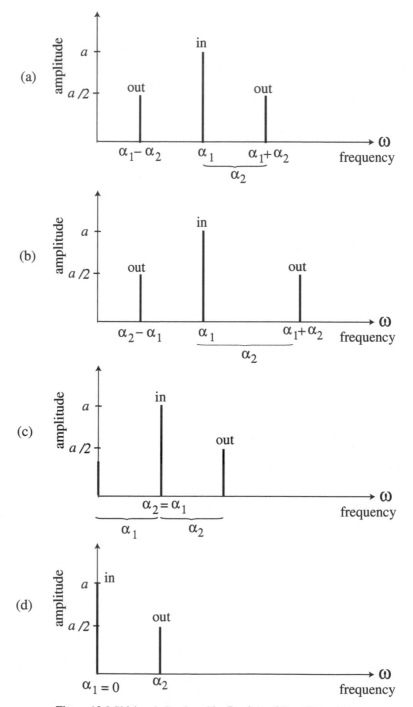

Figure 10.3 Sidebands Produced by Product of Two Sinusoids.

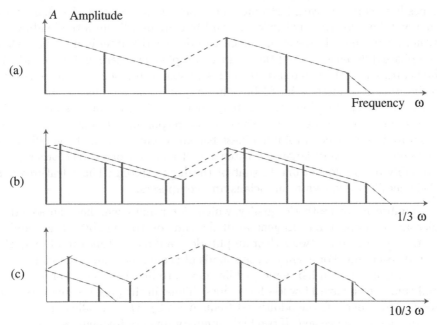

Figure 10.4 Ring Modulations of a Sinusoidal Signal.

The analysis of the three important features of pitch, loudness, and timbre of everyday speech is as follows. The pitch mainly depends on the frequency of vibrations of the air. The higher the frequency is, the higher or sharper the perceived sound is. The physiological process in the simplest terms is that the hair cells in the resonance region (a very limited part of the basilar membrane in the inner ear) are activated by a given frequency. Any multiple of this frequency by a given factor will naturally shift the resonance region by a corresponding amount. For example, the factor 2 associated with the octave will result in a shift of 3.5–4 mm. As Roederer [1979] has shown that although a complex tone composed of superposition of different frequencies activates the hair cells in different regions of the basilar membrane, the combined effect is perceived by the ear's subsequent 'sharpening' process as one ensemble of a definite pitch. The details of the 'MIDI pitch', are given in §4.4.5.

The amplitude of vibrations of the air defines the loudness of the sound, whether it is loud or soft. A larger amplitude of oscillations transmits more acoustic energy into the air and produces a louder sound. However, regardless of amplitude, loudness depends on frequency as well, as the human ear is very sensitive in a range of frequencies around 3500 Hz and completely insensitive to frequencies less than 20 Hz or higher than 20,000 Hz. Consider the following two cases of two pure tones: (i) If the tones are of the same frequency and are played simultaneously, the resulting amplitude is not necessarily the sum of the two respective amplitudes. If the two waveforms generated by these two tones have constructive interference (i.e., they are in phase, see §8.2), a louder tone will be perceived, not twice as loud but about 1.3 times the loudness of either single tone. However, in the case of destructive interference (i.e., when the phase difference is 180°) the resulting amplitude is zero and no sound will be heard. (ii) If the two tones are of closely spaced frequencies, then one tone of intermediate pitch is heard but the amplitude is modulated and creates beats. If the spacing between the frequencies of the two tones is large (> 15 Hz), then these tones are heard since the corresponding regions in the ear are separated. The issue of total loudness becomes complicated. If the difference in frequencies is very large, the listener will focus on one or the other tone, probably on the louder one to that of the higher pitch. These features and their details are known to musicians and composers.

Timbre is an aesthetic quality which determines whether the sound is pleasant or unpleasant. In general, it depends on the number of harmonics, on the relationship between their amplitudes, and on the time evolution of the sound spectrum. For example, the sounds of the human voice, the violin, and the piano are rich in harmonics, while those produced by the recorder, flute, and tuning fork have a few less harmonics. Thus, in the presence of a very loud tone of frequency f, the additional frequencies $2f, 3f, \ldots$, called the *natural harmonics*, are perceived. If two high intensity tones of frequencies f and g are sounded simultaneously, then the additional frequencies $f+g, f-g, f+2g, f-2g, 2f+g, 2f-g, \ldots$, called the *combination tones*, are perceived due to a phenomenon of nonlinear distortion in the ear. But of all these combinations, only the first ones are loud enough to be actually heard (Prestini [2004]). Further detailed analysis of this kind is available in Roederer [1979].

Technically, every sound can be decomposed by Fourier analysis and conversely reconstructed by Fourier resynthesis (see §4.6, and Davis and Hersh [1981]).

10.5.2 Music Bars. In musical notation, a bar (or measure) is a segment of time defined by a given number of beats, with each beat assigned a particular note value. Recall that in music theory the beat denotes a basic unit of time known as the pulse of the beat level. In particular, the beat can refer to different types of related concepts including tempo, rhythm, and groove, all three of which are defined below. A typical representation of different types of bar-lines is shown in Figure 10.5.

The word 'bar' is more common in British English, while the word 'measure' is more common in American English[2]. In the musical community both terms are equally well understood and are used interchangeably in interna-

---

[2] Originally the word 'bar' meant the vertical lines drawn to mark the musical units and not the bar-like (i.e., rectangular) dimensions of a typical measure of music. In British

tional usage. Examples are 'bars 9–16' or 'mm 9–16'. Note that the abbreviation 'bb' is reserved for beats only; bars must be referred to by name in full. Thus, the first metrically complete measure within a musical piece is called 'bar 1' or 'mm 1'. When the music piece begins with an anacrusis (an incomplete measure at its head), the following measure is 'bar 1' or 'mm 1'.

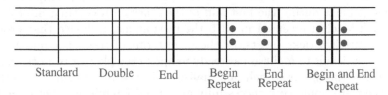

Standard   Double   End   Begin   End   Begin and End
                      Repeat   Repeat   Repeat

Figure 10.5 Types of Bar-Lines.

A double bar-line (or double bar) consists of two single bar-lines drawn close together, thus separating two sections in a piece, or a bar-line followed by a thicker bar-line (as in Figure 10.1), indicating the end of a piece or movement. Note that the term 'double bar is not a type of bar (or measure) but a type of bar-line. Again, a repeat sign (or repeat bar-line) may look like the music end, but it has two dots, one above the other. It indicates that the section of music that is prior to it must be repeated. The beginning of the repeated passage can be marked by a 'begin repeat' sign, but if it is not so marked, repeat is understood to be from the beginning of the piece. However, if this begin-repeat sign appears at the beginning of a staff, it does not act as a bar-line because there is no bar prior to it; it only indicates the beginning of the passage to be repeated.

In music with regular meter, the bars indicate a periodic accent in the music, regardless of its duration. However, in music using mixed meters, bar-lines are generally, but not strictly, used to indicate the beginning of rhythmic note groups. Bars and bar-lines also indicate rhythmic grouping of beats within and between bars and between phrases, and on higher levels such as meter.

A musical piece consists of several bars of the same length. In modern musical notation, the number of beats in each bar is specified at the beginning of the score by a top number of a time signature (such as 3/4) and by a bottom number that indicates the note value of the beat (for example, the beat has 1/4 note value). This time signature also provides the information as to which beats within the bar are more important. For example, in 4/4 time the first beat is strongly accented, compared to the second and fourth beats.

A *tempo* (Italian for 'time') is the speed or pace of a given musical piece.

English these vertical lines are called 'bars', too, but the term 'bar-line' is used to make the distinction clear. In American English, the word 'bar' stands for the vertical lines and nothing else.

Tempo is an important element of most musical compositions. It is measured in terms of beats per minute (BPM), that is, by assigning a particular note value as the beat (e.g., a quarter note or crotchet), and the marking that indicates a certain number of these beats must be played per minute. The greater the tempo, the larger the number of beats that must be played each minute, which means the piece is played faster. The mathematical tempo markings became popular during the first half of the nineteenth century, after Johann Nepomuk Maelzel invented in 1815 the metronome, a device that produces regular, metrical ticks (beats, or clicks) in terms of beats per minute. It is used by musicians to maintain a steady tempo in the play, or to work on issues of irregular timing, or to help internalize a clear sense of timing. Beethoven used the metronome very successfully and in 1817 published metronomic indications for his eight symphonies he had created at that time.

In the modern age of electronics, BPM has become an extremely precise measure. Music sequences use the BPM system to denote tempo, which is very significant in electronic dance music, as it was in the classical. DJs need an accurate knowledge of a tune's BPM to produce beat matching.

According to the Oxford English Dictionary [1971: 2537], a *rhythm* is any regular recurring motion that means a movement marked by the regulated succession of strong and weak elements, or of opposite or different conditions. This is a phenomenon of regular occurrence or pattern in time, and denotes periodicity or frequency of anything from microseconds to million of years. Mathematically, it is a periodic function of time (frequency).

Groove in music is the sense of propulsive rhythmic feel or sense of 'swing' created by the interaction of the music played by a band's rhythm-section (drums, electric bass or double bass, guitar, and keyboards). This word is related to dance, and is often used to describe the aspect of certain music that makes one want to move, dance, or 'groove'.

Rhythmic music is characterized by a repeating (i.e., periodic) sequence of stressed and unstressed beats, often designated as 'strong' and 'weak', and divided into bars organized by time and tempo indicators.

In order to define a musical meter, one needs to understand the poetic foot described in Appendix M. Similar to different types of feet in poetry, the feet-groups in music are called 'pulse-groups', which are defined by taking the accented beat as the first and counting the pulse until the next accented beat (MacPherson [1930: 5]; Scholes [1977]). Most complex meters are divided into a sequence of simpler ones consisting of double and triple pulses. In general, metrical rhythm, measured rhythm, and free rhythm are different classes of rhythms. Metrical rhythm is the most common in western music, in which each time value (tempo) is a multiple or fraction of a fixed unit (beat), and in which normal accents repeat regularly, providing systematic groupings. This aspect can be compared with the poetic rhythm discussed in Appendix M. In the case of the measured rhythm, each time value is a multiple or fraction of a specified time unit but these time values do not necessarily recur regularly; this case is also known as the additive rhythm. In free rhythm there is neither, for example, as in chants.

The structure of music includes meter, tempo, and all types of rhythms that generates temporal regularity. It has different levels which are defined as follows: (i) metric level, which may be distinguished as the beat level at which pulses are heard, is the basic time unit of a musical piece; and (ii) division levels are faster levels, and multiple levels are slower levels. Thus, a rhythmic unit is a duration pattern, periodic in nature, which occupies a time period equivalent to a pulse or pulses at the metric level.

10.5.3 Timbre. In music, timbre or tone quality is the quality of a musical note, sound or tone that distinguishes different types of sound production. In other words, timbre makes a particular musical instrument sound different from the other, even when they have the same pitch and loudness. Different musical instruments can be distinguished based on their timbre. The quality of sound (or timbre) is a very subtle feature. A trumpet and a violin playing the exact same note with the same pitch can be easily distinguished with eyes closed, because their timbres are different. The American Standards Association defines timbre as "... that attribute of sensation in terms of which a listener can judge that two sounds having the same loudness and pitch are dissimilar."

Timbre is also known as 'tone color'. This nomenclature is based on the sound of musical instruments, which may be described as 'bright', 'dark', 'warm', 'harsh', or other terms. There are also colors of noise, such as 'white', 'pink', 'red (brownish)' and 'gray'. For example, white noise refers to a signal whose spectrum has equal power within any equal interval of frequencies, a name that was taken from the concept of 'white light'. These terms are used in audio engineering. The color of a noise signal, produced by a stochastic process, generally defines a broad characteristic of its power spectrum. The concept of color in noise is similar to that of timbre in music, thus representing the spectrum of music.

A decomposition of timbre into component attributes has been described by Schouten [1968: 42] as being determined by at least five major acoustic parameters, as follows: (i) The range between tonal and noiselike character; (ii) the spectral envelope; (iii) the time envelope in terms of rise, duration, and decay (ADSR: attack, decay, sustain, release); (iv) the changes of both spectral envelope (formant–glide) and fundamental frequency (micro-intonation); and (v) the prefix, or onset of a sound, quite dissimilar to the ensuing lasting vibration.

According to Robert Erickson (1917–1997) in his work [1975], the subjective experience and related physical phenomena, based on the above five subjective and objective attributes, are given in Table 10.2 (see Shere [1995], also Sethares [1998])).

Table 10.2 Attributes of Timbre.

Subjective	Objective
Tonal character, usually pitched	Periodic sound
Noisy, with or without some tonal character, including rustle noise	Noise, including random pulses characterized by the rustle time (the mean interval between pulses)
Coloration	Spectral envelope
Beginning/ending	Physical rise and decay time
Coloration glide or formant glide	Change of spectral envelope
Microintonation	Small change (one up and down) in frequency
Vibrato	Frequency modulation
Tremolo	Amplitude modulation
Attack	Prefix
Final sound	Suffix

10.5.4 Harmonics. In music, the pattern of the harmonic series (§1.4) is shown in Figure 10.6. It presents a few terms of the harmonic series of a string, which are written as reciprocals $\left(\frac{2}{1}\right)$ and $\left(\frac{1}{2}\right)$.

The pitched musical instruments are often based on an approximate harmonic oscillator such as a string or a column of air, which oscillates at numerous frequencies simultaneously. At these resonant frequencies, waves travel in both directions along the string or air column, reinforcing and canceling each other to form standing waves. Interaction with the surrounding air causes audible sound waves, which travel away from the instrument. Because of the typical spacing of the resonances, these frequencies are mostly limited to integer multiples, or harmonics, of the lowest frequency, and such multiples form the harmonic series.

String center

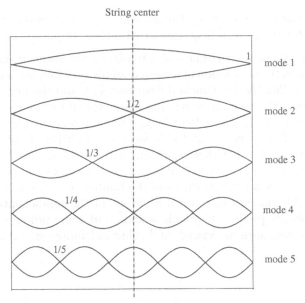

Figure 10.6 Harmonic Series of a String.

Although the contribution of total sound arising from overtones varies from instrument to instrument, from note to note on the same instrument, and even on the same note when a player produces that note differently, e.g., by blowing a little harder, the power due to overtones becomes less pronounced in some cases and the timbre generates a very pure sound to it.

The existence of overtones is a natural phenomenon. Consider the guitar and how it is played. The string is bound at both ends. If it vibrates in a standing waveform, those bound ends will definitely be nodes. When plucked, the string will vibrate in any standing wave condition in which the ends of the string are nodes and the length of the string is some integer numbers of half wavelengths of the initial mode's wavelength, as shown in Figure 10.6, in which the first five modes of a vibrating string are defined.

The musical pitch of a note is usually perceived as the lowest *partial* present (a name for the fundamental frequency), which may be the one created by vibration over the full length of the string or air column, or a higher harmonic chosen by the player. A partial is any of the sinusoids by which the complex tone is described. The musical *timbre* of a steady tone from such an instrument is determined by the relative strength of each harmonic.

A harmonic (or a *harmonic partial*) is any of a set of partials that are whole number multiples of a common fundamental frequency. This set includes the fundamental which is a whole number multiple of itself (1 times itself). The *inharmonicity* is a measure of the deviation of a partial from the closest ideal harmonic, typically measured in *cents* for each partial. Any complex tone can be described as a combination of many simple periodic waves (sinusoids) or *partials*, each with its own frequency of vibration, amplitude and phase.

In a musical note, the richness of a musical instrument or sound is sometimes defined in terms of a sum of a number of distinct frequencies. The lowest frequency is called the *fundamental frequency* (f), and the pitch it generates is used to name the note. However, the fundamental frequency does not necessarily remain the dominant frequency. The dominant frequency is, in fact, the frequency that is heard, and it is always a multiple of the fundamental frequency (Figure 10.7). Other significant frequencies are called *overtones* of the fundamental frequency, which may include harmonics and partials. Recall that harmonics are integer multiples of the fundamental frequency. There are also some subharmonics at integral divisions of the fundamental frequency. Most instruments produce harmonic sounds, but some produce partials and inharmonic tones, such as cymbals and other indefinite-pitched instruments.

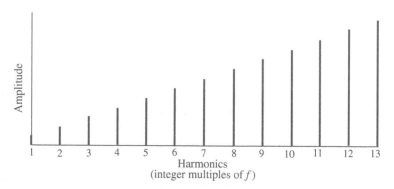

Figure 10.7 Harmonic Spectrum.

In an orchestral note the sound is a combination of 440 Hz, 880 Hz, 1320 Hz, and so on. Each instrument in an orchestra produces a different combination of these frequencies, and also of overtones. The sound waves of different frequencies overlap and combine, and the overall amplitude characterizes sound of each instrument.

The primary contributors to timbre (or quality) of the musical instrument are harmonic content, attack and decay, and vibrato. For sinusoidal tones, the harmonic content and the number and relative intensity of the upper harmonics present in the sound are very important. Some musical instruments have overtones which are not harmonics of the fundamental. But it is always possible to characterize a periodic waveform in terms of harmonics, using the Fourier transform. An example is presented in Figure 10.8, where part (a) displays the harmonic content in a waveform of the sound signal as a function of time, and (b) displays when time is transformed (using Fourier transform) to frequency, showing the amplitude of the harmonic content as a function of frequency.

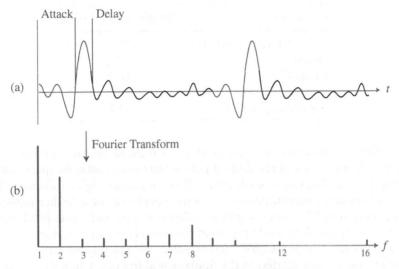

Figure 10.8 Fourier Transform of Harmonic Content.

The term 'vibrato' is defined as a periodic change in the pitch of the tone, and 'tremolo' indicates periodic change in amplitude or loudness of the tone. Thus, vibrato is a frequency modulation (FM) and tremolo an amplitude modulation (AM) of the tone. In fact, in instrumental or vocal music both are usually present to some extent. A moderate amount of vibrato is considered desirable in human voice; it adds a richness to the voice and can be used for expression. It is vibrato that helps the human ear detect the difference in timbre when the harmonic content of a sustained sound from a voice or wind instrument is reproduced precisely. The synthesized music becomes more realistic if some type of vibrato and/or tremolo is added to produce a more realistic tone.

The 'attack' and 'decay' contribute to the quality or timbre of the sound of a musical instrument, for example in a plucked guitar string. The plucking action gives it a sudden attack characterized by a rapid rise to its peak amplitude, while decay is comparatively long and gradual. The ear is sensitive to these attack and decay rates, and uses them to identify the instrument producing music. In the case of a cymbal struck by a stick, the attack is almost instantaneous while the decay is very long.

Sound intensity and its levels are required to hear sounds at different frequencies. Sound intensity in different orchestral instruments measured at 10 m away is provided in Table 10.3

Table 10.3 Intensity of Orchestral Sounds.

Instrument	Intensity Level (dB)
Violin (Quick movement)	34.8
Clarion	76.0
Trumpet	83.9
Cymbals	98.8
Bass drum (loudest)	103

Modes of vibrations are produced by a musical instrument when it is played. The frequency of the desired note is the fundamental frequency, which is caused by the first mode of vibration. However, many higher modes of vibrations always occur simultaneously. The frequencies of these higher modes are called *overtones*. The process goes as follows: the second mode produces the first overtone, the third mode produces the second overtone, and so on. There is one exception: in percussion instruments, like xylophones and marimbas, the overtones are not related to the fundamental frequency in a simple way; in other instruments the overtones are related to the fundamental frequency in a harmonic mode. In fact, harmonics are overtones which are integer multiples of the fundamental frequency.

10.5.5 Musical Scales. The Piano has eight scales C_1 through C_8. All keys are equally spaced, and if all keys are struck in succession from left to right from any one C key to the next C key, it will play the sound of "Do – Re – Me – Fa – So – La – Ti – Do"[3]. The diatonic scale for C_4 is shown in Figure 10.9, in which the white keys from C_4 to B_4 represent the major diatonic scale, and the black keys are intermediate tones. For example, the black key between D_4 and E_4 is of higher frequency (sharper) than D_4 and of lower frequency (flatter) than E_4. Hence, this note is simultaneously known as either 'D_4 sharp' ($D_4^{\#}$) or 'E_4 flat' (E_4^b). The frequency of sound produced by each of these keys is marked on the keys in this figure. All these five sharps and flats with other seven notes give the complete *chromatic scale*.

One can easily tell if two strings are different in frequency by even as little as a fraction of a Hertz. When sound waves interfere, they reinforce each other in constructive interference or subtract each other in destructive interference (§8.2). When sound waves are produced from two sources, like two guitar strings, constructive interference would correspond to a sound louder than each individually, while destructive interference would correspond to a quieter

---

[3] The diatonic scale for the harmonium in India from one C to the next C goes as "Sa – Re – Ga – Ma – Pa – Dha – Ni – Sa", which has the following meaning: *saragama* (rhythm) + *pa* =va (and) + *dhani* = *dhvani* (sound); the last Sa is the pause and repeat back marker: Sa–Ni–Dha–Pa–Ma–Ga–Re–Sa; this last Sa is the pause and repeat forward marker, and this goes on.

sound, or complete silence if the two had the same amplitude. Similarly, the frequency of the beat is the difference in the frequency of the individual waveforms. The beat frequency would decrease as the two frequencies cause the beats to get closer to each other. The beat frequency disappears when the two frequencies are identical.

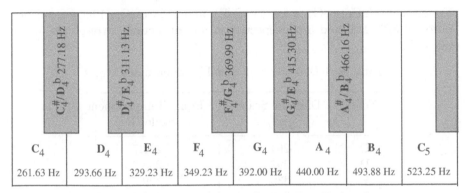

Figure 10.9 Chromatic Scale for C_4 on a Piano Keyboard.

Besides the consideration of beats when choosing frequencies for a musical scale, the issue of *critical bands* is also important. When the frequencies are almost the same, they cannot be distinguished. In this case an average frequency is heard. For example, if two frequencies were 450 Gz and 460 Hz, the average frequency of 455 Hz will be heard beating 10 times per second. However, if one frequency is kept fixed at 450 Hz, and the other raised slowly, they will reach a point when the two frequencies would still be indistinguishable, but the beat frequency would become too high to be easily identified, and the total sound spectrum would become rough. This is known as *dissonance* and it would continue until the higher frequency can be distinguished from the fixed lower one. Thereafter, raising the higher frequency will start decreasing the dissonance.

The development of musical scales is covered in detail in Berg and Stork [1982: Chapter 9], and in Askill [1979: Chapter 5]. Another excellent resource on the science of music is Rossing et al. [2002].

10.5.6 Tempered Scale. The frequencies of the *diatonic scale* are derived from the 'Pythagoras law' which states that the ratios between the fundamental frequencies must be equal to simple ratios of small integers for the corresponding sounds to be musical. Thus, the ratios between the frequencies of the notes in the diatonic scale are as follows:

C	D	E	F	G	A	B	C
1,000	1,125	1,250	1,333	1,500	1,667	1,875	2,000
1/1	9/8	5/4	4/3	3/2	5/3	15/8	2/1

Notice that a fifth (C–G), a fourth (C–F), and a third (C–E) have frequency ratios of 3/2, 4/3, and 5/4, respectively; and an octave corresponds to the ratio of 2/1.

Table 10.4 Diatonic and Equal Temperament Scales.

Note	Diatonic Scale (Hz)	Equal Temperament Scale (Hz)
C	264	261.7
D	297	293.7
E	330	329.7
F	352	349.2
G	396	392.0
A	440	440.0
B	495	493.9
C	528	523.3

The diatonic scale, however, has many limitations. In practice, all musical notes are raised or lowered for a musical piece, say, by one tone. This will then modify the frequency ratios of the third, fourth, and subsequent intervals. For example, if D is of frequency 297 Hz, then in order to maintain the 3/2 ratio of the C–G fifth while raising both notes to one tone to the D–A fifth, the A must have a frequency of 445.5 Hz. If the transposed music were played like this, it would be completely out of tone since the ear is very sensitive to the ratios between the frequencies. The problem ·has been solved by maintaining the 2/1 ratio of the octave and making the ratio between consecutive keys (black and white, on a piano keyboard) identical in all cases. This ratio is called the *semitone* and is equal to $\sqrt[12]{2} = 1.059456$, since there are 12 semitones in one octave. This has resulted in the *equal temperament scale*, which is presented in Table 10.4, along with the diatonic scale.

The history of the note A above the middle C is very interesting. In the 17th century the A had frequency 402.9 Hz. In the 18th century Handel adopted the A of 422.5 Hz, and in the 19th century in France the A had a frequency value of 425 Hz. However, in 1939, during an international music conference, the A above the middle C was assigned the frequency 440 Hz.

The physics of sound and hearing is classified into different fields, as the following tree diagram shows.

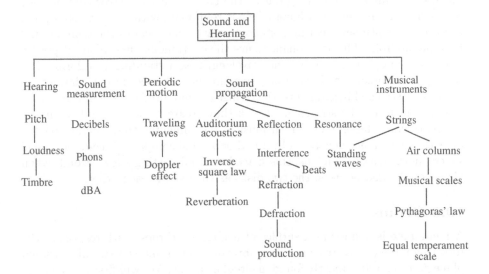

Sound waves, like light waves, are spherical and propagate with a velocity v that is defined using wavelength λ, frequency f and period T as $\lambda = vT = v/f$. Sound waves are reflected by obstacles like light waves or generally electromagnetic waves, resulting in echoes which are used by bats, toothed whales, and sea lions for predaceous practice in the dark. Almost every characteristic of sound and hearing mentioned in this diagram has been discussed throughout the book.

10.5.7 Resonance. The resonance and standing waves are both important in the structure of musical instruments. Most physical systems have many natural frequencies, but each has a low fundamental frequency (first mode) or a higher frequency of vibration (second mode). The structure and function of musical instruments, set into resonance, depend on standing waves, which occur whenever two waves with equal frequency and wavelength move through a medium so that they perfectly superpose on each other. This will happen only when the length of the medium is equal to some integer multiple of half the wavelength of the waves. In the case of standing waves there are parts of the medium that remain motionless (the nodes) and parts where the wave moves back and forth between maximum positive and minimum negative amplitude (the antinodes).

To create music, the physical structure of musical instruments is fixed into a standing wave pattern. For example, in the case of a trumpet the player creates a sound wave by allowing a burst of air into it, which is mostly reflected back when it reaches the end of the trumpet. If the player stops blowing air, the sound waves reflect back and forth between the two ends of the trumpet, but die out quickly by leaking out at the end. So if the player continuously blows the air, the reflected burst of air is reinforced with the new burst, and the process continues creating a standing wave that escapes the trumpet and becomes audible. This is resonance since the continuous burst of air at precise frequency, and hence at the same wavelength, is a standing waveform. The same process holds for all wind pipe instruments, like the flute or clarinet. In the case of stringed instruments, the strings are mounted on a hollow box with a hole, and when the strings are struck they create resonance and the music is generated. This is true for all musical instruments: they operate on the physics of resonance generated by a real or virtual compact three-dimensional space. In the case of cymbals, this space is virtual, lasting a few seconds when air is compressed between the two disks as they strike each other.

10.6 Drums

A membrane is defined as a surface of negligible stiffness with respect to the restoring force due to tension. The vibrating membrane is a two-dimensional plane surface, and is mostly found in circular drums. We will first consider the problem of vibrations of a stretched circular membrane of radius a fastened on its circumference, which can be referred to as a 'circular drum', subject to the conditions that the membrane is initially distorted into a given form which has circular symmetry about the axis through the center perpendicular to the plane of the boundary, and then allowed to vibrate. Recall that a function of the coordinates of a point has circular symmetry about an axis when its value is not affected by rotating the point through an angle about the axis, and a surface has circular symmetry about an axis when it is a surface of revolution about the axis. Using spherical coordinates, the distortion u due to the symmetry is independent of ϕ, and is governed by the equation

$$\frac{\partial^2 u}{\partial t^2} = c^2 \left(\frac{\partial^2 u}{\partial r^2} + \frac{1}{r} \frac{\partial u}{\partial r} \right), \tag{10.9}$$

where c is the velocity of sound in vacuum, subject to the conditions $u(r,0) = f(r)$, $u_t(r,0) = 0$, and $u(a,t) = 0$. Using Bernoulli's separation method, let $u(r,t) = R(r)T(t)$. Substituting in Eq (10.9) we get $RT'' = c^2 T \left(R'' + \frac{1}{r} R' \right)$. Since the left side of this equation does not involve r while the right side does not contain t, each side must be a constant, say $-m^2$, where m is a yet undetermined constant. Then this equation can be written as

$$\frac{T''}{c^2 T} = \frac{1}{R} \left(R'' + \frac{R'}{r} \right) = -m^2,$$

which is equivalent to the following two equations:

$$T'' + m^2 c^2 T = 0, \tag{10.10}$$

$$R'' + \frac{1}{r} R' + m^2 R = 0. \tag{10.11}$$

The particular solutions of Eq (10.10) are $T = \cos mct$ and $T = \sin mct$. To solve Eq (10.11), set $r = x/m$. Then it becomes

$$\frac{d^2 R}{dx^2} + \frac{1}{x} \frac{dR}{dx} + R = 0. \tag{10.12}$$

This equation was first studied by Fourier while considering the cooling of a cylinder and is therefore sometimes called Fourier's equation. We will find a series solution for Eq (10.12): assume that R can be expressed in the form of an infinite series in powers of x, i.e., take $R = \sum\limits_{k=0}^{\infty} a_k x^k$. Then, substituting it in Eq (10.12), we obtain

$$\sum_{k=0}^{\infty} \left[k(k-1) a_k x^{k-2} + k a_k x^{k-2} + a_k x^k \right] = 0.$$

This equation must be true for all values of x. The coefficient of any power of x, say of x^{k-2}, must vanish, which gives $k(k-1) a_k + k a_k + a_{k-2} = 0$, or $k^2 a_k + a_{k-2} = 0$, or

$$a_{k-2} = -k^2 a_k. \tag{10.13}$$

If $k = 0$, then $a_{k-2} = 0$, $a_{k-4} = 0$, and so on. Thus, starting with $k = 0$, and (10.13) gives the recurrence relation

$$a_k = -\frac{a_{k-2}}{k^2},$$

whence we get $a_2 = -\frac{a_0}{2^2}$, $a_4 = \frac{a_0}{2^2 \cdot 4^2}$, $a_6 = -\frac{a_0}{2^2 \cdot 4^2 \cdot 6^2}$, and so on. Hence,

$$R = a_0 \left[1 - \frac{x^2}{2^2} + \frac{x^4}{2^2 \cdot 4^2} - \frac{x^6}{2^2 \cdot 4^2 \cdot 6^2} + \cdots \right],$$

where a_0 is arbitrary, provided the series is convergent. If we take $a_0 = 1$, then $R = J_0(x)$, where $J_0(x)$ is the Bessel function of order zero, and the corresponding series is convergent for all real or purely imaginary values of x since the series

$$J_0(r) = 1 - \frac{r^2}{2^2} + \frac{r^4}{2^2 \cdot 4^2} - \frac{r^6}{2^2 \cdot 4^2 \cdot 6^2} + \cdots,$$

which composed of the moduli of x, is convergent for all $r = |x|$. Using the ratio test for this series, we find that the ratio $\left|\dfrac{a_{n+1}}{a_n}\right| = \dfrac{r^2}{4n^2} \to 0$ as $n \to \infty$. Hence, $J_0(x)$ is absolutely convergent. The function $J_0(x)$ is called a *cylindrical harmonic* or Bessel function of order zero. For Bessel functions, see Appendix K.

Note that Eq (10.12) was obtained by substituting $x = mr$; thus,

$$R = J_0(mr) = 1 - \frac{(mr)^2}{2^2} + \frac{(mr)^4}{2^2 \cdot 4^2} - \frac{(mr)^6}{2^2 \cdot 4^2 \cdot 6^2} + \cdots .$$

Hence, either $u(r,t) = J_0(mr)\cos mct$ or $u(r,t) = J_0(mr)\sin mct$ is a particular solution of Eq (10.9). But since $u(r,t) = J_0(mr)\cos mct$ satisfies the prescribed second condition for any value of m, in order to satisfy the first condition m must be taken such that $J_0(ma) = 0$, i.e., m must be a zero of $J_0(ma)$. It is known that $J_0(x)$ has an infinite number of real positive zeros (see Abramowitz and Stegun [1972]), any one of which can be obtained to any required degree of approximation. Let these zeros be $x_1, x_2, x_3, \ldots$. Then, if $\dfrac{x_1}{a} = m_1$, $\dfrac{x_2}{a} = m_2$, $\dfrac{x_3}{a} = m_3$, and so on, we get a solution of Eq (10.9) as

$$u(r,t) = \sum_{k=1}^{\infty} A_k J_0(m_k r) \cos m_k ct, \tag{10.14}$$

where A_k are constants. This solution satisfies the second and third prescribed conditions. Now set $t = 0$ in (10.14), giving $u(r,0) = \sum_{k=1}^{\infty} A_k J_0(m_k r)$. If $f(r)$ can be expressed as a series of this form, the solution of the problem is obtained by substituting the coefficients of that series for A_k, $k = 1, 2, \ldots$. About the drum problem, Kac [1966] asked the question: "Can one hear the shape of a drum?" This means one should answer the question whether two drums of different shapes and struck in their centers have the same eigenvalues [Protter, 1987]. This question has been resolved negatively by Gordon, Webb and Wolpert [1992].

10.6.1 Vibrating Rectangular Membrane. The problem of a vibrating rectangular membrane is defined as follows: solve the partial differential equation $u_{tt} = c^2 (u_{xx} + u_{yy})$ in the rectangle $R = \{(x,y) : 0 < x < a,\ 0 < y < b\}$, subject to the condition $u = 0$ on the boundary of R for $t > 0$, and the initial conditions $u(x,y,0) = f(x,y)$, $u_t(x,y,0) = g(x,y)$. The solution is

$$u(x,y,t) = \sum_{m,n=1}^{\infty} \left(A_{mn} \cos \lambda_{mn} t + B_{mn} \sin \lambda_{mn}\right) \sin \frac{m\pi x}{a} \sin \frac{n\pi y}{b},$$

where
$$A_{mn} = \frac{4}{ab} \int_0^b \int_0^a f(x,y) \sin \frac{m\pi x}{a} \sin \frac{n\pi y}{b} \, dx \, dy,$$

and $\qquad B_{mn} = \dfrac{4}{ab\lambda_{mn}} \displaystyle\int_0^b \int_0^a g(x,y) \sin \dfrac{m\pi x}{a} \sin \dfrac{n\pi y}{b} \, dx \, dy,$

for $m, n = 1, 2, \ldots$; and the eigenvalues are $\lambda_{mn} = c\pi \sqrt{\dfrac{m^2}{a^2} + \dfrac{n^2}{b^2}}$. For details on circular and rectangular membranes, see Kythe [2011: 118, 182, 208].

10.6.2 Frequency Modulation. This type of modulation, discussed in Appendix E, consists in changing a simple sound $\cos{(\omega_c t)}$, called the carrier, where ω_c is the carrier frequency, by a modulating signal, as in $s(t) = \cos{(\omega_c t + I \sin \omega_m t)}$, where $I \geq 0$ is called the *modulation index*, and ω_m the *modulation frequency*. The resulting sound has a frequency sum of ω_c and ω_m. Then the modulating signal becomes (see Ziemer and Tranter [1994])

$$s(t) = \sum_{k=-\infty}^{\infty} J_k(I) \cos\left[(\omega_c + k\omega_m)\, t\right], \qquad (10.15)$$

where $J_k(t)$ is the Bessel function of order $k = 0, \pm 1, \pm 2, \ldots$. Thus the modulated sound is the sum of harmonics having frequencies $\omega_c + k\,\omega_m$, centered at ω_c and spaced ω_m apart. There are interesting results that can be derived for values of I: for $I - 0$, all $J_k(0) - 0$ except for $k - 0$ when $J_k(0) - 1$, in this case the modulated signal reduces to the carrier wave itself. Next, for $I > 0$, if $\omega_m = \omega_c$, then a harmonic spectrum is obtained. Since $\sin(-\theta) = -\sin\theta$, the negative frequencies can be made positive by changing the sign of the amplitude. Note that the frequencies $(k+1)\omega_c$, $k \geq 0$, have amplitudes given by $J_k(I) - J_{k-2}(I) = J_k(I) + (-1)^{k+1} J_{k+2}(I)$, and the value $k = -1$ gives the constant term. The larger I becomes, the higher harmonics get more energy and the sound gets tinnier (lacking depth). Again, for a fixed I, the values of $J_k(I)$ decrease rapidly as k becomes larger than I. If $\omega_m = 2\omega_c$, only the odd harmonics appear, generating clarinet-like sounds. On the other hand, if $\omega_m = \sqrt{2}\,\omega_c$, the inharmonic sounds appear as in bells, drums and gongs. The spectrum of a frequency modulated sound with $\omega_m = \omega_c$ for values of $I = 0, 1, 2, 3$ is given in Figure 10.10.

If efficiency is the only criterion then many musical notes can be synthesized by assembling an attack, a suitable duration, and a decay. This simple technique fails if high fidelity is desired. In that case, musical sounds must be harmonic or quasi-periodic (as in the case of violin, cello and clarinet) and inharmonic or aperiodic (as in bass drum). Harmonic sounds can be represented by Fourier series with large periods and a large number of Fourier coefficients. They are also expressed as small variations of periodic sounds such as

$$x(t) = \sum_{k=1}^{M} A_k(t) \cos\left[2\pi(\omega_c + k\omega_m)\, t\right], \qquad (10.16)$$

where $A_k(t)$ denote the amplitudes of the harmonics, which vary slowly with time, M is the number of harmonics, and $f_k(t)$ are small frequency deviations with respect to the integer multiples of the fundamental ν (i.e., the pitch), that changes slowly with time. On the other hand, aperiodic sounds can also be represented by (10.16), but the required number of spectral components would be so large that Fourier analysis would not be an appropriate tool to use.

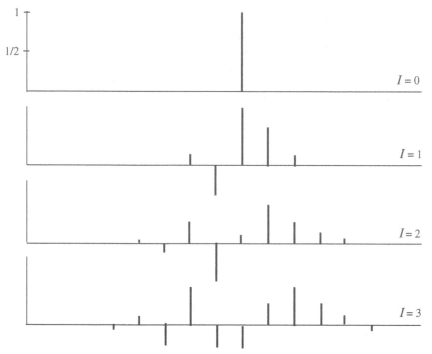

Figure 10.10 Spectrum of a Frequency Modulated Sound.

The waveform represented by (10.16) and to be recognized by devices that check frequency and amplitude of each harmonic must be transformed into the discrete form:

$$x(n) = \sum_{k=1}^{M} A_k(n) \cos\left[2\pi n(\omega_c + k\omega_m)\,T\right], \quad n = 0, 1, \ldots, N-1, \quad (10.17)$$

where N is the number of samples, and T is the interval between successive samples. In practice, the individual (isolated) tone is first digitized; then the pitch ν, assumed to be almost constant over time, is determined very accurately; and lastly, the amplitudes $A_k(n)$ and the frequency deviations $f_k(n)$ of the harmonic are calculated for all harmonics M which are implicitly defined by $A_k(n)$. The summation is terminated when the amplitude reduces to a negligibly small quantity. Thus, formula (10.17) is used to reproduce sound by driving a number of oscillators, although this may require a vast number of data with the sampling rate of 25,600 Hz for each harmonic. This process is called the *additive synthesis*, which can be compared with the subtractive synthesis for a signal defined by (10.15).

10.7 Music Synthesizers

Generally, classical music is described as experimenting with pitch, and contemporary art music as experimenting with timbre. In modern times a new generation of composers who possess a sound knowledge of signal processing techniques are creating a highly refined musical art using music synthesizers. This has been for the most part demanded by sound production in the television and film industries, where it has been synchronized with the action on the screen.

Historically, one of the earliest electric musical instruments was invented in 1876 by an American electrical engineer Elisha Gray. The oscillations of this instrument, known as the *musical telegraph*, were created and transmitted over a telegraph line by electromagnets. In 1897, Thaddeus Cahill invented Teleharmonium (or Dynamophone) using dynamo; it was capable of additive synthesis. However, the first electronic synthesizers were Theremin (1920), Ondes Martenot (1928), and Trautonium (1929). These instruments used *heterodyne circuit* to create audio frequency, and they had very limited sound synthesis capability.

The audio oscillators, audio filters, envelope controllers, and various effect units were invented during the 1930s and 1940s. For example, the *Hammond Novachord* (1939) had an electronic keyboard that used a frequency divider for sound generation, with vibratos, filter, resonator network, and a dynamic envelope controller. Other synthesizers of this period were: Hammond Solovox (1940), Ondioline (1941), Clavioline (1947), and Clarivox (1952). In Japan, Yamaha developed *Magna organ* in 1935, which had a multi-timbre keyboard based on electrically blown free reeds with pickups. In 1951–1952, RCA produced the *Electronic Music Synthesizer*, which was, in fact, more of a composition instrument, since it did not produce sounds in real time.

In 1959–1960, Harald Bode produced a modular synthesizer and sound processor. But the first commercially available modern synthesizer was produced by Robert Moorg in 1965. By the early 1980s, the musical instrument digital interface (MIDI) began to integrate and synchronize synthesizers and other electronic instruments for use in musical composition. The synthesizers started using computer software and other plugins in the 1990s. Eventually, music synthesizers produced a considerable effect on 20th century music.

MIDI is a technical standard in computerized music. It allows a wide variety of electronic musical instruments, computers and other related devices to connect and interact with one another. A single MIDI link can connect up to sixteen channels of information, each of which can be routed to separate devices. MIDI carries messages that specify notation, pitch and velocity, control signals for volume, vibrato, audio panning cues, and clock signals to set and synchronize tempo between multiple devices. MIDI technology was standardized in 1983 by a committee of music industry representatives. It is maintained by the MIDI Manufacturers Association (MMA). In 2012, Ikutaro Kakehashi and Dave Smith were rewarded a Technical Grammy Award for the development of MIDI in 1983.

Music synthesizers use modulation to synthesize waveforms with an extensive overtone spectrum using a small number of oscillators. Modulation has been introduced in Appendix E. MIDI's impact on the music industry has been very rapid. Detailed information on the MIDI, computer-digitized music and music synthesizers can be found in Manning [1994], Shuker [1994], McCutchen [1999], Holmes [2003], and Haas [2010].

11

Computerized Axial Tomography

Applications of the Radon transform to different types of computerized axial tomographic or computed tomographic imaging establish a close relationship to the Fourier transform. The applications go beyond the human body: they are being used in scanning portions of three-dimensional bodies in fields other than medicine, such as optics, oceanography, seismology, geology, oil industry, and others.

11.1 Introduction

Tomography is a compound word derived from Greek $\tau o \mu \iota'$ or $\tau o \mu o \varsigma$ meaning cut or slice. It is a noninvasive imaging technique which is used to visualize internal structures of an object without superposing over-lying and under-lying structures that usually destroy conventional projection images. For example, in a conventional chest radiograph, the heart, lungs, and the ribs are all superimposed on the same film, whereas in a computerized axial tomography (CAT), or simply a computed tomography (CT), and in positron emission tomography (PET), a slice produces each organ in its actual three-dimensional positron. Tomography has applications in many scientific fields, including physics, chemistry, astronomy, geophysics, and medicine. Whereas x-ray CT is the most familiar application, tomography can be performed even in medicine using other imaging modalities including ultrasound, magnetic resonance, nuclear medicine, and microwave techniques.

Mathematically, tomography is based on the Radon transform (Appendix C), developed by Johann Radon (1887–1956). The first application of this transform to radioastronomy was made by Bracewell [1956] and Bracewell and Riddle [1967]. Bracewell was able to obtain images of the Sun in the microwave region, although the available antenna at that time could measure the solar activity only in narrow strips. However, using several measurements of these narrow strips, he used the 'strip sums' in different directions and obtained an image of the Sun and determined those regions that emit microwave radiation.

In medicine, the pioneering work on tomography was done by William Oldendorf [1961] and D. Kuhl and E. Edwards [1963]. The mathematical and physical theory for CT scan was established by the physicist Allan MacLeod Cormack [1963, 1964]. In 1970, the English engineer Godfrey Newbold Hounsfield developed a head scanner. This scanner was later used at Atkinson Morley Hospital in Wimbledon, where the first clinical test was performed in 1972. Since then different companies have developed whole body x-ray scanners in Europe and the U.S. Cormack and Hounsfield shared the Nobel Prize in Medicine in 1979 for their diagnostic technique. Bracewell's original method used backprojection (see Appendix C) and inverted the Radon transform by convolution and Fourier transform. This technique is used by most scanning machines, as described in Natterer [1986], Engl et al. [1996], and Helgason [1980]. An overview is available in Herman and Lewitt [1979].

In all cases the aim is to estimate from measurements the distribution of a particular physical quantity in the object. The physical quantities that can be reconstructed are: (a) CT, where the distribution of linear attenuation coefficient in the slice is imaged; (b) nuclear medicine where the distribution of the radiotracer administered to the patient in the slice is imaged; and (c) ultrasonic diffraction tomography where the distribution of refractive index in the slice is imaged.

Under certain conditions, the measurements made in each modality can be converted into samples of the Radon transform of the distribution, which is then reconstructed. For example, in CT after dividing the measured photon counts by the incident photon counts and taking the negative logarithm we obtain samples of the Radon transform of the linear attenuation map. The Radon transform and its inverse provide the mathematical basis for reconstructing tomographic images from the measured projection or scattering data.

In all applications, the generated data, often called *projections*, are indirect, incomplete, and noisy. Consider data of the form

$$D_{a,b} = \int_{L_{u_a,\theta_b}} f(x,y)\,dx\,dy + \sigma\,z_{a,b}, \qquad (11.1)$$

where $L_{u,\theta} = \{(x,y) : x\cos\theta + y\sin\theta = u\}$ is a line defined by the polar coordinates (u,θ), f is the object under consideration positioned at a point (x,y), and the set $\{z_{u,\theta}\}$ represents a Gaussian white noise process with variance unity. The sets $\{u_a\}_{a=1}^{N_{\text{pos}}}$ and $\{\theta_b\}_{b=0}^{N_{\text{ang}}-1}$, where $N \in \mathbb{N}$, are the discrete samplings of position and angle in the interval $[-1,1]$ and $[0,\pi)$, respectively. A schematic representation of the sampling process for $N_{\text{ang}} = 2$, $\theta = 0$, $\pi/4$, and $N_{\text{pos}} = 9$ per angle is shown in Figure 11.1.

In tomography, an image is reconstructed from its projections along different directions. The basic mathematical approach consists in measuring a different physical quantity in each tomographic modality. Thus, for example,

(i) in CT, the number of x-ray photons transmitted through the patient along individual projection lines;

(ii) in nuclear medicine, the number of photons emitted from the patient along individual projection lines; and

(iii) in ultrasound diffraction tomography, the amplitude and phase of scattered waves along a particular line connecting the source and the detector.

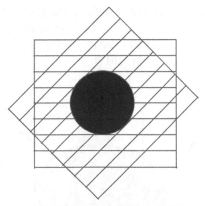

Figure 11.1 A Tomographic Sampling Pattern.

Tomographic images are produced by the resolution of an inverse problem, and more precisely by the inversion of the Radon transform. All tomographic techniques aim at reconstructing the object f from the observed data $\{D_{a,b}\}$. In medical imaging, the machines typically use Fourier-based methods, such as filtered backprojection (FBP), together with maximum likelihood methods, expectation-maximization (EM) algorithm, iterative methods, and orthogonal series methods. However, in tomography we can observe the object only indirectly. The line integrals

$$\mathcal{R}\{f(\rho,\theta)\} \equiv \int_{L_{\theta,\rho}} f(x,y)\,dx\,dy \qquad (11.2)$$

correspond to the two-dimensional Radon transform of f. The reconstruction method, as suggested by Kolaczyk [1996], consists of two parts: first the inversion of the Radon transform in such a way as to obtain not f but the noisy wavelet coefficients of f, and then to apply wavelet shrinkage to the coefficients in order to remove noise[1]. Details of the Radon transform technique are available in Appendix C.

In tomography an image is reconstructed from measurements of its integral over straight lines on a plane, called *projections*. Since the set of projections is the Radon transform of the image, the reconstruction problem is equivalent to the inversion of the Radon transform. Let $f(x,y) \in C^\infty$, with compact

---

[1] For an introduction to wavelets, see Kythe and Schäferkotter [2005].

support, be a two-dimensional image, called the *object*. Its parallel projection $p_\theta(r)$ in a direction θ, considered in §C.2, is defined as

$$p_\theta(r) = \int_{-\infty}^{\infty} \int_{-\infty}^{\infty} f(x,y)\, \delta(x\cos\theta + y\sin\theta - r)\, dx\, dy, \qquad (11.3)$$

where δ is the Dirac delta distribution. Then the Radon transform $\mathcal{R}\{f(r,\theta)\} = p_\theta(r)$, $\theta \in [0,\pi]$ and $r \in (-\infty, +\infty)$, is presented in Figure 11.2. In fact, the two-dimensional Radon transform is given by Eq (C.8). The original inversion formula was developed by Radon [1917].

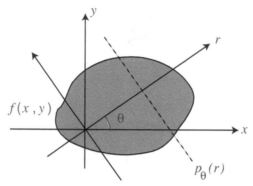

Figure 11.2 Parallel Projection $p_\theta(r)$.

11.2 Types of Tomography

There are different types of tomographic imaging. A list is provided in §11.4. We will discuss in detail some of the simple ones that have provided a platform for more advanced tomographic imaging techniques. Most of the imaging techniques depend on the Fourier transform and convolution methods.

11.2.1 Straight-Ray Imaging. The simplest types of tomographic images are produced by illuminating an object of interest using some form of energy, like x-rays, microwaves, or ultrasound, provided that the object permits the energy pass through it. This process is known as the *straight-ray propagation*. The corresponding measurements involve either the amplitude or the time of approach of the received signal. Such measurements are then used to estimate a line integral of the object's attenuation coefficient or its refractive index. In cases where the energy does not travel in straight line, certain algebraic techniques or diffraction tomographic methods are used.

11.2.2 Series Expansion Method. One of the reconstruction methods is the series expansion method. It originates from the expression and the resolution of the inverse Radon transform problem in the discrete case (§C.4). Let the image be represented by its coordinates f_i in some basis. Then the problem can be expressed as finding the solution of the system of equations

$$\mathbf{R}\mathbf{f} = \mathbf{p},\tag{11.4}$$

where $\mathbf{f}$ is the vector of the image coordinates f_i, $\mathbf{p}$ is the vector of data, and $\mathbf{R}$ is a two-dimensional matrix representing an approximation of the Radon transform. Since the matrix $\mathbf{R}$ is very large, the system (11.4) is generally solved by iterative methods. One of the most-used series expansion methods is the algebraic reconstruction technique (ART), since its implementation and interpretation is simple because only one new projection is processed at each iteration. The only drawback is that this technique may diverge if the data is inconsistent due to noisy projections.

Another useful method is to solve the least square system

$$\left(\mathbf{R}^T\mathbf{R} + \lambda\mathbf{W}\right)\mathbf{f} = \mathbf{R}^T\mathbf{p},\tag{11.5}$$

where λ is a regularization parameter, $\mathbf{W}$ is a matrix that introduces some a priori values in the system. Note that $\mathbf{R}$ and $\mathbf{R}^T$ represent the discrete Radon transform and backprojection operator, respectively. In the case of regular sampled data, the reconstruction becomes a two-dimensional filtering of the backprojection data.

11.2.3 Sinogram. The sinogram is a discrete sampling of the Radon transform. Each row of the sinogram contains the line integral of f, of the form (C.3), at a fixed angle θ and such that different rows correspond to different, yet equally spaced, angles. According to the Fourier slice theorem (§3.7), the one-dimensional Fourier transform of $R(\rho, \theta)$ with respect to the variable θ is the line through the two-dimensional Fourier transform of f, i.e.,

$$\mathcal{F}_1\left(R(\rho, \theta)\right) = \mathcal{F}_2(f)(\alpha\theta) = F(\alpha\theta).\tag{11.6}$$

Using this result and the polar Fourier inversion formula (as in (11.7)), we can construct the Radon transform $R(\rho, \theta)$ using the Fourier reconstruction method, commonly known as the *filtered backprojection*. The function $f(x, y)$ is essentially bandlimited on a circle of radius r. According to Olson [1996: 271], using Eq (11.6) and a polar Fourier inversion formula, the Radon transform $R(\rho, \theta)$ can be related to $f(\mathbf{x})$, where $\mathbf{x} = (x, y, z)$, by the reconstruction formula

$$f(\mathbf{x}) = \int_{S_1}\int_{\mathbb{R}} F(\alpha\theta)|\alpha|\, e^{i\,(\mathbf{x}\cdot\alpha\theta)}\, d\alpha\, d\theta$$
$$= \int_{S_1}\int_{\mathbb{R}} F_1\left(R(\rho, \theta\cdot\mathbf{x})(\alpha)|\alpha|\right)\, e^{i\,(\mathbf{x}\cdot\alpha\theta)}\, d\theta,\tag{11.7}$$

where S_1 denotes the selected slice of the object $f(\mathbf{x})$. Although this formula yields the original function from the Fourier transform, the drawback in this application is that the Fourier transform must be sampled on a radial grid and interpolation must be applied in order to use an FFT on a rectangular grid.

11.2.4 B-Scan Imaging. Sometimes the object's physical restraints do not permit the straight-ray transmission tomography. Examples of such cases involve cardiovascular imaging using ultrasound. In these cases, large impedance discontinuities at tissue-bone or air-tissue interfaces impede measurements of the transmitted signal. Most medical ultrasonic imaging is, therefore, performed using reflected signals, which leads to B-scan imaging and reflection tomography. This methods provides a constructive quantitative cross-section image from the reflection data. It leads to a technique where both transmission as well as reflection tomography is performed using the same imaging instruments and collecting the data. The same technique also works for the B-scan imaging.

We will first discuss the B-scan imaging ('B' for the beam) where a small beam of ultrasonic energy illuminates the object and produces an image that shows the reflected signal as a function of time and direction of the beam. Assume that the inhomogeneous nature of the object is represented by a spatial isotropic scattering function $s(x, y)$, defined at a point (x, y), and called the *object reflectivity function*. This function provides a measure of the portion of the local transmitted field that is reflected back to the receiver. Since it involves the scattering process which also represents the direction of the illumination and the direction in which the reflection is measured, we will avoid the rigorous analysis and confine ourselves to the localized region of imaging. As shown in Figure 11.3(a), a B-scan is simply an example of radar imaging in which the beam is confined to a narrow region along a line. Since the amplitude of the field is not decaying in this region, it can be represented as a function of a single variable along this line. As the illuminating wave propagates in short pulses, it will provide a direct mapping between the time at which a portion of the reflected waves is received and the distance it traveled into the object.

Mathematically, the received wavefront is a convolution of the input waveform $p(t)$ (which represents a pulse) and the object's reflectivity. Then the incident field can be represented as

$$\psi_i(x, y) = \begin{cases} p\left(t - \dfrac{x}{c}\right) & \text{for } y = 0, \\ 0 & \text{elsewhere,} \end{cases} \tag{11.8}$$

where c is the wave speed. This function is a pulse that propagates along the x-axis with speed c and is orthogonal to the face of the transducer. Figure 11.3(b) represents the pressure along the lines $t = t_n$ for $n = 1, 2, \dots, N$, which is orthogonal to the x-axis. At some interior point (x, y) of the object, a portion of the incident field $\psi_i(x,y)$ is scattered (reflected) back toward the transducer. This means that the amplitude $a(x,y)$ of the scattered field at the scatterer point is approximately equal to

$$a(x,0) = s(x,0)\, p\!\left(t - \frac{x}{c}\right).\qquad(11.9)$$

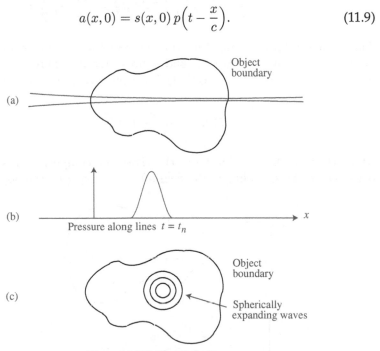

Figure 11.3 B-Scan Imaging.

Because of the propagation distance involved and attenuated, the wave is delayed by x/c. Figure 11.3(c) shows the spherically expanding waves produced by a single scatterer at an interior point of the object. Thus, the reflected field is diverging. By the law of conservation of energy, the amplitude attenuation as a result of expanding waveform is proportional to $1/\sqrt{x}$. This implies that the energy density will decay as $1/x$. Then, integrating over the boundary of a circle enclosing the scatterer point, the total energy will remain invariant regardless of the radius of the circle. Thus, the field F_s due to the scattering at x is given by

$$\psi_s = s(x,0)\, p\!\left(t - \frac{x}{c} - \frac{x}{c}\right)\frac{1}{\sqrt{x}}.\qquad(11.10)$$

The total field $F_s(t)$ at the receiver is obtained by integrating Eq (11.10) with respect to x:

$$\psi_s(t) = \int p\!\left(t - 2\frac{x}{c}\right)\frac{s(x,0)}{\sqrt{x}}\,dx.\qquad(11.11)$$

Taking the simplest case of a pulse as the Dirac δ-function, the scattered field can be approximated by

$$\psi_s(t) = \int \delta\left(t - 2\frac{x}{c}\right) \frac{s(x,0)}{\sqrt{x}}\, dx = \sqrt{\frac{c}{2t}}\, s\left(\frac{tc}{2}, 0\right). \qquad (11.12)$$

This result establishes a direct relation between the scattered field at t and the object's reflectivity at $x = tc/2$, and is presented in Figure 11.4. Let $\hat{s}(x,0)$ denote an estimate of the reflectivity function $s(x,0)$ at $x = tc/2$. Then

$$\hat{s}(x,0) = \sqrt{\frac{4x}{c^2}}\, F_s\left(\frac{2x}{c}\right), \qquad (11.13)$$

where the term $4x/c^2$ is known as the *time gain compensation* since it compensates for the spreading of the field after scattering by the object.

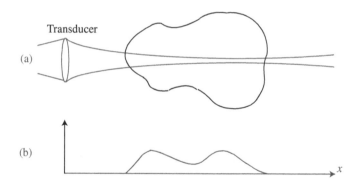

Reflected field measured at transducer

Figure 11.4 Pulse-Illuminated Field.

The B-scan involves two parameters: the duration of the incident impulse and the width of the beam. There are two kinds of resolutions: one known as the *range resolution* which is determined by the duration of the pulse, and the other the *lateral resolution* which is controlled by the width of the beam. For example, the range resolution is determined from (11.11), which gives the field measured at the point $(0,0)$ due to a single scatterer of unit strength at $x = x_0$ as

$$\psi_s(t) = \frac{1}{\sqrt{x_0}}\, s\left(t - \frac{2x_0}{c}\right).$$

Substituting this into Eq (11.13), we get the image of the object's reflectivity as

$$\hat{s}(x,0) = \sqrt{\frac{4x}{c^2}}\, \psi_s\left(\frac{2x}{c}\right) = \sqrt{\frac{4x}{c^2 x_0}}\, s\left(\frac{2x}{c} - \frac{2x_0}{c}\right). \qquad (11.14)$$

Obviously, from (11.14) we see that an incident pulse of t seconds duration provides an estimate that is tc units wide. If we define a modified reflectivity function s^* by

$$s^*(x, y) = \frac{1}{\sqrt{x}} s(x, y),$$

then the scattered field at the origin can be written as the convolution

$$\psi_s(t) = \int p\left(t - \frac{2x}{c}\right) s^*(x, 0) \, dx,$$

which in the Fourier domain becomes

$$\Psi_s(\omega) = P(\omega) S^*\left(\frac{2\omega}{c}, 0\right). \tag{11.15}$$

Thus, the Fourier transform of the received field by $P(\omega)$ can be used to determine $S^*\left(\frac{2\omega}{c}, 0\right)$ from the formula (11.15), or from

$$S^*\left(\frac{2\omega}{c}, 0\right) = \frac{\Psi_s(\omega)}{P(\omega)}. \tag{11.16}$$

However, this formula does not work in most cases because there can be frequencies for which $P(\omega) = 0$, and this causes instabilities in the division in (11.16).

11.2.5 Reflection Tomography. The instabilities encountered in the B-scan techniques can be eliminated using reflection tomography. Whereas in the straight-ray transmission tomography both narrowband and broadband signals are used, reflection tomography uses only the broadband signals (short pulses). Thus, the important question is to compare the reflection tomography with the B-scan imaging. Reflection tomography is based on the evaluation of line integrals of the object reflectivity function. Consider a single point transducer illuminating an object with a very wide fan-shaped beam. As noted above, if the incident field is an impulse in the time domain, then the received signal at time t represents the total of all reflections at a distance tc from the transducer. All points at the same distance from the transmitter/receiver lie on a circle. Thus, the reflection tomography will measure line integrals over circular arcs, as shown in Figure 11.5, where the incident field covers the entire object. Then, if the transducer is moved over a plane, or alternatively on a sphere wrapped around the object, we will be able to collect enough line integrals to reconstruct the entire object. This method was developed by

Norton and Linzer [1979a,b].

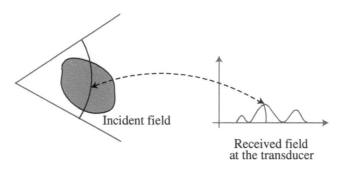

Figure 11.5 Circular Transducer Arrays.

Assuming that the transducer is set in the plane $x = 0$, a plane wave transducer when excited by the waveform $p(t)$ generates a field that is equal to

$$\psi_i(x, y, t) = p\left(t - \frac{x}{c}\right), \quad x > 0, \tag{11.17}$$

At the receiver, the situation becomes a bit complicated. If $\psi_r(x, y, t)$ is a scattered field, the generated signal is proportional to the integral of this field. Omitting the constant of proportionality, the electrical received signal $p(t)$ is given by

$$p_r(t) = \int \psi_r(0, y, t) \, dy. \tag{11.18}$$

If the field at points away from the transducer is known, the received waveform can be described provided that we know how the waves propagate back to the transducer. For this purpose, first we assume that there exists a line of reflectors at $x = x_0$ which reflect a portion $f(x_0, y)$ of the field. The scattered field at the line $x = x_0$ can then be defined as the product of the incident field and the reflectivity parameter:

$$\psi_s(x_0, y, t) = \psi_i(x_0, y, t) f(x_0, y) = p\left(t - \frac{x}{c}\right) f(x_0, y). \tag{11.19}$$

Next, the Fourier transform of the field will yield the field at the transducer by propagating each plane wave to the transducer face. This is accomplished by finding the spatial and temporal Fourier transform of the field at the line of reflectors, thus giving

$$\Psi_s(f_y, \omega) = \int_{-\infty}^{\infty} \int_{-\infty}^{\infty} \psi_s(x_0, y, t) \, e^{-i f_y y} \, e^{i \omega t} \, dy \, dt, \tag{11.20}$$

so that the function $\Psi_s(f_y, \omega)$ represents the amplitude of the plane wave propagating with direction vectors $\left(-\sqrt{(\omega/c)^2 - f_y^2}, \, f_y\right) \mathbf{j}$. Note that in Eq

(11.20) the transform has a phase factor of $e^{-i f_y y}$ in the spatial domain, and a factor $e^{i \omega t}$ in the temporal domain, where f_x, f_y denote the frequency in x and y direction. With this plane wave expansion, each plane wave can now be propagated to the transducer face.

For example, consider an arbitrary plane wave $\psi(x, y) = e^{i (f_x x + f_y y)}$. In view of (11.18), it can be shown that the electrical signal is zero for all plane waves for $f_y \neq 0$, since $\int_{-\infty}^{\infty} e^{i f_y y} dy = \delta(f_y)$. In the case when $f_y = 0$, the plane waves traveling orthogonal to the transducer face will be delayed due to the propagation distance x_0. This represents a factor of $e^{i \omega (x_0/c)}$ in the frequency domain. Since the electrical response due to a unit amplitude plane wave is $p_r(\omega, f_y) = \delta(f_y) e^{i \omega (x_0/c)}$, we can add each plane wave at frequency ω in (11.20), yielding the total electrical response due to the scattered fields from the plane $x = x_0$ and obtain in the Fourier domain

$$P_r(\omega) = \Psi_s(0, \omega) e^{i \omega (x_0/c)}, \tag{11.21}$$

or, in the time domain

$$p_r(t) - \frac{1}{2\pi} \int_{-\infty}^{\infty} \Psi_s(0, \omega) e^{i \omega (x_0/c)} e^{-i \omega t} d\omega. \tag{11.22}$$

By substituting (11.17), (11.20), and (11.19) into (11.22), we obtain the received signal

$$p_r(t) = \frac{1}{2\pi} \int_{-\infty}^{\infty} \int_{-\infty}^{\infty} \int_{-\infty}^{\infty} \psi_s(x_0, y, \tau) e^{i \omega (x_0/c)} e^{-i f_y y} e^{i \omega \tau} e^{-i \omega t} dy \, d\tau \, d\omega, \tag{11.23}$$

or, equivalently,

$$p_r(t) = \frac{1}{2\pi} \int_{-\infty}^{\infty} \int_{-\infty}^{\infty} \int_{-\infty}^{\infty} \psi_i(x_0, y, \tau) f(x_0, y) e^{i \omega (x_0/c)} e^{i \omega \tau} e^{-i \omega t} dy \, d\tau \, d\omega,$$

$$= \frac{1}{2\pi} \int_{-\infty}^{\infty} \int_{-\infty}^{\infty} \int_{-\infty}^{\infty} p\left(\tau - \frac{x_0}{c}\right) f(x_0, y) e^{i \omega (x_0/c)} e^{i \omega \tau} e^{-i \omega t} dy \, d\tau \, d\omega. \tag{11.24}$$

By interchanging the order of integration, we finally get the measured signal due to a single line of the scatterer at $x = x_0$ as

$$p_r(t) = p\left(t - \frac{2x_0}{c}\right) \int_{-\infty}^{\infty} f(x_0, y) dy. \tag{11.25}$$

This signal is similar to that of the B-scan, since the transmitted pulse is convoluted with the reflectivity of the object, as in the B-scan, with the only difference that in each case the reflectivity is summed over a portion of the object illuminated by the incident field. In fact, by using a common signal source and adding all electrical signals, we use an array of transducers to generate a plane wave for reflection tomography, as shown in Figure 11.6.

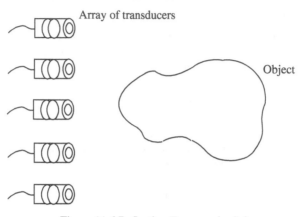

Figure 11.6 Reflection Tomography Scheme.

Limitations to reflection tomography are mostly due to the fact that it does not provide information about an object at low frequencies. However, this problem can be rectified to a certain extent by extrapolation using the Gerchberg-Papoulis algorithm available in Gerchberg [1974] and Papoulis [1975]. It is an iterative process that combines the Fourier transform with independent domain (spatial) constraints. This algorithm works as follows: suppose the reflection tomography yields $F_0(u, v)$ as an estimate of the Fourier transform of an object's cross-section. Its inverse $f_0(x, y)$ will then be the image. The function $F_0(u, v)$ is known in a doughnut-shaped region of the (u, v)-space, which we denote by D_f. If $f(x, y)$ is the true cross-section and $F(u, v)$ the true corresponding Fourier transform, we run into the following problem:

$$F(u, v) = \begin{cases} F_0(u, v) & \text{if } (u, v) \in D_f, \\ ? & \text{elsewhere.} \end{cases}$$

Now, if we impose the constraint that the object is spatially limited, i.e.,

$$f(x, y) = \begin{cases} ? & \text{if } (x, y) \in D_s, \\ 0 & \text{elsewhere,} \end{cases}$$

where D_s denotes the maximum known object cross-section, then the inverse Fourier transform of the known data $F_0(u, v)$ will provide a reconstruction that will not be spatially limited. The Gerchberg-Papoulis algorithm simply finds a reconstruction $f^*(x, y)$ that satisfies the space constraint and has a Fourier transform $F^*(u, v)$ that is equal to the one provided by the reflection tomography. Hence, this algorithm works as follows: let $F_0(u, v)$ be an initial estimate. Then a better estimate of the object is determined by finding the IFT of $F_0(u, v)$ and setting the first iteration as

$$f_1(x, y) = \begin{cases} \text{IFT}\{F_0(u, v)\} & \text{if } (x, y) \in D_s, \\ 0 & \text{elsewhere.} \end{cases}$$

The next iteration is obtained by taking Fourier transform of $f_1(x, y)$ and then constructing a composite function in the frequency domain as

$$F_1(u, v) = \begin{cases} F_0(u, v) & \text{if } (u, v) \in D_f, \\ FT\{f_1(x, y)\} & \text{elsewhere.} \end{cases}$$

Then we construct the next iterate $f_2(x, y)$ by first taking IFT of $F_1(u, v)$ and setting to zero any values that are outside the region D_s. This iterative process is continued to yield $f_3, f_4, \ldots$ until the difference between two successive approximations is less than the prescribed tolerance. The flow chart of this algorithm is given in Figure 11.7.

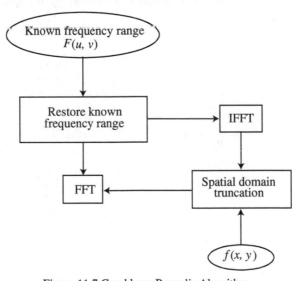

Figure 11.7 Gerchberg-Papoulis Algorithm.

11.3 Magnetic Resonance Imaging

This technique, also known as MRI, deals with a noninvasive medical imaging modality which has become an important part of diagnostic radiology complementing the older techniques like ultrasound and x-ray computed tomography (CT). The MRI process has the capability to capture images of motion, temperature effects, and chemical effects. It provides high-contrast imaging of soft tissues without using contrast agents. However, it is limited in resolution and speed of image acquisition. All techniques used in MRI encode information about an object under examination in radio frequency signals generated by the stimulated precession of nuclear magnetic moments only in a portion of the object. This encoding process provides measurement of projections of functions representing various sample properties.

The difference between MRI and other techniques needs some clarification. The ultrasound process, though inexpensive, is done in real time, but it cannot be used for structures that are deep inside the object, or where the air or bone surrounds the structure of the object. In medical tomography, ultrasound is mainly used in the heart or the abdomen. The x-ray CT techniques provide excellent spatial resolution, but fail to provide soft tissue contrast, which is determined by the electron density of the object. On the other hand, MRI contrast is determined by the relaxation parameters of the water in the tissue. For example, according to Droege et al. [1983], whereas the white matter of the brain has a 12% lower CT number than the gray matter, the white matter has a 14% higher signal than gray matter in MRI examination. As compared to the CT, the MRI gets the images at arbitrary positions and orientations, although MRI is sometimes restricted by its spatial resolution and long imaging time.

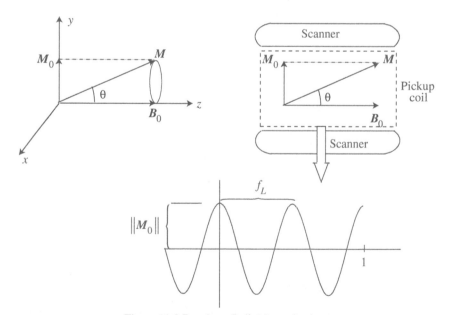

Figure 11.8 Resultant Bulk Magnetization.

The nuclear magnetic resonance (NMR) plays an important role in a typical imaging technique. An object under examination, e.g., a human body, is placed in a strong homogeneous magnetic field $\mathbf{B}_0$ inside the scanner, with components B_x, B_y, B_z along the x, y and z direction, respectively, so that we have $\mathbf{B}_0 = B_x\mathbf{i} + B_y\mathbf{j} + B_z\mathbf{k}$, where $\mathbf{i}, \mathbf{j}, \mathbf{k}$ are the unit vectors along the coordinate axes. Let the z-axis be taken along the scanner axis. As the object is placed in the scanner's field, it causes an alignment of the magnetic moments of the hydrogen nuclei in the object along the direction of the magnetic field $\mathbf{B}_0$. It creates a resultant bulk magnetization $\mathbf{M}$ in the form of a vector sum of the aligned nuclear magnetic moments $\{M_x, M_y, M_z\}$. According to Healy and Weaver [1996: 300], the bulk magnetization vector $\mathbf{M}$ can be excited by a radio-frequency electromagnetic pulse, which results in its precession at a characteristic frequency, known as the *Larmor frequency*, marked as f_L in Figure 11.8, around the external field $\mathbf{B}_0$. A 'pickup coil' detects a component of the resulting oscillating magnetic field and produces a signal which carries the information about the properties of the object. In this process, the frequency of the precession of $\mathbf{M}$ and of the resulting output signal is proportional to the strength of the external magnetic field, such that $f_L = \gamma\|\mathbf{B}_0\|$, where γ is the proportionality constant. Note that f_L is the natural resonant frequency of the spin system that circulates the object entirely. The term 'magnetic resonance' in NMR means that the spin system absorbs and emits radiation at the characteristic frequency f_L determined by the magnetic field $\mathbf{B}_0$. For example, $f_L \approx 60$ MHz for hydrogen nuclei in a typical scanner.

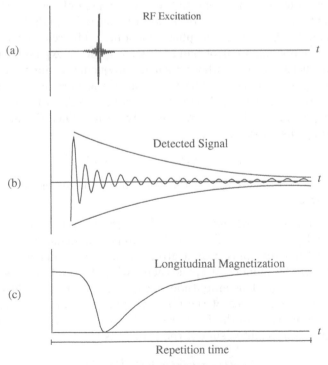

Figure 11.9 Sequence of MR Measurements.

A typical sequence of operations of MRI measurements is shown in Figure 11.9 in three time lines. Figure 11.9(a) presents the beginning of this sequence when an RF excitation at the Larmor frequency f_L is used as the input to excite the spins in the object. These spins generate a decaying sinusoid (i.e., a free induction decay, FID) signal (shown in (b)) at the frequency f_L. The decay is caused partly by the eventual realignment of the spins with the static field, as indicated by the recovery in the longitudinal magnetization (as in (c)). This recovery determines the recovery time which must be allowed before the spin is excited again. However, during this intermittent period no signal measurements are available because there is another faster decay occurring due to the loss of coherence among the spins.

All nuclei generate signals in the object at the corresponding Larmor frequency after excitation. Thus, the output signal exhibits the properties of the object in its entirety. For a successful imaging, the variation of the property of interest must be mapped, in most cases by determining a map of the weighted spin density over the entire object. This can be accomplished by perturbing the external magnetic field so as to create variation in its strength throughout the object. As a result the nuclei at different positions will experience a slightly different field and have a Larmor frequency that depends on that position.

In order to map tissue properties across a region of the object, there are two different ways to encode information on the RF signal: (i) selective excitation, and (ii) frequency and phase encoding. The selective excitation is obtained by supplying a narrowband RF excitation, as explained above. The stimulated nuclei are those whose resonant frequency is almost the same as that of the excitation signal. The variation of the Larmor frequency across the excited region is provided by special gradient coils in the scanner. The natural frequencies of the spin vary linearly with z since the magnetic field $\mathbf{B}_0$ acts along the z-axis. Thus,

$$f_L = f_L(z) = \gamma\left(B_z + G_z z\right), \tag{11.26}$$

where G_z is a constant reflecting the gradient of the magnetic field strength along the z-axis.

At the end of the RF excitation, the z gradient is turned off and the output signal is measured. This yields a sinusoidal signal, whose amplitude contributed from the spins near a given plane $z = z_0$ depends on the two quantities: (i) the number of spins there, $d(z_0)\Delta z$, and (ii) the excitation amplitude at $z = z_0$. This amplitude $A(z_0)$ is determined by the quantity of the RF excitation provided at the point z_0, which is approximately equal to the Fourier component of the RF pulse at frequency $\gamma\left(B_z + G_z\right)$. Hence, the output signal is a weighted sum of the spin density:

$$s_A(t) = \int A(z)d(z)\, e^{i\gamma B_z t}\, dt. \tag{11.27}$$

As an example, to reconstruct the density in z let us choose a basis of amplitude profiles using selective excitation encoding. If we are interested in mapping the variation of spin density along the z-axis of the object, let us selectively excite a succession of slabs along that axis and record the signal strength from each slab. Assuming that profiles are ideal boxcar functions $\phi_k(z)$ partitioning the z-axis into slabs of width Δz:

$$\phi_k(z) = \begin{cases} 1 & \text{if } k\Delta z < z < (k+1)\Delta z, \\ 0 & \text{otherwise,} \end{cases}$$

these functions correspond to sinc-like RF excitation applied with gradient G_z turned on. This gradient is turned off at the end of the RF pulse, and at that time the output signal measures are proportional to

$$s_k(t) = \int \phi_k(z)d(z)\, e^{i\gamma B_z t}\, dt \quad \text{successively for } k = 1, 2, \ldots, N. \quad (11.28)$$

In the case of frequency encoding, if we are interested in mapping the spin density in the three-dimensional space, we must do the following: after initiating the selective excitation along the z-axis, as described above, the magnetic field strength must be made to vary linearly in one direction across the excited region, say the x direction, during the time interval when the output signal is being measured. Then the Larmor frequency in the excited region of the object will vary linearly with x, yielding $f_L = f_L(x) = \gamma\,(\mathbf{B}_0 + F_x x)$, where G_x is a constant for the gradient of the magnetic field strength along the x-axis, as shown in Figure 11.10. The demodulated signal output with the phase profile $\Phi(z)$ in this case is

$$s(t) = \int \int \int \Phi(z)d(x, y, z)\, e^{i\gamma G_x x t}\, dx\, dy\, dz, \quad (11.29)$$

where Φ is the amplitude of the selective excitation.

In the case of phase encoding, suppose there is a linear phase variation along the y-axis of the excited slice before the frequency encoding and signal measurement occur. The signal output is then a modified form of the signal output defined by (11.28), i.e., it is given by

$$s_\theta(t) = \int \int \int \Phi(z)\rho(x, y, z)\, e^{i\gamma G_x x t}\, e^{i\theta y}\, dx\, dy\, dz, \quad (11.30)$$

where θ is the constant phase gradient. As an example, if the phase profile $\Phi(z)$ is associated with a thin slice at $z = z_0$, then the output signal provides the values of the two-dimensional spatial Fourier transform $\mathcal{F}\{\rho_{z_0}(\omega_x, \omega_y)\}$ of the spin density of that slice along one line segment through the two-dimensional spatial frequency plane $\omega_x = \gamma G_x t$, $\omega_y = \theta$, where t varies over the time interval taken by the measurement.

11.3.1 Nonlocality of Radon Inversion. Fourier analysis is based on the fact that the function f can only be supported on a localized region if its Fourier transform is very smooth or possesses finitely many smooth derivatives. Thus, the function cannot be well-localized if the Fourier transform is discontinuous or has a discontinuous derivative. The reconstruction formula (11.30) is seriously flawed because the inverse Fourier transform of a function like $|\alpha|b(\alpha)$ is not locally supported.

One reason for this problem is that the function $|\alpha|$ is not differentiable at the origin, and the cutoff introduces discontinuities at the points $\pm r$ in the neighborhood of the origin. According to Olson [1996: 273], the Fourier reconstruction must be replaced by the so-called *filtered backprojection* method, which is another reconstruction technique. Assuming that the function f is essentially bandlimited on a circle of radius r, we introduce a characteristic function $b(\alpha)$ (which is a window) into Eq (11.7) and get

$$
\begin{aligned}
f_b(\mathbf{x}) &= \int_{S_1} \int_{\mathbb{R}} F_1\big(R(\rho, \theta \cdot v\mathbf{x})\big)\,(b(\alpha)|\alpha|)\, e^{i\,\alpha(\mathbf{x}\cdot\theta)}\, d\alpha\, d\theta \\
&= \int_{S_1} R(\rho, \theta \cdot v\mathbf{x}) \star F_1^{-1}(b(\alpha)|\alpha|)\, d\theta \\
&= \int_{S_1} \int R(\theta, t) F_1^{-1}(b(\alpha)|\alpha|)\,(\theta \cdot \mathbf{x} - t)\, dt\, d\theta.
\end{aligned}
\tag{11.31}
$$

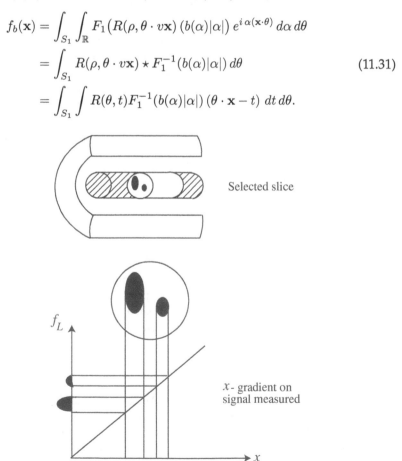

Selected slice

f_L

x- gradient on
signal measured

x

Figure 11.10 Frequency Encoding.

In this formula the window $b(\alpha)$ must satisfy two conditions: (i) it must be chosen to agree with the essential bandlimit of f so that f_b provides a good approximation to f, and (ii) the window must represent a mollification of the inverse Radon transform, which remains an unbounded operator without the addition of the window (Natterer [1986: 222]).

A simple algorithm for this reconstruction process is as follows: (i) Filter the line integrals, at each angle, using the convolution filter generated by $|\alpha|b(\alpha)$; (ii) then backproject the result obtained in (i) for every angle θ, as shown in Figure 11.11. This will yield an accurate reconstruction of the image. A useful window is $b(\alpha) = \chi_{(-r,r)}(\alpha)$, where r is chosen to agree with the essential bandlimit of f, and $\chi_{(-r,r)}$ is the characteristic function on the interval $(-r, r)$.

Figure 11.11(a) shows the data gathered just from two angles, although in practice this is done for several hundreds of angles. Then the data is filtered and backprojected as shown in Figure 11.11(b).

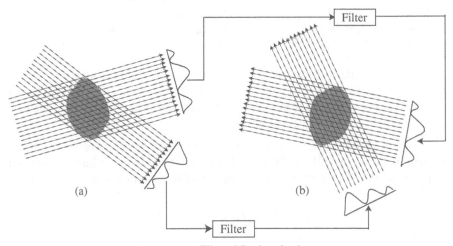

Figure 11.11 Filtered Backprojection.

11.4 Modern Tomography

The modern efforts in tomography consist in gathering projection data from all possible different directions and analyzing that data through a tomographic reconstruction software. Details of the algorithm and the related software are available in Herman [2009]. To obtain a tomographic image, different types of signal acquisition are used, as listed in Table 11.1.

Modern usage of the term 'volume imaging' is becoming more prevalent than the word 'tomography'. However, in most cases in clinics, the output in the form of two-dimensional slice image is more desirable. There are many reconstruction algorithms available these days. Most of them fall under two categories: (i) filtered backprojection (FBP), and (ii) iterative reconstruction (IR). These procedures are, however, prone to yield mostly inaccurate results; they represent a compromise between accuracy and computation time. Although FBP requires fewer computational resources, IRs generally produce fewer errors in reconstruction at a higher computation time (cost).

Table 11.1 Types of Signal Acquisition.

Energy Source	Type of Tomogram
X-rays	CT and angiography
Gamma rays	SPECT
Radio-frequency waves	MRI
Electron-positron annihilation	PET
Electrons	Electron tomography or 3D TEM
Muons	Muon tomography
Ions	Atom probe
Magnetic particle	Magnetic particle imaging

Although MRI and ultrasound generate cross-sectional images, they do not acquire data from different sources. In MRI the spatial information is obtained using magnetic fields, but in ultrasound it is obtained by focusing and targeting a pulsed ultrasound beam.

A recent development is a new technique called *synchrotron x-ray tomographic microscopy* (SRXTM) that allows for detailed three-dimensional scanning of fossils. Different types of tomographic techniques are presented in Table 11.2.

Table 11.2 Types of Tomograms.

Name of Tomography	Source of Data	Abbreviation
Atom probe tomography	Atom probe	APT
Computed tomography	Visible light	CTIS
Imaging spectrometer	Spectral imaging	
Confocal microscopy	Laser scanning	LSCM
Laser confocal screening	Confocal microscopy	
Cyro-electron	Cyro-electron	Cyro-ET
Electric capacitance	Electric capacitance	ECT
Electrical resistivity	Electric resistivity	ERT
Electric impedance	Electric impedance	EIT

(continued)

Name of Tomography	Source of Data	Abbreviation
Electron tomography	Electron attenuation/scatter	EI
Functional magnetic resonance	Magnetic resonance	fMRI
Magnetic induction	Magnetic induction	MIT
Magnetic resonance	Nuclear magnetic moment	MRI or MRT
Neutron tomography	Neutron	
Ocean acoustic tomography	Sonar	
Optical coherence tomography	Interferometry	OCT
Optical diffusion tomography	Absorption of light	ODT
Optical projection	Optical microscope	OPT
Photoacoustic imaging	Photoacoustic spectrometry	PAT
Positron emission	Positron emission	PET
Positron emission	Positron emission & x-ray	PET-CT
Quantum tomography	Quantum state	
Single photon emission CT	Gamma rays	SPECT
Seismic tomography	Seismic waves	
Thermoacoustic imaging	Photoacoustic spectroscopy	TAT
Ultrasound-modulated optical	Ultrasound	UOT
Ultrasound transmission	Ultrasound	
X-ray tomography	X-ray	CT, CTScan
Zeeman-Doppler imaging	Zeeman effect	

Modern research focusses on discrete tomography and geometric tomography. They deal with the reconstruction of objects that are discrete, such as crystals, or homogeneous. This research again centers around the reconstruction methods.

11.4.1 Nuclear Medicine. In nuclear medicine, radioactive substances are used in the diagnosis and treatment of disease. The procedure involves combining radionuclides with other elements to form chemical compounds known as *radiopharmaceuticals*, which, once administered to the patient, localize to specific diseased organs or cells. Diagnostic medical imaging is performed after the radiopharmaceuticals are taken internally. The imaging procedure, which is unlike a diagnostic x-ray, is to capture and form images from the radiation emitted by the radiopharmaceuticals. The following are some of the imaging techniques involved in diagnostic nuclear medicine:

(i) 2-D scintigraphy, also known as *scint*, in which the internal radionuclides are used to create two-dimensional images, as explained in Taylor et al. [2000].

(ii) 3-D SPECT, which is a three-dimensional tomographic technique that uses gamma camera data from several projections and is then reconstructed in different planes (see Mark et al. [2007]).

(iii) Positron emission tomography (PET) uses coincidence detection for imaging. A maximum intensity projection (MIP) of a whole-body PET is usually performed after an intravenous injection of 371 MBq of 18F-FDG (one hour prior to imaging).

Nuclear medicine tests are different from other imaging techniques. In these diagnostic tests the aim is to see the physiological function of the system as opposed to the traditional anatomical imaging such as CT or MRI. Nuclear medicine is organ- or tissue-specific. Besides, nuclear medicine studies the whole-body scans or PET/CT scans, gallium scans, indium white blood cell scans, and octreotide scans. However, there are risks from low-level radiation exposures, which are not yet properly understood. A universally adopted cautious approach is to keep all human radiation exposures 'As Low As Reasonably Practicable' (ALARP).

A

Tables of Fourier Transforms

The Fourier transform of a signal $x(t)$ and the inverse Fourier transform are defined as

$$\mathcal{F}(\omega) \equiv X(\omega) = \int_{-\infty}^{\infty} x(t)\, e^{-i\omega t}\, dt, \quad \mathcal{F}^{-1}\{X(\omega)\} \equiv x(t) = \frac{1}{2\pi} \int_{-\infty}^{\infty} X(\omega) e^{i\omega t}\, d\omega,$$

or

$$X(f) = \int_{-\infty}^{\infty} x(t)\, e^{-2\pi i f t}\, dt, \quad x(t) = \int_{-\infty}^{\infty} X(f)\, e^{2\pi i f t}\, dt,$$

where, as pointed in the footnote in §3.4.1, ω is the radian frequency, and f the cyclic frequency ($\omega = 2\pi f$).

A.1 Complex Fourier Transform Pairs

	$x(t)$	$\mathcal{F}\{x(t)\} = X(f)$
1.	$x'(t)$	$if\, X(f))$
2.	$x^{(n)}(t)$	$(-if)^n\, X(f))$
3.	$x(at), \quad a > 0$	$\dfrac{1}{a} X\left(\dfrac{f}{a}\right)$
4.	$x(t - a)$	$e^{iaf}\, X(f)$
5.	$\delta(t)$	1
6.	$\delta(t - a)$	e^{-iaf}
7.	$e^{-at} H(\pm t), \quad a > 0$	$\dfrac{1}{a \pm if}$
9.	$e^{-a|t|}, \quad a > 0$	$\dfrac{2a}{a^2 + f^2}$

	$x(t)$	$\mathcal{F}\{x(t)\} = X(f)$
10.	$e^{-at^2}, \ a > 0$	e^{-af^2}
11.	$e^{i\,at}$	$2\pi\,\delta(f - a)$
12.	$c, \ -\infty < t < \infty$	$2\pi c\,\delta(f)$
13.	$\dfrac{1}{|t|}$	$A - 2\log|f|, \ A$ const
14.	$H(t)$	$\dfrac{1}{if} + \pi\,\delta(f)$
15.	$H(t \pm a)$	$\dfrac{e^{\pm iaf}}{if} + \pi\,\delta(f)$
16.	$H(t) - H(-t)$	$-\dfrac{2i}{f}$
17.	$H(t + a) - H(t - a)$	$\dfrac{2\sin af}{f}$
18.	$|t|\,e^{-a|t|}, \ a > 0$	$\dfrac{2\left(a^2 - f^2\right)}{\left(a^2 + f^2\right)^2}$
19.	$\mathrm{sgn}(t)$	$\dfrac{2}{if}$
20.	$t^{-n-1}\,\mathrm{sgn}(t)$	$\dfrac{(-if)^n}{n!}\,(A - 2\log|f|), \ A$ const
21.	$(t - a)^{-1}$	$-i\pi e^{-i\,af}\,\mathrm{sgn}(f)$
22.	t^{-n}	$\dfrac{(-i)^n \pi f^{n-1}}{(n-1)!}\,\mathrm{sgn}(f)$
23.	$(t - a)^{-n}$	$(-i)^n \pi\, e^{-i\,af}\,\dfrac{(-if)^{n-1}}{(n-1)!}\,\mathrm{sgn}(f)$
24.	$\sin(at)$	$\dfrac{\pi}{i}\,[\delta(f - a) - \delta(f + a)]$
25.	$\cos(at)$	$\pi\,[\delta(f - a) + \delta(f + a)]$
26.	$\log(|t|)$	$-\dfrac{\pi}{|f|}$
27.	$\mathrm{rect}(at) = H(t + a) - H(t - a)$	$\dfrac{2\sin(af)}{f}$
28.	$\mathrm{rect}(t/2)$	$\mathrm{sinc}\left(\dfrac{f}{2}\right)$
29.	$\mathrm{tri}(t/2)$	$\mathrm{sinc}^2\left(\dfrac{f}{2}\right)$
30.	$\mathrm{comb}(t) = \mathrm{comb}\left(f - \dfrac{m}{T}\right)$	$\displaystyle\sum_{n=-\infty}^{\infty} e^{-i\pi fmt}$

A.2 Fourier Cosine Transform Pairs

	$x(t)$	$\mathcal{F}_c\{x(t)\} = X_c(f)$
1.	$x(at)$	$\dfrac{1}{a} X_c\left(\dfrac{f}{a}\right)$
2.	e^{-at}, $a > 0$	$\dfrac{2a}{a^2 + f^2}$
3.	$e^{-a^2 t^2}$	$\dfrac{\sqrt{\pi}}{a} e^{-f^2/(4a^2)}$
4.	$t^{-\alpha}$, $0 < \alpha < 1$	$\dfrac{\pi f^{\alpha-1}}{\Gamma(f)} \sec\left(\dfrac{\pi\alpha}{2}\right)$
5.	$\dfrac{a}{t^2 + a^2}$, $a > 0$	πe^{-af}, $\quad f > 0$
6.	$H(a - t)$	$\dfrac{2\sin af}{f}$
7.	$e^{-t} \operatorname{sinc}(t)$	$\arctan\left(\dfrac{2}{f^2}\right)$
8.	$t^{-\alpha}$, $0 < \alpha < 1$	$\dfrac{\pi f^{\alpha-1}}{\Gamma(\alpha)} \sec\left(\dfrac{\pi\alpha}{2}\right)$
9.	$\dfrac{1 - t^2}{(1 + t^2)^2}$	$\pi f e^{-f}$
10.	$\left(\dfrac{1}{a} + t\right) e^{-at}$	$\dfrac{4a^2}{(a^2 + f^2)^2}$
11.	$\dfrac{a}{a^2 + t^2}$, $a > 0$	πe^{-af}, $f > 0$
12.	$\sin(at^2)$, $a > 0$	$\sqrt{\dfrac{\pi}{2a}} \left[\cos(f^2/4a) - \sin(f^2/4a)\right]$
13.	$\cos\left(at^2\right)$, $a > 0$	$\sqrt{\dfrac{\pi}{2a}} \left[\cos(f^2/4a) + \sin(f^2/4a)\right]$
14.	$\log\left(1 + \dfrac{a^2}{t^2}\right)$, $a > 0$	$\dfrac{2\pi}{f}\left(1 - e^{-af}\right)$
15.	$\log\left(\dfrac{a^2 + a^2}{b + t^2}\right)$, $a, b > 0$	$\dfrac{2\pi}{f}\left(e^{-bf} - e^{-af}\right)$

A.3 Fourier Sine Transform Pairs

	$x(t)$	$\mathcal{F}_s\{x(t)\} = X_s(f)$
1.	e^{-at}, $a > 0$	$\dfrac{2f}{a^2 + f^2}$
2.	$t\,e^{-at}$, $a > 0$	$\dfrac{4af}{(a^2 + f^2)^2}$
3.	$t^{\alpha-1}$, $0 < \alpha < 1$	$2f^{-\alpha}\Gamma(\alpha)\sin\left(\dfrac{\pi\alpha}{2}\right)$
4.	$t^{-1/2}$	$\sqrt{\dfrac{2\pi}{f}}$, $f > 0$
5.	$\dfrac{t}{t^2 + a^2}$, $a > 0$	$\pi\,e^{-af}$
6.	$\dfrac{t}{(t^2 + a^2)^2}$, $a > 0$	$\dfrac{f}{a}\,e^{-af}$
7.	$t^{-\alpha}$, $0 < \alpha < 2$	$\sqrt{2\pi}\,\Gamma(1-\alpha)\,f^{\alpha-1}\cos\left(\dfrac{\alpha\pi}{2}\right)$
8.	$H(a-t)$, $a > 0$	$\dfrac{2}{f}\,(1 - \cos af)$
9.	$\arctan(t/a)$	$\dfrac{\pi}{f}\,e^{-af}$
10.	$\operatorname{erfc}(at)$, $a > 0$	$\dfrac{2}{f}\left[1 - e^{-f^2/(4a^2)}\right]$
11.	$J_0\left(a\sqrt{t}\right)$, $a > 0$	$\dfrac{2}{f}\cos\left(\dfrac{a^2}{4f}\right)$
12.	$\dfrac{J_0(at)}{t}$, $a > 0$	$\begin{cases} 2\sin\left(\dfrac{f}{a}\right), & 0 < f < a \\ \pi, & a < f < \infty \end{cases}$

B

Hilbert Transforms

B.1. Hilbert Transforms

The Hilbert transform of a signal $x(t)$ is the convolution of the signal $x(t)$ with $\dfrac{1}{\pi t}$ and is defined as

$$\mathcal{H}\{x(t)\} \equiv X_H(z) = x(t) \star \frac{1}{\pi t} = \frac{1}{\pi} \oint_{-\infty}^{\infty} \frac{x(t)}{t-z}\, dt = \frac{1}{\pi} \oint_{-\infty}^{\infty} \frac{x(t-z)}{z}\, dt. \quad \text{(B.1)}$$

Note that the integral in (B.1) is improper since the integrand has a singularity and the integration ranges over an infinite region. It can be evaluated as a Cauchy principal value (p.v) integral whenever that value exists, using the formula

$$\mathcal{H}\{x(t)\} = \frac{1}{\pi} \lim_{\varepsilon \to 0+} \left(\int_{t-1/\varepsilon}^{t-\varepsilon} \frac{x(t)}{t-z}\, dt + \int_{t+\varepsilon}^{t+1/\varepsilon} \frac{x(t)}{t-z}\, dt \right). \quad \text{(B.2)}$$

B.1.1. Properties. The Hilbert transform of a time-domain signal $x(t)$ is another time-domain signal $X_H(z)$. If $x(t)$ is real-valued, so is $X_H(z)$. The Hilbert transform is linear, and we have $x(t) \rightleftharpoons X_H(z)$. If $\mathcal{H}\{x(t)\} = c$ (const), then $X_H(z) = 0$, since by definition (B.1)

$$\mathcal{H}\{c\} = \lim_{\varepsilon \to 0+} \left(\frac{1}{\pi} \int_{-1/\varepsilon}^{-\varepsilon} \frac{c}{\tau}\, d\tau + \int_{\varepsilon}^{1/\varepsilon} \frac{c}{\tau}\, d\tau \right)$$

$$= \lim_{\varepsilon \to 0+} \left(-\frac{1}{\pi} \int_{\varepsilon}^{1/\varepsilon} \frac{c}{\tau}\, d\tau + \frac{1}{\pi} \int_{\varepsilon}^{1/\varepsilon} \frac{c}{\tau}\, d\tau \right) = \lim_{\varepsilon \to 0+} \left(\frac{1}{\pi} \int_{\varepsilon}^{1/\varepsilon} \left(\frac{c}{\tau} - \frac{c}{\tau} \right) d\tau \right) = 0.$$

Using linearity property, we get $\mathcal{H}\{x(t) + c\} = \mathcal{H}\{x(t)\} + \mathcal{H}\{c\} = X_H(z)$. It is an example of an ideal differentiator, i.e., a Hilbert transform 'loses' DC effects.[1]

---

[1] This refers to the phenomenon of a direct current crossing from the insulator in the absence of any external electromagnetic force. The DC current is proprtional to $\sin(\psi)$, where ψ is the phase difference across the insulator taking the values between the current $-I_c$ and I_c. It is defined by the Josephson or weak-link current-phase relation $I(t) = I_c \sin(\psi(t))$ (see Josephson [1962], Feynman et a. [1965]). Brian D. Josephson received Physics Nobel Prize in 1973.

Properties of the Hilbert transform include:

(i) $\mathcal{H}\{x(t+a)\} = X_H(z+a)$ (time-shifting property);

(ii) $\mathcal{H}\{x(at)\} = \mathrm{sgn}(a)\,X_H(az)$, $a \neq 0$;

(iii) $\mathcal{H}\{x(-at)\} = -X_H(-az)$;

(iv) $\mathcal{H}\{x'(t)\} = \dfrac{d}{dz}X_H(z)$;

(v) $\mathcal{H}\{tx(t)\} = zX_H(z) + \dfrac{1}{\pi}\displaystyle\int_{-\infty}^{\infty} x(t)\,dt$;

Convolution satisfies the property: $\mathcal{H}\{x_1(t) \star x_2(t)\} = x_1(t) \star \mathcal{H}\{x_2(t)\}$.
The Hilbert transform of the derivative of a signal $x(t)$ is defined as $\mathcal{H}\left\{\dfrac{d}{dt}x(t)\right\}$
$= \dfrac{d}{dt}\mathcal{H}\{x(t)\}$ (by Property (iv)). This result is obtained using Leibniz's rule,
which states that

$$\frac{d}{dc}\int_{a(c)}^{b(c)} f(x,c)\,dx = \int_{a(c)}^{b(c)} \frac{\partial}{\partial dc}f(x,c)\,dx + f(b,c)\frac{d}{dc}b(c) - f(a,c)\frac{d}{dc}a(c).$$

Thus,

$$\frac{d}{dt}\mathcal{H}\{x(t)\} = \frac{1}{\pi}\int_{-\infty}^{\infty}\frac{x(t-u)}{u}\,du = \frac{1}{\pi}\int_{-\infty}^{\infty}\frac{x'(t-u)}{u}\,du = \mathcal{H}\{x'(t)\}.$$

The Hilbert transform is related to the Fourier transform via the Fourier
transform of the signal $x(t) = 1/(\pi t)$. We know that in the cyclic frequency
domain

$$\mathcal{F}\{1/(\pi t)\} = -i\,\mathrm{sgn}(f) = \begin{cases} -i & \text{if } f > 0, \\ 0 & \text{if } f = 0, \\ i & \text{if } f < 0. \end{cases}$$

Using the notation of Chapter 3, let $x(t) \rightleftharpoons X(f)$. Since $X_H = x(t) \star \dfrac{1}{\pi t}$,
the Fourier convolution theorem gives $\hat{X}(f) = -i\,\mathrm{sgn}(f)\,X(f)$, where $\hat{X}(f)$
denotes vthe Fourier transform of X_H. The Hilbert transform is defined in the
frequency domain, and it does not change the magnitude $|X(f)|$; it changes
only the phase. The Fourier transform values at positive (or negative) frequencies are multiplied by $-i$ (or i) corresponding to a phase change of $-\pi/2$
(or $\pi/2$), respectively. In other words, if $X(f) = a + ib$, then

$$\hat{X}(f) = \begin{cases} -b + ia & \text{if } f < 0, \\ b - ia & \text{if } f > 0. \end{cases}$$

Since $|X(f)| = |\hat{X}(f)|$, both $X(f)$ and $\hat{X}(f)$ have the same energy spectral
density. Thus, e.g., if $X(f)$ is bandlimited to some value v Hz, so is $\hat{X}(f)$.
This means, that both $X_H(t)$ and $x(t)$ have the same energy.

If $x(t)$ is real-valued, then $X(-f) = \overline{X(f)}$, i.e., $X(f)$ has Hermitian symmetry. But then $\hat{X}(-f) = -i \operatorname{sgn}(-f)X(-f) = \overline{-i \operatorname{sgn}(f)X(f)} = \hat{X}(f)$, where the line over a quantity represents its complex conjugate, so $\hat{X}(f)$ also has Hermitian symmetry. Further, let $x(t)$ be a real-valued signal. If $x(t)$ is even, i.e., $x(-t) = x(t)$, then $X(f)$ is real-valued, whereas if $x(t)$ is odd, i.e., $x(-t) = -x(t)$, then $X(f)$ is purely imaginary. Thus, if $X(f)$ is real-valued, then $\hat{X}(f)$ is purely imaginary-valued, and conversely. This means that if $x(t)$ is even, then $X_H(t)$ is odd, and if $x(t)$ is odd, then $X_H(t)$ is even.

If $x(t)$ is a real-valued signal, then $x(t)$ and $X_H(t)$ are orthogonal, as shown by the inner product

$$\langle x(t), X_H(t) \rangle = \int_{-\infty}^{\infty} x(t)\overline{X_H(t)}\, df = \int_{-\infty}^{\infty} X(f)\overline{\hat{X}(f)}\, df$$

$$= \int_{-\infty}^{\infty} X(f)\overline{[-i \operatorname{sgn}(f)X(f)]}\, df$$

$$= \int_{-\infty}^{\infty} i\,|X(f)|^2 \operatorname{sgn}(f)\, dt = 0,$$

where, since $|X(f)|^2$ is an even function of f and so $|X(f)|^2 \operatorname{sgn}(f)$ is an odd function of f, the integral is zero.

Let $x(t)$ be a signal such that its Fourier transform $X(f) = 0$ for $|f| \geq \alpha$, and let $y(t)$ be another signal such that $Y(f) = 0$ for $|f| < \alpha$. Then

$$\mathcal{H}\{x(t)y(t)\} = X(f) \star \hat{Y}(f). \tag{B.3}$$

The Hilbert transform of an amplitude modulated signal $x(t)$ is given by (see Hahn [1996])

$$\mathcal{H}\{x(t)\cos(2\pi f_c t + \phi)\} = \left[x(t) \star \frac{\cos(2\pi f_c t + \phi)}{\pi t}\right]\cos(2\pi f_c t + \phi)$$

$$+ \left[x(t) \star \frac{\sin(2\pi f_c t + \phi)}{\pi t}\right]\sin(2\pi f_c t + \phi). \tag{B.4}$$

For proofs of these results, see Hahn [1996].

B.2. Inverse Hilbert Transforms

First note that

$$\mathcal{H}^{-1}\{\hat{X}(f)\} = (-i \operatorname{sgn}(f))^2\, X(f) = -\operatorname{sgn}^2(f)\, X(f), \tag{B.5}$$

where $\operatorname{sgn}^2(f) = 1$ for $f \neq 0$. Thus, in general, $\mathcal{H}\{\mathcal{H}\{x(t)\}\} = -x(t)$ except in the case when $X(f)$ has a singularity at $f = 0$ (e.g., a delta function, which at $f = 0$ generates a nonzero DC offset (i.e., offsetting a signal from zero) and, therefore, is not covered by the Hilbert transform). Let

$x_{\text{mean}}(t) = \lim_{T \to \infty} \int_{-T/2}^{T/2} x(t)\,dt$ define the mean value of $x(t)$ on $[-T/2, T/2]$, provided this limit exists. Then assuming that $x_{\text{mean}}(t) \neq 0$, we recover $x(t)$ from $X_H(t)$. Thus, the inverse Hilbert transform is obtained by applying the Hilbert transform again, and negating the result:

$$x(t) = -\mathcal{H}\{X_H(t)\} = -X_H(t) \star \frac{1}{\pi t}. \tag{B.6}$$

In general, we have

$$x(t) = -X_H(t) \star \frac{1}{\pi t} + c, \tag{B.7}$$

for some constant c. Zero-mean signal $x(t)$ and $X_H(t)$ are known as a *Hilbert transform pair*, and every such pair has the *dual* pair $X_H(t)$ and $-x(t)$. Note that $\mathcal{H}\{\delta(t)\} = \frac{1}{\pi z}$; $\mathcal{H}\left\{\frac{1}{\pi t}\right\} = -\delta(z)$; $\mathcal{H}\left\{\frac{1}{t}\right\} = -\pi\delta(z)$; $\mathcal{H}\{\sin t\} = \cos(z)$; $\mathcal{H}\{\cos t\} = -\sin(z)$.

B.3. Discrete Hilbert Transforms

The discrete Hilbert transform can be related to the discrete Fourier transform (DFT), which is described in §3.4. Defining DFT as in (3.53), let $x(n)$ be the real-valued input data and $X(k)$ the output frequency components for $k = 0, 1, \ldots, N-1$ and $n = 0, 1, 2, \ldots, N-1$. Then the properties of the frequency components are as follows:

(i) $X(k) = \overline{X(N-k)}$ for $k = 1, \ldots, \frac{N}{2} - 1$. $\tag{B.8}$

(ii) Let $X_1(k)$ be a new set of frequency components, for $k = 0, 1, \ldots, N-1$. Then

$$X_1(0) = \frac{1}{2}X(0),$$

$$X_1(k) = X(k) \quad \text{for } k = 1, \ldots, \frac{N}{2} - 1,$$

$$X_1(N/2) = \frac{1}{2}X(N/2),$$

$$X_1(k) = 0 \quad \text{for } k = \frac{n}{2} + 1, \ldots, N-1. \tag{B.9}$$

(iii) The new data $x_1(n)$ in the time domain can be obtained from the inverse DFT (IDFT) of $X_1(k)$ as

$$x_1(n) = \frac{1}{n} \sum_{k=0}^{N-1} X_1(k)\, e^{2\pi i\, kn/N}. \tag{B.10}$$

Thus, if there are N real-valued input data, they will yield N complex-valued data. This is the result of the padding of the $X_1(k)$ values with zeros as shown in Eq (B.9). Note that padding with zeros in the frequency domain is the same as interpolating in the time domain.

This method, used in MATLAB, generates N complex data from N real input data. However, the new data so obtained may increase processing time without improving receiver's performance. A different method suggested in Tsui [2000: 143], which is a modified MATLAB approach, generates only $N/2$ complex output data. With digitized real input data, the sampling frequency $f_s \approx 2.5 \, \Delta f$ is used and the input signal is aliased close to the center of the baseband. This should yield a very small frequency component $X(N/2)$. The complex data is then obtained by the following three steps:

STEP 1: Write the DFT equation (3.53) as above in the MATLAB approach and take the FFT of this input signal.

STEP 2: Obtain the new $X_1(k)$ as

$$X_1(k) = X(k) \quad \text{for } k = 0, 1, \dots, \frac{N}{2} - 1. \tag{B.11}$$

This step keeps only half of the frequency components.

STEP 3: Obtain the new data in time domain as

$$x_1(n) = \frac{2}{N} \sum_{k=0}^{N/2-1} X_1(k) \, e^{4\pi nki/N}. \tag{B.12}$$

This step yields $N/2$ complex data in the time domain, as desired. These complex data contain the same information as the N real data and cover the same length of time. Thus, the equivalent sampling rate of the complex data is $f_{s1} = f_s/2$ which matches reasonably well with the Nyquist sampling rate of $F_{s1} = \Delta f$ for complex data.

Next, the problem is to change the complex data into real data. A common I-Q converter design to make the input frequency (IF) zero is shown in Figure B.1. If it is used, the center frequency of the input signal is determined by the Doppler shift, using the following four steps:

STEP 1: Take the DFT of $x(n)$ to yield $X(k)$ using (3.53), i.e.,

$$X(k) = \sum_{n=0}^{N-1} x(n) \, e^{-2\pi nki/N}, \quad k = 0, 1, \dots, N - 1, \tag{B.13}$$

where $x(n)$ is complex.

STEP 2: Generate a new set of frequency components $X_1(k)$ from $X(k)$ as

$$X_1(k) = X\left(\frac{N}{2} + k\right) \quad \text{if } k = 0, \dots, \frac{N}{2} - 1,$$

$$X_1(k) = X\left(-\frac{N}{2} + k\right) \quad \text{if } k = \frac{N}{2}, \dots, N - 1. \tag{B.14}$$

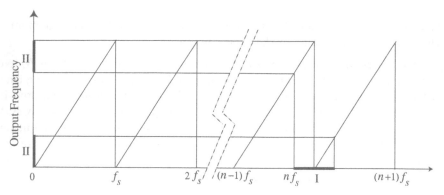

Legend: I : Input bandwidth;   II : Output bandwidth.

Figure B.1 Frequency Aliasing for Complex Data.

In Figure B.1, the lower input frequency (IF) is converted into a higher output frequency and conversely. Thus, the design in this figure shifts the center of the input signal from zero to $f_s/2$. If the IF is not at zero frequency, a different shift is required. But no shift is required if the IF of the I-Q channels is at $f_s/2$. This step can, therefore, be omitted because the input signal will not split into two separate bands.

STEP 3: Generate another frequency component for $X_1(k)$ as

$$X_1(N) = 0,$$
$$X_1(N + k) = \overline{X_1(N - K)} \quad \text{for } k = 1, \ldots, N - 1. \tag{B.15}$$

Eqs (B.14) and (B.15) together yield $2N$ frequency components for $k = 0, \ldots, 2N - 1$.

STEP 4: Using FFT the new data in the time domain is computing by the formula

$$x_1(n) = \frac{1}{2N} \sum_{k=0}^{2N-1} X_1(k)\, e^{nk\pi i/N}. \tag{B.16}$$

Thus, the N complex data generate $2N$ real data which contain the same amount of information over the same time period. Hence, $f_{s1} = 2f_s$, i.e., the equivalent sampling is doubled. This is confirmed from actual complex data collected from satellites with I-Q channels of zero IF, as reported in Tsui [2000: 145–146].

C

Radon Transforms

Radon transforms are used in computerized tomography which is studied in Chapter 11. Radon [1917] established a remarkable result that the set of all projections along any line L in the plane uniquely determines the distribution of a function $f(x, y)$ under general assumptions that f is continuous and has compact support. Consider a density distribution f given by a small mass centered at a fixed point x_0 in the plane. Then at a point x_0 along the x-axis the projection P of the distribution along the y-axis is given by

$$P\{f(x_0)\} = \int_{-\infty}^{\infty} f(x_0, y) \, dy.$$ As x_0 varies along the x-axis, a projection $P_1\{f(x)\}$ is obtained. A similar situation occurs along the y-axis with a point y_0 yielding another projection $P_2\{f(y)\}$. The problem of uniquely locating a point mass at (x, y) is completely solved if both of these projections are known. However, in the case of two point masses the two projections are insufficient. Radon's contribution in problems of this nature not only provides the solution but also paves the way to computerized tomography.

C.1 Radon Transform

The Radon function computes projections of an image matrix along specified directions. Since a projection of a two-dimensional function $f(x, y)$ is a set of line integrals, the Radon function computes line integrals from multiple sources along parallel lines (or paths, called *beams*) in a certain direction, such that the beams are set one pixel apart from each other. The Radon function takes multiple, parallel-beam projections of the image from different angles by rotating the source about the center of the image. A parallel-beam projection at a specified rotation angle θ is shown in Figure C.1.

Note that in Figure C.1(a) the line integral of $f(x, y)$ in the vertical direction is the projection of $f(x, y)$ onto the x-axis, whereas the integral in the horizontal direction is the projection of $f(x, y)$ onto the y-axis. The horizontal and vertical projections for a simple two-dimensional function $f(x, y)$, which is a rectangle, is shown in Figure C.1(b), where the projections are calculated using the angle θ.

Let θ denote the angle by which the (x, y) coordinate system is rotated such that the positive x axis coincides with the positive x'-axis. If (x, y) are the coordinates of a point relative to the (x, y) system, let (x', y') denote the new coordinates of the same point relative to the (x', y') coordinate system. The relation between these two coordinate systems is defined by

$$\begin{Bmatrix} x \\ y \end{Bmatrix} = \begin{bmatrix} \cos\theta & \sin\theta \\ -\sin\theta & \cos\theta \end{bmatrix} \begin{Bmatrix} x' \\ y' \end{Bmatrix}. \tag{C.1}$$

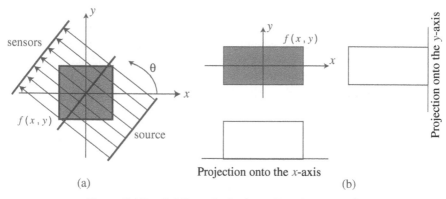

Figure C.1 Parallel-Beam Projection at Rotation Angle θ.

The Radon transform of $f(x, y)$ is the line integral of f parallel to the y'-axis and is defined as

$$R_\theta(x', y') = \int_{-\infty}^{\infty} f\left(x'\cos\theta - y'\sin\theta, x'\sin\theta + y'\cos\theta\right) dy', \tag{C.2}$$

The geometry of the Radon transform is presented in Figure C.2.

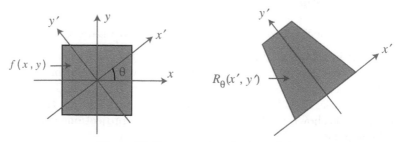

Figure C.2 Geometry of Radon Transform.

In view of (C.2), the Radon transform of a two-dimensional function $f(x, y)$ with compact support that includes the origin is defined as the set of projections along an angle θ, $0 \leq \theta < \pi$, as

$$\mathcal{R}\{f(x,y)\} \equiv R(\rho,\theta) = \int_{-\infty}^{\infty} \int_{-\infty}^{\infty} f(x,y)\,\delta(x\cos\theta + y\sin\theta, \rho)\,dx\,dy$$

$$= \int_{-\infty}^{\infty} f(\rho\cos\theta - k\sin\theta, \rho\sin\theta + k\cos\theta)\,dk, \qquad (C.3)$$

where the δ-function converts the two-dimensional integral to a line integral dk along the line $x\cos\theta + y\sin\theta = \rho$. The transformed function $R(\rho,\theta)$ is called the *sinogram* of $f(x,y)$ because a point δ-function in f transforms a sinusoidal line δ-function into $R(\rho,\theta)$.

A projection is formed by combining a set of line integrals. The simplest projection, which is a collection of parallel ray integrals (i.e., constant c), is shown in Figure C.3 (a), and a simple fan beam projection in Figure C.3(b).

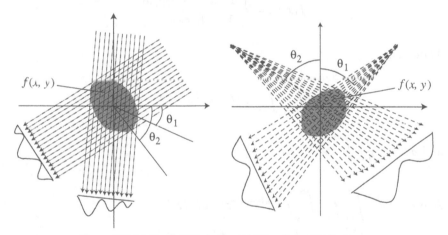

Figure C.3 (a) Parallel Projection; (b) Fan Beam Projection.

C.2 Inverse Radon Transform

The inverse Radon transform (also known as the *backprojection*) is defined as

$$\mathcal{R}^{-1}\{f(x,y)\} = \int_0^{\pi} R(x\cos\theta + y\sin\theta, \theta)\,d\theta. \qquad (C.4)$$

Geometrically, this operation simply propagates the measured sinogram back into the image space along the projection paths, as shown in Figure C.4.

Note that for a point source at the origin, $\delta(x,y)$, which is the intensity of the backprojection image, decays like $1/r$. It is shown in Figure C.4 as the surface plot of the backprojection image of a point source. Thus, we have

$$R^{-1}\{f(x,y)\} = \frac{1}{r} \star f(x,y). \qquad (C.5)$$

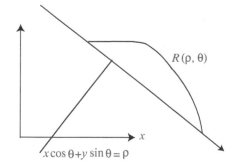

Figure C.4 Geometrical Interpretation of Inverse Radon Transform.

Example C.1. (Square) Let the square be defined as

$$f(x,y) = \begin{cases} 1 & x, y \in [-a, a] \\ 0 & \text{otherwise.} \end{cases}$$

Then using

$$R(\rho, \theta) = \int_{-\infty}^{\infty} \int_{-\infty}^{\infty} f(x,y)\, \delta\left(y - (\theta + \rho x)\right) dx\, dy,$$

and $\delta(x) = \dfrac{1}{2\pi} \displaystyle\int_{-\infty}^{\infty} e^{-ikx}\, dx$, we get

$$
\begin{aligned}
R(\rho, \theta) &= \frac{1}{2\pi} \int_{-a}^{a} \int_{-a}^{a} \int_{-\infty}^{\infty} e^{-i\,k[y-(\theta+\rho x)]}\, dk\, dx\, dy \\
&= \frac{1}{2\pi} \int_{-\infty}^{\infty} e^{i\,k\theta} \left[\int_{-a}^{a} e^{-i\,ky}\, dy \int_{-a}^{a} e^{i\,k\rho x}\, dx \right] dk \\
&= \frac{1}{2\pi} \int_{-\infty}^{\infty} e^{i\,k\theta} \frac{1}{-i\,k} \left[e^{-i\,ky} \right]_{-a}^{a} \frac{1}{i\,k\rho} \left[e^{i\,k\rho x} \right]_{-a}^{a} dk \\
&= \frac{1}{2\pi} \int_{-\infty}^{\infty} e^{i\,k\theta} \frac{1}{k^2\rho} \left[-2i\,\sin(ka) \right] \left[2i\,\sin(k\rho a) \right] dk \\
&= \frac{2}{\pi\rho} \int_{-\infty}^{\infty} \frac{\sin(ka)\sin(k\rho a)\, e^{i\,k\theta}}{k^2}\, dk \\
&= \frac{4}{\pi\rho} \int_{0}^{\infty} \frac{\sin(ka)\sin(k\rho a)\cos(k\theta)}{k^2}\, dk \\
&= \frac{2}{\pi\rho} \int_{0}^{\infty} \frac{\sin\left(k(\theta+a)\right) - \sin\left(k(\theta-a)\right)}{k^2}\, \sin(k\rho a)\, dk \\
&= \frac{2}{\pi\rho} \left\{ \int_{0}^{\infty} \frac{\sin\left(k(\theta+a)\right)\sin(k\rho a)}{k^2}\, dk - \int_{0}^{\infty} \frac{\sin\left(k(\theta-a)\right)\sin(k\rho a)}{k^2}\, dk \right\}.
\end{aligned}
$$

Using the formula $\int_0^\infty \dfrac{\sin(ax)\sin(bx)}{x^2}\, dx = \dfrac{1}{2}\pi\, \mathrm{sgn}(a,b)\, \min\left(|a|, |b|\right)$ (see Gradshteyn and Ryzhik [2000: Eq (3.741.3)]), we obtain

$$R(\rho, \theta) = \frac{1}{\rho}\left\{ \mathrm{sgn}\left((\theta + a)\rho a\right) \min\left(|\theta + a|, |\rho a|\right) - \mathrm{sgn}\left((\theta - a)\rho a\right) \min\left(|\theta - a|, |\rho a|\right)\right.$$

$$= \frac{|a + a\rho - \theta| + |a + a\rho + \theta| - |a - a\rho - \theta| - |a - a\rho + \theta|}{2\rho}. \tag{C.6}$$

The projection of the square is shown in Figure C.5. ∎

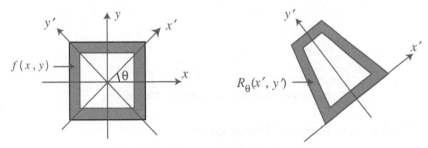

Figure C.5 A Square under Radon Transform.

C.3 2-D Radon Transform

The solution of the inverse Radon transform is based on the Fourier slice theorem (FST, §3.7), which relates $F(v_x, v_y)$ of 2-D Fourier transform (FT) of $f(x, y)$ and $R(v, \theta)$ of 1-D FT of $R(\rho, \theta)$. Another version of Theorem 3.10 is as follows.

Theorem C.1. (FST) *The following result holds:*

$$R(\rho, \theta) = F(\rho \cos\theta, \rho \sin\theta). \tag{C.7}$$

PROOF. Using the two-dimensional version of (C.3) we get

$$R(\rho, \theta) = \int_{-\infty}^{\infty} \int_{-\infty}^{\infty} \int_{-\infty}^{\infty} f(x, y)\, \delta(x\cos\theta + y\sin\theta - \rho)\, e^{-i\omega\rho}\, dx\, dy\, d\rho$$

$$= \int_{-\infty}^{\infty} \int_{-\infty}^{\infty} f(x, y)\, e^{i\omega(x\sin\theta + y\cos\theta)}\, dx\, dy = F(u, v), \tag{C.8}$$

where $u = \rho\cos\theta$ and $v = \rho\sin\theta$. ∎

This theorem means that the value of 2-D FT of $f(x, y)$ along a line at an inclination θ is given by the 1-D FT of $R(\rho, \theta)$ which is the projection profile of the sinogram acquired at angle θ. This theorem is geometrically represented in Figure C.6. Thus, with a sufficient number of projections, $R(\rho, \theta)$ can fill the (u, v)-plane to generate $F(u, v)$. In the Fourier domain, Eq (C.7) becomes

$$R(\rho, \theta + \pi) = R(-\rho, \theta). \tag{C.9}$$

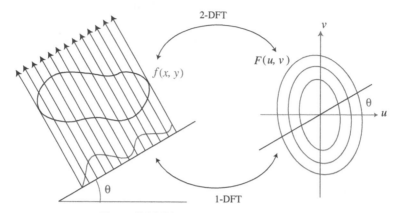

Figure C.6 2-Dimensional Slice Theorem.

C.4 Discrete Radon Transform

As shown in previous sections dealing with the continuous two-dimensional case as defined by Eq (C.3), the two-dimensional FT of f is recovered in polar coordinates from the slices, and an inverse two-dimensional FT recovers f. Other references for continuous reconstruction of the Radon transform are Natterer and O'Malley [2001] and Natterer and Wübbeling [2001], and algorithms for discrete Radon transform (DRT) and fast discrete Radon transforms (FDRT) are available in Beylkin [1987], Kelley and Madisetti [1993], Götz and Druckmüller [1996], Brady [1998], Brandt et al. [1999], although no single algorithm can claim to be final and complete. A description of three-dimensional segmented human anatomy is available in Zubal et al. [2004]. Now, we will discuss an algorithm for a fast and accurate IDRT as developed in Press [2006].

Since FFT approximates the one-dimensional continuous transform in $O\left(N \log N\right)$ operations and the two-dimensional in $O\left(N^2 \log N\right)$ operations, the DRT is fast, of order $O\left(N^2 \log N\right)$ for an $N \times N$ image, because it uses only addition, and not multiplication or interpolation with the backprojection. Most of the fast DRT algorithms are based on the discretization of the Fourier slice properties of the continuous case. As noted in Beylkin [1987], these fast algorithms are formulated using the FFT for the fast solution of the sets of linear equations, like the block-circulant forms. On the other hand, slow DRTs and IDRTs approximate with at least $O\left(N^3\right)$ operations, and therefore will not be considered here. Another important feature of any fast DRT algorithm is the accuracy which depends on the discrete representation of $f(x, y)$ and the choice of the DRT algorithm approximating a continuous Radon transform. For example, the assumption of periodicity of $f(x, y)$ along one axis, or severely bandwidth limited $f(x, y)$, generally provides a well-defined reconstructed image in and around the center but shows small tails outside. In terms of speed and accuracy, the DRT algorithms of Götz and Druckmüller [1996] and of Brady [1998] generate an exact inverse that can be realized by an iteratively fast algorithm.

A discrete approximation of lines is described by considering an image of an $N \times N$ array of intensity values f_{ji}, $0 < i, j < N$, where $N = 2^m$ and $m \in \mathbb{Z}^+$. A set of digital lines, called d-lines, that transect the image, passing exactly through one array point (i, j) in each column of the array and parameterized by integers h and s, is denoted by $\{D_N(h, s)\}$. Thus, the d-line $D_N(h, s)$ connects two array points: $(i, j) = (0, h)$ and $(i, j) = (N - 1, h + s)$, where h and s denote the intercept and the rise, respectively, of the d-line.

We will consider only the case $0 \leq s \leq N - 1$, which corresponds to the slopes in $[0, \pi/4]$ of the d-line. Then the D-lines are defined recursively in terms of d-lines on half-wide images:

$$D_N(h, 2s) = D_{N/2}(h, s) \cup D_{N/2}(h + s, s),$$
$$D_N(h, 2s + 1) = D_{N/2}(h, s) \cup D_{N/2}(h + s + 1, s), \qquad \text{(C.10)}$$

where $\cup$ is an operator joining left and right halves. The d-line $D_N(h, s)$ approximates a continuous line with intercept h and slope $s/(N-1)$. According to Brady [1998], its maximum vertical deviation from the continuous line is at most $\frac{1}{6} \ln N$. The recursion and some of d-lines for $N = 4$ are shown in Figures C.7(a) and (b), respectively, where slight offsets in the lines are for the sake of clarity. All d-lines pass through integer lattice points, one in each column.

The DRT in one quadrant, designated I: $[0, \pi/4]$,[1] is defined as the sum of image intensities over the array points on a d-line:

$$R^{\mathrm{I}}(h, s) = \sum_{(i,j) \in D_N(h,s)} f_{ji}, \qquad \text{(C.11)}$$

such that $f_{ji} = 0$ if i or j is outside the range $[0, N-1]$. The recursive definition (C.10) of the d-lines induces a recursive computation of the DRT components. This is done by associating each partial d-line with its corresponding partial sum of function values, as follows: two half-images of the partial sums are converted to one full image in a single upward sweep, as shown in Figure C.7, which uses one row of scratch domain. The value of $R^{\mathrm{I}}(h, s)$ of an image is computed by carrying out $(\ln N)$ sweeps, first combining adjacent pairs of

---

[1] There are three other quadrants, namely, II: $[\pi/4, \pi/2]$, III: $[-\pi/2, -\pi/4]$, and IV: $[-\pi/4, 0]$. They are discussed in the sequel.

columns, then pair of pairs, and so forth in $(\ln N)$ sweeps.

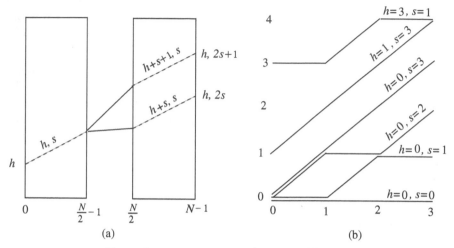

Figure C.7 (a) Recursion; (b) d-Lines for $N = 4$.

    The intercept h takes some negative values so as to include all d-lines which intersect the partial images, or the final full image. When a sweep produces a partial image of width N, the most negative value of the intercept h is $-n+1$, since a d-line at $\pi/4$ with this intercept will intersect the single pixel in the lower right of the partial image. Thus, the sums are computed for i and j on each partial image such that all the following three conditions are satisfied for a total of $nN + \frac{1}{2} n(n-1)$ cases in each partial image:

(a) $0 \le i \le n-1$,

(b) $-n+1 \le j \le N-1$,

(c) $0 \le i_j$.

The partial images are summed in each sweep, where all the $(\ln N)$ sweeps give a total of $(N^2 - \frac{1}{2} N) \ln N + N(N-1) = O\left(N^2 \ln N\right)$ additional operations, without any multiplication or other floating point arithmetic. In the remaining three quadrants, the DRT is defined by appropriately flipping the original image, and then repeating the above algorithm for Quadrant I. Thus,

$$R^{\mathrm{II}}\{f_{ji}\} = \mathcal{R}^{\mathrm{I}}\{f_{ij}\} \quad \text{for Quadrant II: } [\pi/4, \pi/2],$$
$$R^{\mathrm{III}}\{f_{ji}\} = \mathcal{R}^{\mathrm{I}}\{f_{i,N-1-j}\} \quad \text{for Quadrant III: } [-\pi/2, -\pi/4],$$
$$R^{\mathrm{IV}}\{f_{ji}\} = \mathcal{R}^{\mathrm{I}}\{f_{N-1,i}\} \quad \text{for Quadrant IV: } [-\pi/4, 0]. \qquad \text{(C.12)}$$

C.4.1. DRT and Sinogram. The DRT as defined above from a discrete image can also be identified with values sampled from the continuous sinogram of a continuous image. If a sinogram $R(\rho, \theta)$ is scaled such that it is nonzero for $-\frac{1}{2} \leq \rho \leq \frac{1}{2}$, and defined for $0 \leq \theta \leq \pi$, then the sampled positions are given by

$$\theta_s = \arctan \frac{s}{N-1}, \quad \rho_{hs} = \cos \theta_s \left(\frac{h}{N} - \frac{1}{2} + \frac{1}{2(N-1)} \right), \qquad \text{(C.13)}$$

which yield

$$R^{\mathrm{I}}(h, s) = \cos \theta_s \, R \left(\rho_{hs}, \theta_s \right),$$
$$R^{\mathrm{II}}(h, s) = \cos \theta_s \, R \left(\rho_{hs}, \pi - \theta_s \right),$$
$$R^{\mathrm{III}}(h, s) = \cos \theta_s \, R \left(1 - \rho_{hs}, \pi/2 - \theta_s \right),$$
$$R^{\mathrm{IV}}(h, s) = \cos \theta_s \, R \left(1 - \rho_{hs}, \pi/2 + \theta_s \right), \qquad \text{(C.14)}$$

where it is assumed, as before, that $R(\rho, \theta) = 0$ for arguments outside the above-mentioned ranges. The factors $\cos(\theta_s)$ scale the transform values so that the DRT always sums one value in each of the N columns, independent of the angle, while the sinogram is a true line integral of length proportional to $\sec \theta_s$.

C.4.2. Inverse DRT: Continuous Case. In the case of a two-dimensional continuous Radon transform, the backprojection, which is fast and exact, is defined as

$$\mathcal{R}^{-1}\{R\} \equiv f(x, y) = \int_{-\infty}^{\infty} \int_0^{\pi} R(\rho, \theta) \, \delta(x \cos \theta + y \sin \theta - \rho) \, d\rho \, d\theta,$$
$$= \int_0^{\pi} R(x \cos \theta + y \sin \theta, \theta) \, d\theta. \qquad \text{(C.15)}$$

This integral is computed over all directions of all the projections that pass through a point (x, y). The recursive definition of d-lines, given by Eq (C.10), induces a fast backprojection algorithm as found in Brady [1998], by reversing the sweeps shown in Figure C.8. Every element of left-half and right-half partial images is assigned the sum of two full-image elements that corresponds with the d-line recurrence. Thus, if $F_n(h, s)$ denotes a partial reverse transformation of an image of width n, then

$$F_{n/2}^{\mathrm{L}}(h, s) = F_n(h, 2s) + F_n(h, 2s + 1),$$
$$F_{n/2}^{\mathrm{R}}(h, s) = F_n(h, 2s) + F_n(h - 1, 2s + 1),$$

where the superscripts L and R refer to left and right half-width images. It will take $(\ln N)$ sweeps to obtain the backprojection in each quadrant. The full backprojection is an appropriate sum over the four quadrants:

$$F_{ji} = \frac{1}{4(n-1)} \left(F_{ji}^{I} + F_{ij}^{II} + F_{i,N-1-j}^{III} + F_{N-1-j,i}^{IV} \right).$$

This result is exactly the sum of the discrete projections along d-lines that pass through an image pixel.

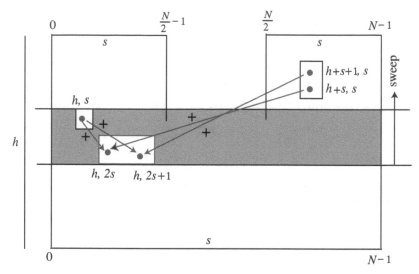

Figure C.8 Computation of DRT.

C.4.3. IDRT as Approximate IFT.

An approximate inverse of the DRT can be obtained by Fourier methods. It requires no interpolation, or application of a ramp or other filter, for manipulating out of the Fourier domain. However, while the inverse so obtained is exact it is ill-conditioned. Let $a \equiv s/(N-1)$ in Eq (C.13). Then $\tan\theta = a$. This means that the number of nonzero values of $R(h, s)$ in a column of constant s varies as $N(1+a)$, as in Figure C.8. Hence, we zero-pad each column, symmetrically at both ends, into a vector of length $2N$, and take FFT of each column. Since the data are real, and negative frequencies are complex conjugates of the positive ones, we consider only the N nonnegative frequency components (excluding the components at the Nyquist frequency).

This process implies that these components are mapped directly onto the N values along the d-line in two-dimensional frequency domain with the same value of s. However, the formal inverse (IDRT) so obtained is exact it is ill-conditioned. To see this, note that the distance represented by the $N(1+a)$ components in a column is not $N(1+a)$ but rather $N(\sin\theta + \cos\theta) = N(1 + a)\left(1+a^2\right)^{-1/2}$, as seen in Figure C.9, such that the sampling interval is $\Delta = \left(1+a^2\right)^{-1/2}$. Thus, the component $j \geq 0$ represents the frequency

$f_j = \dfrac{j}{2N\Delta} = \dfrac{j}{2N}\left(1 + a^2\right)^{-1/2}$, which means that there is no scaling problem.

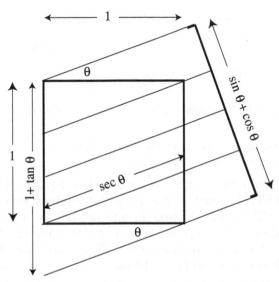

Figure C.9 Trigonometry of a Square and its Projection at an Angle θ.

Many d-lines can hit the same frequency cell, particularly near the origin. To avoid this situation, each cell is assigned the mean of all components that hit it, which is a discrete approximation to the Fourier slice theorem. This will guarantee that every cell gets at least one hit. Then the two-dimensional IFFT provides an approximation of the original image f_{ji}. In fact, by reversing the sweep procedure of Figure C.9 it is possible to derive a formal algebraic inverse from the data in each quadrant, but then using the Fourier slice theorem this quadrant-by-quadrant inverse may turn out to be very ill-conditioned since for each quadrant it lacks information on 75% of the Fourier domain.

C.4.4. Relationship with Fourier Transform. The Fourier transform of one and two variables defined in Chapter 3, and the Radon transforms are related in the following sense. Define the Radon transform (C.3) as

$$\mathcal{R}_\theta\{f(s)\} = R\{f(\theta, s)\}, \tag{C.16}$$

such that the Radon transform R_θ takes the Fourier transform in the s variable. Then the Fourier slice theorem states that

$$\mathcal{R}\{f(\theta)\} = R\{f(\sigma)\} = F\left(\sigma \mathbf{u}(\theta)\right) \tag{C.17}$$

where the unit vector $\mathbf{u}(\theta) = (\cos\theta, \sin\theta)$. Thus, the 2-D Fourier transform of the initial function is the one variable Fourier transform of the Radon transform of that function. The commutative diagram of this relationship is presented in Figure C.10.

$$f(x, y) \xleftarrow{\quad \text{2-D IFT} \quad} F(v_x, v_y)$$

$$\uparrow \text{FST}$$

$$R(\rho, \theta) \xrightarrow{\quad \text{1-D IFT} \quad} R(v, \theta)$$

Figure C.10 Commutative Diagram of Fourier Slice Theorem.

C.4.5. Tomographic Reconstruction Methods. Among many such methods, the two most useful methods are

(i) DIRECT FOURIER RECONSTRUCTION. Once $F(v_x, v_y)$ is obtained from $R(\rho, \theta)$ using FST, the function $f(x, y)$ can be obtained by applying IFT to $F(v_x, v_y)$. If FFT is used, the issue of interpolation is important. For this purpose the values of $F(v_x, v_y)$ should be available at rectangular grid points. However, the values generated from the FST are available at a polar grid. This requires Fourier-domain interpolation, as shown in Figure C.11.

(ii) The filtered backprojection algorithm (FBP) is derived as follows. We have

$$\begin{aligned}
f(x, y) &= \int_{-\infty}^{\infty} \int_{-\infty}^{\infty} F(v_x, v_y) \, e^{-2\pi i \, v_x x} \, e^{-2\pi i \, v_y y} \, dv_y \, dv_x \\
&= \int_{0}^{2\pi} d\theta \int_{0}^{\infty} v \, F(v \cos\theta, v \sin\theta) \, e^{-2\pi i \, vx \cos\theta} \, e^{-2\pi i \, vy \sin\theta} \, dv \\
&= \int_{0}^{2\pi} d\theta \int_{0}^{\infty} v \, R(v, \theta) \, e^{-2\pi i \, v(x \cos\theta + y \sin\theta)} \, dv
\end{aligned}$$

Since

$$\begin{aligned}
&\int_{\pi}^{2\pi} d\theta \int_{0}^{\infty} v \, R(v, \theta) \, e^{-2\pi i \, v(x \cos\theta + y \sin\theta)} \, dv \\
&= \int_{0}^{\pi} d\theta \int_{0}^{\infty} v \, R(-v, \theta) \, e^{-2\pi i \, v(x \cos\theta + y \sin\theta)} \, dv, \quad \text{using (C.9)} \\
&= \int_{0}^{\pi} d\theta \int_{-\infty}^{0} (-v) \, R(v, \theta) \, e^{-2\pi i \, v(x \cos\theta + y \sin\theta)} \, dv,
\end{aligned}$$

we finally obtain

$$\begin{aligned}
f(x, y) &= \int_{0}^{\pi} d\theta \int_{-\infty}^{\infty} |v| \, R(v, \theta) \, e^{-2\pi i \, v(x \cos\theta + y \sin\theta)} \, dv \\
&= \int_{0}^{\pi} R'(x \cos\theta + y \sin\theta, \theta) \, d\theta, \quad\quad\quad\quad\quad \text{(C.18)}
\end{aligned}$$

where

$$R'(\rho, \theta) = \int_{-\infty}^{\infty} |v| R(v, \theta) \, e^{-2\pi i \, v\rho} \, dv = R(\rho, \theta) \star b(\rho).$$

Here, $b(\rho)$ is the so-called *ramp filter*[2] to $R(\rho, \theta)$, defined in the frequency domain as $B(v) = |v|$. Hence, $f(x, y)$ can be obtained by backprojection of $R'(\rho, \theta)$ using Eq (C.4). The filtered projection profile $R'(\rho, \theta)$ is obtained by applying the ramp filter $b(\rho)$. The FBP algorithm is mostly used in medical applications. The flow of the FBP algorithm is presented in Figure C.11.

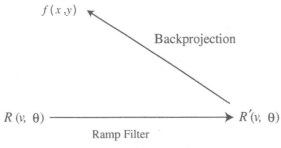

Figure C.11 FBP Algorithm.

C.4.6. Practical Issues. (i) In the case of insufficient sampling, a reduction in the number of projections is required for the purpose of (a) reducing scanning time; (b) reducing noise; (c) reducing motion artifact; and (d) reducing patient dose.

(ii) Aliasing: in the case of insufficient radial sampling, which occurs when there is a sharp intensity change caused by, e.g., bones.

(iii) Incomplete/Missing data: a part of data cannot be acquired due to physical or instrumental limitations, such as limited angles, or in CT metal blocking radiation.

(iv) Motion artifact, caused by patient's movement during data acquisition, such as respiration and heartbeat.

(v) Noise, where the photon detection is a stochastic process: lower noise level can be obtained by increasing the radiation dose, and/or the acquisition time, where the perception of structures depends on the contrast, size, and noise level.

(vi) Finite resolution: finite detector size limits the resolution of the acquired data, producing FBP imaging of noisy and blurred sinograms.

(vii) Partial beam effect and beam hardening can occur.

---

[2] Ramp filters belong to the family of backprojection filters.

D

Maxwell's Equations and Solitons

We will define Maxwell's equations, and derive the sine-Gordon equation in four different cases. Various solitons, i.e., traveling wave type solutions, of this equation are constructed and studied.

D.1 Maxwell's Equations

Maxwell's equations are defined as

$$\nabla \cdot \mathbf{E} = 4\pi \left(q_i n_i + q_e n_e \right), \quad \nabla \cdot \mathbf{B} = 0,$$

$$\frac{\partial \mathbf{B}}{\partial t} + c \left(\nabla \times \mathbf{E} \right) = 0, \quad -\frac{\partial \mathbf{E}}{\partial t} + c \left(\nabla \times \mathbf{B} \right) = 4\pi \left(q_i n_i \mathbf{u}_i + q_e n_e \mathbf{u}_e \right),$$

$$\text{(D.1)}$$

with the equation of continuity and the equation of state defined, respectively, by

$$\frac{\partial n_j}{\partial t} + \nabla \cdot (n_j \mathbf{u}_j) = 0, \quad p_j = n_j T_j, \tag{D.2}$$

where $\mathbf{E}$ is the electric field, $\mathbf{B}$ the magnetic field, T the product of the Boltzmann constant and the temperature, q the charge, m the mass, c the speed of light, and $\mathbf{u}$ a unit-magnitude vector. The classical theory for the propagation of light based on Maxwell's equations produced a unified treatment of electric and magnetic fields. The law of conservation of charge is related to Maxwell's equations. Let $\mathbf{H}$ and $\mathbf{J}$ represent the magnetic intensity and current density, respectively. If $\mathbf{X}$ represents $\mathbf{E}$, $\mathbf{H}$, or $\mathbf{J}$ in a conducting medium, respectively, then they satisfy the equation

$$\nabla^2 \mathbf{X} = \mu \epsilon \mathbf{X}_{tt} + \mu \sigma \mathbf{X}_t, \tag{D.3}$$

where μ is the magnetic inductive capacity, ϵ the electric inductive capacity, and σ the electrical conductivity. For details see Courant and Hilbert [1953].

An electromagnetic wave can be regarded as a modulated, circularly polarized plane wave traveling along the x-axis such that electric field $\mathbf{E}(x, t)$ is represented as

$$\mathbf{E}(x, t) = A(x, t) \left(\mathbf{j} \cos \theta + \mathbf{k} \sin \theta \right), \tag{D.4}$$

where $A(x,t)$ is the amplitude, $\theta = (kx - \omega t) + \phi(x)$, k being the refractive index determined by the linear dispersion relation of the medium, and $\phi(x)$ the phase that varies slowly compared to the carrier wave, i.e.,

$$\left|\frac{\partial A}{\partial x}\right| \ll |kA|, \quad \left|\frac{\partial A}{\partial t}\right| \ll |\omega A|, \tag{D.5}$$

with similar results holding for $\phi(x)$. The electric field $\mathbf{E}$ can be determined from Maxwell's equations (D.1): it can be shown to satisfy the wave equation

$$\frac{\partial^2 \mathbf{E}}{\partial x^2} - \left(\frac{n}{c}\right)^2 \frac{\partial^2 \mathbf{E}}{\partial t^2} = \left(\frac{4\pi}{c^2}\right) \frac{\partial^2 \mathbf{P}}{\partial t^2}, \tag{D.6}$$

where $\mathbf{P}(x,t)$ is the polarization of the medium caused by the electromagnetic wave. If the equation for $\mathbf{P}$ is known, then the two unknowns, amplitude A and phase ϕ, can be determined from (D.4) and (D.6). In the case when the medium is constituted of N two-level atoms and the transition frequencies ω of individual atoms are different but are distributed around the frequency ω_0 of the incident wave field with the spectral density function $f(\Delta\omega)$ normalized by the conditions $\Delta\omega = |\omega - \omega_0|$ and $\int_{-\infty}^{\infty} f(\Delta\omega)d(f(\Delta\omega)) = 1$, then the polarization of the medium becomes

$$\mathbf{P}(x,t) = N \int_{-\infty}^{\infty} f(\Delta\omega)\mathbf{p}(\Delta\omega, x, t)\, d(f(\Delta\omega)), \tag{D.7}$$

where $\mathbf{p}(\Delta\omega, x, t)$ is the polarization of a two-level atom induced by the electromagnetic wave such that it satisfied the slow-variation conditions of the type (D.5). This implies that the polarization $\mathbf{p}$ consists of two parts, one being in the phase with the carrier wave of the electric field $\mathbf{E}(x,t)$, and the other out of phase, so that it can be represented as

$$\mathbf{p}(\Delta\omega, x, t) = \nu_1(\Delta\omega, x, t)\,(\mathbf{j}\cos\theta + \mathbf{k}\sin\theta) + (\Delta\omega, x, t)\,(\mathbf{j}\cos\theta - \mathbf{k}\sin\theta), \tag{D.8}$$

where $\nu_{1,2}$ are the modulations of the electric field and are slowly varying functions of x and t. These modulations can be determined from the Schrödinger equation for a two-level atom in the incident electric field. According to Polyanin and Zaitsev [2004], the equations for $\nu_{1,2}$ are

$$\frac{\partial \nu_1}{\partial t} = \nu_2 \Delta\omega,$$

$$\frac{\partial \nu_2}{\partial t} = -\nu_1 \Delta\omega - \left(\frac{\kappa^2}{\omega}\right) AW, \quad \kappa = \frac{2p}{\hbar}, \tag{D.9}$$

$$\frac{\partial W}{\partial t} = \nu_2 A\omega,$$

where p is the magnitude of the electric dipole moment, $\hbar = 1.054 \times 10^{-34}$ joules-sec is the Planck's constant, and W is the energy of the two-level atom such that $W = -\frac{1}{2}\hbar\omega_0$ in the general case and $W = -\frac{1}{2}\hbar\omega$ in the excited state. Eqs (D.6) and (D.9) form a closed system for $\nu_{1,2}$. Since $\nu_{1,2}$ and $\mathbf{E}$ are slowly varying functions for the modulations of $\mathbf{p}$, the equations for the amplitude A and phase ϕ are obtained by substituting Eqs (D.4), (D.7), and (D.8) in (D.6) and using the conditions (D.5) as

$$\frac{\partial A}{\partial x} + \left(\frac{n}{c}\right)\frac{\partial A}{\partial t} = -\frac{2\pi\omega N}{nc}\int_{-\infty}^{\infty}\nu_2 f(\Delta\omega)\,d(\Delta\omega),$$

$$A\frac{\partial\phi}{\partial x} = \frac{2\pi\omega N}{nc}\int_{-\infty}^{\infty}\nu_1 f(\Delta\omega)\,d(\Delta\omega). \tag{D.10}$$

Thus, Eqs (D.9) for $\nu_{1,2}, W, A$ and ϕ define the self-induced transparency phenomena of pulsed coherent light.

D.2 sine-Gordon Equation

In the particular case when all transition frequencies of N two-level atom are identical, i.e., when $f(\Delta\omega) = \delta(\Delta\omega)$ and the frequency ω of the electromagnetic wave is equal to that frequency, i.e., when $(\Delta\omega) = 0$, Eqs (D.9) reduce to the sine-Gordon equation if we set $\nu_1 = 0$ and $\phi = 0$ because of the first equation in (D.9) and second equation in (D.10) and, writing ν_2 as simply ν, we obtain

$$\frac{\partial\nu}{\partial t} = -\left(\frac{\kappa^2}{\omega}\right)AW,$$

$$\frac{\partial W}{\partial t} = \omega_0\nu A,$$

$$\frac{\partial A}{\partial x} + \frac{n}{c}\frac{\partial A}{\partial t} = -\frac{2\pi\omega N}{nc}\nu. \tag{D.11}$$

Introducing a new function $u(x,t)$ by

$$u(x,t) = \kappa\int_{-\infty}^{t}A(x,\tau)\,d\tau, \tag{D.12}$$

and using the boundary condition $W(x,t)\big|_{t=-\infty} = -\frac{1}{2}\hbar\omega_0 = W_0$, Eqs (D.11) yield the following forms for the functions $\nu(x,t)$ and $W(x,t)$:

$$\nu(x,t) = p\sin(u(x,t)), \quad W(x,t) = W_0\cos(u(x,t)). \tag{D.13}$$

At this state all two-level atoms are in the ground state. Using Eq (D.12) and the first equation in (D.13), the third equation in (D.11) reduces to the sine-Gordon equation

$$u_{xx} + \frac{1}{c'}u_{tt} = -\gamma^2\sin u, \tag{D.14}$$

where $c' = c/n$ and $\gamma^2 = (\pi N\omega_0\hbar)/(nc)$. With a suitable transformation of variables, Eq (D.14) can be written as

$$u_{xx} - u_{tt} = \sin u. \tag{D.15}$$

Setting $v(x,t) = \tan\left(\dfrac{u}{4}\right)$, and using the trigonometric identity $\sin u = \dfrac{4v(1+v^2)}{(1+v^2)^2}$, Eq (D.15) reduces

$$(1+v^2)\left(v_{xx} - v_{tt} - v\right) - 2v\left(v_x^2 - v_t^2 - v^2\right) = 0,$$

which can be solved using Bernoulli's method of separation of variables by setting $v(x,t) = \dfrac{X(x)}{T(t)}$ and using the identity $\sin 4\theta = \dfrac{4\tan\theta\left(1 - \tan^2\theta\right)}{\left(1 + \tan^2\theta\right)^2}$, which gives

$$\left(X^2 + T^2\right)\left(\frac{X_{xx}}{X} + \frac{T_{tt}}{T}\right) - 2\left(X_x^2 + T_t^2\right) = X^2 - T^2.$$

If we differentiate this equation with respect to x and t, the equations are separated into

$$\frac{1}{XX_x}\left(\frac{X_{xx}}{X}\right) = \frac{1}{TT_t}\left(\frac{T_{tt}}{T}\right) = -4k^2, \tag{D.16}$$

where $-4k^2$ is the separation constant. Each ordinary differential equation in (D.16) can be integrated twice to give

$$X'^2 = -k^2X^4 + aX^2 + b, \quad T^2 = k^2T^4 + cT^2 + d,$$

which, when substituted in (D.16), yields $a - c = 1$ and $b + d = 0$. Then, setting $a = m^2$ and $b = n^2$, we get

$$X_x^2 = -k^2X^4 + m^2X^2 + n^2, \quad T_t^2 = k^2T^4 + (m^2 - 1)T^2 - n^2, \tag{D.17}$$

where m and n are integration constants. In general, Eqs (D.17) can be solved in terms of elliptic functions. There are four particular cases of interest, which we will discuss below.

CASE 1: $k = n = 0$ and $m > 1$. Eqs (D.17) become

$$X - x = \pm mX, \quad T_t = \pm\sqrt{m^2 - 1}\,T, \tag{D.18}$$

which has the solutions

$$X(x) = a_1\,e^{\pm mx}, \quad T(t) = a_2\,e^{\pm\sqrt{m^2-1}\,t}, \tag{D.19}$$

where $a_{1,2}$ are integration constants. This gives

$$u(x,t) = 4\arctan\left[\alpha\exp\left(\pm\frac{x \pm Ut}{\sqrt{1-U^2}}\right)\right],\qquad(\text{D.20})$$

where $\alpha = a_1/a_2$ and $U = \dfrac{\sqrt{m^2-1}}{m}$ or $m = (1-U^2)^{-1/2}$ are constants.
Note that one of the solutions in (D.20) is of the form

$$u(x,t) = 4\arctan\left[\alpha\, e^{m(x-Ut)}\right].\qquad(\text{D.21})$$

This represents the *soliton* solution of the sine-Gordon equation. Another form of the solution (D.21) is

$$u(x,t) = 4\arctan\left[\alpha\, e^{m(x-Ut)}\right] = 4\,\text{arccot}\left[\frac{1}{\alpha}e^{m(x-Ut)}\right],\qquad(\text{D.22})$$

which represents the *antisoliton* solution of the sine-Gordon equation. The functions u_x and u_t defined by

$$u_x(x,t) = \pm 2m\,\text{sech}\left[m(x \pm Ut) + \log\alpha\right],$$
$$u_t(x,t) = \pm 2\sqrt{m^2-1}\,\text{sech}\left[m(x \pm Ut) + \log\alpha\right],\qquad(\text{D.23})$$

represent solitary wave solutions, where the minus sign in each represents *antisoliton* solution.

CASE 2: $k = 0$, $m^2 > 1$ and $n \neq 0$. Eq (D.17) gives the solutions after integration in terms of hyperbolic functions as

$$X(x) = \pm\frac{n}{m}\,\text{sech}\,(mx + a_1),\quad T(t) = \frac{n}{\sqrt{m^2-1}}\cosh\left(\sqrt{m^2-1}\,t + a_2\right),\qquad(\text{D.24})$$

where $a_{1,2}$ are integration constants. Then the solution is given by

$$u(x,t) = \pm\arctan\left[\frac{U\sinh(mx + a_1)}{\cosh\left(\sqrt{m^2-1}\,t + a_2\right)}\right].\qquad(\text{D.25})$$

Note that this solution does not depend on n. In particular, if $a_1 = a_2 = 0$, the solution (D.25) becomes

$$u(x,t) = \pm 4\arctan\left[\frac{U\sinh(mx)}{\cosh\left(\sqrt{m^2-1}\,t\right)}\right],\quad 0 < U^2 < 1.\qquad(\text{D.26})$$

This solution, found by Perring and Skyrme [1962], describes the interaction of two equal solitons by expressing $\sinh(mx)$ as the sum of two exponentials. Noting that $\cosh(mUt) \sim \frac{1}{2}e^{mU|t|}$ as $t \to \pm\infty$, we have

$$u(x,t) \sim U\, e^{m(x+Ut)} - U\, e^{-m(x-Ut)}\quad\text{as } t \to -\infty,$$

which holds uniformly for all x and represents two distinct solitons approaching each other with equal and opposite speed U, as shown in Figure D.1(a). In fact, in the limit as $x \to -\infty$ and $t \to -\infty$, the solution (D.26) becomes

$$u(x,t) \sim -4\arctan\left[U\,e^{-m(x-Ut)}\right].\tag{D.27}$$

This solution represents a pulse moving in the positive x direction from -2π to 0 as x passes through the value Ut.

On the other hand, as $x \to +\infty$ and $t \to -\infty$, the solution has the form

$$v(x,t) \sim 4\arctan\left[U\,e^{-m(x+Ut)}\right].\tag{D.28}$$

This represents a pulse traveling in the negative x direction as $u(x,t)$ increases from 0 to 2π while x passes through the value $-Ut$. At $t=0$ the two pulses undergo an interaction, as shown in Figure D.1(b). As $t \to +\infty$, we have

$$v(x,t) \sim -U\exp\left(-\frac{x+Ut}{\sqrt{1-U^2}}\right) + U\exp\left(-\frac{x-Ut}{\sqrt{1-U^2}}\right),$$

which as $x \to -\infty$ and $t \to +\infty$ yields the asymptotic solution

$$u(x,t) \sim -4\arctan\left[U\exp\left(-\frac{x+Ut}{\sqrt{1-U^2}}\right)\right].\tag{D.29}$$

As $x \to +\infty$ and $t \to +\infty$, this asymptotic solution becomes

$$u(x,t) \sim 4\arctan\left[U\exp\left(\frac{x-Ut}{\sqrt{1-U^2}}\right)\right].\tag{D.30}$$

Here, u varies from -2π to 2π as x goes from $-\infty$ to $+\infty$; the related solution is called a 4π-pulse.

The only noticeable interaction as $t \to +\infty$ is a longitudinal displacement of each soliton. To see this, note that the solution satisfies

$$\pm U\,e^{\pm m(x+Ut)} = \pm e^{\pm m(x+Ut\pm\gamma/2)},$$

where

$$\gamma = 2\sqrt{1-U^2}\,\log\left(U^{-1}\right)$$

denotes the displacement of each soliton which is decreased by the interaction. The above soliton solution yields, as before, the functions u_x and u_t. For example, for $U > 0$ we have

$$u_x(x,t) = \frac{4U\cosh(mx)\cosh(mUt)}{\sqrt{1-U^2}\left[\sinh^2(mx)+\cosh^2(mUt)\right]}.\tag{D.31}$$

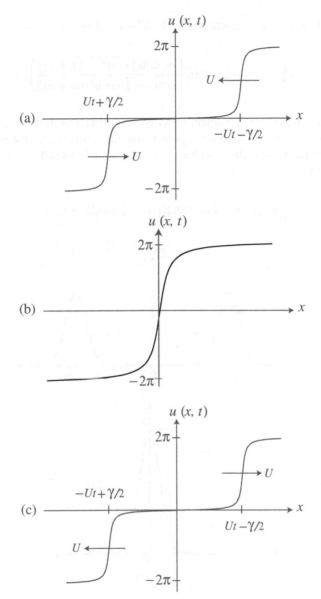

Figure D.1 Interaction of Two Equal Solitons.

This represents the interaction of two hump-shaped solitons presented in Figure D.2, where (a) represents the case $t \to -\infty$; (b) for $t = 0$, and (c) for $t \to +\infty$ with amplitude $2U/\sqrt{1 - U^2}$.

CASE 3: $k \neq 0$, $n = 0$, and $m^2 > 1$. The solution $u(x,t)$ is

$$u(x,t) = -4\arctan\left[\frac{m\sinh\left(\sqrt{m^2-1}\,t + a_2\right)}{\sqrt{m^2-1}\,\cosh(mx + a_1)}\right]. \qquad (D.32)$$

This solution represents a soliton moving toward the boundary and reflected back as an antisoliton, thereby representing an interaction of a soliton with an antisoliton. In the limit as $m \to 1$, the solution (D.32) reduces to the asymptotic form

$$u(x,t) \sim -4\arctan\left[(t + a_2)\,\mathrm{sech}(x + a_1)\right]. \qquad (D.33)$$

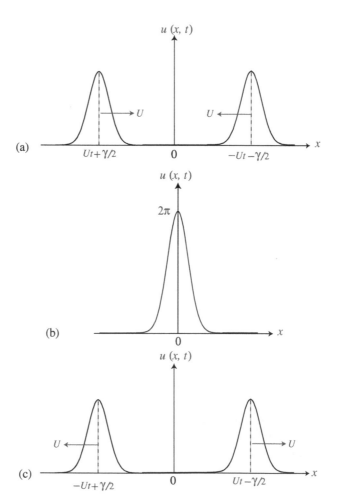

Figure D.2 Interaction of Hump-Shaped Solitons.

CASE 4: $k \neq 0$, $n = 0$, and $m^2 < 1$. This solution is given by

$$u(x,t) \sim -4 \arctan \left[\frac{m \sinh(\omega t + a_2)}{\sqrt{1 - m^2} \cosh(mx + a_1)} \right]$$

$$= -4 \arctan \left[\frac{\sinh(\omega t + a_2)}{U \cosh(mx + a_1)} \right], \tag{D.34}$$

where $\omega = \sqrt{1 - m^2} = mU$. This solution, known as the *breather solution* of the sine-Gordon equation, represents a pulse-type soliton which for fixed x becomes a periodic function of time (i.e., a sinusoid) with frequency ω. This solution is shown in Figure D.3 for $m = 0.8$. The case $m \ll 1$ yields a small-amplitude breather solution which can be obtained from (D.34) by expanding the arctan function for small m and keeping only the first term in this expansion. It is given by

$$u(x,t) \sim 2i\, m \operatorname{sech}(mx) \exp\left[i\left(t - \tfrac{1}{2}m^2\right)t\right] + \cdots = w(x,t)\, e^{it}, \tag{D.35}$$

where $w(x,t) = 2i\, m \operatorname{sech}(mx)\, e^{itm^2/2}$, such that $w(x,t)$, after neglecting the term $w_{tt} = O(m^2)$, satisfies the one dimensional Schrödinger nonlinear equation

$$2i\, w_t - w_{xx} - |w|^2 w = 0. \tag{D.36}$$

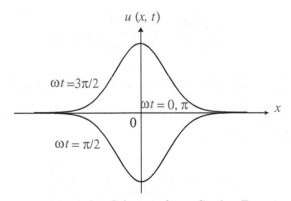

Figure D.3 Breather Solution of sine-Gordon Equation.

D.3 K-dV Equation

A soliton is a self-reinforcing solitary wave or pulse that maintains its shape while it travels at constant speed. Solitons are caused by cancellation of nonlinear and dispersive effects in the medium. Thus, a soliton wave appears when dispersion balances refraction. For example, the dispersion relation for a water wave in an ocean at depth h is $\omega^2 = gh \tanh ah$, where ω is the radian frequency, g the acceleration due to gravity, and a the amplitude. Thus,

$\omega = \left(gh\tanh ah\right)^{1/2} \approx ca\left(1 - \frac{1}{6}a^2h^2\right)$, where $c = \sqrt{gh}$. In 1895, Korteweg-deVries developed the following nonlinear equation for long dispersive water waves in a channel of depth h as

$$\eta_t + c_0\left(1 + \frac{3}{2}\frac{\eta}{h}\right)\eta_x + \sigma\eta_{xxx} = 0, \tag{D.37}$$

where $\eta(x,t)$ determines the free surface elevation of the waves, c_0 is the shallow water wave speed, and σ is a constant for fairly long waves. The dispersive term η_{xxx} in this K-dV equation allows solitary and periodic waves that are not found in shallow water wave theory. Eq (D.37) is solved in the form $\eta(x,t) = hf(X), X = x - Ut$, where U is a constant wave velocity. Substituting this form into Eq (D.37) yields $\frac{1}{6}h^2f''' + \frac{3}{2}ff' + \left(1 - \frac{U}{c}\right)f' = 0$, $\sigma = ch^2/6$, which after integrating twice gives

$$\frac{1}{3}h^2f'^2 + f^3 + \left(1 - \frac{U}{c}\right)f^2 + 4Af + B = 0, \tag{D.38}$$

where A and B are integration constants. Since we seek a solitary wave solution with boundary conditions $f, f', f'' \to 0$ as $|X| \to \infty$, we have $A = B = 0$. With these values of A and B, we integrate Eq (D.38) and obtain

$$X = \int_0^f \frac{df}{f'} = \sqrt{\frac{h^2}{3}}\int_0^f \frac{df}{f\sqrt{\alpha - f}},$$

where $\alpha = 2\left(\frac{U}{c} - 1\right)$. The substitution $f = \alpha\,\mathrm{sech}^2\,\theta$ reduces this result to $X - X_0 = \left(\frac{4h^2}{3\alpha}\right)^{1/2}\theta$, where X_0 is an integration constant. Hence, the solution for $f(X)$ is given by

$$f(X) = a\,\mathrm{sech}^2\left[\sqrt{\frac{3\alpha}{4h^2}}\,(X - X_0)\right]. \tag{D.39}$$

This solution increases from $f = 0$ as $X \to -\infty$ such that it attains a maximum value $f = f_{\max}$ at $X = 0$, and then decreases to $f = 0$ as $X \to \infty$, as shown in Figure D.4. This also implies that $X_0 = 0$ so that the solution (D.39) becomes $f(X) = \alpha\,\mathrm{sech}^2\left[\sqrt{\frac{3\alpha}{4h^2}}\,X\right]$. Hence the final solution is

$$\eta(x,t) = \eta_0\,\mathrm{sech}^2\left[\sqrt{\frac{3\eta_0}{4h^2}}\,(x - Ut)\right], \tag{D.40}$$

where $\eta_0 = \alpha h$. This is called a solitary wave solution of the K-dV equation, also known as a *soliton* (named so by Zabusky and Kruskal [1965]).

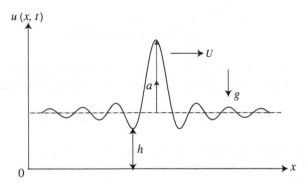

Figure D.4 A Soliton.

Since $\eta > 0$ for all x, the soliton is a self-reinforcing wave that propagates in the medium without change of shape with velocity $U = c\left(1 + \dfrac{\eta_0}{2h}\right)$, which is directly proportional to the amplitude η_0. The width $\sqrt{4h^3/(3\eta_0)}$ is inversely proportional to $\sqrt{\eta_0}$, which means that the soliton propagates to the right with a velocity U that is directly proportional to the amplitude. Therefore, the taller solitons travel faster and are narrower than the shorter (or slower) ones. They can overtake the shorter ones and, surprisingly, they emerge from this interaction without change of shape. These waves are stable and can travel very long distances. A sech-envelope soliton for water waves is shown in Figure D.5.

Figure D.5 A Soliton for Water Waves.

D.4 Nonlinear Schrödinger Equation

Besides the K-dV equation, soliton solutions can also be obtained from the nonlinear Schrödinger equation $i\psi_t + \frac{1}{2}w''\psi_{xx} = 0$, which after using a Taylor expansion for $w = w(k, a^2)$ of the form

$$w = w_0 + (k - k_0)\left(\frac{\partial^2 w}{\partial k^2}\right)_{k=k_0} + \tfrac{1}{2}(k - k_0)^2\left(\frac{\partial^2 w}{\partial k^2}\right)_{k=k_0} + \left(\frac{\partial w}{\partial |a|^2}\right)_{|a|^2=k_0},$$

where $w_0 = w(k_0)$, reduces to

$$i\left(a_t + w_0' a_x\right) + \frac{1}{2}\,w_0'' a + xx + \gamma |a|^2 a = 0,$$

where $\gamma = -\left(\dfrac{\partial w}{\partial |a|^2}\right)_{|a|^2 = k_0}$ is a constant. The amplitude satisfies the normalized nonlinear Schrödinger equation

$$i\, a_t + \frac{1}{2}\, w_0'' a_{xx} + \gamma |a|^2 a = 0. \tag{D.41}$$

The wave modulation is stable if $\gamma w_0'' < 0$ and unstable if $\gamma w_0'' > 0$. In this case again, writing the nonlinear Schrödinger equation in the standard form

$$i\psi_t + \psi_{xx} + \gamma |\psi|^2 \psi = 0, \quad -\infty < x < \infty, \ t \geq 0, \tag{D.42}$$

we seek a solution of the form $\psi(x,t) = f(\xi)\, e^{i(mX - at)}$, $\xi = x - Ut$ for some finite f and constant wave speed U, and m and n are constants. The solution of Eq (D.42) is

$$f(\xi) = \sqrt{\frac{2\alpha}{\gamma}}\, \text{sech}^2 \left[\sqrt{\alpha}\,(x - Ut)\right], \quad \alpha > 0. \tag{D.43}$$

This represents a soliton that propagates without change of shape but constant velocity U. Unlike the K-dV equation, the amplitude and velocity of this wave are independent parameters. The soliton exists only for the unstable case ($\gamma > 0$), which means that small modulations of the unstable wave train lead to a series of solitons. Since $\alpha > 0$ for all ξ, the soliton is a wave of elevation which is symmetric about $\xi = 0$. It propagates in the medium without change of shape with velocity $U = c_0 \left(1 + \frac{1}{2}\alpha\right)$, which is directly proportional to the amplitude $\sqrt{2\alpha/\gamma}$. The solitary wave propagates to the right with velocity U which has the width inversely proportional to the square root of the amplitude. This means, just as with the K-dV equation, that taller solitons are narrower but travel faster than the shorter (slower) ones. They can overtake the shorter ones and emerge from the interaction without change of shape (see Kythe and Kythe [2012: §17.8.1]).

Besides the K-dV and nonlinear Schrödinger equations, dispersion and nonlinearity effects in the coupled nonlinear Schrödinger equation and the sine-Gordon equation also interact and generate permanent and localized solitary waves. These waves were noticed by John Scott Russell in 1834, who called them the 'wave of translation'. Solitons occur in fiber optics, which prompted Robin Bullough in 1973 to propose the idea of a soliton-based transmission system in optical telecommunications. For applications of discrete solitons in coding theory, see Kythe and Kythe [2012: 368ff].

E

Modulation

The concept and types of modulation, and waveshaping methodology is discussed. This appendix is useful in almost all aspects of signal processing including computerized music.

E.1 Definition

In ordinary speech, modulation is an inflection of the tone or pitch of the voice, generally used to convey meaning. In music, it is a change from one musical key to another. It is the process of varying any one of the following three properties: amplitude (volume), frequency (pitch), or phase (timing) of a carrier signal using a modulating signal which typically contains information to be transmitted. A low frequency signal is used to vary any one of these properties to obtain a modulated signal. Usually a high-frequency sinusoid waveform is used as a carrier signal, while the square pulse is sometimes preferred. Modulation is, therefore, a process of transmitting a message signal consisting of a digital bit stream or an analog audio signal inside another signal (carrier signal) that can be physically transmitted.

A device used for modulation is known as a *modulator*, and for inverse operation a *demodulator* ('demod' for short). A device that does both of these operation is called a *modem* (**mo**dulator-**dem**odulator).

The effect of *digital modulation* is to transfer a sequence of digital bits over an analog bandpass channel (e.g., public switched telephone network, limited to the frequency range of 300 to 3400 Hz, or a limited radio frequency band). The *analog modulation* transfers an analog baseband (or lowpass) signal (e.g., an audio signal or TV signal) over an analog bandpass channel at a different frequency. Both analog and digital modulations facilitate frequency division multiplexing (FDM), where several lowpass information signals are transferred simultaneously over the same shared medium, using separate bandpass channels.

The FDM technique divides the total bandwidth available in a communication medium into a series of nonoverlapping frequency sub-bands, each of which is used to carry a separate signal. It allows a single medium to be shared by many signals. Cable TV uses this technique to transmit many television channels simultaneously on a single cable.

The *digital baseband modulation*, also known as *line coding*, is used to transfer a sequence of digital bits over a baseband channel. The *pulse modulation* methods are used to transfer a narrowband analog signal, e.g., in phone calls over a wideband baseband channel, or a sequence of bits over another digital transmission system.

Music synthesizers use modulation to synthesize waveforms with an extensive overtone spectrum using a small number of oscillators. For more on this subject see Chapter 10, and Sudakshina [2010: 163–184].

Modulation of a sinusoid is used to transform a baseband message signal into a passband signal, for example low-frequency audio signal into a radio-frequency (RF) signal in radio communications, cable TV signals or the public switch-board telephone networks. Electrical signals can only be transmitted over a limited passband frequency spectrum with nonzero lower and upper cutoff frequencies. The advantage in using a sinusoid carrier is to keep the frequency content of the transmitted signal as close as possible to the center frequency (carrier frequency) of the passband.

Let ω_c, ω_m, and r denote the carrier frequency, modulation frequency, and index of modulation, respectively. Then, in the case of a frequency-modulated sinusoid, frequency varies sinusoidally at some angular frequency ω_m, about a central frequency ω_c such that the instantaneous frequencies vary on the interval $((1-r)\omega_c, (1+r)\omega_c)$, where r controls the depth of the variation and ω_m controls the frequency of variation. A similar formulation is used if the phase of the carrier sinusoid is modulated sinusoidally, since the instantaneous frequency is the rate of change of phase and the rate of change of a sinusoid is another sinusoid. The phase modulation is analyzed as follows: Assuming that the modulating oscillator and the waveguide are both sinusoidal, and the carrier and modulation frequency remain fixed in time, the signal is defined by

$$x(n) = \cos\left(a\cos(n\omega_m) + n\omega_c\right),$$

where the parameter $a > 0$ replaces the earlier parameter r and becomes the new index of modulation which controls the range of frequency variation relative to the carrier frequency ω_c. If $a = 0$, there is no frequency variation and the signal $x(n)$ reduces to $x(n) = \cos n\omega_c$ which is the unmodulated carrier sinusoid. Thus, the resulting spectrum is defined by the signal

$$x(n) = \cos(n\omega_c) \cdot \cos\left(a\cos(n\omega_m)\right) - \sin(n\omega_c) \cdot \sin\left(a\cos(n\omega_m)\right),$$

which is the product of the two harmonic spectra, one of which consists of the waveshaping outputs $\cos\left(a\cos(n\omega_m)\right)$ and the other consists of the outputs $\sin\left(a\cos(n\omega_m)\right)$, such that the former have even harmonics at $0, 2\omega_m, 4\omega_m, \ldots$

and the latter odd harmonics at $\omega_m, 3\omega_m, 5\omega_m, \ldots$, each spectrum multiplied by the sinusoid of the carrier frequency ω_c. The entire spectrum is centered at the carrier frequency ω_c with sidebands at both even and odd multiples of the modulation frequency ω_m, contributed respectively by the cosine and sine waveshaping terms. The index a of modulation, as it changes, controls the relative strength of the various partials which are located at the frequencies $\omega_c + m\omega_m$, where $m = \ldots, -2, -1, 0, 1, 2, \ldots$. Further, suppose that the two frequencies are multiples of some common frequency ω, i.e., $\omega_m = k\omega$ and $\omega_c = m\omega$, where k and m are relatively prime. Then, since the two periodic signals are multiplied, the resulting product will repeat at a frequency of the common submultiple of ω. But if no common multiple of ω exists, or if the only submultiples are smaller than any discernible pitch, the result will be inharmonic. A block diagram for frequency modulation (FM) (a) in the classic form, and (b) in a phase modulation is shown in Figure E.1, where a is the index of modulation.

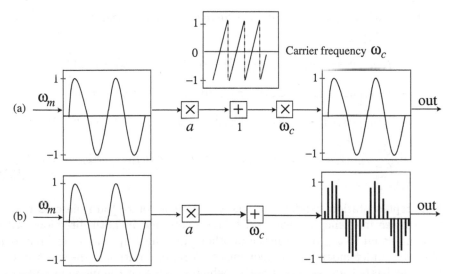

Figure E.1 Frequency Modulation in (a) Classic Form; (b) in Phase Modulation Form.

E.2 Phase Modulation and FM

Phase modulation is obtained by splitting the carrier oscillator into its phase and cosine lookup components. In this case the signal has the form

$$x(t) = \cos\left(a\cos(n\omega_m) + n\omega_c\right), \tag{E.1}$$

where a is the (angular) index of modulation, and ω_m and ω_c are the (angular) modulation and carrier frequency, respectively. Thus, applying the formula $\cos(u + v) = \cos u \cos v - \sin u \sin v$ to (E.1), and using the expansions (see

Abramowitz and Stegun [1972: 361])

$$\cos(z \cos \theta) = J_0(z) + 2 \sum_{k=1}^{\infty} (-1)^k J_{2k}(z) \cos(2k\theta),$$

$$\sin(z \cos \theta) = 2 \sum_{k=0}^{\infty} (-1)^k J_{2k+1}(z) \cos \left((2k+1)\theta \right),$$

and

$$\cos z = J_0(z) - 2J_2(z) + 2J_4(z) - 2J_6(z) + \cdots ,$$
$$\sin z = 2J_1(z) - 2J_3(z) + 2J_5(z) - 2J_7(z) + \cdots ,$$

the signal spectrum becomes

$$
\begin{aligned}
x(t) &= \cos \left(a \cos(n\omega_m) \right) \cos(n\omega_c) - \sin \left(a \cos(n\omega_m) \right) \sin(n\omega_c) \\
&= J_0(a) \, \cos(n\omega_c) \\
&\quad + J_1(a) \Big[\cos \left(n(\omega_m + \omega_c) + \pi/2 \right) + \cos \left(n(\omega_m - \omega_c) + \pi/2 \right) \Big] \\
&\quad + J_2(a) \Big[\cos \left(n(\omega_m + 2\omega_c) + \pi \right) + \cos \left(n(\omega_m - 2\omega_c) + \pi \right) \Big] \\
&\quad + J_3(a) \Big[\cos \left(n(\omega_m + 3\omega_c) + 3\pi/2 \right) + \cos \left(n(\omega_m - 3\omega_c) + 3\pi/2 \right) \Big] + \cdots \\
&= \sum_{k=1}^{\infty} J_k(a) \Big[\cos \left(n(\omega_c + k\omega_m) + k\pi/2 \right) + \cos \left(n(\omega_c - k\omega_m) + k\pi/2 \right) \Big].
\end{aligned}
$$
$$\text{(E.2)}$$

This representation signifies that the components of $x(t)$ are centered about the carrier frequency ω_c with sidebands extending in both directions with spacing equal to ω_m between one another, and the amplitudes are functions of a, thus independent of the frequencies. A representation of the spectrum of the signal $x(t)$ is given in Figure E.2, which shows that the phase modulation is of the form of ring modulated waveshaping.

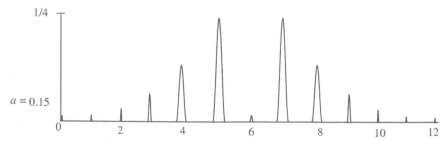

Figure E.2 Phase Modulation as Ring-Modulated Waveshaping.

On the other hand, frequency modulation is not restricted to purely sinusoidal carrier or modulation oscillators. An example is to effect a phase
modulation on the phase modulation spectrum itself. Thus, let a_1 and a_2 be
two indices of modulation with respective modulation frequencies ω_{m_1} and
ω_{m_2}, respectively. Then the signal has the waveform

$$x(n) = \cos\left(n\omega_c + a_1 \cos(n\omega_{m_1}) + a_2 \cos(n\omega_{m_2})\right), \qquad (E.3)$$

and thus, the spectrum of the signal $x(t)$ becomes

$$x(t) = \sum_{k=1}^{\infty} J_k(a_1)\Big[\cos\left(n(\omega_c + k\omega_{m_1}) + k\pi/2 + a_2 \cos(nk\omega_{m_2})\right)$$

$$- \cos\left(n(\omega_c + k\omega_{m_1}) + k\pi/2 + a_2 \cos(nk\omega_{m_2})\right)\Big]. \qquad (E.4)$$

The right and left sidebands are again centered about ω_c, described by the
first and the second cosine terms for $k \neq 0$ inside the square brackets in
(E.4), with frequencies $\omega_c + k\omega_{m_1} + \nu\omega_{m_2}$ for $\nu = 0, \pm 1, \pm 2, \ldots$, amplitude
$J_k(a_1)J_k(a_2)$, and phase $(k + \nu)\pi/2$.

E.3 Waveshaping

This is an alternative method of modulating a signal. In waveshaping, a signal
is passed through a suitably chosen nonlinear function. A function f, called
the *transfer function*, is used to distort the input waveform into a different
(new) shape, which depends on the shape and amplitude of the input wave.
The amplitude of the input waveform, called the *waveshaping index*, is very
significant as it affects the shape of the output waveform. This transformation
method can be used repeatedly. Thus, it is advisable to include a leading
amplitude control as part of the waveshaping operation, which is shown in
Figure E.3. A small index sometimes results in very little distortion so much
so that the input and output resemble each other. Thus, a larger index always
leads to a more distorted output.

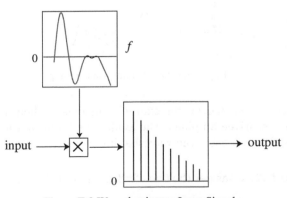

Figure E.3 Waveshaping an Input Signal.

Waveshaping is a nonlinear function of its input signals, whereas ring modulation is linear. However, by considering the input signals in waveshaping as separate linear processes, we can analyze the nonlinear case if we consider the action of all input components together with the interactions among them. Although the results may sometimes become very complex, we will explain the waveshaping process by some examples.

Example E.1. To show the effect of the magnitude of the index, let the transfer function f behave like a *clipping function*. The index (input magnitude) has a marked effect on the output waveform, where the clipping function passes its input to the output unchanged as long as it stays in the interval $(-0.3, 0.3)$. That is, the output remains the same so long as the input value is up to at most $|0.3|$. But when input increases outside this interval, the output stays within it, and the larger the amplitude of the signal, the more severe the effect of the clipping action. This phenomenon is presented in Figure E.4 where the input is a decaying sinusoid, and the output starts from an almost-square waveform and ends in a pure sinusoid, where (a) is the input, a decaying sinusoid; (b) the waveshaping function, which clips its input to the interval $(-0.3, 0.3)$; and (c) shows the output. ∎

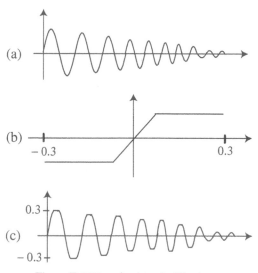

Figure E.4 Waveshaping via Clipping.

This effect is well known; for example, in music when an instrument is played on an overdriven amplifier, the louder the input is, the more distorted the output is. In such cases the waveshaping generates distortion.

Example E.2. Consider the transfer function $f(x) = x^2$, and a sinusoidal input function $x(n) = a\cos(n\omega + \phi)$. The result of the composition $f \circ x$ is

$$(f \circ x)(n) = f(x(n)) = \frac{a^2}{2}(1 + \cos(2n\omega) + 2\phi). \tag{E.5}$$

The amplitude $a = 1$ amount to ring modulating the sinusoid by another sinusoid of the same frequency, resulting in the output function a DC sinusoid (zero-frequency) plus a sinusoid with twice the original frequency. Unlike the ring modulation, in this waveshaping example the amplitude of the output grows as square of the input. This example is presented in Figure E.5, where (a) is the input; (b) the transfer function; and (c) the result, with twice the original frequency. ∎

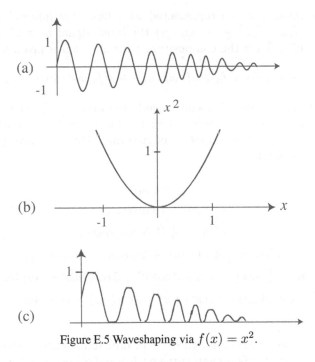

Figure E.5 Waveshaping via $f(x) = x^2$.

Example E.3. Using the same transfer function as in Example E.2, consider a composite of two sinusoids with amplitudes a and b and angular frequencies α and β, respectively. If the initial phase terms are omitted, and if we set $x(n) = a\cos(\alpha n) + b\cos(\beta n)$, then the composite $f \circ x$ is given by

$$f(x(n)) = \frac{a^2}{2}(1 + \cos(2\alpha n)) + \frac{b^2}{2}(1 + \cos(2\beta n))$$
$$+ ab\left[\cos((\alpha + \beta)n)) + \cos((\alpha - \beta)n))\right]. \tag{E.6}$$

The first two terms in (E.6) correspond to the case when the two sinusoidal signals are sent separately; the third term is twice the product of the two input terms, which corresponds to the middle term in the expansion $f(x + y) = (x+y)^2 = x^2 + 2xy + y^2$. This effect is called *intermodulation*, which becomes more dominant as the number of input signal increase. That is, if there are k sinusoids in the input, then there are only k terms corresponding to k signals sent separately, but there are $k(k - 1)/2$ intermodulation terms. ∎

In general, in the case of a periodic input, the periodicity still remains in effect no matter how complex the waveshaping is. Thus, consider a signal with period p, such that $x(n + p) = x(n)$. For simplicity, let the index be $a = 1$. Then the composition $f \circ x$ is given by $f(x(n + p)) = f(x(n))$. In some cases the output may repeat at a submultiple of p and yield subharmonics of the input (see §E.4).

Let the input $f(x)$ be represented as a finite (or infinite) power series $f(x) = f_0 + f_1 x + f_2 x^2 + \cdots$, and let the input signal be $x(n) = \cos(n\omega)$ of amplitude unity. Then the composition $f \circ x$ at $x = a$ is given by

$$f(ax(n)) = f_0 + af_1 \cos(n\omega) + a^2 f_2 \cos(n\omega) + \cdots,$$

where the terms of the series will depend on a lower or higher value of a, i.e., a lower value ($a < 1$) will result in larger value of initial terms while a higher value ($a > 1$) in the larger value of later terms. Thus, we have (using Pascal triangle for coefficients):

$$1 = \cos(0),$$
$$x(n) = \cos(n\omega),$$
$$x^2(n) = \tfrac{1}{2} [1 + \cos(2n\omega)],$$
$$x^3(n) = \tfrac{1}{4} [\cos(-n\omega) + 2\cos(n\omega) + \cos(n\omega)],$$
$$x^4(n) = \tfrac{1}{8} [\cos(-2n\omega) + 3\cos(0) + 3\cos(2n\omega) + \cos(4n\omega)],$$
$$x^5(n) = \tfrac{1}{16} [\cos(-3n\omega) + 4\cos(-n\omega) + 6\cos(n\omega) + 4\cos(3n\omega) + \cos(5n\omega)],$$
$$\text{(E.7)}$$

and so on. By the central limit theorem of probability theory, the kth row can be approximated by a Gaussian curve with standard deviation σ proportional to $\sqrt{k}$. The negative frequency terms in (E.7) can be combined with the positive terms.

Example E.4. Consider the waveshaping functions of the general form $f(x) = x^k$ which are assumed to be in phase and the spectra spread out as k increases. If the function f has a series expansion of the form $f = f_0 + f_1 x + f_2 x^2 + \cdots$, the higher power terms in this expansion will carry relatively more weight under a higher index of modulation. Let the index of modulation be a. Then the various x^k terms are multiplied by $f_0, af_1, a^2 f_2, \ldots$. In order to rearrange the terms in the above expansion as a function of the index a, let f_0 be the largest term for $0 < a < 1$. Then f_0 will be overtaken by the next more rapidly growing term af_1 for $1 < a < 2$. This will then be overtaken by the next term $a^2 f_2$ for $2 < a < 3$, and this process will continue until the kth term $a^k f_k$ is overtaken for $k < a < k + 1$. This will require that $f_1 = f_0, 2f_2 = f_1, 3f_3 = f_2, \ldots, kf_k = f_{k-1}, \ldots$. Thus, if we take $f_0 = 1$, we get $f_1 = 1$, $f_2 = 1/2$, $f_3 = 1/6$, and in general, $f_k = \dfrac{1}{k!}$. Since these are the coefficients of the power series for $f(x) = e^x$, this type of waveshaping uses the exponential function. However, since the exponential function grows faster as x increases, a sinusoid of amplitude a provides, under the exponential transfer function, a maximum output of e^a which occurs whenever the phase is zero. To obtain the peak of unit height, we use

$$f\left(a\cos(\omega n)\right) = \frac{e^{a\cos(\omega n)}}{e^a} = e^{a\cos(\omega n)-1}. \qquad \text{(E.8)}$$

The spectra of waveshaping using an exponential transfer function for $a = 0, 4$, and 10 are presented in Figure E.6.

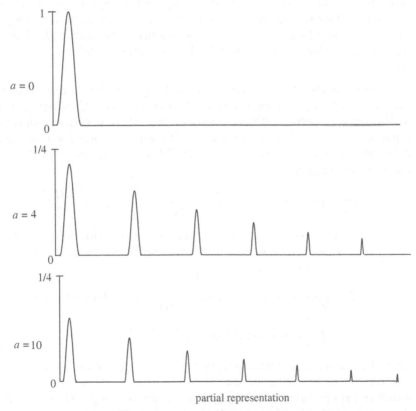

partial representation

Figure E.6 Waveshaping Using an Exponential Transfer Function.

E.4 BPSK Modulation

Phase-shift keying (PSK) is a digital modulation that conveys data by changing, or modulating, the phase of a carrier signal waveform. It uses a finite
number of distinct phases, each assigned a unique pattern of binary digits
(bits), where each phase encodes an equal number of bits, and each pattern
of bits forms the symbol that is represented by the corresponding phase. The
demodulator, which is designed specially for the symbol-set used by the modulator, determines the phase of the received signal and maps it back to the
symbol it represents, thus recovering the original data.

Binary phase-shift keying (BPSK, also called 2PSK) is the simplest form of
PSK. It uses two phases which are separated by 180° (hence the name 2PSK).
The position of the constellation points is immaterial, but they can be at 0°
and 180°. This modulation is the most robust one among all the PSKs because it takes the highest level of noise (distortion) to make the demodulator
reach an incorrect decision. However, it can only modulate at 1 bit/symbol
as in Figure E.7 and so it cannot be used for high data-rate applications.
BPSK is functional equivalent to 2-QAM modulation, or quadrature amplitude modulation which is both an analog and a digital scheme, conveying
two analog message signals or two digital bit sequences, by changing (modulating) the amplitude of two carrier waves using the amplitude-shift keying
digital modulation scheme or amplitude modulation (AM) analog modulation
scheme.

In terms of the Shannon theorem capacity (§5.1), the channel capacity $\mathfrak{C}$ is
measured in bits per channel use (modulated symbol). Consider an additive
white Gaussian noise (AWGN) channel (Example 5.3) with one-dimensional
input $y = x + n$, where x is a signal with average energy variance E_s and
n is the Gaussian with variance $N_0/2$. Then the capacity of AWGN with
unconstrained input is

$$\mathfrak{C} = \max_{p(x)}\{I(X;Y)\} = \frac{1}{2}\ln\left(1 + \frac{2E_s}{N_0}\right) = \frac{1}{2}\ln\left(1 + \frac{2RE_b}{N_0}\right). \tag{E.9}$$

This capacity is achieved by a Gaussian input x. However, this is not a
practical modulation. If we consider antipodal BPSK modulation, then $X = \pm\sqrt{E_s}$, and the capacity is

$$\mathfrak{C} = \max_{\mathfrak{p}(x)}\{I(X;Y)\} = I(X;Y)\Big|_{\mathfrak{p}(x):\mathfrak{p}=1/2} = H(Y) - H(N)$$

$$= \int_{-\infty}^{\infty} \mathfrak{p}(y)\log_2 \mathfrak{p}(y)\,dy - \frac{1}{2}\log_2(\pi e N_0), \tag{E.9}$$

since the maximum is attained when two signals are equally likely. The integral term in (E.9) is computed numerically using FFT defined by the convo-
lution $\mathfrak{p}_X(y) \star \mathfrak{p}_N(y) = \int_{-\infty}^{\infty} \mathfrak{p}_X(u)\mathfrak{p}_N(y-u)\,du$, where $\mathfrak{p}$ is the probability for
the transmitted signal to be received with error (bias).

F

Boolean and Bitwise Operations

The distinction between Boolean logical and bitwise operations is important, although the XOR (exclusive or) operation in both is the same. The truth tables for Boolean logical operations with only two values, 'true' and 'false', usually written T and F, or 1 and 0 in the case of the binary alphabet $\{0, 1\}$, are given in Table F.1 for most commonly used operators AND, OR, XOR, XNOR, IF-THEN, and THEN-IF. The operator NOT is defined by NOT $0=1$, and NOT $1=0$. The others are:

Table F.1. Boolean Logical Operators.

p	q	AND	OR	XOR	XNOR	IF-THEN	THEN-IF
0	0	0	0	0	1	1	1
0	1	0	1	1	0	1	0
1	0	0	1	1	0	0	1
1	1	1	1	0	1	1	1

The bitwise operation carried out by operator XOR is defined as follows: XOR (denoted by $\oplus$). This bitwise operator, known as the *bitwise exclusive-or*, takes two bit patterns of equal length and performs the logical XOR operation on each pair of the corresponding bits. If two bits are different, then the result is 1; but if they are the same, then the result is 0. Thus, for example, $0101 \oplus 0011 = 0110$. In general, if x, y, z are any items, then (i) $x \oplus x = 0$, (ii) $x \oplus 0 = x$, (iii) $x \oplus y = y \oplus x$, and (iv) $(x \oplus y) \oplus z = x \oplus (y \oplus z)$. In C programming languages, the bitwise XOR is denoted by $\oplus$.

The bitwise XOR operation is the same as addition mod 2. The XOR function has the following properties, which hold for any bit values (or strings) a, b, and c:

PROPERTY 1. $a \oplus a = 0$; $a \oplus 0 = a$; $a \oplus 1 =\sim a$, where $\sim$ is bit complement; $a \oplus b = b \oplus a$;

$a \oplus (b \oplus c) = (a \oplus b) \oplus c$; $a \oplus a \oplus a = a$, and if $a \oplus b = c$, then $c \oplus b = a$ and $a \oplus a = b$.

PROPERTY 2. As a consequence of Property 1, given $(a \oplus b)$ and a, the value of the bit b is determined by $a \oplus b \oplus a = b$. Similarly, given $(a \oplus b)$ and b, the value of a is determined by $b \oplus a \oplus b = a$. These results extend to finitely many bits, say a, b, c, d, where given $(a \oplus b \oplus c \oplus d)$ and any 3 of the values, the missing value can be determined. In general, for the n bits $a_1, a_2, \ldots, a_n$, given $a_1 \oplus a_2 \oplus \cdots \oplus a_n$ and any $(n-1)$ of the values, the missing value can be easily determined.

PROPERTY 3. A string $\mathbf{s}$ of bits is called a *symbol*. A very useful formula is

$$\mathbf{s} \oplus \mathbf{s} = \mathbf{0} \quad \text{for any symbol } \mathbf{s}. \tag{F.1}$$

A Hadamard matrix $\mathfrak{H}_n$ is an $n \times n$ matrix with elements $+1$ and -1 such that $\mathfrak{H}_n \mathfrak{H}_n^T = n I_n$, where I_n is the $n \times n$ identity matrix, since $\mathfrak{H}_n^{-1} = (1/n) \mathfrak{H}_n^T$. Also, since $\det(\mathfrak{H}_n \mathfrak{H}_n^T) = n^n$, we have $\det(\mathfrak{H}_n) = n^{n/2}$, although a theorem of Hadamard that imparted the name to the matrix $\mathfrak{H}_n$ states that $\det(\mathfrak{H}_n) \leq n^{n/2}$ (see Kythe and Kythe [2012: 183, Theorem 9.4.1]). The multiplication of the elements 1 and -1 is defined in Table F.2.

Table F.2. Bitwise XOR.   Multiplication (voltages).

p	q	XOR	$\|$	p	q	product
0	0	0	$\|$	-1	-1	1
0	1	1	$\|$	-1	1	-1
1	0	1	$\|$	1	-1	-1
1	1	1	$\|$	1	1	1

G

Galois Field

A finite field can be regarded as a finite vector space with two operations, denoted by $+$ and $*$, which are not necessarily ordinary addition and multiplication. Under the operation $+$ all elements of a field form a commutative group whose identity is denoted by 0 and inverse of a by $-a$. Under the operation $*$, all elements of the field form another commutative group with identity denoted by 1 and the inverse of a by a^{-1}. Note that the element 0 has no inverse under $*$. There is also a *distributive identity* that links $+$ and $*$, such that $a * (b + c) = (a * b) + (a * c)$ for all field elements a, b and c. This identity is the same as the *cancellation property*: If $c \neq 0$ and $a * c = b * c$, then $a = b$. The following groups represent fields:

1. The set $\mathbb{Z}$ of rational numbers, or the set $\mathbb{R}$ of real numbers, or the set $\mathbb{C}$ of complex numbers, are all infinite fields under ordinary addition and multiplication.

2. The set F_p of integers mod p, where p is a prime number, can be regarded as a group under $+$ (ordinary addition followed by remainder on division by p). All its elements except 0 form a group under $*$ (ordinary multiplication followed by remainder on division by p), with the identity 1, while the inverse of an element $a \neq 0$ is not obvious. However, it is always possible to find the unique inverse x using the extended Euclidean algorithm, according to which, since p is prime and $a \neq 0$, the $\gcd(p, a) = 1$.

G.1 Galois Field Arithmetic

There exists a unique finite field with p^n elements for any integer $n > 1$. This field is denoted by $GF(p^n)$ and is known as a *Galois field*. In particular, for $p = 2$, it has 2^n elements for $n > 1$. The field of binary numbers, having strings of 0s and 1s, is denoted by F_{2^m}. Since the digital data is transmitted in multiples of eight bits (2^3), we can regard each message as having the values from a field of 2^m bits, where $m > 0$ is an integer. The field of our choice must be F_{2^m}, which has the number of elements 2^m. This can be generalized to a field F_{p^m}, where p is prime. However, the field F_{2^m} has a serious drawback, namely that it does not always meet the criterion for inverses. For example, F_{2^4} is not a field with multiplication as defined in F_p. This situation forces us to look for a different arithmetic for multiplication, such as the polynomial arithmetic, since $F_{2^m} \simeq F_2[x]/ < g(x) >$, where $g(x)$ is a minimal polynomial with root α. The new arithmetic replaces the non-prime 2^m with a 'prime polynomial' which is irreducible $p(x)$, and gives us an extension field $GF(2^m)$.

We construct the extension field $GF(2^m)$ of $F_2[x]$ using the above isomorphism and find a minimal polynomial with a root in F_{2^4}. Since we need 16 elements in $GF(2^m)$, the degree of this polynomial must be 4 because $2^4 = 16$. In addition, we require another condition to be fulfilled: This minimal polynomial must be primitive in $GF(2^4)$. Thus, every element in F_{2^4} must be expressible as some power of $g(x)$. Under these two restrictions, there are only two choices that may be made from primitive polynomials among all possible nonzero elements of $GF(2^4)$, namely $1 + x^3 + x^4$ and $1 + x + x^4$, of which either may be used.

Table G.1 Field of 16 Elements.

(a) Generated by $1 + x + x^4$ (b) Generated by $1 + x^3 + x^4$

Power	4-bit Form	Polynomial	4-bit Form	Polynomial
0	0000	0	0000	0
1	0001	1	0001	1
α	0010	α	0010	α
α^2	0100	α^2	0100	α^2
α^3	1000	α^3	1000	α^3
α^4	0011	$\alpha + 1$	1001	$\alpha^3 + 1$
a^5	0110	$\alpha^2 + \alpha$	1011	$\alpha^3 + \alpha + 1$
α^6	1100	$\alpha^3 + \alpha^2$	1111	$\alpha^3 + \alpha^2 + \alpha + 1$
α^7	1011	$\alpha^3 + \alpha + 1$	0111	$\alpha^2 + \alpha + 1$
α^8	0101	$\alpha^2 + 1$	1110	$\alpha^3 + \alpha^2 + \alpha$
α^9	1010	$\alpha^3 + \alpha$	0101	$\alpha^2 + 1$
α^{10}	0111	$\alpha^2 + \alpha + 1$	1010	$\alpha^3 + \alpha$
α^{11}	1110	$\alpha^3 + \alpha^2 + \alpha$	1101	$\alpha^3 + \alpha^2 + 1$
α^{12}	1111	$\alpha^3 + \alpha^2 + \alpha + 1$	0011	$\alpha + 1$
α^{13}	1101	$\alpha^3 + \alpha^2 + 1$	0110	$\alpha^2 + \alpha$
α^{14}	1001	$\alpha^3 + 1$	1100	$\alpha^3 + \alpha^2$

The process to generate the elements is as follows: Start with three initial elements of the extension field and perform all arithmetic operations mod $g(x) = 1 + x^3 + x^4$ (which we have chosen). The initial elements are $0, 1, \alpha$. Raising α to successive powers identifies α^2 and α^3 as members of the extension field; note that α^4 is not an element of F_{2^4}. Since $g(\alpha) = 0 = 1 + \alpha^3 + \alpha^4$, we have the identity $\alpha^4 = 1 + \alpha^3$,[1] and we use this identity to reduce each power greater than 3. This identity can be used to reduce every power of α, which will build an entire field $GF(2^4)$. These fields of 16 elements, generated by two primitive polynomials $1 + x + x^4$ and $1 + x^3 + x^4$, are presented below in Table G.1.

An irreducible polynomial $f(x)$ of degree m is said to be *primitive* if the smallest positive integer n for which $f(x)$ divides $x^n + 1$ is $n = 2^m - 1$. A table for primitive polynomials over F_2 is available in Abramowitz and Stegun [1972]. An irreducible polynomial can be easily tested for primitiveness using the fact that at least one of its roots must be a primitive element. For example, to find the $m = 3$ roots of $f(x) = 1 + x + x^3$, we will list the roots in order. Clearly, $\alpha^0 = 1$ is not a root since $f(\alpha^0) = 1$. Using Table G.2, we check if α^1 is a root: Since $f(\alpha) = 1 + \alpha + \alpha^3 = 1 + \alpha^0 = 0 \pmod 2$, so α is a root. Next, we check α^2. Since $f(\alpha^2) = 1 + \alpha^2 + \alpha^6 = 1 + \alpha^0 = 0$, so α^2 is a root. Then, we check α^3: Since $f(\alpha^3) = 1 + \alpha^3 + \alpha^9 = 1 + \alpha^3 + \alpha^2 = \alpha + \alpha^2 = \alpha^4 \neq 0$, so α^3 is not a root. Finally, we check α^4: Since $f(\alpha^4) = 1 + \alpha^4 + \alpha^{12} = 1 + \alpha^4 + \alpha^5 = 1 + \alpha^0 = 0$, so α^4 is a root. Hence the roots of $f(x) = 1 + x + x^3$ are $\{\alpha, \alpha^2, \alpha^4\}$, and since each of these roots will generate seven nonzero elements in the field, each of them is a primitive element. Since we only need one root to be a primitive element, the given polynomial is primitive.

Example G.1. Consider the primitive polynomial $f(x) = 1 + x + x^3$ $(m = 3)$. There are $2^3 = 8$ elements in the field defined by $f(x)$. Solving $f(x) = 0$ will give us its roots. But the binary elements (0 and 1) do not satisfy this equation because $f(0) = 1 = f(1) \pmod 2$. Since, by the fundamental theorem of algebra, a polynomial of degree m must have exactly m zeros, the equation $f(x) = 0$ must have three roots, and these three roots lie in the field $GF(2^3)$. Let $\alpha \in GF(2^3)$ be a root of $f(x) = 0$. Then we can write $f(\alpha) = 0$, or $1 + \alpha + \alpha^3 = 0$, or $\alpha^3 = -1 - \alpha$. Since in the binary field $+1 = -1$, α^3 is represented as $\alpha^3 = 1 + \alpha$. Using $\alpha^3 = 1 + \alpha$, we get $\alpha^4 = \alpha * \alpha^3 = \alpha * (1 + \alpha) = \alpha + \alpha^2$ and $\alpha^5 = \alpha * \alpha^4 = \alpha * (\alpha + \alpha^2) = \alpha^2 + \alpha^3 = \alpha^2 + 1 + \alpha = 1 + \alpha + \alpha^2$, (using $\alpha^3 = 1 + \alpha$). Further, $\alpha^6 = \alpha * \alpha^5 = \alpha * (1 + a + a^2) = \alpha + \alpha^2 + \alpha^3 = 1 + \alpha^2$; and $\alpha^7 = \alpha * \alpha^6 = \alpha * (1 + \alpha^2) = \alpha + \alpha^3 = 1 = \alpha^0$. Thus, the eight finite field elements of $GF(2^3)$ are $\{0, \alpha^0, \alpha^1, \alpha^2, \alpha^3, \alpha^4, \alpha^5, \alpha^6\}$, which are similar to a binary cyclic code. ∎

Some formulas useful for Galois field arithmetic are as follows:

$$+1 = -1; \quad \alpha^i = -\alpha^i \text{ for } i = 0, 1, \dots, k;$$

$$\alpha^i \alpha^j = \alpha^{i+j} \text{ for any } i \text{ and } j \ (i, j = 0, 1, \dots, k);$$

$$1\alpha^i = \alpha^0 \alpha^i = \alpha^i; \quad \alpha^i + \alpha^i = 0 \text{ (by formula (F.1))}; \qquad \text{(G.1)}$$

$$2\alpha^i = \alpha^i + \alpha^i = 0; \quad 2n\alpha^i = n(2\alpha^i) = 0 \text{ for } n = 1, 2, \dots.$$

---

[1] Recall that in a binary field, addition is equivalent to multiplication.

Addition and multiplication tables for the above eight field elements of $GF(2^3)$ are given in Tables G.2 and G.3. Note that in some tables 1 and α are written as α^0 and α^1, respectively.

Table G.2 (Addition.)

+	1	α	α^2	α^3	α^4	α^5	α^6
1	0	α^3	α^6	α	α^5	α^4	α^2
α	α^3	0	α^4	1	α^2	α^6	α^5
α^2	α^6	α^4	0	α^5	α	α^3	1
α^3	α	1	α^5	0	α^6	α^2	α^4
α^4	α^5	α^2	α	α^6	0	1	α^3
α^5	α^4	α^6	α^3	α^2	1	0	α
α^6	α^2	α^5	1	α^4	α^3	α	0

Table G.3 (Multiplication.)

$\times$	1	α	α^2	α^3	α^4	α^5	α^6
1	1	α	α^2	α^3	α^4	α^5	α^6
α	α	α^2	α^3	α^4	α^5	α^6	1
α^2	α^2	α^3	α^4	α^5	α^6	1	α
α^3	α^3	α^4	α^5	α^6	1	α	α^2
α^4	α^4	α^5	α^6	1	α	α^2	α^3
α^5	α^5	α^6	1	α	α^2	α^3	α^4
α^6	α^6	1	α	α^2	α^3	α^4	α^5

The Gauss elimination method is also used in binary arithmetic operations in a Galois field if $m \le 16$. The traditional Gauss elimination method, as found in numerical analysis, is performed on the infinite set of real numbers $\mathbb{R}$. However, in $GF(2^m)$, especially where multiplication/division is performed, the traditional rules do not apply. The operation of division is not defined over all elements in $GF(2^m)$; for example, $\frac{3}{2}$ is not defined in $GF(4)$, and thus, Gauss elimination becomes unsolvable in many cases. For positive integer m, the $GF(2^m)$ field contains 2^m elements that range from 0 to $2^m - 1$. A C code for this method is available in Kythe and Kythe [2012].

G.2 LFSR Encoder

Consider the RS(7, 3) code, for which the number of correctable errors is 2, and its generator polynomial is

$$\mathbf{g}(x) = \alpha^3 + \alpha x + x^2 + \alpha^3 x^3 + x^4. \tag{G.2}$$

For the purpose of discussing the encoding we take specifically a nonbinary 3-symbol data $\underbrace{010}_{\alpha} \underbrace{110}_{\alpha^3} \underbrace{111}_{\alpha^5}$, and we multiply the data polynomial $\mathbf{d}(x)$ by $x^{n-k} = x^4$, which gives

$$x^4 \mathbf{d}(x) = x^4 \left(\alpha + \alpha^3 x + \alpha^5 x^2 \right) = \alpha x^4 + \alpha^3 x^5 + \alpha^5 x^6.$$

Then, by dividing $x^4 \mathbf{d}(x)$ by $\mathbf{g}(x)$ we get the quotient and remainder, respectively, as $\mathbf{q}(x) = \alpha^4 + x + \alpha^5 x^2$, and $\mathbf{p}(x) = 1 + \alpha^2 x + \alpha^4 x^2 + \alpha^6 x^3$. Thus, the codeword polynomial is

$$\mathbf{c}(x) = \sum_{i=0}^{6} c_i x^i = 1 + \alpha^2 x + \alpha^4 x^2 + \alpha^6 x^3 + \alpha x^4 + \alpha^3 x^5 + \alpha^5 x^6$$

$$= (100) + (001)x + (011)x^2 + (101)x^3 + (10)x^4 + (110)x^5 + (111)x^6. \tag{G.3}$$

A check on the accuracy of the encoded codeword can be implemented by using the following criterion: The roots of the generator polynomial $\mathbf{g}(x)$ must be the same as those of codeword $\mathbf{c}(x)$, since $\mathbf{c}(x) = \mathbf{d}(x)\mathbf{g}(x)$.

We will assume that during transmission the 7-symbol codeword gets corrupted with two-symbol errors. Let the error polynomial $\mathbf{e}(x)$ have the form $\mathbf{e}(x) = \sum_{i=0}^{6} e_i x^i$, and suppose that these two errors consist of one parity error, say a 1-bit error in the coefficient of x^3 such that α^6 (101) changes to α^2 (001), and one data-symbol error, say a 2-bit error in the coefficient of x^4 such that α (010) changes to α^5 (111), i.e., the two-symbol error can be represented as

$$\mathbf{e}(x) = 0 + 0x + 0x^2 + \alpha^2 x^3 + \alpha^5 x^4 + 0x^5 + 0x^6$$

$$= (000) + (000)x + (000)x^2 + (001)x^3 + (111)x^4 + (000)x^5 + (000)x^6. \tag{G.4}$$

Thus, the received word polynomial $\mathbf{w}(x)$ is represented as the sum of the transmitted codeword polynomial $\mathbf{c}(x)$ and the error polynomial $\mathbf{e}(x)$: $\mathbf{w}(x) = \mathbf{c}(x) + \mathbf{e}(x)$, which is

$$\mathbf{w}(x) = (100) + (001)x + (011)x^2 + (100)x^3 + (101)x^4 + (110)x^5 + (111)x^6$$

$$= 1 + \alpha^2 x + \alpha^4 x^2 + x^3 + \alpha^6 x^4 + \alpha^3 x^5 + \alpha^5 x^6. \tag{G.5}$$

There are four unknowns: 2 error locations and 2 error values, so four equations are needed to determine the two error locations L_1, L_2, we use syndrome computation, and find that the error locator polynomial is given by

$$\mathbf{L}(x) = 1 + L_1 x + L_2 x^2 = 1 + \alpha^6 x + x^2, \tag{G.6}$$

which after some algebraic simplification gives the errors locations at α^3 and α^4. Next, we determine the error polynomial as $\mathbf{e}(x) = \alpha^2 x^3 + \alpha^5 x^4$. Finally, the received codeword polynomial becomes

$$\hat{\mathbf{c}}(x) = \mathbf{c}(x) + \hat{\mathbf{e}}(x) = \mathbf{c}(x) + \mathbf{e}(x) + \hat{\mathbf{e}}(x), \tag{G.7}$$

where

$$\mathbf{w}(x) = 1 + \alpha^2 x + \alpha^4 x^2 + x^3 + \alpha^6 x^4 + \alpha^3 x^5 + \alpha^5 x^6,$$

$$\hat{\mathbf{e}}(x) = \alpha^2 x^3 + \alpha^5 x^4,$$

$$\hat{\mathbf{c}}(x) = 1 + \alpha^2 x + \alpha^4 x^2 + (1 + \alpha^2) x^3 + (\alpha^6 + \alpha^5) x^4 + \alpha^3 x^5 + \alpha^5 x^6$$

$$= 1 + \alpha^2 x + \alpha^4 x^2 + \alpha^6 x^3 + \alpha x^4 + \alpha^3 x^5 + \alpha^5 x^6$$

$$= \underbrace{(100) + (001)x + (011)x^2 + (101)x^3}_{\text{4 parity symbols}} + \underbrace{(010)x^4 + (110)x^5 + (111)x^6}_{\text{3 data symbols}}.$$

This gives the corrected decoded data symbols as $\underbrace{010}_{\alpha} \; \underbrace{110}_{\alpha^3} \; \underbrace{111}_{\alpha^5}$. For details of calculations, see Kythe and Kythe [2012: 268–273].

The parity-symbol polynomial, which has been generated above by purely algebraic rules in the Galois field, is electronically determined easily by the built-in circuitry of an $(n-k)$-stage shift register is shown in Figure G.1. Since we are considering an RS $(7,3)$ code, it requires a linear feedback shift register (LFSR) circuit. The multiplier terms taken from left to right correspond to the coefficients of the generator polynomial in Eq (G.2) which is written in low order to high order.

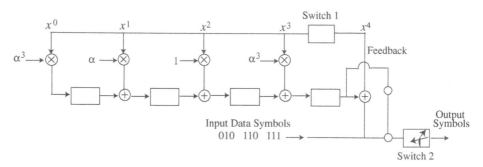

Figure G.1 LFSR for RS$(7,3)$ Code.

In this example, the RS$(7,3)$ codeword $\mathbf{c}(x)$, defined by (G.3), has $2^m - 1 = 2^3 - 1 = 7$ symbols (bytes), where each symbol is composed of $m = 3$ bits. Each stage in the shift register of Figure G.1 will hold a 3-bit symbol; and since each coefficient is specified by 3 bits, it can take one of $2^3 = 8$ values. The operation of the encoder controlled by the LFSR circuit will generate codewords in a systematic manner according to the following steps:

STEP 1: Switch 1 is closed during the first $k = 3$ clock cycles; this allows shifting the data symbols into the $n - k = 4$-stage shift register.

STEP 2: Switch 2 is in the drawn position during the first $k = 3$ clock cycles; this allows simultaneous transfer of data symbols directly to an output register.

STEP 3: After transfer of the k-th (third) data symbol to the output register, switch 1 is opened and switch 2 is moved to the up position.

STEP 4: The remaining $n - k = 4$ clock cycles clear the parity symbols contained in the shift register by moving them to the output register.

STEP 5: The total number of clock cycles is equal to $n = 7$, and the output register contains the codeword polynomial $\mathbf{c}$ defined by Eq (G.3).

The data symbols $\underbrace{010}_{\alpha}\ \underbrace{110}_{\alpha^3}\ \underbrace{111}_{\alpha^5}$ move to the LFSR from right to left, where the rightmost symbol is first symbol and the rightmost bit is the first bit. The operation of the first $k = 3$ shift of the encoding circuit of Figure G.1 are given below, where Tables G.2 and G.3 are used.

Clock Cycle	Input Queue	Register Contents	Feedback
0	$\alpha\ \alpha^3\ \alpha^5$	0, 0 0 0	α^5
1	$\alpha\ \alpha^3$	$\alpha\ \alpha^6\ \alpha^5\ \alpha$	1
2	α	$\alpha^3\ 0\ \alpha^2\ \alpha^2$	α^4
3	$-$	$1\ \alpha^2\ \alpha^4\ \alpha^6$	$-$

At the end of the third cycle, the register contents are the four parity symbols $1, \alpha^2, \alpha^4, \alpha^6$, as shown in the above table. At this point, switch 1 is opened, switch 2 is toggled to the up position, and the 4 parity symbols are shifted from the register to the output. This suggests that the output (encoded) codeword can be written as

$$\mathbf{c}(x) = \sum_{i=0}^{6} c_i x^i = 1 + \alpha^2 x + \alpha^4 x^2 + \alpha^6 x^3 + \alpha x^4 + \alpha^3 x^5 + \alpha^5 x^6,$$

which is the same as (G.3), where the first four terms correspond to register contents after cycle 3, and the last three terms to the data input. ∎

H

GPS Geometry

H.1 Earth

The Earth is an ellipsoid, and as such it has two latitudes: (i) the geocentric latitude L_c, and (ii) the geodetic latitude L which is used in day-to-day maps. Consider the case of four satellites, and let (x_i, y_i, z_i), $i = 1, 2, 3, 4$, denote the location of these four satellites, and let (x_u, y_u, z_u) be unknowns. Then Eq (6.3) for the pseudo-range ρ_i can be written as

$$\rho_i = \sqrt{(x_i - x_u)^2 + (y_i - y_u)^2 + (z_i - z_u)^2} + b_u. \tag{H.1}$$

Differentiating (H.1) we get the differential

$$\begin{aligned}
\delta\rho_i &= \frac{(x_i - x_u)\delta x_u + (y_i - y_u)\delta y_u + (z_i - z_u)\delta z_u}{\sqrt{(x_i - x_u)^2 + (y_i - y_u)^2 + (z_i - z_u)^2}} + \delta b_u \\
&= \frac{(x_i - x_u)\delta x_u + (y_i - y_u)\delta y_u + (z_i - z_u)\delta z_u}{\rho_i - b_u} + \delta b_u,
\end{aligned}$$

in which we can regard the quantities $\delta x_u, \delta y_u, \delta z_u$, and δb_u as the only unknowns, if we treat the quantities x_u, y_u, z_u, and b_u as known by assuming them to be the initial set of values for these quantities. Now, Eq (H.1) can be written in matrix form as

$$\begin{Bmatrix} \delta\rho_1 \\ \delta\rho_2 \\ \delta\rho_3 \\ \delta\rho_4 \end{Bmatrix} = \begin{bmatrix} \alpha_{11} & \alpha_{12} & \alpha_{13} & 1 \\ \alpha_{21} & \alpha_{22} & \alpha_{23} & 1 \\ \alpha_{31} & \alpha_{32} & \alpha_{33} & 1 \\ \alpha_{41} & \alpha_{42} & \alpha_{43} & 1 \end{bmatrix} \begin{Bmatrix} \delta x_u \\ \delta y_u \\ \delta z_u \\ \delta b_u \end{Bmatrix}, \tag{H.2}$$

where

$$\alpha_{i1} = \frac{x_i - x_u}{\rho_i - b_u}, \quad \alpha_{i2} = \frac{y_i - y_u}{\rho_i - b_u}, \quad \alpha_{i3} = \frac{z_i - z_u}{\rho_i - b_u}.$$

This equation can be written as $\delta\rho = \alpha\,\delta x$, which has the solution $\delta x = \alpha^{-1}\delta\rho$, α^{-1} represents the inverse of the matrix α. This solution can be represented in detail as

$$\begin{Bmatrix} \delta x_u \\ \delta y_u \\ \delta z_u \\ \delta b_u \end{Bmatrix} = \begin{bmatrix} \alpha_{11} & \alpha_{12} & \alpha_{13} & 1 \\ \alpha_{21} & \alpha_{22} & \alpha_{23} & 1 \\ \alpha_{31} & \alpha_{32} & \alpha_{33} & 1 \\ \alpha_{41} & \alpha_{42} & \alpha_{43} & 1 \end{bmatrix}^{-1} \begin{Bmatrix} \delta\rho_1 \\ \delta\rho_2 \\ \delta\rho_3 \\ \delta\rho_4 \end{Bmatrix}. \tag{H.3}$$

However, such a solution does not usually provide correct results, unless it is used iteratively.

If there are n satellites, $n > 4$, we must have $i = 1, 2, \ldots, n$ in Eq (H.2). If we linearize this equation, we get

$$
\begin{Bmatrix} \delta\rho_1 \\ \delta\rho_2 \\ \delta\rho_3 \\ \delta\rho_4 \\ \vdots \\ \delta\rho_n \end{Bmatrix} = \begin{bmatrix} \alpha_{11} & \alpha_{12} & \alpha_{13} & 1 \\ \alpha_{21} & \alpha_{22} & \alpha_{23} & 1 \\ \alpha_{31} & \alpha_{32} & \alpha_{33} & 1 \\ \alpha_{41} & \alpha_{42} & \alpha_{43} & 1 \\ & \vdots & & \\ \alpha_{n1} & \alpha_{n2} & \alpha_{n3} & 1 \end{bmatrix} \begin{Bmatrix} \delta x_u \\ \delta y_u \\ \delta z_u \\ \delta b_u \end{Bmatrix}, \qquad \text{(H.4)}
$$

or, in the simpler form, $\delta\rho = \boldsymbol{\alpha}\,\delta\mathbf{x}$. Although $\boldsymbol{\alpha}$ is not a square matrix, it cannot be inverted directly. But since (H.4) is linear, we use the least-square method, and get the solution of Eqs (H.4) as

$$
\delta\mathbf{x} = \left[\boldsymbol{\alpha}^T\boldsymbol{\alpha}\right]^{-1}\boldsymbol{\alpha}^T\delta\rho, \qquad \text{(H.5)}
$$

and thus, determine the values of δx_u, δy_u, δz_u, and δb_u. Step-by-step details of this method are given in Tsui [2000: 30].

H.2 User Location

The location of the user is computed in spherical coordinate system, where this location is usually given in latitude, longitude, and altitude. The Earth's latitude is from $-90°$ to $90°$ with the equator at $0°$. The longitude is from $-180°$ to $180°$ with the Greenwich meridian at $0°$. The altitude is the height above Earth's surface. Since Earth is not a perfect sphere, the distance r from the center of Earth to the user's location is

$$
r = \sqrt{x_u^2 + y_u^2 + z_u^2}. \qquad \text{(H.6)}
$$

as in Figure H.1, so that the latitude L_c, longitude L, and the altitude h are

$$
L_c = \arctan\left(\frac{z_u}{\sqrt{x_u^2 + y_u^2}}\right), \quad L = \arctan\left(\frac{y_u}{x_u}\right), \quad h = r - r_e, \qquad \text{(H.7)}
$$

where r_e is the radius of an ideal spherical Earth or its average radius. Since Earth is an ellipsoid, and not a sphere, these quantities must be modified accordingly. An ellipse with semi-major and semi-minor axis a_e and b_e, respectively, and the foci $2c_e$ apart, has the equation $x^2/a_e^2 + y^2/b_e^2 = 1$, with $a_e^2 - b_e^2 = c_e^2$, where the eccentricity e_e is defined as

$$
e_e = \frac{c_e}{a_e} = \frac{\sqrt{a_e^2 - b_e^2}}{a_e}, \quad \text{or} \quad \frac{b_e}{a_e} = \sqrt{1 - e_e^2}.
$$

The ellipticity e_p is defined as $e_p = \dfrac{a_e - b_e}{a_e}$, where, for Earth, $a_e = 6378137 \pm 2$ m; $b_e = 6356752.3142$ m; $e_e = 0.0033528106674$, according to Department of Defense [1987] and Riggins [1996].

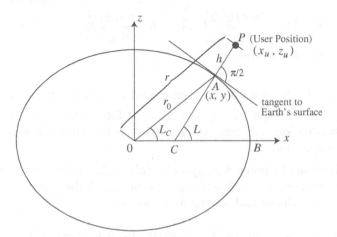

Figure H.1 Geocentric and Geodetic Latitudes.

H.3 Altitude

Using Figure H.2, the altitude h can be found from the triangle OPA as

$$r^2 = r_0^2 - 2r_0 h \cos\left(\pi - D_0\right) + h^2 = r_0^2 + 2r_0 n h \cos d_0 + h^2,$$

where r_0 is the distance from Earth's center to the point on Earth's surface under the user location.

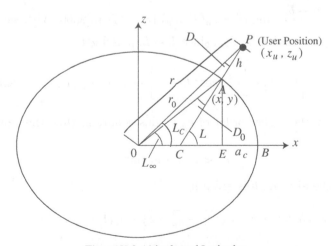

Figure H.2 Altitude and Latitudes.

The amplitude of r is given by

$$r = \sqrt{(r_0 + h)^2 - 2r_0 h(1 - \cos D_0)} = (r_0 + h)\left[1 - \frac{2r_0 h(1 - \cos D_0)}{(r_0 + h)^2}\right]^{1/2}$$

$$\approx r_0 + h - \left[1 - \frac{2r_0 h D_0^2/2}{(r_0 + h)^2}\right] = r_0 + h - \frac{r_0 h D_0^2}{2(r_0 + h)}$$

$$= r_0 + h - \text{error term}, \tag{H.8}$$

where, since D_0 is very small, we have used $1 - \cos D - 0 \approx \frac{1}{2} D_0^2$, and the angle D_0 is in radians. Using the data for Earth, i.e., $h = 100$ km, and $r_0 = r_e \approx 6368$ km, the error term in (H.8) is less than 0.6 km, which is a good approximation.

The location of a point $A(x, y)$ on Earth's surface is, in fact, on the ellipse, and thus, we have $x = r_0 \cos L_{co}$ $y - r_0 \sin L_{co}$. Substituting these in the equation of the ellipse and solving for r_0 we get

$$r_0^2 \left(\frac{\cos^2 L_{co}}{a_e^2} + \frac{\sin^2 L_{co}}{b_e^2}\right) = r_0^2 \frac{b_e^2 \cos^2 L_{co} + a_e^2(1 - \cos^2 L_{co})}{a_e^2 b_e^2} = 1,$$

or

$$r_0^2 = \frac{a_e^2 b_e^2}{a_e^2 \left[1 - \left(1 - \frac{b_e^2}{a_e^2}\right)\cos L_{co}\right]} = \frac{b_e^2}{1 - e_e^2 \cos L_{co}},$$

or

$$r_0^2 = b_e\left(1 + \frac{1}{2}e_e^2 \cos^2 L_{co} + \cdots\right). \tag{H.9}$$

Since $e_p = \dfrac{a_e - b_e}{a_e}$, and $e_e^2 = e_p(2 - e_p)$ (as in Tsui [2000: 37]), we replace b_e by a_e, e_e by e_p, and L by L_{co}, since $L \approx L_{co}$, and get

$$r_0 \approx a_e(1 - e_p)\left[1 - \left(e_p - \frac{1}{2}e_p^2\right)\cos^2 L + \cdots\right] \approx a_e(1 - e_p)(1 + e_p - e_p \sin^2 L + \cdots).$$

If we neglect the higher order of e_p in this equation, then the value of r_0 is given by

$$r_0 \approx a_e(1 - e_p \sin^2 L), \tag{H.10}$$

and thus, the altitude h is given by

$$h \approx r - r_0 = \sqrt{x_u^2 + y_u^2 + z_u^2} - a_e(1 - e_p \sin^2 L). \tag{H.11}$$

H.4 Sidereal Day

A sidereal day differs slightly from a solar day. As we know, a solar day that we use has 24 hours. But a sidereal day is the time between two successive transits of the Sun across our local meridian. A sidereal day is the time that Earth takes to make one revolution. The difference between these two concepts is shown in Figure H.3, though a little magnified. A solar day is slightly larger than a sidereal day; their difference is about 4 minutes ($24 \times 60/365$) per day. The mean sidereal day is 23 hr 56 min 4.09 sec. The difference from an apparent day is 3 min 55.91 sec. Each satellite rotates around Earth twice in a sidereal day, i.e., one rotation each half a sidereal day, which is 11 hr, 58 min, 2.05 sec. Obviously, there is always a satellite approximately at the same location at the same time each day.

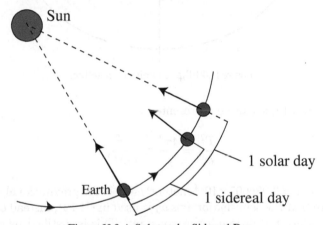

Figure H.3 A Solar and a Sidereal Day.

H.5 Kepler's Laws

The three Kepler's laws (§9.1) apply to the motion of the GPS satellites. The orbit of a satellite SV is elliptic with the Earth at one of its foci F, as in Figure H.4, where the semi-major axis a_s and the semi-minor axis b_s, $a_s > b_s$, are such that they are very close to each other so that the actual orbit of the satellite is very close to a circle; c_s is the focal length of the ellipse, and the angle α is called the actual anomaly. This figure is a representation of Kepler's second law. Let t represent the satellite's position at time t, t_p the time when the satellite passes the perigee, A the area contained between the ellipse and the lines $t = t$, $t = t_p$, and T the period of the satellite. Then

$$\frac{t - t_p}{A} = \frac{T}{\pi a_s b_s},\tag{H.12}$$

where $\pi a_s b_s$ is the area of the ellipse. Eq (H.12) states that the time to sweep the area A is proportional to the time T to sweep the entire area of the ellipse.

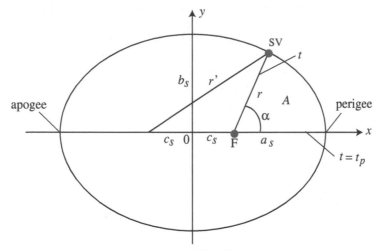

Figure H.4 Elliptic Orbit of a Satellite.

Kepler's third law can be represented as

$$\frac{T^2}{a_s^3} = \frac{4\pi^2}{\mu} \equiv \frac{4\pi^2}{GM},$$

(H.13)

where $GM = \mu = 3.986005 \times 10^{14}$ meters^3/sec^2 is the gravitational constant of Earth. Note that the semi-major axis a_s is used in (H.12), instead of the mean distance r_s, since the ratio a_s/r_s is constant. This is justified by considering the relation $\pi a_s b_s = \pi r_s^2$, which gives $a_s/r_s = r_s/b_s$. Since a_s, b_s and r_s are constant, so is the ratio a_s/r_s.

I

Gold Codes

I.1 Definition

A Gold code, also known as Gold sequence, and named after Robert Gold [1967, 1968], is a type of binary sequence used in telecommunication CDMA (code division multiple access) and GPS. CDMA is a channel access method. It should not be confused with the mobile phone standards called CDMA1 and CDMA2000, which is the 3G version of CDMA1. A set of Gold code sequences consists of $2^n - 1$ sequences, each one with a period of $2^n - 1$. The steps to generate a set of Gold codes are as follows: Choose two maximum length sequences of the same length $2^n - 1$, such that their absolute cross-correlation is less than or equal to $2^{(n+2)/2}$, where n is the size of the LSFR used to generate the maximum length sequence. The set of the $2^n - 1$ XOR of the two sequences in their various phases (i.e., translated into all relative positions) is a set of Gold codes. The highest absolute cross-correlation in the set of codes is $2^{(n+2)/2} + 1$ for even n and $2^{(n+1)/2} + 1$ for odd n. The XOR of two Gold codes from the same set is another Gold code in some phase. Within a set of Gold codes about half of the codes are balanced — the numbers of zeros and ones differ by only one (see Holmes [2007: 100]). Let m denote the number of LFSR elements, l the sequence length, s the number of l sequences, and t the number of Gold cross-correlation of Gold sequences. Then the data in Table I.1 provides the number t, where 'Max l' denotes the maximum cross-correlation of the l sequence and the last column is the normalized cross-correlation of Gold sequence.

Thus, Gold sequences are constructed by XOR-ing two l sequences of the same length. If the length is $l = 2^m - 1$, one uses two LFSR, each of length $2^m - 1$. If the LFSR are chosen correctly, Gold sequences have better cross-correlation properties than maximum length LSFR sequences. Gold has shown that for certain well-chosen l-sequences, the cross-correlation only takes on three possible values, namely, -1, $-t$, or $t-2$, where t depends only on the length of the linear feedback shift register (LFSR) used. Two such sequences are called *preferred sequences*. For a LFSR with m memory elements,

$$t = \begin{cases} 2^{(m+1)/2} + 1 & \text{if } m \text{ is odd,} \\ 2^{(m+2)/2} + 1 & \text{if } m \text{ is even.} \end{cases} \tag{I.1}$$

Thus, a Gold sequence is formally an arbitrary phase of a sequence in the set $G(u, v)$ defined by

$$G(u, v) = \left\{ u, v, u \oplus v, u \oplus Tv, u \oplus T^2 v, \dots, u \oplus T^{(N-1)} v \right\}, \tag{I.2}$$

where T^k denotes the operator which shifts vectors cyclically to the left by k places, and u, v are the l sequences of period generated by different binary polynomials.

Table I.1 Cross-Correlation of Gold Sequence.

m	$2^m - 1$	l	Max l	t	$t/(2^m - 1)$
3	7	2	0.71	5	0.71
4	15	2	0.60	9	0.60
5	31	6	0.35	9	0.29
6	63	6	0.36	17	0.27
7	127	18	0.32	17	0.13
8	255	16	0.37	33	0.13
10	1023	60	0.37	65	0.06
12	4095	144	0.34	129	0.03

I.2 m-Sequences

We will discuss generation of preferred pairs m-sequences for Gold codes. These are spreading sequences in a spread spectrum system, which can be generated using diversified codes like m-sequences, Gold Codes, Kasami Codes, Walsh Codes, and others. Compared to m-sequences (i.e., maximum length PN sequences), Gold codes have the worst auto-correlation properties but they have better cross-correlation properties. The sequences associated with Gold codes are generated by binary addition (modulo 2) , or XORing, of cyclic shifted versions of two m-sequences of length $N = 2^n - 1$. If two m-sequences of cyclic shift t_1 and t_2 are used in Gold code generation, the generated Gold code will be unique for each combination of t_1 and t_2. Thus, a large number of Gold codes are generated with just two m-sequences.

When two m-sequences are randomly chosen for generating Gold codes, the cross-correlation property of the generated Gold code may not be quite satisfactory. Since Gold codes are generated using preferred pairs of m-sequences, both of their good cross-correlation and auto-correlation properties are satisfied.

The Gold sequence generator block uses two PN sequence generator blocks. It generates the preferred pair of sequences, and then XORs these sequences to produce the output sequence. A list of pairs of two preferred polynomials P_1 and P_2 of degree n, with the value $N = 2^n - 1$, is provided in Table I.2.

Table I.2 Pairs of Preferred Polynomials P_1 and P_2.

n	N	P_1	P_2
5	31	[5 2 0]	[5 4 3 2 0]
6	63	[6 6 0]	[6 5 2 1 0]
7	127	[7 3 0]	[7 3 2 1 0]
9	511	[9 34 0]	[9 6 4 3 0]
10	1023	[10 3 0]	[10 8 3 2 0]
11	2047	[11 2 0]	[11 8 5 2 0]

For example, the vectors [5 2 0] and [1 0 0 1 0 1] both represent the polynomial $x^5 + x^2 + 1$. The parameters of the initial states [1] and [2] are vectors that specify the initial values of the registers corresponding to P_1 and P_2, respectively. These parameters must satisfy the following conditions: (i) The vectors of all elements of initial states [1] and [2] must be binary numbers; (ii) the length of the vector of initial states [1] must be equal to the degree of P_1, and so for the length of the vector of the initial states [2] which must be equal to the degree of P_2; and (iii) the initial state of at least one of the registers must be nonzero (i.e., at least one element of the initial state vectors must be nonzero) so that the block can generate a nonzero sequence. Thus, the sequence index parameter specifies which sequence in the set $G(u, v)$ of the Gold sequences the block outputs. The range of the sequence index is $[-2, -1, 0, 1, 2, \ldots, 2^n - 2]$. The output sequence corresponding to each sequence index is presented in Table I.3.

Table I.3 Sequence Index and Output Sequence.

Sequence Index	Output Sequence
-2	u
-1	v
0	$u \oplus v$
1	$u \oplus Tv$
2	$u \oplus T^2v$
$\ldots$	$\ldots$
$2^n - 2$	$u \oplus T^{2^n-2}v$

A method of choosing the preferred pairs for Gold code generation is as follows (Gold, 1967]):

STEP 1. Take an m-sequence d_1 of given length $N(= 2^n - 1)$, where n is the number of the LFSR registers.

STEP 2. Decimate the m-sequence by a decimation factor q. This yields the second preferred sequence $d_2 = d_1(q)$. If the value of q is chosen according to the following three conditions, then the two sequences d_1 and d_2 will be preferred pairs:

(i) n is odd or $\mathrm{mod}(n, 4) = 2$;

(ii) q is odd and either $q = 2^k + 1$ or $q = 2^{2k} - 2k + 1$ for an integer k;

(iii) $\gcd(n, k) = \begin{cases} 1 & \text{when } n \text{ is odd,} \\ 2 & \text{when } \mathrm{mod}(n, 4) = 2. \end{cases}$

I.3 Gold Code Generator

Many Gold code sequences can be generated by shifting two m-sequences that are preferred pairs and XORing them together. If d_1 and d_2 are preferred pair m-sequences, then a set of Gold sequences is defined as

$$\mathrm{gold}(d_1, d_2) = \{d_1, d_2, d_1 + d_2, d_1 + T d_2, d_1 + T^2 d_2, \dots, d_1 + T^n d_2\}, \quad \text{(I.3)}$$

where T^k is the cyclic shift by k bits, $k = 1, 2, \dots, n$.

The theoretical cross-correlation of Gold sequences is three-valued, given by

$$\left\{ -\frac{1}{N} t(n), -\frac{1}{N}, \frac{1}{N} (t(n) - 2) \right\}, \quad \text{(I.4)}$$

where $t(n)$ is defined, as in (I.1), by

$$t(n) = \begin{cases} 2^{(n+1)/2} + 1 & \text{if } n \text{ is odd,} \\ 2^{(n+2)/2} + 1 & \text{if } n \text{ is even.} \end{cases} \quad \text{(I.5)}$$

To generate Gold code sequences, use Viswanathan's `genGoldCodes.m` available at http://gaussianwaves.com (Viswanathan [2013]).

J

Doppler Effect

The Doppler frequency shift is caused by the satellite motion both on the carrier frequency and on the C/A code. The angular velocity $d\theta/dt$ and the speed v_s of a satellite is calculated using the approximate radius of the satellite orbit as $\dfrac{d\theta}{dt} \approx 1.458 \times 10^{-4}$ rad/s, and $v_s = r_s\dfrac{d\theta}{dt} \approx 3874$ m/s. where r_s is the average radius of the satellite orbit. Since the time difference between a solar day and the sidereal day is 3 min 55.91 sec, the satellite will travel approximately 914 km. With reference to the surface of Earth with the satellite toward the zenith, the corresponding angle is about 0.045 rad or 2.6°. If the satellite is close to the horizon, this angle is about 0.035 rad or 2°. This means that the satellite position changes about 2 to 2.6 degrees per day at the same time with respect to a fixed point on Earth's surface.

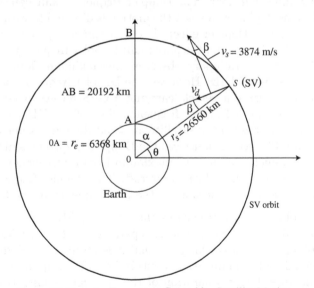

Figure J.1 Doppler Frequency Caused by Satellite Motion.

Figure J.1 shows the circular orbit of an SV around Earth, where the velocity component v_d causes the Doppler effect. The Doppler frequency is caused by the satellite velocity component v_d toward the user, i.e.,

$$v_d = v_s \sin \beta. \tag{J.1}$$

In the triangle OAS, the law of cosines gives

$$(AS)^2 = r_e^2 + r_s^2 - 2r_e r_s \cos \alpha = r_e^2 + r_s^2 - \sin \theta,$$

while the law of sine gives $\dfrac{\sin \beta}{\sin \alpha} = \dfrac{\sin \beta}{\cos \theta} = \dfrac{r_e}{AS}$. Substituting these results in (J.1) we find that the velocity component v_d is given by

$$v_d = \frac{v_s r_e \cos \theta}{\sqrt{r_e^2 + r_s^2 - 2r_e r_s \sin \theta}},$$

where $r_e = 6368$ km is the average Earth radius, and r_s is the average radius of the satellite orbit.

When $\theta = \pi/2$, the Doppler velocity is zero. Doppler velocity occurs when the satellite is at the horizon. A high-speed military aircraft has speed of about 2078 miles per hour. The Doppler frequency shift caused by a land vehicle is often negligible even when the motion is directed toward the satellite to produce maximum Doppler effect. For L1 frequency $f = 1575.42$ MHz as modulated by the C/A signal, the maximum Doppler frequency shift is about 4.9 KHz. Thus, for a stationary observer the maximum Doppler frequency shift can be taken as ± 5 KHz. However, the Doppler frequency shift becomes significant in the case of a vehicle carrying a GPS receiver that moves at a high speed. To create a Doppler frequency shift of ± 5 KHz by the vehicle alone, it must move toward the satellite at about 2078 miles per hour. Thus, the design of a GPS receiver must use the Doppler frequency shift of ± 5 KHz for a low-speed vehicle, and use the maximum Doppler frequency shift of ± 10 KHz for a high-speed vehicle like an aircraft. The Doppler frequency shift on the C/A code is very small because of the low frequency of the C/A code.

If the sampling frequency (SF) of the data is 5 MHz, i.e., if the data is sampled every 5 MHz, each sample is separated by 200 ns which is known as the sampling time. In tracking a signal it is required to align the locally generated signal and the input signal within half the sampling time or about 100 ns. Otherwise, larger separation of the two signals loses tracking. The chip time of the C/A code is 977.5 ns or $1/(1023 \times 10^6)$ sec. It takes 156.3 ms ($1/6.4$) to shift one cycle or 977.5 ns. Thus, it will take 16 ms ($100 \times 156.3/977.5$) to shift 100 ms. In a high-speed vehicle, a selection of a block of the input data must be checked ever 16 ms to ensure their alignment with the locally generated data. The noise on the signal effects this alignment, and therefore, the input data must be adjusted to every 20 ms for slow-speed vehicles and to 40 ms for high-speed vehicles. This analysis clearly implies that the adjustment of input data depends on the sampling frequency: Higher SF shortens the adjustment time because the sampling time is short.

There are two methods to calculate the Doppler frequency rate: (i) Estimate the average rate of change of the Doppler frequency, and (ii) determine the maximum rate of change of the Doppler frequency. The average rate of change of the Doppler frequency δf_{dr} is given by $\delta f_{dr} \approx 0.54$ Hz/s, whereas its maximum rate of change is $\delta f_{dr}\big|_{max} = 0.936$ Hz/s. This value, being very small, is taken on the order of 1 Hz, giving the update rate of almost 1 sec even when the Doppler frequency is at its maximum changing rate. This analysis shows that the Doppler frequency rate of change, caused by the satellite motion, is very low, and as such it does not create any serious problem in tracking a GPS signal. The input adjustment rate due to Doppler frequency on the C/A code is about 20 ms. In the case when the carrier frequency of the tracking loop has a bandwidth of the order of 1 Hz and the receiver accelerates at 7 g, the tracking frequency needs be updated about every 2.8 ms (1/360) because of the change in the carrier frequency. Since there is always noise present in the carrier frequency, this adjustment is extremely essential. Further details are available in Tsui [2000: 51 ff].

K

Bessel Functions

In the study of subtractive synthesis and frequency modulation, let the carrier signal be defined by ω_c. Then the modulating signal $s(t) = \sin(\omega_c t + I \sin \omega_m t)$, where I is the modulation index and ω_m the modulation frequency. As shown in §E.2, the signal $s(t)$ is defined as

$$s(t) = \sum_{k=-\infty}^{\infty} J_k(I) \sin\left[(\omega_c + k\omega_m)t\right],$$

where $J_k(t)$ is the Bessel function of first kind and order k. This brings us to define different kinds of Bessel functions because of their importance in our study.

K.1 Bessel Functions of the First Kind

The solutions of Bessel's equation of order ν

$$x^2 \frac{d^2 y}{dx^2} + x \frac{dy}{dx} + (x^2 - \nu^2) y = 0, \tag{K.1}$$

where ν is an integer, are the Bessel functions $J_\nu(x)$ of the first kind and $Y_\nu(x)$ of the second kind (also known as Weber's functions) and the Hankel functions $H_\nu^{(1)}(x)$ and $H_\nu^{(2)}(x)$, defined for each $\nu \geq q0$ as

$$J_\nu(x) = \sum_{n=0}^{\infty} \frac{(-1)^n}{n!(n+\nu)!} \left(\frac{x}{2}\right)^{2n+\nu}, \quad -\infty < x < \infty, \tag{K.2}$$

$$Y_\nu(x) = \frac{J_\nu(x)\cos(\nu\pi) - J_{-\nu}(x)}{\sin(\nu\pi)}, \tag{K.3}$$

$$H_\nu(1)(x) = J_\nu(x) + i\, Y_\nu(x) = i\, \csc(\nu\pi) \left[e^{-\nu\pi i} J_\nu(x) - J_{-\nu}(x)\right], \tag{K.4}$$

$$H_\nu^{(2)}(x) = J_\nu(x) - i\, Y_\nu(x) = i\, \csc(\nu\pi) \left[J_{-\nu}(x) - e^{\nu\pi i} J_\nu(x)\right]. \tag{K.5}$$

These power series solutions are obtained by the Frobenius method, details of which are available in textbooks on ordinary differential equation, e.g., Boyce and DiPrima [1992]. The infinite series (K.2) is uniformly convergent and can be differentiated or integrated term-by-term. The differentiation and integration formulas are as follows (see Abramowitz and Stegun [1972: 372 ff]):

$$J_\nu(-x) = (-1)^\nu J_\nu(x),$$

$$\frac{d}{dx}[x^\nu J_\nu(x)] = x^\nu J_{\nu-1}(x),$$

$$\frac{d}{dx}[x^{-\nu} J_\nu(x)] = -x^{-\nu} J_{\nu+1}(x),$$

$$x J_\nu'(x) = \nu J_\nu(x) - x J_{\nu+1}(x), \tag{K.6}$$

$$x J_\nu'(x) = -\nu J_\nu(x) + x J_{\nu-1}(x),$$

$$\int_0^x t^\nu J_{\nu-1}(t)\, dt = x^\nu J_\nu(x).$$

In particular,

$$J_0'(x) = -J_1(x), \quad \text{and} \quad \int_0^x t J_0(t)\, dt = x J_1(x).$$

The integral representations for integer ν are

$$J_\nu(x) = \frac{1}{\pi} \int_0^\pi \cos(x \sin\theta - \nu\theta)\, d\theta, \tag{K.7}$$

$$Y_\nu(x) = \frac{1}{\pi} \int_0^\pi \sin(x \sin\theta - \nu\theta)\, d\theta. \tag{K.8}$$

where for $\nu = 0$ we have

$$J_0(x) = \frac{1}{\pi} \int_0^\pi \cos(x \sin\theta)\, d\theta = \frac{2}{\pi} \int_0^{\pi/2} \cos(x \cos\theta)\, d\theta. \ x > 0 \tag{K.9}$$

$$Y_0(x) = \frac{4}{\pi^2} \int_0^{\pi/2} \cos(x \cos\theta) \left[\gamma + \ln(2x \sin^2\theta)\right] d\theta = -\frac{2}{\pi} \int_0^\pi \cos(x \cosh t)\, dt, \ x > 0. \tag{K.10}$$

Note that J_{2k} is an even function and J_{2k+1} an odd function. Moreover, $J_0(0) = 1$, but $J_\nu(0) = 0$ for $\nu \geq 1$. In fact, J_ν has a zero of multiplicity ν at $x = 0$. The integral representation (K.7) shows that $\|J_\nu(x)\| \leq 1$ for all real x and $\nu \geq 0$.

K.2 Modified Bessel Functions

The modified Bessel functions $I_{\pm\nu}(x)$ of the first kind and $K_{\pm\nu}(x)$ of the second kind are solutions of the differential equation $x^2 u'' + x u' - (x^2 + \nu^2)\, u =$

0. They are defined as

$$I_\nu(x) = e^{-\nu\pi i/2}\, J_\nu\big(x\, e^{pii/2}\big), \quad -\pi < \arg\{x\} \leq \pi/2, \qquad \text{(K.11)}$$

$$K_\nu(x) = \frac{\pi}{2}\frac{I_{-\nu}(x) - I_\nu(x)}{\sin(\nu\pi)}. \qquad \text{(K.12)}$$

In particular, $I_{-\nu}(x) = I_\nu(x)$, and $K_{-\nu}(x) = K_\nu(x)$. The integral representations are:

$$I_0(x) = \frac{1}{\pi}\int_0^\pi e^{\pm x \cos\theta}\, d\theta = \frac{1}{\pi}\int_0^\pi \cosh(x\cos\theta)\, d\theta, \qquad \text{(K.13)}$$

$$K_0(x) = \frac{1}{\pi}\int_0^\pi e^{\pm x \cos\theta}\left[\gamma + \ln(2x\sin^2\theta)\right]\, d\theta. \qquad \text{(K.14)}$$

Kelvin functions ber, bei, ker, kei, which are solutions of the differential equation $x^2 u'' + x u' - \left(i\, x^2 + \nu^2\right) u = 0$ are available in Abramowitz and Stegun [1972: 379 ff]; and Bessel functions for fractional order are given in Abramowitz and Stegun [1972: 437 ff]

K.3 Airy Functions

Airy functions are defined in Abramowitz and Stegun [1972: 446 ff]. There are three pairs of solutions of the differential equation $u'' - xu = 0$, namely, (i) Ai(x) and Bi(x); (ii) Ai$\big(x\,e^{2\pi i/3}\big)$, and (iii) Ai$(x)$ and Ai$\big(x\,e^{-2\pi i/3}\big)$. The function Ai$(x)$ is defined in terms of the Bessel functions as follows:

$$\text{Ai}(x) = \frac{\sqrt{x}}{3}\left[I_{-1/3}(y) - I_{1/3}(y)\right] = \frac{1}{\pi}\sqrt{\frac{x}{3}}\,K_{1/3}(x), \qquad \text{(K.15)}$$

$$\text{Ai}(-x) = \frac{\sqrt{x}}{3}\left[J_{1/3}(y) - J_{-1/3}(y)\right]$$

$$= \frac{1}{\pi}\sqrt{\frac{x}{3}}\left[e^{\pi i/3}H_{1/3}^{(1)}(y) + e^{pii/3}H_{1/3}^{(2)}(y)\right], \qquad \text{(K.16)}$$

$$\text{Ai}(x) = \frac{x}{3}\left[I_{-2/3}(y) - I_{2/3}(y)\right] = \frac{1}{\pi}\frac{xc}{\sqrt{(3)}}K_{2/3}(x), \qquad \text{(K.17)}$$

K.4 Hankel Transforms

In two-dimensional polar coordinate system we set

$$r = \sqrt{t^2 + \tau^2}, \ \ t = r\cos\theta, \qquad \theta = \arctan\left(\frac{\tau}{t}\right), \ \ \tau = r\sin\theta,$$

$$\rho = \sqrt{f^2 + g^2}, \ \ f = \rho\cos\phi, \ \ \phi = \arctan\left(\frac{g}{f}\right), \ \ g = \rho\sin\phi. \qquad \text{(K.18)}$$

Then the two-dimensional Fourier transform (3.29) in the polar coordinate system, denoted by $\mathcal{F}\{x(t,\tau)\} = X(\rho,\phi)$, is defined as

$$X(\rho,\phi) = \int_0^\infty x(r,\theta)\,dr \int_0^{2\pi} e^{-i\,2\pi r\rho\cos(\theta-\phi)}\,d\theta. \qquad \text{(K.19)}$$

Consider the Bessel function J_0 of the first kind and order zero, and use the identity

$$J_0(a) = \frac{1}{2\pi}\int_0^{2\pi} e^{-i\,a\cos(\theta-\phi)}\,d\theta. \qquad \text{(K.20)}$$

Then substituting (K.20) into (K.19), the transform X is independent of ϕ, and we get

$$X(\rho) = 2\pi \int_0^\infty rx(r,\theta)J_0(2\pi r\rho)\,dr. \qquad \text{(K.21)}$$

This is known as the *Fourier-Bessel transform* $\mathcal{B}$, or simply the *Hankel transform*, of order zero $\mathcal{H}_0$, which is defined as follows. If we denote $\mathbf{r} = (t,\tau)$ and $\boldsymbol{\phi} = (f,g)$, the two-dimensional Hankel transform pair is defined by:

$$\mathcal{H}_0\{x(t,\tau)\} = \hat{x}(f,g) = \int_{-\infty}^\infty \int_{-\infty}^\infty e^{-i\,(\boldsymbol{\phi}\cdot\mathbf{r})}x(t,\tau)\,dt\,d\tau,$$

$$\mathcal{F}^{-1}\{\hat{x}(f,g)\} = x(t,\tau) = \int_{-\infty}^\infty \int_{-\infty}^\infty e^{i\,(\boldsymbol{\phi}\cdot\mathbf{r})}\hat{x}(f,g)\,df\,dg,$$

$$\text{(K.22)}$$

or in polar coordinates, since $(t,\tau) = r(\cos\theta,\sin\theta)$ and $(f,g) = \rho(\cos\phi,\sin\phi)$, we find that $\boldsymbol{\phi}\cdot\mathbf{r} = r\rho\cos(\theta-\phi)$, and thus,

$$\hat{x}(\rho,\phi) = \int_0^\infty r\,dr \int_0^{2\pi} e^{-i\,r\rho\cos(\theta-\phi)}\,x(r,\theta)\,d\theta. \qquad \text{(K.23)}$$

In case $x(r,\theta) = x(r)\,e^{i\,n\theta}$, we make a change of variable $\theta - \phi = \alpha - \pi/2$, and then (K.23) reduces to

$$\hat{x}(\rho,\phi) = \int_0^\infty rx(r)\,dr \int_0^{2\pi+\phi_0} e^{-i\,n\phi_0 + i\,(n\alpha - r\rho\sin\alpha)}\,d\alpha, \qquad \text{(K.24)}$$

where $\phi_0 = \pi/2 - \phi$. The operational properties of the Hankel transform of order zero are: (i) $\hat{x}(ar) = \dfrac{1}{a^2}\hat{x}(\rho/a)$ (scaling); and (ii) $\int_0^\infty rx(r)y(r)\,dr = \int_0^\infty \hat{x}(\rho)\hat{y}(\rho)\,d\rho$ (Parseval's relation). Some useful Hankel transform pairs of order zero are given in Table K.1.

Table K.1. Hankel Transform Pairs of Order Zero.

	$x(r)$	$\hat{x}(\rho) = \int_0^\infty r J_0(\rho r) x(r)\, dr$
1.	$\dfrac{1}{r}\, \delta(r)$	1
2.	e^{-ar}	$a\left(a^2 + \rho^2\right)^{-3/2}$
3.	$\dfrac{e^{-ar}}{r}$	$a\left(a^2 + \rho^2\right)^{-1/2}$
4.	$H(a - r)$	$\dfrac{a}{k}\, J_1(a\rho)$
5.	$\left(a^2 - r^2\right) H(a - r)$	$\dfrac{4a}{\rho^3}\, J_1(a\rho) - \dfrac{2a^2}{\rho^2}\, J_0(a\rho)$
6.	$a\left(a^2 + r^2\right)^{-3/2}$	$e^{-a\rho}$
7.	$\dfrac{1}{r}\cos(ar)$	$\left(\rho^2 - a^2\right)^{-1/2} H(\rho - a)$
8.	$\dfrac{1}{r}\sin(ar)$	$\left(a^2 - \rho^2\right)^{-1/2} H(a - \rho)$
9.	$\dfrac{1}{r^2}\left(1 - \cos ar\right)$	$\cosh^{-1}\left(\dfrac{a}{\rho}\right) H(a - \rho)$
10.	$\dfrac{1}{r}\, J_1(ar)$	$\dfrac{1}{a} H(a - \rho),\quad a > 0$
11.	$Y_0(ar)$	$\dfrac{2}{\pi}\left(a^2 - \rho^2\right)^{-1}$
12.	$K_0(ar)$	$\left(a^2 + \rho^2\right)^{-1}$
13.	$\left(r^2 + a^2\right)^{-1/2}$	$\dfrac{1}{\rho}e^{-a\rho}$
14.	$\dfrac{\sin r}{r^2}$	$\begin{cases} \dfrac{\pi}{2}, & \rho < 1, \\[2mm] \arcsin\dfrac{1}{\rho}, & \rho > 1 \end{cases}$
15.	$\mathrm{circ}(r)$	$\dfrac{1}{\rho}\, J_1(2\pi\rho)$

L

Heisenberg's Uncertainty Principle

This uncertainty principle is based on the simple premise, known as the Heisenberg principle (Heisenberg [1925/1930: 20]), that asserts a fundamental limit to the simultaneous precision between position and momentum. The two variables, position x and momentum p, behave in a manner such that the more precisely the position x is determined, the less precisely the momentum p can be known. According to Kennerd [1927] and Weyl [1928], the standard deviations σ_x of position and σ_p of momentum are related by the inequality

$$\sigma_x \, \sigma_p \geq \hbar/2, \tag{L.1}$$

where $\hbar$ is the Planck's constant.

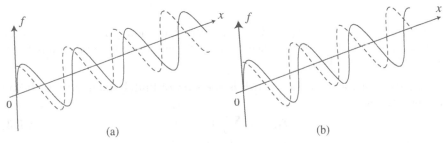

Figure L.1 (a) Plane Wave, (b) Wave Packet $f(x,t)$.

The uncertainty principle is inherent in all wave-like systems. Among different forms, there are at least two mathematical forms: the wave mechanics, and the matrix mechanics form. In the wave mechanics format, the uncertainty principle arises because the wave functions in the two corresponding bases are Fourier transform of each other, thus making position and momentum as conjugate variables, called the *Fourier conjugates*. The variance of Fourier conjugates exists in the case of sound waves; for example, a pure tone is a sharp spike at a single frequency so that its Fourier transform gives the sinusoidal shape of the sound wave in the time domain. In quantum mechanics, the two Fourier conjugates take the form of a particle wave such that momentum is its Fourier conjugate that satisfies the de Broglie relation $p = \hbar k$, where k is the wavenumber. Thus, the function $f(x, t) = A e^{i(px - \omega t)}$, represents a plane wave, and the finite sum $f(x, t) = \sum_n A_n e^{i(p_n x - \omega_n t)}$, represents a wave packet, both in the (x, t)-plane, as shown in Figure L.1, where the solid curve represents the real part and the dotted curve the imaginary part of $f(x, t)$. The probability of finding a particle at a given point x is spread all over the waves. As the amplitude A (or A_n) increases above zero, the wave curvature reverses sign, so the amplitude begins to decrease again, and the process repeats. In both cases, the position of the particle is determined by the wave function $f(x, t)$.

In the time-independent case, the wave function of a single-mode plane wave with wavenumber k_0, or momentum p_0, is given by

$$f(x) \propto e^{i k_0 x} = e^{i p_0 x / \hbar}. \tag{L.2}$$

Then the probability of finding a particle between a and b is

$$\mathfrak{p}\,[a \le x \le b] = \int_a^b |f(x)|^2 \, dx. \tag{L.3}$$

Note that the quantity $|f(x)|^2$ is a normal distribution. This means that the particle position is extremely uncertain in the sense that it could be anywhere along the wave packet.

Consider a wave function which is the sum of finitely many waves. It can be written as

$$f(x) \propto \sum_n A_n e^{i p_n x / \hbar}, \tag{L.4}$$

where A_n is the relative overall amplitude of the mode p_n. The representation (L.4) can be written in the limiting case when the wave function is an integral over all possible modes, i.e.,

$$f(x) = \frac{1}{\sqrt{\hbar}} \int_{-\infty}^{\infty} \phi(p) \, e^{i p x / \hbar} \, dp, \tag{L.5}$$

where $\phi(p)$ is the amplitude of all the above-mentioned modes. The function $f(x)$ in (L.5) is called the *wave function in the momentum mode*. Obviously, in view of the definition (3.1), the function $\phi(p)$ in Eq (L.5) is the Fourier transform of $f(x)$, and x and p are Fourier conjugates. However, when all these wave functions are summed up, the momentum p becomes less precise as it becomes a combination of many different momenta.

There is one way to obtain the precise position and, thus a precise momentum, by considering the quantity $|f(x)|^2$. since it represents a probability density function of position. By calculating its standard deviation, the precision of the position x increases since σ_x is reduced by using many plane waves; this decreases the precision of momentum since σ_p increases. In other words, σ_x and σ_p have an inverse relationship. This is the uncertainty principle, and its exact limit is the *Kennerd bound* (L.1). For a proof of this bound, see Kennerd [1927].

In the matrix form of the uncertainty principle, the position and momentum are considered as self-adjoint operators. Define the commutator for a pair of operators $\mathbf{a}$ and $\mathbf{b}$ as $[\mathbf{a}, \mathbf{b}] = \mathbf{ab} - \mathbf{ba}$, the commutator for the position vector $\mathbf{x}$ and $\mathbf{p}$ is $[\mathbf{x}, \mathbf{p}] = i\hbar$. Using the bra-ket notation(see §5.5.5), let $|f\rangle$ be the right eigenstate of position with a constant eigenvalue x_0. This means that $\mathbf{x}|f\rangle = x_0|f\rangle$. Then applying the commutator to $|f\rangle$ we obtain

$$[\mathbf{x}, \mathbf{p}] = (\mathbf{x}\,\mathbf{p} - \mathbf{p}\,\mathbf{x}) = (\mathbf{x} - x_0\mathbf{I}) \cdot \mathbf{p}|f\rangle = i\,\hbar|f\rangle \neq 0, \qquad \text{(L.6)}$$

where $\mathbf{I}$ is the identity operator. This result shows that it is not possible for both a position and a momentum eigenstate to occur simultaneously. Thus, if a particle's position is computed, then the state exists at least momentarily in the position eigenstate, i.e., the state is not in a momentum eigenstate, but exists as a sum of multiple momentum eigenstates. This implies that the momentum is less precise. In this case the standard deviations are defined by

$$\sigma_x = \sqrt{\langle \mathbf{x}^2 \rangle - \langle \mathbf{x} \rangle^2}, \quad \sigma_p = \sqrt{\langle \mathbf{p}^2 \rangle - \langle \mathbf{p} \rangle^2}. \qquad \text{(L.7)}$$

M

Classical Latin Prosody[1]

M.1 Rhythm and Meter

Percy Bysshe Shelley (1792-1822) in *A Defense of Poetry* has said, "In the infancy of society every author is necessarily a poet, because language itself is poetry." A necessary condition to compose a work of musical value is to devise a system of disciplined and linked movement. This system, called rhythm, consists of the reproduction of the sound of human voices, musical instruments and orchestras. Such expressions in metrical art are attained by an arrangement of syllables, which is known as rhythm. The science of meter consists in establishing laws which define the rhythmic forms of poetry and music. Thus, rhythm means a harmonious movement. The classical Latin verse is rhythmic, and the rhythm occurs as a periodic motion.

Rhythm is marked by the stress of voice, i.e., accent. The rhythmic accent is called the *ictus* (beat, blow). The accented part of the syllables is called the *thesis* ($\theta\epsilon'\sigma\iota\varsigma$), and the unaccented the *arsis* ($\alpha'\rho\sigma\iota\varsigma$). Thesis and arsis are Greek terms, meaning putting down and raising of the foot, respectively, in military marching.

Rhythm, when represented in poetry, is embodied in *meter* or measure ($\mu\epsilon'\tau\rho o\nu$). A meter is a system of syllables standing in a predetermined order. Meter is used in two senses: (i) as a definite system or combination of particular verses; and (ii) as a definite portion of a particular verse. The stress-accent is not of supreme importance in Latin meters, but it has an influence on the quality of sound.

The difficulty of learning the rhythm of the Latin verse arises from the fact that Roman poets found it generally desirable that stress-accent and ictus must not coincide. The effect of their falling on different syllables was not only an evenly distribution of emphasis (weights), but also a clear and precise pronunciation of each syllable. This feature produced a valuable result when the quantities were so important. Even when stress-accent and ictus coincide, the weight is concentrated on one syllable, and the metrical rhythm is more

---

[1] Excerpts from author's unpublished work on the mathematical structures in classical Latin poetry.

strongly marked. This effect was used with Hexameters and Pentameters for the purpose of showing the approach of the end of the line. For example, consider

Dāedalus īntere|ā † Cré|tēn | lon|gūmque per|ōsus. _ *Ovid, Metamorphosis*, viii, 183; or the line

ībant | ōbscú|rī só|lā sub|nōcte per|ūmbram. _ *Vergil, Aeneid* vi, 218,

where only the ictus (–) and the natural stress-accent (´) are marked, and † denotes the caesura[2]. Notice that they coincide in the first, fifth, and sixth feet (defined below in §M.2 and M.3). Generally, the poets liked such coincidences in both fifth and sixth feet (the coincidence at the beginning depends on the choice of words) to mark the conclusion of the line, and they are actually avoided altogether in both third and fourth feet, by putting a strong caesura in one of them. When stress-accent and ictus do not coincide, the accentuation of syllables is more regular, and the flow more stately. It is the avoidance of coincidences in the body of the Hexameter that is largely responsible for its stately character.

Thus, meter or measure ($\mu\epsilon'\tau\rho o\nu$) is a system of syllables in a definite order. The unit of measure is the short syllable ($\smile$), and is called mŏra, tempus (time); its value in music is a quaver (= 1/8). The long syllable (–) is the double of the short one; its value in music is a crotchet (= 1/4). Thus, one crotchet is equal to two quavers[3]. The prosodic and musical notations are presented in the following table:

Syllable	Prosodic notation	Musical notation	Value
Short	$\smile$	♪	1/8
Long	–	♩	1/4

A syllable which is not an exact multiple of the standard unit is called an irrational syllable; feet containing such quantities are, therefore, called

---

[2] The conflict of the word-foot and the verse-foot gives rise to caesura which is denoted by †. Caesura means the incision in a verse where a word ends, so as to cut (caedo) the verse-foot into two, and the voice pausing a little. This pause is served by the incision itself, partly to prevent monotony by distributing the masses of the verse. Many scholars call any incision in a verse a caesura. It should not, and does not, occur in the middle of a verse, since in that case the verse would become monotonous.

[3] This does not mean that the Latin poets thought this equality to be mathematically exact, nor even that they made it so in reciting their verses, by keeping a strict musical time. The syllables vary greatly in length. However, it was almost true to make it a practical working rule.

irrational (see below for the definition of a foot, and also §M.3).

In some verses, if two short syllables are used instead of a long one, a *resolution* is said to take place; or if a long syllable is used instead of two short syllables, a *contraction* is said to take place. This system may be compared with music, so that resolution and contraction are represented as follows:

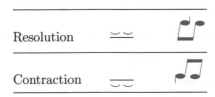

Regarding the graphical representation of the short and long syllables, a simple sinusoidal wave of the form $x(t)$ is employed, where the time t-interval along the abscissa represents the quantity of quavers and crotchets in an exact mathematical manner, such that the tempus of a long syllable is twice that of a short syllable. These graphical forms are presented together with the description of various foot in Figure M.1.

M.2 Feet

A foot (pēs, $\pi\alpha\upsilon'\varsigma$) is a name given to a set of two or more syllables by which the lines of Latin poetry are divided. As elements of orchestrics, meters are called *bars*; as elements of verse they are called *feet*.

The scansion or scanning is the distribution of a verse into its proper feet. As musical strains are composed of equal bars, so verses are composed of equal feet. Consider the lines

Phăsēlŭs īllĕ, quēm vĭdē tĭs, hōspĭtēs,
    ăīt fŭīssĕ nāvĭūm cĕlērrĭmŭs. ‗ *Catallus*, iv, 1-2.

Note that the last syllable of the second line is short. Since there is a natural pause at the end of the line, the effect is the same as if it were long. This often happens in other meters as well.

As we mark these two lines by the quantity of the syllables, we find that short and long syllables alternate regularly, i.e., the group ⌣ – is repeated, as it is technically called, and the accent falls on the long syllable, ⌣ -́. The group ⌣ – is called a 'foot', and we mark the divisions between the feet by vertical lines, thus completing the scansion:

Phăsē|lŭs īl|lĕ, quēm |vĭdē |tĭs, hōs|pĭtēs,
    ăīt |fŭīs|sĕ nā|vĭūm | cĕlēr|rĭmŭs.

M.3 List of Feet

Theoretically, the number of meters is unrestricted. However, practically only those meters are of importance that serve to embody the principal rhythms. The feet of Latin verse are listed below. Those marked by ⋆ are mentioned by Latin grammarians, but they do not occur in the classical Roman verse; the other remaining feet were in use.

- Feet of Two Times:

  Pyrrhic ⋆ (pēs pyrrhichius, $\pi v' \rho \rho \iota \chi \iota o \varsigma$), ⌣ ⌣   lĕgĭt.

- Feet of Three Times: ($\tau \rho o \chi \alpha \iota o \varsigma$)

  Trochee[4] (pēs trŏchaeus, $\tau \rho o \chi \alpha \iota o \varsigma$), – ⌣   lēgĭt.

  Iambus (pēs ĭambus, $\iota \alpha \mu \beta o \varsigma$). ⌣ –   lĕgūnt.

  Tribrach (trībrăchys, $\tau \rho \iota \beta \rho \alpha \chi v \varsigma$), ⌣ ⌣ ⌣   lĕgĕtĕ.

- Feet of Four Times:

  Anapaest (pēs ănăpaestus, $\alpha v \alpha' \pi \alpha \iota \sigma \tau o \varsigma$), ⌣ ⌣ –   lĕgērēnt.

  Spondee (pēs spondēus, $\sigma \pi o v \delta \epsilon \iota \varsigma$), – –   lēgī.

  Proceleusmaticus, ⌣ ⌣ ⌣ ⌣   rĕlĕgĭtŭr.

- Feet of Five Times:

  Cretic (pēs crēticus), – ⌣ –   lēgĕrīnt.

  First paeōn ($\pi \alpha \iota \omega' v$), – ⌣ ⌣ ⌣   lēgĕrĭtĭs.

  Second Paeōn ⋆, ⌣ – ⌣ ⌣   lĕgēntĭbŭs.

  Third Paeōn ⋆, ⌣ ⌣ – ⌣   lĕgĭtōtĕ.

  Fourth Paeōn, ⌣ ⌣ ⌣ –   lĕgĭmĭnī.

  Bacchīus ($\beta \alpha \kappa \chi \epsilon \iota o \varsigma$), ⌣ – –   lĕgēbānt.

  Antibacchīus, – – ⌣   lēgīstĭs.

- Feet of Six Times:

  Iōnicus ā māiōre, – – ⌣ ⌣   cōllēgĭmŭs.

  Iōnicus ā minōre, ⌣ ⌣ – –   rĕlĕgēbānt.

  Chorĭamous (Chŏrēus or Trochee + Iambus), – ⌣ ⌣ –   cōllĭgĕrānt.

  Ditrochee (Trochee + Trochee), – ⌣ – ⌣   cōllĭgūntŭr.

  Diiambus (Iambus + Iambus), ⌣ – ⌣ –   lĕgāmĭnī.

  Antispast ⋆ (Iambus + Trochee), ⌣ – – ⌣   lĕgēbārĭs.

  Molossus ⋆ (connected with $Mo\lambda o \sigma \sigma o \iota'$), – – –   lēgērūnt.

---

[4] Trochee was originally called a Chŏrēus or Chŏrĭus ($\chi o \rho \epsilon \iota o \varsigma$).

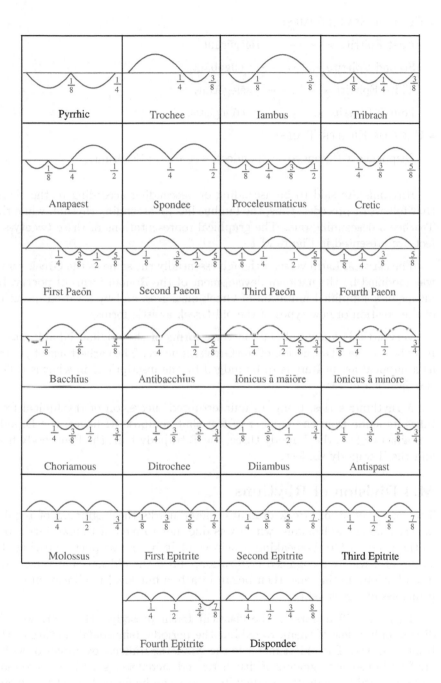

Figure M.1 Graphical Forms of the Twenty-Six Feet.

- FEET OF SEVEN TIMES:

  First Epitrite ⋆ ⌣ – – –   rĕlēgērūnt.

  Second Epitrite ⋆, – ⌣ – –   ēlĭgēbānt.

  Third Epitrite ⋆, – – ⌣ –   sēlēgĕrīnt.

  Fourth Epitrite ⋆, – – – ⌣   cōllēgīstĭs.

- FEET OF EIGHTH TIMES:

  Dispondee ⋆ (Spondee + Spondee), – – – –   sēlēgērūnt.

Rhythms are said to be ascending or descending according to the thesis that follows or precedes; thus, the Iambus has an ascending rhythm, while the Trochee a descending one. The graphical representations of these twenty-six feet are presented in Figure M.1.

The classical Latin verse, although essentially the same as the Greek verse, was modified by the national development of the Roman forms of poetry, by means of a simplified imitation of Greek measures, and by a varied intensity of the creation of new types of the old Greek artistic forms.

Very little is known about the verse forms of the original inhabitants of Italy before the introduction of the Greek influence. The earliest use of poetry as a national art in Italy is to be judged by the inscriptions in what is called the Saturian meter.

A rhythmic series, being an uninterrupted succession of rhythmical feet, takes its name from the number of feet that compose it; hence, the Dipody contains two feet, the Tripody three, the Tetrapody four, the Pentapody five, and the Hexapody six feet.

M.4 Division of Rhythms

The introduction of Greek dramatic meter marks the beginning of regular poetry among the Romans, which was due, not to men of Roman birth, but to the poets of Greek-speaking provinces of Italy. These poets, bearing the stamp of a widely recognized civilization, threw the old national verse back into oblivion. Latin verse then began in a free but loyal modification of the principles of Greek verse.

There are 79 meters in the classical Latin prosody. However, we will discuss only a few of them, to exhibit the periodic (sinusoidal) nature of the Roman poetry. The oldest meter is the Nŭmĕrus Ītălĭcus, examples of which are found in some fragments of ritualistic and sacred songs, and seems to have had no regard to quantity. No definite theory can be formed about this meter since the examples are very few. The Saturian verse (Nŭmĕrus Sāturnĭus) is another old Latin rhythm, examples of which occur in the earlier monuments.

It is divided into two parts, with three theses in each; but the exact metrical composition is much disputed, since the remains are again insufficient to admit of any precise structure.

The Dactylic (heroic) Hexameter was an invention ascribed to the Delphic priests, although this conjecture may be of purely religious origin. The hexameter has existed for centuries before Homer's Iliad assumed its form. The hexameter arose, as is seen from the importance of the caesura in its scheme, from a combination of two short lines, the first of which normally is $-\smile\smile$ | $-\smile\smile$ | $-$, and the second has the same structure but with an anacrūsis and an added syllable at the end as $-\smile\smile$ | $-\smile\smile$ | $-$ ‖ $-$. The Pentameter was formed from this scheme by the omission of the added syllables in the second half, which changed the character of the verse entirely.

Occasionally a Spondee is found in the fifth foot, such as:

tune ille Aeneas, quem Dardanio Anchisae. _ *Verg., Aen., i, 617.*
This line contains a hiatus and a Spondee in the fifth foot instead of the customary Dactyl, a license which Vergil permits himself only in the lines with proper names. Such a verse, in which the fifth foot is a Spondee, is called Sponaic Hexameter. The Sponaic Hexameter produces a special effect. An example, from Horace, is the line:

ınvitum qui servat idem facit occidenti. _ *de Ars Poetica,* 467.

In the Hexameter, in general, the first four feet must be either Dactyls or Spondees; the fifth foot is, with very rare exceptions, a Dactyl, and the sixth a Spondee, although with the usual license in the last syllable of any line where a Trochee may take its place. Either in the third or fourth foot one word must end and another begin; if this break (caesura) occurs after the first syllable of the foot, the line is said to have strong caesura; if after the second, a weak caesura occurs. Thus, the scheme for the Hexameter is

$$-\smile\smile \mid -\smile\smile \mid -\smile\smile \mid -\smile\smile \mid -\smile\smile \mid --$$

or

$$-- \mid -- \mid -- \mid -- \mid -\smile\smile \mid -\smile$$

mūltās | pēr gēn|tēs † ēt | mūltă pĕr | aēquŏră | vēctŭs
ūt tē | pōstrēm |ō † dōn|ārēm mūnĕrĕ | mōrtĭs. _ *Catallus, ci,* 1–2.

In the scansion of the lines, no account is taken of syllables at the close of words ending in a vowel or an *m*, if they are followed immediately by a word commencing with a vowel or an *h*. Such a final syllable is elided, although it was more probably slurred in pronunciation.

The scheme of the Hexameter makes this meter elastic and gives it variable length. This makes possible such onomatopoeic lines as

quādrŭpĕ|dāntĕ pŭ|trēm † sŏnĭ|tū quătĭt | ūngŭlă | cāmpŭm.

īll(ī) īn|tēr sē|sē māg|nā vī|brācchĭă | tōllūnt.

Pure Dactylic lines are rare; the most usual forms of the first four feet of the stichic measure are: (i) DSSS ; (ii) DSDS; (iii) DDSS, and (iv) SDSS, where D represents a Dactyl, and S a Spondee. The most common measures are SSDD, and SDDD.

The Heroic Hexameter, also known as Versus Pȳthĭus (a word connected with Pȳtho, $\Pi v\theta\omega'$), is composed of two Dactylic Tripodies, the second of which ends in a Spondee. Spondees may be substituted for the Dactyls in the first four feet, and in the fifth foot only when a special effect is desired. Such a verse is called Spondaic Hexameter which has been discussed above. The longest Hexameter (17-syllable) contains five Dactyls and one Spondee (or Trochee); the shortest (13-syllable) five Spondees and one Dactyl. This variety in the length of the verse is combined with the greatest number of caesuras, and gives the Hexameter peculiar advantages for continuous composition. The metrical scheme of the Heroic Hexameter is:

$$\underset{\smile\smile}{\prime\prime} \;\Big|\; \overset{\prime}{\underset{\smile\smile}{—}} \;\Big|\; \overset{\prime}{\underset{\smile\smile}{—}} \;\overset{\prime}{}\;\overset{\prime\prime}{\underset{\smile\smile}{—}} \;\Big|\; \overset{\prime}{—} \;\overset{(—\,\smile)}{\smile\smile} \;\Big|\; \overset{\prime}{—}\,—$$

The Hexameter is an important measure in Latin poetry, and so much of its beauty depends on the selection and arrangement of the words, considered as metrical elements, and on the exquisite finish of the rhythm. As a general rule, monosyllables at the end of this meter denote surprise, and the Anapaestic words rapid movement. Again, the Hexameter may be lowered to a conversational tone by large masses of Spondees and free handling of the caesura. The only instance of Spondaic Hexameter is in Horace:

> invitum qui servat idem facit occidenti. _ de Ars Poetica, 467;
> (īnvīt(um) | quīsēr|vāt īd|ēm fā|cīt ŏccĭ|dēntĭ.)

M.5 Roman Concept of Music and Drama

The Greeks originally regarded music as an art over which the Muses presided. A Muse (Mūsa, $Mov\sigma\alpha$) was generally regarded as a goddess of music. However, eventually it got narrowed down by excluding poetry, dancing and other arts, although music was still closely related to them. The ancient Greeks regarded music generally subordinate to verse and kept it limited as a means of expression of their thoughts. One reason for this inferior status of music was probably the use of simple instruments, mainly the lyre and flute. However, this simplicity itself produced the diatonic scale or modes (see §10.5.5), which eventually developed the basic key relationships.

On the other hand, the Romans were somehow indifferent to art. They mostly regarded the artist merely as a superior mechanic (*deux ex machina*). A reason for this indifference was the fact that the lesser arts in Greece were practiced mostly by slaves (metoecs), and the Greek aristocrats classified even the higher artist with these people and called all of them technicians.

Regarding the musical notation, there seems to be an important factor about the comparative value in their expressional power. The musical notation dates back from a time when the alphabetical method of delineation was prevalent among the Greeks who were perhaps pioneer in realizing and evolving a system of the kind. But we learn for certainty from Boethius, a distinguished Roman statesman, philosopher and theologian of the post-classical period (AD 524), that this Greek notation was not adopted by the Romans. The reduction of the scale to the octave is ascribed to St. Gregory, as also the naming of the seven notes, but it is not safe to assume that such an ascription is accurate or final. Indications of a scheme of notation based, not on the alphabet, but on the use of dashes, hooks, curves, dots and strokes are found to exist as early as the sixth century AD, while specimen in illustration of this different method do not appear until the eighth century AD. The origin of these signs, known as neumēs ($\nu\epsilon\upsilon\mu\alpha\tau\alpha$), is the fullstop (puntus), the comma (virga) and the mound or undulating line (clīvus), where the first indicates a short sound, the second a long sound, and the third a group of the two notes. The musical intervals were suggested by the distance of these signs from the words of the text.

Notwithstanding, the Greeks still possessed a peculiar sense of the value of art, and believed that beautiful sights and sounds germinate beautiful thoughts and actions. Plăto in advocating his view of 'music for mind' holds the artist worthy of reverence and praise. Such a feeling is, however, totally foreign to the Roman mind.

In explaining these facts we need not go beyond the Roman character, which was essentially practical and attached to a life of action. Their chauvinism or exaggerated opinion of themselves as a nation is due to their political history. Their success had filled them with the pride in an imperial nation when they encountered the Greeks; they were not prepared to bow down before the genius of a foreign and defeated people. At a later date they grudgingly admitted the excellence of much that they were secretly conscious of being unable to perform themselves. But they could save their national pride by asserting that they had failed to excel not from inability but from choice. Cicero says in (_ Tusc., 1, 1): "meum semper indicium fuit omnia nostros aut invenisse per se sapintibus quam Graecos aut accepta ab illis fecisse meliora, quae quidem digna statuissent, in quibus elaborarent", i.e., 'I have always maintained that in every case where Romans thought it (consistent with their dignity) to make the effort, they have either made brilliant and original discoveries, or else improved upon what they have borrowed.'

If it were not for the immense influence of the Stoic philosophy, this national prejudice might have been greatly modified among the more educated classes. The system of Zēno and Chrȳsippus, which had been acclimatized at Rome by men like Panaeticus, who was the favorite of the Acipionic circle, propagated its cardinal principle of the exaltation of Reason at the expense of the feelings and emotions. This endeavored to suppress the emotions and left no room for art. To understand a scientific fact, such as a proposition of Euclid, a man need only use his powers of reasoning, but to understand a work of art — a beautiful statue, painting, or poem — he must be able to feel as well as reason. Thus, the result of the Stoic influence and of the national pride was to harden the Roman character, and leave art a thing totally without meaning to the ordinary citizen and even to the educated statesman.

Nonetheless, the almost mathematical precision of the Roman verse laid the foundation of music throughout the middle ages and eventually became the source of composition of western masterpiece music. The beauty and wonder of this evolution in music permeates even today.

Bibliography

(Note: First author or single author is cited with last name first.)

Abbe, Ernst Karl. 1873. Beitrage zur Theorie des Mikroskops und der mikroskopischen Wahmehmung. *Archive für Mikroskopischen Anatomie.* 9(1): 413–468.

Abramowitz, M. and I. A. Stegun. 1972. *Handbook of Mathematical Functions.* New York: Dover.

Ackerman, M., M. Ajello, A. Allafort, L. Baldini, J. Ballet, G. Barbiellini, M. G. Baring, D. Bastieri, et al. 2013. Detection of the Characteristic Pion-Decay Signature in Supernove Remnants. *Science* (American Association for the Advancement of Science). 339 (6424): 807–811.

Agnew, D. C. and K. M. Larson. 2007. Finding the repeat times of the GPS constellation. *GPS Solutions.* 11(1): 71–76. New York: Springer-Verlag.

Alcubierre, Miguel. 1994. The warp drive: Hyper-fast travel within general relativity. *Classical and Quantum Gravity.* 11: L73–L77.

Aldroubi, A. and M. Unser (eds.). 1996. *Wavelets in Medicine and Biology.* Boca Raton, FL: CRC Press.

Allen, L. , M. W. Beijersbergen, R. J. C Spreeuw, and J. P. Woerdman. 1992. Orbitalangular momentum of light and the transformation of Laguerre-Gaussian laser model. *Physics Review A.* 45: 8185–8189.

Arden, Jeremy. 2006. *Keys to the Schillinger System. Course A: Basic Principles and Foundations.* Harwich Port, MA: Clock & Rose Press. ISBN 1-59386-031-5.

———— . 1996. Focusing the musical imagination: Exploring in composition the ideas and techniques of Joseph Schillinger. *Ph.D. Thesis.* London: City University.

Arnowitt, R., S. Deser and C. Misner. 1959. Dynamic structure and definition of energy in general relativity. *Physical Review.* 116 (5): 1322–1330.

Askill, John. 1979. *Physics of Musical Sounds.* New York: Van Nostrand.

Attali, Jasques. 1985. *Noise: The Political Economy of Music.* Translated by Brian Massumi. Minneapolis: University of Minnesota Press.

Bancroft, S. 1985. An algebraic solution of the GPS equations. *IEEE Transactions on Aerospace and Electronics Systems.* AES-21: 56–59.

Bandres, Miguel A. and Juio C. Gutierez-Vega. 2004. Gaussian beams. *Opt. Lett.* 29(2): 144–146.

Bate, R. R., D. D. Mueller and J. E. White. 1971. *Fundamentals of Astrody-*

namics. New York: Dover.

Berg, Richard E. and David G. Stork. 1982. *The Physics of Sound.* Engelwood Cliffs: Prentice Hall.

Bernoulli, D. 1753/1755. Reflexions et éclaircissemens sur les nouvelles vibrations des cordes exposées dans les mémoires de l'Académie de 1747 et 1748. *Mémoires de l'Académie Royale de Berlin*, 9: 147–172.

Beylkin, G. 1987. Discrete Radon Transform. *IEEE Trans. ASSP.* 35: 162–172.

Bôchner, M. 1905/1906. Introduction to the theory of Fourier series. *Math. Ann.*, 7: 81–152.

Borre, Kai, Dennis M. Akos, N. Bertelsen, P. Rinder, and Søren H. Jensen. 2007. *A Software-Defined GPS and Galileo Receiver: A Single-Frequency Approach.* Boston, MA: Birkhäuser.

Boyce, W. E. and R. C. DiPrima. 1992. *Elementary Differential Equations.* 5th ed. New York: Wiley.

Braasch, M. and A. Van Dierendonck. 1999. GPS Receiver Architectures and Measurements. *Proceedings of IEEE.* 87: 48–64.

Bracewell, R. N. 1956. Strip integration in radio astronomy. *Austr. Journal of Physics.* 9: 190–217.

———. 1960. Communications from superior galactic communities. *Nature.* 186, No. 4726: 670–671.

———. 1986. *The Hartley Transform.* Oxford: Oxford University Press.

———. 2000. *The Fourier Transform and Its Applications.* New York: McGraw-Hill, 1st ed. 1965, 2nd ed. 1978.

——— and A. C. Riddle. 1967. Inversion of fan-beam scans in radioastronomy. *Astro Phys. Journal.* 150: 427–434.

Brady, M. L. 1998. A fast discrete approximation algorithm for the Radon transform. *SIAM J. Comput.* 27: 107–119.

Brandt, A., J. Mann, M. Brodski and M. Galun. 1999. A fast and accurate multilevel inversion of the Radon transform. *SIAM J. Appl. Math.* 60:437–462.

Brigham, E. Oran. 1974. *The Fast Fourier Transform.* Englewood Cliffs, NJ: Prentice Hall.

Brown, A., M. May, and B. Tanju. 2000. Benefits of Software GPS Receivers for Enhanced Signal Processing. *GPS Solutions.* 4(1).

———, G. Neil, and Keith Taylor. 2000. Modeling and Simulation of GPS Using Software Signal Generation and Digital signal Reconstruction. *Proc. of ION Technical Meeting*, Anaheim, CA, January 2000.

Bunch, Bryan H. and Alexander Hellemans. 2004. *The History of Science and Technology.* Hartcourt, CT: Houghton Mifflin.

Carslaw, H. S. 1925. A historical note on Gibbs' phenomenon in Fourier's series and integrals. *Bull. Amer. Math. Soc.*, 31: 420–424.

Castagnoli, G., S. Bräuer, and M. Herrmann. 1993. Optimization of cyclic redundancy-check codes with 24 and 32 parity bits. *IEEE Trans. Com-*

mun. 41: 883–892.

Chaisson, Eric and S. McMillan. 2008. *Astronomy Today.* San Francisco, CA: Pearson/Addison-Wesley.

Chandrasekhar, S. 1983. *The Mathematical Theory of Black Holes.* New York: Clarendon Press.

Churchill, R. V. 1972. *Operational Methods,* 3rd ed. New York: McGraw-Hill.

―――― and J. W. Brown. 1978. *Fourier Series and Boundary Value Problems.* New York: McGraw-Hill.

Cocconi, Giuseppe and Philip Morrison. 1959. Searching for interstellar communications. *Nature.* 184: 844–846.

Cooley, J. W. and J. W. Tukey. 1965. An algorithm for machine calculation of complex Fourier transform. *Math. Computation.* 19: 287–301.

Cormack, A. M. 1963. Representation of a function by its line integrals, with some radiological applications. *J. Applied Phys.* 34: 2722–2727.

―――― . 1964. Representation of a function by its line integrals, with some radiological applications II. *J. Applied Phys.* 35: 2908–2912.

Couch, L. W., II. 2001. *Digital and Analog Communications Systems,* 6th Ed. Engelwood Cliffs, NJ: Prentice Hall.

Courant, R. and D. Hilbert. 1953/1962. *Methods of Mathematical Physics.* Vol. 1 (1953); Vol. 2 (1962). New York: Interscience.

Davis, H. F. 1963. *Fourier Series and Orthogonal Functions.* Boston, MA: Allyn and Bacon.

Davis, P. J. and R. Hersh. 1981. *The Mathematical Experience.* Boston, MA: Birkhäuser.

d'Alembert, J. le R. 1749. Recherches sur la courbe que forme une corde tendue mise en vibration. *Mémoires de l'Académie Royale de Berlin,* 3: 214–219.

De Sabbata, V. and Z. Zhang (Eds.) 1992. *Black Hole Physics.* Boston, MA: Kluwer Academic Publishers.

Department of Defense. 1987. *World Geodetic System 1984 (WGS-84).* DMA-TR-8350.2. Defense Mapping Agency, September 1987.

Dillinger, Marcus. 2003. *Software Defined Radio: Architecture, Systems and Functions.* New York: Wiley.

Dirac, P. A. M. 1939. A new notation for quantum mechanics. *Mathematical Proceedings of the Cambridge Philosophical Society.* 35: 416–418.

―――― . 1947. *The Principles of Quantum Mechanics.* Oxford: Oxford University Press.

―――― . 1964. *Lectures on Quantum Mechanics.* New York: Belfer Graduate School of Science Monographs Series, Vol. 2, Yeshiva University. Reprinted 2001, Dover Publication, Inc., New York.

Dixon, R. C. 1976. *Spread Spectrum Systems.* New York: Wiley.

Droege, R. T. , S. N. Wiencer, M. S. Rzeszotarski, G. N. Holland, and I. R. Young. 1983. Nuclear magnetic resonance: a gray scale model for head images. *Radiology*. 148: 763–771.

Duru, H. and H. Kleinert. 1979. Solution of the path integral for the H-atom. *Physics Letters B* 84: 185–88.

Edwards, R. E. 1967. *Fourier Series: A Modern Introduction*. New York: Holt, Rinehart and Winston.

Einstein, Albert. 1905. An heuristic viewpoint concerned with the generation and transformation of light. *Ann. Phys.* 322: 132–148.

El-Rabbany, A. 2002. *Introduction to GPS: the Global Positioning System*. Norwood, MA: Artech House.

Emden, R. 1907. *Gaskugeln*. Leipzig: B. G. Teubner.

Engelberg, S. 2007. *Random Signals and Noise: A Mathematical Introduction*. Boca Raton, FL: CRS Press.

Engl, H. W., M. Hanke, A. Neubauer. 1996. *Regularisation of Inverse Problems*. Dordrecht: Kluwer.

Erdélyi, A., W. Magnus, F. Oberhettinger, and F. G. Tricomi. 1954. *Tables of Integral Transforms*, vol. 1. New York: McGraw-Hill.

Erickson, Robert. 1975. *Sound Structures in Music*. University of California Press.

Euler, L. 1798. Disquisito ulterior super seriebus secundum multiplacuiusdam anguli progredientibus. *Nova Acta Academie Scientarum Prtropolitanae*, 11:114–132.

———— . 1765/1767. Sur le mouvement d'une corde, qui au commencement n'a été ébranlée que dans une partie. *Mémoires de l'Académie Royale de Berlin*, 307–334.

Federal Radionavigation Plan (1999). February 2000. Washington, DC: U.S. Department of Transportation and Department of Defense. Available on line from United States Coast Guard Navigation Center.

Feynman, R. P. 1989. *What Do You Care What Other People Think?* New York: Bantam.

———— , R. B. Lighton and M. Sands. 1965. *The Feynman Lectures*. Vol. 3. Reading, MA: Addison-Wesley.

Fieser, Louis F., Mary Fieser, and Srinivasa Rajagopalan. 1948. Absorption spectroscopy and the structure of the diosteroids. *J. Org. Chem.* 13(6): 800-806.

Folger, Tim. 2013. Crazy Far. *National Geographic*. 125th Anniversary Special Issue, January 2013. National Geographic Society, 68–81.

Fourier, J. 1816. Théorie de la chaleur. *Annales de chemie et de physique*, 3: 350–375.

———— . 1978. *The Analytic Theory of Heat*. London: Cambridge University Press. Reprinted by Dover, New York, 1985.

Friedrich, Walter, Paul Knipping, and Max von Laue. 1912. Interferenzerscheinungen bei Röntgenstrahlen. *Ann. Physik.* 41(5): 971–988.

Gallager, R. G. 1962. Low-density parity-check codes. *IEEE Transactions on Information Theory*. 8: 21–28.

———— . 1963. *Low-Density Parity-Check Codes.* Cambridge, MA: MIT Press.

Gerchberg, G. 1974. Super-resolution through error energy reduction. *Opt. Acta.* 21: 709–720.

Ghosh, Pranab. 2007. Rotational and acceleration powered pulsars. *World Scientific.*

Gibbs, J. 1899. Fourier Series. *Nature,* 59: 606. (http://www.worldcat.org/issn/0028-0836).

Global Positioning System Standard Positioning Service Specification. 2nd Edition, June 2, 1995. Available online from United States Coast Guard Navigation Center.

Gold, R. 1967. Optional binary sequences for spread spectrum multiplexing. *IEEE Transactions on Information Theory.* 13: 619–621.

———— . 1968. Maximal Recursive Sequences with 3-valued Recursive Cross-Correlation Functions. *IEEE Trans. Infor. Theory.* 14 : 154-156.

Goodman, J. W. 1968. *Introduction to Fourier Optics.* New York: McGraw-Hill.

Gordon, C., L. Webb and S. Wolpert. 1992. One cannot hear the shape of a drum. *Bull. Am. Math. Society.* 1992: 134–138.

Götz, W. A. and H. J. Druckmüller. 1996. A fast digital Radon transform An efficient means for evaluating the Hough transform. *Pattern Recognition.* 29: 711-718.

Gouy, C. R. 1890. Sur une propriété nouvelle des ondes lumineuses. *Comptes Rendus Acad. Sci. Paris.* 110: 1251.

GPS Joint Program Office. 1997. ICD-GPS-200: GPS Interface Control Document. ARINC Research. Available online from United States Coast Guard Navigation Center.

Gradshteyn, I. S. and I. M. Rhyzik. 2000. *Tables of Integrals, Series, and Products.* 6th ed. San Diego, CA: Academic Press.

Grier, David G. 2000. Adaptive-Additive Algorithm. http://www.physics.nyu.edu/ dg86/cgh2b/node6.html (October 10, 2000).

Griffiths, David J. 1995. *Introduction to Quantum Mechanics.* Upper Saddle River, NJ: Prentice Hall.

Haas, Jeffrey. 2010. *Chapter Three: How MIDI Works.* Indiana University Jacobs School of Music (Web August 13, 2012).

Hahn, S. L. 1996. *The Transforms and Applications Handbook.* (A. Poularakis, ed.) Boca Raton, FL: CRC Press. Ch. 7.

Harris, F. J. 1978. On the use of windows for harmonic analysis with the discrete Fourier transforms. *Proc. IEEE.* 66: 51–83.

Harrison, Edward R. 1978. *Darkness at Night: A Riddle of the Universe.* Harvard University Press.

———— . 1992. The redshift-distance and velocity-distance laws. *Astrophysical Journal,* Part 1. 403: 25–31.

Hawking, S. W. 2009. Does God Play Dice? www.hawking.org.uk.2009-03-14.

Healy, Dennis M., Jr. and John B. Weaver. 1996. Adapted wavelet techniques for encoding magnetic resonance images, in *Wavelets in Medicine and Biology* (A. Aldroubi and M. Unser, eds.). Boston, MA: Birkhäuser.

Heaviside, Oliver. 1892/1893. On operational methods in physical mathematics. Part I and Part II. *Proceedings of the Royal Society, London.* 52: 504–529; 54: 105–143.

Hecht, K. T. 2000. *Quantum Mechanics.* New York: Springer-Verlag.

Heisenberg, W. 1925. *Die Physik der Atomkerne.* Taylor & Francis. English translation: *The Physical Principles of Quantum Theory.* Chicago: University of Chicago Press, 1930.

Helgason, S. 1980. *The Radon Transform.* Boston, MA: Birkhäuser.

Helmholtz, H. 1954. *On the Sensation of Tone.* New York: Dover.

Herman, Gabor T. 2009. *Fundamentals of Computerized Tomography: Image Reconstruction from Projections.* 2nd ed. New York: Springer-Verlag. ISBN 978-1-85233-617-2.

———— R. M. Lewitt. 1979. Overview of image reconstruction from projections. In *Overview of Image Reconstruction from Projections* (G. T. Herman, ed.). Vol. 32. New York: Springer-Verlag.

Heurtley, J. C. 1973. Scalar Rayleigh-Sommerfeld and Kirchhoff diffraction integrals: A comparison of exact evaluations for axial points. *J. Opt. Soc. Am.* 63: 1003.

Holmes, J. K. 2007. *Spread Spectrum Systems for GNSS and Wireless Communications.* Norwood: Artech House.

Holmes, Thom. 2003. *Electronic and Experimental Music: Pioneers in Technology and Composition.* New York: Routledge.

Institute of Navigation. 1980, 1984, 1986, 1993. Global Positioning System monographs. Washington, DC: The Institute of Navigation.

Hooft, G. 2009. Introduction to the Theory of Black Holes. New York: Institute for Theoretical Physics/Spinoza Institute.

Israel, W. 1992. Thermodynamics and Internal Dynamics of Black Holes, in *Black Hole Physics* (V. De Sabbata and Z. Zhang, Eds.) Boston, MA: Kluwer Academic Publishers.

Jackson, J. C. 1972. A critique of Rees's theory of primordial gravitational radiation. *Monthly Notices of the Royal Astronomical Society.* 156: 1–6.

Jones, David L. 2012. Sending signals to submarines. *New Scientist.* 26: 37–41. Revised February 17, 2012. London: Holborn Publishing Group.

Josephson, B. D. 1962. Possible new effects in superconductive tunnelling. *Phys. Lett.* 1: 251.

Kac, M. 1966. Can one hear the shape of a drum? *Am. Math. Monthly.* 74: 1–23.

Kaiser, N. 1987. Clustering in real space and in redshift space. *Monthly Notices of the Royal Astronomical Society,* 227: 1–21.

Kak, A. and M. Slaney. 1988. *Principles of Computerized Tomographic Imaging.* New York: IEEE Press.

Kaku, Michio. 1994. *Hyperspace*. New York: Doubleday.

Kaplan, E. 1990. *Understanding GPS: Principles and Applications*. Norwood, MA: Artech House.

Kelley, B. T. and V. K. Madisetti. 1993. The fast discrete Radon transform — I: Theory. *IEEE Trans. Image Proc.* 2: 382–400.

Kelley, C., J. Cheng, and J. Barnes. 2002. Open source software for learning about GPS. *15th Int. Tech. Meeting of the Satellite Division*, U.S. Inst. of Navigation, Portland, Oregon, September 24-27, 2002.

Kelvin, Lord. 1839–1841. On Fourier's expansions of functions in trigonometric series. *Cambridge Mathematical Journal*, 2: 258–262.

————. 1841–1843. Note on a passage in Fourier's Heat. *Cambridge Mathematical Journal*, 3: 25–27.

————. 1841–1843. On the linear motion of heat. *Cambridge Mathematical Journal*, 3: 170–174; 206–211.

Kennerd, E. H. 1927. Zur Quantenmechanik einfacher Bewgungstypen. *Zeitschrift für Physik*. 44(4–5): 326.

Kirchhoff, G. 1883. Zur Theorie der Lichtstrahlen. *Weidemann Ann.* (2), 18: 663.

Kittel, C. 1971. *Solid State Physics*. New York: Wiley.

Klein, Miles V. and Thomas E. Furtak. 1986. *Optics*. New York: Wiley.

Knight, W. and R. Kaiser. 1979. A simple fixed-point error bound for the fast Fourier transform. *IEEE Trans. Acoustics, Speech and Signal Processing*. 27: 615–620.

Kolacyzk, Eric D. 1996. An application of wavelet shrinkage to tomography, in *Wavelets in Medicine and Biology*, (eds. A. Aldroubi and M. Unser). Boca Raton, FL: CRC Press. pp. 77–92.

Kottler, F. 1965. Diffraction at a black screen. In E. Wolf (ed.) *Progress in Optics*, Vol. IV. Amsterdam: North Holland Publishing Company.

Kovacs, Z., K. S. Cheng and T. Harko. 2009. Can stellar mass black holes be quark stars? *Monthly Notices of the Royal Astronomical Society*. 400(3): 1632–1642.

Krisnikov, S. 2003. The quantum inequalities do not forbid spacetime shortcuts. *Physical Review*. D 67 (10): 104–113.

Krumvieda, K., C. Cloman, E. Olson, J. Thomas, W. Kober, P. Madhani, and P. Axelrad. 2001. A Complete IF Software GPS Receiver: A Tutorial About the Details. *ION GPS-2001*. Salt Lake City, UT, 789–829.

Krzysztof, I. 2007. *Wireless Technologies: Circuits, Systems, and Devices*. Boca Raton, FL: CRC Press.

Kuhl, D. E. and R. Q. Edwards. 1963. Image separation radioisotope scanning. *Radiology*. 80: 653–661.

Kythe, D. K. and P. K. Kythe. 2012. *Algebraic and Stochastic Coding Theory*. Boca Raton, FL: CRC Press.

Kythe, P. K. 1996. *Fundamental Solutions for Differential Operators and Applications*. Boston, MA: Birkhäuser.

————— . 1998. *Computational Conformal Mapping*. Boston, MA: Birkhäuser.

————— , P. Puri and M. R. Schäferkotter. 2003. *Partial Differential Equations and Boundary Value Problems*. 2nd ed. Boca Raton, FL: Chapman & Hall/CRC.

————— , and M. R. Schäferkotter. 2005. *Handbook of Computational Methods for Integration*. Boca Raton, FL: Chapman & Hall/CRC.

————— . 2011. *Green's Functions and Linear Differential Equations: Theory, Applications and Computation*. Boca Raton, FL: CRC Press.

Langen, D. et al. 2002. Implementation of a RISC Processor Core for SoC Designs – FPGA Prototype vs. ASIC Implementation. *Proc. of the IEEE-Workshop: Heterogeneous Reconfigurable Systems on Chip (SoC)*. Hamburg, Germany, April 2002.

Leick, Alfred. 1995. *GPS Satellite Surveying*. 2nd. ed. New York: John Wiley & Sons.

Li, Y. and J. Santanilla. 1995. Existence and nonexistence of positive singular solutions for semilinear elliptic problems with applications to astrophysics. *Diff. and Integral Eqs.* 8: 1369–1383.

Lin, D. and J. Tsui. 1998. Acquisition schemes for software GPS receiver. *ION* GPS-98: 317–325. Nashville, TN, September 15–18, 1998.

Litton, James D., Graham Russel and Richard K. Woo. 1996. Methods and apparatus for digital processing in a global positioning system receiver. *US Patent 5576715*, issued November 16, 1996, assigned to Leica Geosystems.

Longair, M. S. 1994. *High Energy Astrophysics*, Vol. 2. Cambridge University Press.

Loudon, R. 2001. *The Quantum Theory of Light*. New York, NY: Oxford University Press.

Low, Robert J. 1999. Speed limits in general relativity. *Classical and Quantum Relativity*. 16 (2): 543–549.

Luby, M. 2002. LT Codes. *43rd Annual IEEE Symposium on Foundations of Computer Science*. http://ieeexplore.ieee.org/xpl/freeabs_all.jsp?arnumber =1181950

————— . 2002. LT Codes. *Proc. IEEE Symposium on the Foundations of Computer Science*. (FOCS): 271–280.

MacPherson, Stewart. 1930. *Form in Music*. London: Joseph Williams, Ltd.

Mandel, Leonard and Emil Wolf. 1995. *Optical Coherence and Quantum Optics*. Ch. 5: Optical Beams, pp. 267. Cambridge: Cambridge University Press.

Manning, Peter. 1994. *Electronic and Computer Music*. Oxford University Press.

Mark, J. , M. J. Shumate, D. A. Kooby and N. P. Alazraki. 2007. *A Clinician's Guide to Nuclear Oncology: Practical Molecular Imaging and Radionuclide Therapies*. Society of Nuclear Medicine, January 2007.

Matukuma, T. 1938. *The Cosmos*. Tokyo: Iwanami Shoten (in Japanese).

McCutchen, Ann. 1999. *The Muse That Sings: Composers Speak about the*

Creative Process. New York: Oxford University Press.

Michelson, A. A. 1898. *Nature.* 58:544. London.

—— and F. G. Pease. 1921. Measurement of the diameter of α Orionis with the interferometer. *Astrophysics J.* 53: 249–259.

Misra, P. and P. Enge. 2001. *Global Positioning System: Signals, Measurements, and Performance.* Lincoln, MA: Ganga-Jamuna Press.

Moon, Todd K. 1999. *Mathematical Models and Algorithms for Signal Processing.* Englewood Cliffs, NJ: Prentice Hall.

Nallatech. 2002. *Strathnuey PCI Card User Guide,* NT 107-0076, Issue 9, Nallatech Datasheet.

Narlikar, J. V. 1982. *Violent Phenomena in the Universe.* Cambridge: Oxford University Press.

National Imagery and Mapping Agency. 1997. Department of Defense World Geodetic System 1984: Its Definition and Relationship with Local Geodetic Systems. NIMA TR8350.2, Third Edition, 4 July 1997. Bethesda, MD: National Imagery and Mapping Agency. Available on line from National Imagery and Mapping Agency.

Natterer, F. 1986. *The Mathematics of Computerized Tomography.* Chichester, UK: Wiley.

—— and R. O'Malley. 2001. *The Mathematics of Computerized Tomography.* Cambridge, UK: Cambridge University Press.

—— and F. Wübbeling. 2001. *Mathematical Methods in Image Reconstruction.* Philadelphia, PA: SIAM.

Nave, Carl R. 2013. *Cosmic Rays.* Harper Physics Concepts. Georgia State University.

Navstar GPS User Equipment Introduction. 1996. Available on line from United States Coast Guard Navigation Center

Neugebauer, O. E. 1952. *The Exact Sciences in Antiquity.* Princeton, NJ: Princeton University Press.

Ni, W.-M. 1986. Uniqueness, nonuniqueness and related questions of nonlinear elliptic and parabolic equations. *Proc. Symposia Pure Math.* 45: 229–241.

Norton, S. L. and M. Linzer. 1979(a). Ultrasonic reflectivity tomography: Reconstruction with circular transducer arrays. *Ultrason. Imaging.* 1 (2): 154–184.

—— and M. Linzer. 1979(b). Ultrasonic reflectivity tomography: Reconstruction with spherical transducer arrays. *Ultrason. Imaging.* 1 (2): 210–231.

Oberhettinger, F. 1990. *Tables of Fourier Transforms and Fourier Transforms of Distributions.* Berlin: Springer-Verlag.

Oldendorf, W. H. 1961. Isolated flying spot detection of radio density discontinuities displaying the internal structural pattern of a complex object. *IRE Transactions Biomedical Electronics.* 8: 68–72.

Olson, T. 1996. Optimal time-frequency projections for localized tomography, in *Wavelets in Medicine and Biology* (A. Aldroubi and M. Unser, eds.). Boca Raton, FL: CRC Press.

Oppenheim, A. V. 2009. Discrete-Time Signal Processing. 3rd ed. Englewood Cliffs, NJ: Prentice Hall.

———— and R. W. Schäfer. 2009. *Digital Signal Processing*. 3rd ed. Englewood Cliffs, NJ: Prentice Hall.

Oxford English Dictionary. 1971. Compact Edition II. Oxford University Press.

Pace, S., G. Frost, et al. 1995. *The global positioning system: assessing national policies*. Santa Monica, CA: RAND Corp.

Page, D. N. 1992. Black-Hole Thermodynamics, Mass-Inflation and Evaporation, in *Black Hole Physics* (V. De Sabbata and Z. Zhang, Eds.) Boston, MA: Kluwer Academic Publishers.

Pang, Jing. 2003. *Direct Global Positioning System P-Code Acquisition Field Programmable Gate Array Prototyping*. Ph.D. Dissertation, Fritz J. and Dolores H. Russ College of Engineering and Technology, Ohio University. August 2003.

Papoulis, A. 1975. A new algorithm in spectral analysis and band limited extrapolation. *IEEE Trans. Circuits Syst.* CAS-22: 735–742.

Parkinson, Bradford W. and James J. Spilker (eds.) 1996. *Global Positioning System: Theory and Practice*. Volumes I and II. Washington, DC: American Institute of Aeronautics and Astronautics, Inc.

Pauli, M. 1926. Über das Wasserstoffspektrum von Standpunkt der neuen Quantenmechanik. *Zeitschrift für Physik* 36: 336–363.

Peebles, P. J. E. 1972. Black holes are where you find them. *Gen. Rel. Grav.* 3: 63–82.

Perlmutter, S. et al. (The Supernova Cosmology Project). 1999. Measurements of Omega and Lambda from 42 high redshift supernova. *Astrophysical Journal.* 517(2): 565–586.

Perring, J. K. and T. H. R. Skyrme. 1962. A model unified field equation. *Nucl. Phys.* 31: 550–555.

Perutz, M. F. 1964. The hemoglobin molecule. *Scientific American*, November issue: 64–76.

Pfleegor, R. L. and L. Mandel. 1967. Interference of independent photon beams. *Phys. Rev.* 159: 1084–1088.

Pike, John. 2009. *Operational Control Segments* (OCX). Globalsecurity.org. Retrieved December 8, 2009.

Polyanin, Andrei D. and Valentin F. Zaitsev. 2004. *Handbook of Nonlinear Partial Differential Equations*. Boca Raton, FL: Chapman & Hall/CRC Press.

Prasad, R. 1996. *CDMA for Wireless Personal Communications*. Boston, MA: Artech House.

Prestini, E. 2004. *Applied Harmonic Analysis: Models of the Real World*. Boston, MA: Birkhäuser.

Press, William H. 2006. Discrete Radon transform has an exact, fast inverse and generalizes to operations other than sums along lines. Proceedings of the National Academy of Sciences. 103 (51): 19249–19254. Doi: 10.1073/pnas.0609228103.

————, Saul A. Tekolsky, William T. Vettering, and Brian P. Flannery. 2007. *Numerical Recipes: The Art of Scientific Computing.* 3rd ed. (C++ code). Cambridge University Press.

Proakis, John G. 1995. *Digital Communications.* 3rd ed. New York: McGraw-Hill.

Protter, M. H. 1987. Can one hear the shape of a drum? Revisited. *SIAM Review.* 29: 185–197.

Puckette, M. 2007. *The Theory and Techniques of Electronic Music.* Singapore: World Scientific Publishing Company.

Puthoff, H. E. 1996. SETI, the velocity-to-light limitation, and the Alcubierre warp drive. An interesting overview. *Physics Today.* 9: 156–158.

Rabiner, L. and B. Gold. 1975. *Theory and Applications of Digital Signal Processing.* Englewood Cliffs, NJ: Prentice Hall.

Radon, Johann. 1917. Über die Bestimmung von Funktionen durch ihre Integralwerte längsgewisser Manningfaltigkeiten. *Berichte Über die Verhandlungen der Königlich-SächsischenAkademie der Wissenschaften zu Leipzig, Mathematisch-Physische Klasse* (Reports on the Proceedings of the Royal Saxonian Academy of Sciences at Leipzig, Mathematical and Physical Section). Leipzig: Teubner. 69: 262–277. Translated by P.C. Parks (1986): "On the determination of functions from their integral values among certain manifolds", *IEEE Transactions on Medical Imaging.* 5:170–176. http//:dx. doi.org/10.1109/ TMI.1986.4307775. PMID 18244000 (http://www.ncbi. nlm.nih.gov/pubmed/18244009).

Reed, Jeffrey H. 2002. *Software Radio: A Modern Approach to Radio Engineering.* Pittsburgh: Prentice Hall.

Riggins, R. 1996. Navigation using the global positioning system. Chapter 6, classnotes. Wright Peterson AFB, OH: Air Force Institute of Technology.

Robel, Axel. 2005. *Adaptive-Additive Synthesis of Sound Technical.* University of Berlin, Germany: Einsteinufer 17.(http://i2pi.com/PAPERS/music-dsp/adaptive-additive-synthesis-of.pdf).http://ieeexplore.ieee.org/iel5 /10376 /32978/101109TSA2005858529.pdf?arnumber=101109STA2005858529.

Roederer, J. C. 1979. *Introduction to the Physics and Psychophysics of Music.* Berlin: Springer-Verlag.

Roerdink, J. B. T. M. 2001. Tomography.(http://www.encyclopediaofmath. org/index.php?title=t/t092980), in Michael Hazewikel, *Encyclopedia of Mathematics.* Springer-Verlag, ISBN 978-1-55608-010-4.

Rossing, Thomas D., Richard F. Moore and Paul A. Wheeler. 2002. *The Science of Sound.* 3rd ed. Reading, MA: Addison-Wesley.

Sagan, C. and F. Drake. 1975. The search for extraterrestrial intelligence. *Scientific American.* 232 (May): 80–89.

Sagan, Hans. 1989. *Boundary and Eigenvalue Problems in Mathematical Physics.* New York: Dover.

Saleh, B. E. A. and M. C. Teich. 1991. *Fundamentals of Photonics.* New York: Wiley.

Scholes, Percy. 1977. *Metre and Rhythm.* 10th ed., 6th printing. The Oxford Companion to Music. London: Oxford University Press.

Schwarzschild, K. 1916. Über das Gravitationsfeld eines Massenpunktes nach der Einsteinschen Theorie. *Sitzungsberichte der Deutschen Akademie der Wissenschaften zu Berlin, Klasse für Mathematik, Physik, und Technik.* pp. 189.

———— .1916. Über das Gravitationsfeld einer Kugel aus inkompressibler Flussigkeit nach der Einsteinschen Theorie. *Sitzungsberichte der Deutschen Akademie der Wissenschaften zu Berlin, Klasse für Mathematik, Physik, und Technik.* pp. 424.

Schwinger, J. 1951. *The Theory of Quantized Fields.* Harvard University U.S. Department of Energy: Atomic Energy Commission.

Sellaro, T. L., A. K. Ravindra, D. B. Stolz and S. F. Badylak. 2007. Maintenance of hepatic sinusoidal endothelial cell phenotype in vitro using organ-specific extracellular matrix scaffolds. *Tissue Eng.* 13 (9): 2301–2310.

Sethares, William. 1998. *Tuning, Timbre, Spectrum, Scale.* New York: Springer.

Shannon, C. E. 1948. A mathematical theory of communication. *Bell System Tech. J.* 27: 379–423.

Shapiro, S. L. and S. A. Teukolsky. 1992. Black holes, star cluster, and naked singularities: numerical solution of Einstein's equations. *Phil. Trans. Royal Soc., Series A.* 340: 365–390.

Shepp, L. A. and B. F. Logan. 1974. The Fourier reconstruction of a head section. *IEEE Trans. Nucl. Sci.* NS21(3): 21–43.

Shere, Charles. 1995. *Thinking Sound Music: The Life and Work of Robert Erickson.* Berkeley, CA: Fallen Leaf Press.

Shillinger, Joseph. 2005. *The Schillinger System of Musical Composition.* 2 Vols. Harwich Port, MA: Clock & Rose Press. ISBN 1-59368-028-5.

Shokrollahi, A. 2006. Raptor codes. *IEEE Transactions on Information Theory.* IT-52: 2551–2567.

Shouten, J. F. 1968. The Perception of timbre, in *Reports of the 6th International Congress of Acoustics,* Tokyo, GP-6-2, 6 vols, Y. Kohasi (ed.). Amsterdam: Elsevier.

Shuker, Roy. 1994. *Understanding Popular Music.* London: Routledge.

Siegman, A. E. 1973. Hermite-Gaussian functions of complex argument as optical-beam eigenfunctions. *Journal of the Optical Society of America.* 63: 1093–1094.

———— . 1986. *Lasers.* Mill Valley, CA: University Science.

Smoller, J. A., A. G. Wasserman, S.-T. Yau and J. B. McLeod. 1993. Smooth static solutions of Einstein/Yang–Mills equations. *Commun. Math. Phys.*

151: 303–325.

Sneddon, I. N. 1957. *Partial Differential Equations.* New York: McGraw-Hill.

———— .1978. *Fourier Transforms and Their Applications.* Berlin: Springer-Verlag.

Soifer, V. K. and L. Doskolovich. 1997. *Iterative Methods for Diffractive Optical Elements Computation.* Bristol, PA: Taylor & Francis.

Sommerfeld, A. 1896. Mathematische Theorie der Diffraction. *Math. Ann.* 47: 317.

Spilker, J. J. 1978. GPS signal structure and performance characteristics. *Navigation,* Institute of Navigation. 25: 121–146.

Sparkle, L. S. and J. S. Gallagher. 2000. Galaxies in the Universe: An Introduction. Princeton,NJ: Princeton University Press.

———— 1996. GPS signal structure and theoretical performance, in *Global Positioning System: Theory and Applications* by B. W. Parkinson and J. J. Spilker Jr., Chapter 3. Washington, DC: American Institute of Aeronautics and Astronautics.

Steel, W. H. 1986. *Interferometry.* Princeton, NJ: Cambridge University Press.

Stein, E. M. and G. Weiss. 1971. *Introduction to Fourier Analysis on Euclidean Spaces.* Princeton, NJ: Princeton University Press.

Strang, G. and K. Borre. 1997. *Linear Algebra, Geology, and GPS.* Cambridge: Wellesley.

Sudakshina, Kundu. 2010. *Analog and Digital Communications.* Pearson Education India.

Svelto, Orazio. 2010. *Principles of Lasers.* 5th ed. New York: Springer-Verlag.

Talbot, H.F. 1836. Facts Relating to Optical Science. *Philos. Mag.* 9: 401–407.

Taylor, A., D. M. Schuster and N. Naomi Alazraki. 2000. *A Clinicians' Guide to Nuclear Medicine.* 2nd ed. Society of Nuclear Medicine.

Taylor, C. A. and H. Lipson. 1964. *Optical Transforms.* London: Bell.

Terras, A. 1985. *Harmonic Analysis on Symmetric Spaces and Applications.* New York: Springer-Verlag.

Titschmarsh, E. C. 1937. *Introduction to the Theory of Fourier Integrals.* Oxford, UK: Oxford University Press (Clarendon).

———— .1958. *Eigenfunction Expansions,* Part 2. Oxford, UK: Oxford University Press.

Tsui, James Bao-Yen. 2000. *Fundamentals of Global Positioning System Receivers: A Software Approach.* New York: John Wiley & Sons.

UIUC Histology Subject 589. 2007. Report available at the website *https//:histo. life.illinois.edu/ histo/atlas/oimages.php?oid=589*

Van der Broeck, Christian. 2000. Alcubierre's warp drive: Problems and Prospects. *AIP Conference Proceedings.* 504: 1105–1110.

Viswanathan, Mathuranathan. 2013. *Simulation of Digital Communication*

Systems Using Matlab. eBook. ISBN 9781301525089.

Viterbi, A. 1995. *CDMA: Principles of Spread Spectrum Communication*. Reading, MA: Addison-Wesley.

Wald, R. M. 1992. Black Holes and Thermodynamics, in *Black Hole Physics* (V. De Sabbata and Z. Zhang, Eds.). Boston, MA: Kluwer Academic Publishers.

Walker, J. S. 1988. *Fourier Analysis*. Oxford, UK: Oxford University Press.

Watson, J. D. and F. H. C. Crick. 1953. A structure for deoxyribosenucleic acid. *Nature*. 171: 737; Genetical implications of the structure of deoxyribosenucleic acid. *Nature*. 171: 964.

Weinberger, H. F. 1965. *A First Course in Partial Differential Equations*. New York: John Wiley.

Wells, David (ed). 1989. *Guide to GPS Positioning*. Fredericton, NB, Canada: Canadian GPS Associates.

Weyl, Herman. 1928. Gruppentheorie under Quantenmechanik. Leipzig: Herzel.

Wesson, Paul . 1991. Olbers' paradox and the spectral intensity of the extragalactic background light. *The Atrophysical Journal*. 367: 399–406.

White, H. 2007. Successful defense of published paper 'A Discussion of Space-Time Metric Engineering', part of PH. D. candidacy process in Physics at Rice University, 2007.

———— . 2011. Encore of Warp Field Mechanics 101, technical presentation requested by AIAA, Houston Chapter, Gilruth Center, Houston, TX 2011.

———— . 2011. Warp Field Mechanics 101. Technical presentation given at the 100 Year Starship Symposium, Orlando, FL, 2011, http://ntrs.nasa.gov/archive/ nasa/casi.ntrs.nasa.gov/20110015936_2011016932.pdf; also http:// ntrs.nasa.gov/search.jsp?R=20110015936#.UwNqQBS_exA.email.

———— and W. Davis. 2006. The Alcubierre Warp Drive in Higher Dimensional Spacetime. *Proceedings of Space Technology and Applications International Forum* (STAIF 2006), American Institute of Physics Conference Proceedings, Melville, New York.

Wicker, S. B., and V. K. Bhargava (Eds.). 1999. *An Introduction to Reed-Solomon Codes*. New York: Institute of Electrical and Electronics Engineers.

Wieidermann, E. . 1888. Über Fluorescenz und Phosphorescence, I, Abhandlung. *Annalen der Physik*. 34: 446–463.

Winold, Allen. 1975. *Rhythm in Twentieth-Century Music*, in *Aspects of Twentieth-Century Music*, Gary Wittich (ed). Englewood Cliffs, NJ: Prentice Hall.

Wolf, E. and E. W. Marchand. 1964. Comparison of the Kirchhoff and Rayleigh-Sommerfeld theories of diffraction at an aperture. *J. Opt. Soc. Am.* 54: 587.

Wollack, E. J. 2010. *Cosmology: The Study of the Universe. Universe 101: Big Bang Theory*. NASA. Archived from the original on May 14, 2011.

Woods, Thomas E., Jr. 2005. *How the Catholic Church Built Western Civilization.*Washington DC: Regnery Publishing.

Xilinx. 2000. *High-Performance 1024-Point Complex FFT/IFFT*, V1.0.5, Xilinx Product Specification Datasheet, July 2000.

————— . 2002. *Virtex* ᵗᵐ*-E 1.8 V Field Programmable Gate Arrays*, DS022-1 (v2.3). Xilinx Product Specification, July 2002.

Zanotti, C., U. Caretta, C. Grimaldi and G. Colombo. 1992. Self-sustained oscillatory burning of solid propellants: Experimental results. *Progress in Astronautics and Aeronautics.* 143: 399–439.

————— and P. Giuliani. 1993. Experimental and numerical approach to the frequency response of solid propellants. *IV International Symposium on Flame Structure*, August 1992, Novosibirsk, Siberia. Published in *Fizika Gorreniya i Vzryva*, 3: 36–41.

Ziemer, R. E. and W. H. Tranter. 1995. *Principles of Communications, Systems, Modulation, and Noise.* 4th ed. New York: Wiley.

Zubal, I. G., C. R. Harrell, E. O. Smith, Z. Rattner, G. Gindi, and P. B. Hoffer. 2004. Computerized 3-dimensional segmented human anatomy. *Med. Phys.* 21: 299–302.

Zabusky, N. J. and M. D. Kruskal. 1965. Interaction of 'solitons' in a collisionless plasma and recurrence of initial states. *Phys. Rev. Letters.* 15: 240–243.

Zygmund, A. 1968. *Trigonometric Series.* Cambridge, MA: Cambridge University Press.

Index

Milton Keynes UK
Ingram Content Group UK Ltd.
UKHW021919071024
449327UK00022B/1681

9 780367 378578